P9-AFP-784

COSMIC DEBRIS

COSMIC DEBRIS

Meteorites in History

JOHN G. BURKE

UNIVERSITY OF CALIFORNIA PRESS
Berkeley / Los Angeles / London

University of California Press
Berkeley and Los Angeles, California

University of California Press, Ltd.
London, England

Library of Congress Cataloging in Publication Data

Burke, John G.
Cosmic debris.

Bibliography: p.
Includes index.
1. Meteorites—History. I. Title.
QB755.B87 1986 523.5′1 85–28841
ISBN 0–520–05651–5 (alk. paper)

Printed in the United States of America

1 2 3 4 5 6 7 8 9

To the memory of

PEGGY PORTER BURKE

Contents

Acknowledgments

I am grateful to the John Simon Guggenheim Foundation for the award of a fellowship in 1979–80, which permitted me to visit archives and major meteorite collections in both Europe and the United States. I also thank the Smithsonian Institution for providing a stipend during my extended visit to review the Paneth collection of meteorite literature and other archival material in the Division of Meteorites at the National Museum of Natural History. In addition, I appreciate the financial support for this study made available by The College of Letters and Science, UCLA.

John T. Wasson of the Institute of Geophysics and Planetary Physics, UCLA, first suggested that writing a history of meteoritics would be a worthwhile and rewarding undertaking. He has encouraged and supported me throughout the project, and members of his UCLA research group, in particular Derek Sears (now at the University of Arkansas) and Alan E. Rubin, have made numerous helpful suggestions and furnished offprints of relevant material. Also at UCLA, Wayland D. Hand, director of the Center for the Study of Comparative Folklore and Mythology, generously guided me to sources on the folklore of meteors and meteorites, for which I owe him thanks.

A. C. Bishop, Keeper of Mineralogy at the British Museum (Natural History), gave me free access to the meteorite archives there. I profited greatly from discussions with Andrew Graham, Alex Bevan, and the late Max Hey. I thank Peter Embrey especially for referring me to several crucial historical sources, and I am very grateful to Robert Hutchison for his continual help during my visit and for arranging the loan of several photographs that appear in this book.

I also thank Gero Kurat, Director of the Mineralogical-Petrographic Department of the Natural History Museum, Vienna, for allowing me open access to all records and correspondence there concerning meteorites, and also for arranging the loan of a number of photographs that appear herein. I am also grateful to Alfred Kracher (now at Iowa State University) for guiding me through the rich archival resources of the department and for seeing to it that I profited from the many amenities present in the city of Vienna.

At the Division of Meteorites, National Museum of Natural History, Smithsonian Institution, I was fortunate to talk on several occasions with Brian Mason, who also read and commented on the early drafts of some chapters. I also had the opportunity to meet with Kurt Fredriksson, who kindly loaned me several photographs. Twila Thomas helped me at every turn during my visit to the division. I also acknowledge the great debt I owe to Roy S. Clarke,

Jr., who has been a constant source of help and encouragement during the writing of this book. He read the early drafts of the entire manuscript and offered many comments and suggestions. He made available much unpublished material and arranged the loan of numerous photographs from the files of the Museum of Natural History, which are printed herein.

Michael J. Crowe, University of Notre Dame, referred me to several of the sources I used in chapter 5 in describing the controversy in the 1870s and 1880s concerning the possibility of life in meteorites. I thank him for his careful reading of the chapter and for his helpful comments. John C. Greene, University of Connecticut, graciously sent his notes on Thomas Jefferson's references to meteorites. Barbara Narendra at the Peabody Museum, Yale University, generously shared the results of her research on the activities of Benjamin Silliman with respect to meteorites. Michelle Aldrich at the AAAS archives kindly provided a copy of the little book about meteorites written in 1859 by Mrs. G. H. Silliman.

I thank many persons for their ready responses to my inquiries and requests: Vagn F. Buchwald, Technical University of Denmark; Bernard S. Finn, Museum of American History, Smithsonian Institution; Elbert A. King, University of Houston; Charles F. Lewis, Center for Meteorite Studies, Arizona State University; Ursula Marvin, Harvard-Smithsonian Center for Astrophysics; Robert P. Multhauf, Museum of American History, Smithsonian Institution; Lettie S. Multhauf, Washington, D.C.; Bartholomew Nagy, University of Arizona; Edward J. Olsen, Field Museum of Natural History; Blyth Robertson, Energy, Mines and Resources, Ottawa; John L. Rosenfeld, UCLA; Cyril S. Smith, Massachusetts Institute of Technology; George W. Wetherill, Carnegie Institute of Washington; and John A. Wood, Harvard-Smithsonian Center for Astrophysics.

Numerous librarians and archivists have aided my research. I am particularly obliged to: Eileen Brunton and Roland T. W. Atkins, Mineralogy Library, British Museum (Natural History); William A. Deiss and Harry G. Heiss, Smithsonian Institution Archives; Karl Kabelac and Mary Huth, Department of Rare Books and Special Collections, University of Rochester; Anna McCosland, Interlibrary Loan, University of Washington; and Brenda Corbin; U.S. Naval Observatory Library, Washington, D.C.

I made extensive use of two publications that proved invaluable. Harrison S. Brown's *A Bibliography of Meteorites,* listing about 8,600 citations to the scientific literature, relieved me of innumerable hours of library work. Vagn F. Buchwald's three-volume *Handbook of Iron Meteorites* contains, in addition to detailed scientific descriptions of the then-known meteoritic irons, thorough historical accounts of many irons and of the folklore and legends surrounding them. I would be remiss not to thank these authors for their contributions to this book.

Finally, I take sole responsibility for any errors that appear herein.

Port Townsend, Washington J. G. B.
October 1985

Introduction

Tens of millions of meteoroids, solid bodies from outer space, enter the Earth's atmosphere each year. They are pieces of stone, iron, or stony iron conglomerates, and they range in mass from fractions of a gram to hundreds of kilograms. The smallest are incinerated immediately, emitting little if any visible light. Somewhat larger pieces appear as meteors or shooting stars, brief luminous streaks in the night sky, which result from their incandescence. Meteorites are meteoroids that survive passage through the atmosphere and fall to earth. These larger chunks of matter signal their arrival by appearing as huge fiery masses or fireballs, which light up a broad area of the sky at night and sometimes in daylight. Their extremely rapid transit through the atmosphere creates sonic booms that people far distant from the scene can hear. Fireballs that detonate are called bolides. Sometimes the explosion causes complete disintegration of the meteoroid, so that only dust particles remain to drift slowly to earth. In other instances, the explosion yields a few fragments or a shower of several thousand pieces, which become meteorites. Very rarely, a huge mass weighing thousands of tons penetrates the Earth's atmosphere, and upon impact with the ground vaporizes almost completely, producing a massive crater, such as Barringer Meteorite Crater in Arizona.

Meteorites have undoubtedly been falling to earth for eons. Scientists make a distinction between meteorite falls and finds. A fall is a meteorite, which observers actually witnessed and the fragments of which were recovered very soon after the event. A find is a meteorite whose fall was not seen, but which was later recognized as a meteorite because of its chemical composition, its mineral constituents, or its structure. A large number of the meteorite finds are irons or stony irons, both because they have a greater resistance to weathering and because their presence has seemed unusual to farmers, hunters, or primitive peoples who have encountered them. It requires keen observation, however, to discriminate between a terrestrial rock and a meteoritic stone, which, particularly if it is friable, disintegrates over time.

Each meteorite has a name—usually that of the town, village, or geographical landmark closest to where it fell or was found. Since the distribution of meteorites is worldwide, a list of their names recalls the index of an atlas: Pultusk (Poland), Sikhote-Alin (USSR), Ogi (Japan), Queen's Mercy (South Africa), or Dingo Pup Donga (Australia). Before there was universal agreement on nomenclature, some meteorites had several names, which caused confusion. For example, Babb's Mill is a meteoritic iron that was found in 1842 in Greene County, Tennessee, and was named alternatively Blake's Iron, Greene County, or Troost's Iron. Herein, when a former name appears in a quotation, I place the standard name in brackets following it—for example, Otumpa [Campo del Cielo], or the Pallas iron [Krasnojarsk]. Meteorites only occasionally carry the name of the individuals who found them; an example is Lutschaunig's Stone, recognized by A. Lutschaunig in 1861 at Copiapo, Chile. Recently, because of the large number of finds on the barren Antarctic continent, meteorites receive an identifying letter-and-number combination. ALHA81005, for example, designates the fifth specimen found in 1981 at Allan Hills. It so happens that this particular meteorite was the first one recognized by scientists as having been catapulted from the surface of the Moon into an orbit that eventually crossed that of the Earth.

The first chapter and part of the second are devoted to a description of the underlying causes of eighteenth-century scientific disbelief or skepticism concerning the fall of meteorites, and to a review of the sequence of events that led to scientific acceptance of the phenomenon. One reason for my detailed analysis is that I hope thereby to counter misinformation and to correct errors, which through repetition have become rooted in accounts of or references to the actions or attitudes of eighteenth-century individual scientists or scientific bodies with respect to meteorite falls and finds.

A more important reason is that the episode furnishes a remarkable illustration of how a theory can dominate the minds of even the best scientists. This observation is, of course, not a new one. Historical accounts of the revolution in astronomy and physics that led to the rejection of the Aristotelian-Ptolemaic cosmos and to its replacement by that of Copernicus, Galileo, and Newton have clearly demonstrated how tenaciously scientists can cling to a familiar theory. With respect to the fall of meteorites, however, there were no religious beliefs that were possible influencing factors. The prevailing theory was that, apart from the celestial bodies, interplanetary or cosmic space did not contain any matter except the subtle and invisible aether. Therefore, neither stones nor irons could originate there. A subsidiary eighteenth-century theory affirmed what was true—namely, that stones or minerals could not be generated in the Earth's atmosphere. Taken together, then, theory held that stones could not and did not fall from the heavens. It is also noteworthy that acceptance of falls occurred within a relatively brief period of time, when evidence of falls accumulated.

Even at the present time, meteorites are enigmatic objects. Scientists who study them are intrigued not only about their place of origin but also about the conditions that prevailed there and gave rise to their characteristic chemical and mineralogical compositions and structures. Meteorites have been investigated in various ways by scientists from a number of different disciplines. Karl Reichenbach in 1858 called attention to the fact that a meteorite is simultaneously a cosmological, astronomical, physical, geological, chemical, mineralogical, and meteorological object—that is to say, a meteorite can be viewed from many different perspectives. It is significant, then, that the various hypotheses proposed about the origin and history of meteorites have usually been intimately associated with the disciplines of the scientists advancing them. Mineralogists, for example, have elicited evidence from meteorites, which they have interpreted as favorable to one theory, and have discounted other evidence which astronomers believed convincingly supported an alternative theory. Only in the past three decades have scientists studying meteorites been equipped, because of their broad educational backgrounds, to consider meteorites from many of the perspectives mentioned by Reichenbach. The matter is important; viewing a natural or social phenomenon holistically is more likely to yield the most satisfactory explanation.

Scientists who investigate meteorites also face practical problems. The difficulties that nineteenth-century scientists encountered in completing reasonably accurate chemical analyses and in identifying the minerals in meteorites have now been largely overcome. However, as scientists have begun to study meteorites much more intensively during the past thirty years, new problems have arisen. I have described several of the obstacles in some detail, because the methods and techniques that scientists employ in efforts to arrive at accurate and reproducible results are sometimes overlooked. Painstaking (and often boring) experimentation, observation, and data gathering are, in my view, important components of science.

New instruments or instrumental techniques have had a distinct role in advancing knowledge about meteorites. I have not described these in detail, but trust that I have emphasized their significance. In the nineteenth century, the invention of the polarizing microscope and the technique of preparing thin sections of meteorites for detailed study provided mineralogists with the capability of identifying mineral constituents and structures. Moreover, improved telescopes aided astronomers both in the determination of cometary orbits and in the discovery of more and more asteroids. In the twentieth century, X-ray diffraction techniques, the mass spectrometer, the electron microprobe, and high-speed computers are among the numerous instruments and techniques that have significantly advanced scientific understanding of meteorites.

The recovery and collection of meteorites has been an important activity ever since scientists accepted the fact that they fell. I have described in chapter 6 the beginnings and development in the nineteenth century of several of the major museum collections in Europe and the United States. From the outset there was a lively market in meteorites, and it seems clear that shrewd Yankee dealers usually got the best of the bargain in transactions with their European cousins. Meteorite collections became symbols of civic pride and national prestige. High prices brought wails of anguish from scientists, who felt that the importance of meteorites for scientific research was becoming subservient to their preservation as precious relics.

How peoples at various times and in different places interpreted the appearance of meteors and fireballs and the fall and finding of meteorites is another important subject. Folklore and legends reflect cosmographical and religious beliefs, but the origins of some practices associated with the phenomena are buried in the distant past and still require clarification. We know that meteorites inspired intense fear among some peoples and were revered and worshiped by others. Yet in some primitive societies peoples used the irons to manufacture jewelry, tools, and weapons.

I am fully aware that I have given scant attention to some important aspects of meteoritics and have completely omitted mention of others. Some scientists, for example, have devoted a great amount of time to the study of the physics of the fall of meteorites through the atmosphere, how the attitude of the meteorite affects its loss of mass through ablation, and how crusts form. These investigations were of crucial importance in the design of reentry shields for space vehicles. Meteorite impact theory is another subject that deserves much more extensive treatment. I can only plead limitations of space for my failure in this respect.

Meteorites are fascinating objects that were formed during the birth of the Solar System. I hope that what I have included in this book will give readers a better understanding of how scientists and nonscientists alike have regarded and studied them.

1

Disbelief

Most eighteenth-century natural philosophers either did not believe or doubted that meteorites fell. This denial of the occurrence of an unusual but nonetheless genuine natural phenomenon has been an embarrassment to scientists, particularly because it persisted for about a century and therefore implies a serious error of judgment on the part of the scientific community. One explanation that appears to have gained general currency in scientific articles and texts about meteorites is that Enlightenment natural philosophers—especially the members of the Académie Royale des Sciences in Paris—were, in their justifiable desire to eradicate popular superstitions, overzealous, dogmatic, and too hasty in their dismissal of eyewitness reports.[1] The apparent parallel between the attitude of many twentieth-century scientists toward reports of unidentified flying objects or sightings of giant sea serpents and that of Enlightenment scientists toward meteorites has not escaped notice and has provoked defensive postures.[2] Indeed, this analogy has caught the attention of sociologists of science, one of whom has turned it to good account in a general critique of the reasons why the scientific community has failed to assess properly reports of anomalous events.[3] All explanations to date of Enlightenment skepticism about the fall of meteorites share one defect: they fail to describe eighteenth-century ideas about meteorology or the origin of minerals, and they do not give an adequate account of chemical knowledge and practice at that time. However naive the then prevailing theories as to the causes of atmospheric phenomena may appear by twentieth-century standards, eighteenth-century scientists had reasons for their disbelief—reasons apart from attributing reports of the fall of stones to superstition, exaggeration, or plain idiocy.

Four related topics require investigation. We should first review eighteenth-century theories of the causes of meteors and fireballs, and trace the sources of these ideas. Then we should examine theories as to how minerals formed in the earth and thunderstones in the atmosphere. Next, we should determine why scientists failed to make the connection between fireballs and meteorites, and why they were unable to distinguish meteorites from terrestrial stones and irons. Finally, we should summarize, insofar as the evidence permits, the attitudes of eighteenth-century scientists toward reports of actual meteorite falls.

THE LEGACY OF ARISTOTLE'S *METEOROLOGICA*

The word *meteors* in the eighteenth century did not have the meaning it has at present. In his *Lexicon Technicum* (1704), John Harris wrote that:

> METEORS . . . are various impressions made upon the Elements, exhibiting them in different forms, . . . for the most part, they appear high in the Air, and they are either Fiery, Airy, or Watery. *Fiery Meteors,* are such as consist of a Fat, Sulphureous kindled Smoak, whereof there are several kinds: as *Ignis Fatuus* [will-of-the-wisp], *Trabs* [beam], *Ignis Pyramidalis* [aurora], *Draco Volans* [shooting star], *Capra Saltans* [leaping goat], *Thunder* and *Lightning,* etc. *Airy Meteors,* are such as consist of Flatuous and Spirituous *Exhalations,* as *Winds,* etc. *Watery Meteors,* consist of Vapours, or Watery Particles, by the Action of Heat separated from each other and variously modified as *Rain, Dew,* etc.

Meteors, then, were transitory atmospheric phenomena caused by the interaction in the air of terrestrial vapors or exhalations and consisting of elemental air, fire, or water. Their various manifestations depended on the particular kind of mixing process that had occurred in the atmosphere. Beams and goats were different types of fireballs. The former exhibited a broad, somewhat rectangular swath of flame across the sky. The latter appeared to be almost spherical and seemed to jump or swerve in its path.

This early eighteenth-century description and explanation of meteors stems directly from the meteorological theory established by Aristotle in the fourth century B.C. Aristotelian astronomical and physical theory had been destroyed by the scientific revolution of the sixteenth and seventeenth centuries. The planets with their satellite moons were shown to revolve in elliptical orbits around the sun in accordance with the law of universal gravitation. The spheres, supposedly holding the planets and the fixed stars in their daily orbits around the Earth, vanished, and the Earth with its Moon was assigned its proper place as the third planet distant from the Sun. However, Aristotle's celestial element—the quintessence or aether—remained, both to transmit light from the Sun and the stars and as the medium by which the gravitational force was exerted throughout the universe. With a few exceptions, Aristotle's teachings about phenomena occurring in the area between the Earth and the Moon—the sublunary region—were still respected. One major exception was the theory of comets, which Aristotle believed were meteors occurring in the higher regions of the atmosphere. Seneca disputed this notion in the first century A.D., but it was not until the late sixteenth century that observations, especially by Tycho Brahe, demonstrated conclusively that comets were astronomical phenomena. Another change in thinking had to do with the Milky Way, which Aristotle held was a sublunary appearance. A few medieval scholars, such as Ananis Shirakatsi of Armenia in the seventh century and Albertus Magnus in the thirteenth, considered the Milky Way as a collection of stars, and the telescopic observations of Galileo proved their ideas to be correct.

Aristotle's views of sublunary phenomena, drawn principally from the teachings of Heracleitus of Ephesus, are united in the *Meteorologica,* written about 340 B.C. and considered authentic by most Aristotelian specialists. The work was well known in classical times and late antiquity. Books I through III were translated into Latin from the Arabic by Gerard of Cremona in the late twelfth century, and into Latin from the Greek by William of Moerbeke in the mid-thirteenth century. Its popularity thereafter is evidenced by the fact that almost three hundred manuscripts of the work are preserved in collections dating from the late thirteenth to the fifteenth century.[4] In the early modern period, scientists were fully conversant with the *Meteorologica;* its influence was pervasive and continuing.

In the *Meteorologica,* Aristotle described just about every conceivable terrestrial and atmospheric phenomenon, and explained the cause of each. The classical four elements—earth, air, fire, and water—or, more precisely, the qualities that they embodied, lay at the basis of his teaching. The four opposing qualities were dry, moist, hot, and cold, and their intermixture produced hot-dry (fire), hot-moist (air), cold-moist (water), and cold-dry (earth). Ideally, each of the elements occupied its own sphere below the Moon. The celestial revolutions, however, and the heat of the Sun, which generated exhalations from the Earth, caused the intermixture of the elements and produced the changing phenomena both of the atmosphere and of the Earth. Aristotle's writing style is pedantic and didactic. Nevertheless, it exudes self-assurance and confidence. His descriptions had just enough verisimilitude to make them palatable and convincing to naturalists for many centuries.

Aristotle explained that two kinds of exhalations arose from the Earth. One was a vapor, emanating from the water within and upon the Earth, while the other was dry, windy, and smoky, and came from the crevices in the Earth. The latter, called "fire," rose to the top of the sublunary region, "immediately beneath the circular celestial motion," and was inflammable. "Wherever then conditions are most favorable," Aristotle wrote, "this composition bursts into flame when the celestial revolution sets it in motion." What appeared in the sky when the material ignited depended upon the quantity and location of the material. If it extended both lengthwise and breadthwise, then one saw a "beam"; if lengthwise only, then "torches" or "goats" appeared. A "goat" threw off sparks; a "torch" did not. If the exhalation was separated into small parts, which were scattered about, then shooting stars resulted.

Sometimes, Aristotle continued, fiery meteors appeared much lower in the sublunary region, but originated from a different cause. Occasionally, the air contracted because of the cold, and the resulting pressure forced out heat, producing meteors that took a course similar to that of a projectile in motion. All these fiery meteors took place below the lunar sphere, Aristotle emphasized, "a proof of which is the fact that the speed of their movement is comparable to that of objects thrown by us, which seem to move faster than the stars and sun and moon because they are close to us." The dry exhalations also caused lightning and thunder.[5]

Certain natural phenomena known to the Greeks supported the theory of exhalations. Smoke and steam issue from volcanoes, fumaroles and hot springs give off odorous gases, and combustible gases occur in mines. Moreover, fireballs do appear to be composed of combustible substances, and lightning bolts do cause fires. It is not surprising, then, that Aristotle's theory of exhalations was accepted by most ancient writers who described meteors. Thus Seneca, though denying that comets were sublunary phenomena, followed Aristotle closely in other respects. Describing fireballs he wrote: "Our sight does not discern their passing but believes the entire path is on fire wherever they fly. The speed of their transit is so great that its stages are not observable. Only movement as a whole is grasped."[6] Pliny the Elder, about A.D. 75, stated that the sublunary region consisted of superior air and terrestrial exhalations, and from there emanated hail, hoar frost, the rains, the tempests, and the whirlwinds. Once in a while, he added, torches suddenly blazed forth, of which two kinds could be distinguished: the beautiful simple torches, which made long tails by burning in their anterior parts; and the bolides, which, burning from end to end, traced an extended wake of fire.[7]

Infrequently, there were alternative explanations of meteors. For example, in reporting the ideas of the atomist Epicurus, Diogenes Laertes, in the third century A.D., agreed that shooting stars might result when certain parts were expelled from a mixture of fire and air or when confined wind burst forth and ignited. But they might be caused, he wrote, by the mutual friction of the stars, or they might be "due to the meeting of atoms capable of generating fire."[8] However, the Aristotelian tradition proscribed both disorder in the celestial regions and atomistic explanations of phenomena, so such speculations did not provoke commentaries.

When we turn to late seventeenth- and early eighteenth-century explanations of fiery meteors, we find that the terrestrial exhalations still remain, although their characteristics have changed owing to contemporary chemical theory. John Woodward in 1695 declared that when the Sun's power was great, it raised exhalations "out through the Mouths of Caverns, and through the Ordinary Cracks and Pores of the Earth, mounting them up, along with the Watery Exhalations, into the Atmosphere, especially Sulphur, Nitre, and the other more light and active Minerals; where they form *Meteors;* and are particularly the Cause of *Thunder* and of *Lightning*."[9] In the *Opticks,* Sir Isaac Newton wrote that "sulphureous Steams, at all times when the Earth is dry, ascending into the Air, ferment there with nitrous Acids, and sometimes taking fire cause Lightning and Thunder and Fiery Meteors."[10] The "sulphur" and "nitre" here mentioned were not ordinary sulfur and niter. "Sulphur," mercury, and salt were the three "principles" in the chemical tradition inaugurated by Paracelsus in the early sixteenth century. "Sulphur" was the principle of combustibility, variously characterized as moist, oily, clammy, fatty, and light. "Nitre" was a volatile salt, present in living things and chemically active in the vapor form.[11]

In Paris, the academician Nicolas Lemery attempted to demonstrate the existence of sulfurous exhalations and their role in the production of fiery meteors by drawing an analogy from experiments.[12] In the heat of summer, he thoroughly mixed equal quantities of iron filings and pulverized sulfur in a large jar, added water, placed the container in a hole in the earth, and covered it with a rag and some soil. In about eight hours, the earth above the jar swelled, became hot, and developed fissures from which hot sulfurous fumes emerged. Soon these ignited, enlarging the crevices and dispersing a blackish yellow powder.

Lemery declared that the identical process took place within the earth as occurred in his experiment. Violent fermentation of iron and sulfur caused earthquakes, and the subterranean fires created by the sulfurous exhalations, if extensive and powerful enough, produced volcanoes on land and water spouts at sea. The sulfurous exhalations rising into the air became mixed with the clouds, and when greatly compressed, they ignited and appeared as lightning. Lemery anticipated the questions of how the sulfurous exhalations could be ignited in clouds formed of water, and why the flames were not extinguished when compressed. Sulfur, he explained, was an oily substance, not affected by water. It could be ignited in water and burn therein, just like camphor and other sulfurous materials. He granted that a portion of the vapor might be extinguished after a great detonation, but maintained that the more subtle portion of the sulfur—that more disposed to motion—continued to ignite.

To explain the source of the "sulphur" within the earth, Lemery performed another experiment. He poured some *esprit de vitriol* (concentrated sulfuric acid) into a flask, diluted it with water, and warmed it gently. Then he gradually added iron filings. Soon there was an ebullition of white vapors that, when exposed to a lighted candle, ignited and also exploded. An important conclusion to be drawn from this experiment, Lemery wrote, was that "the iron contains a great deal of sulphur . . . ; necessarily then the sulphur of the iron filings having been rarefied and expanded by the spirit of vitriol is excited as an exhalation very susceptible to fire." Certainly, Lemery was misled by the confused state of contemporary chemical knowledge. He considered the vapor evolved in this experiment to be sulfurous exhalations; other mid-eighteenth-century chemists thought it was phlogiston; and eight decades later adherents of the new chemistry identified it as hydrogen gas. In any case, many scientists believed it to be the cause of meteors and fireballs.

In the mid-eighteenth century, scientists eliminated two types of fiery meteors, the aurora borealis and lightning, from the list of those supposedly caused by terrestrial exhalations. An aurora in March 1707, which was seen over most of central Europe, aroused scientific interest, and over 160 auroras were reported from 1716 to 1732.[13] William Whiston in England and J. P. Maraldi in France adhered to tradition and explained them in accordance with the exhalation theory. There were, however, now alternative theories. Edmund Halley proposed that auroras resulted from the flow of luminous magnetic effluvia from the Earth's polar regions. J. J. Dortous de Mairan supposed that the solar atmosphere extended outward from the Sun to such a great distance that it sometimes enveloped the Earth. When this occurred, the differing densities of the solar and terrestrial atmospheres produced auroras. Leonhard Euler made an analogy between auroras and the tails of comets, which, he wrote, glowed by reflected sunlight. Similar to comets, the Earth and the other planets had tails, whose particles accumulated at the poles and glowed by reflected sunlight at times when clouds did not absorb the solar rays.

Meanwhile, a number of scientists investigating electricity had postulated the electrical nature of lightning, and in 1752 Benjamin Franklin's proposals and experiments proved this hypothesis to be true. Within a year John Canton suggested that the aurora was a passage of "electrical fire from positive to negative clouds at a great distance, through the upper parts of the atmosphere, where the resistance is least."[14] Though Canton's suggestion was far off the mark, the notion that the aurora was a manifestation of electricity gained ground. Moreover, the idea was immediately extended to include meteors. In 1753, for example, Franklin asked his friend John Perkins of Boston to send him his thoughts on the cause of shooting stars. Perkins replied: "As to shooting stars . . . I imagine them to be passes of electric fire from place to place in the atmosphere. . . . Electric fire in our globe is always in action, sometimes ascending, descending, or passing from region to region. I suppose it avoids the dry air, and therefore we never see these shoots ascend."[15] Perkins's reply demonstrates how rapidly hypotheses could be fashioned to offer explanations of phenomena in terms of a new discovery.

From mid-century onward, then, many scientists considered that meteors and fireballs were unusual manifestations of atmospheric electricity. Others thought that they were closely associated with auroras, which were, in turn, dependent in some manner upon electricity or magnetism. These ideas became strong competitors to the explanation of meteors as ignited terrestrial exhalations. But electricity did supply a vital ingredient to the exhalation theory, for it served as the most probable cause of the ignition of the atmospheric vapors.

THE GENESIS OF MINERALS AND THUNDERSTONES

The contents of the Aristotelian cosmos and the Newtonian one were almost exactly the same. The vast differences between the two lay in other areas: the distances, relative positions, and motions of the celestial bodies; the anisotropy of Aristotelian space versus the isotropy of Newtonian space; the function of the aether; and the fact that change occurred in the Newtonian cosmos—for example, stellar novae. But with the exceptions of the change in the status of comets from sublunar to astronomical phenomena, and the identification of the constituents of the Milky Way as stars, the contents of interplanetary space were identical. There were the stars, the Sun and its planets and their satellites, and the aether. There was nothing else.

Robert Hooke in his *Micrographia* (1665) succinctly expressed the prevailing conviction that no small bodies existed in interplanetary space. Hooke attempted to reproduce experimentally how the lunar craters that he had observed telescopically might have been formed. He dropped some bullets into a mixture of clay and found that they produced cavities similar to the lunar craters. However, he concluded that there was no probability that lunar craters would be formed by the impact of objects, "for it would be difficult to imagine whence these bodies should come."[16]

Newton ruled out the existence of any but "the great bodies, Fixed Stars, Planets, & Comets" in the cosmos,[17] stating that "to make way for the regular and lasting Motions of the Planets and Comets, it's necessary to empty the Heavens of all Matter, except perhaps some very thin Vapours, Steams, or Effluvia, arising from the Atmospheres of the Earth, Planets, and Comets."[18] Gravitational theory dictated that each planet with any accompanying satellites was a self-contained entity, as if wrapped in a cocoon, and that interplanetary space was free of all matter except an invisible aether.

Newton's ideas were reiterated throughout the eighteenth century. Euler, for example, reaffirmed that the universe contained only fixed stars, planets, and comets.[19] The discovery of Uranus in 1781 by Sir William Herschel merely added another planet to the Solar System; it did not alter conceptions of the contents of interplanetary space. French scientists termed it *le vide planétaire*—the planetary void. Ernst Chladni described this idée fixe in 1794 in a paragraph that was both bitter and plaintive:

> Now . . . the statement that in the reaches of space there exist in addition to the celestial bodies many smaller aggregations of matter appears to many so incredible that they repudiate [my] entire theory [of meteorites]. The incredibility of this statement is only pretended; it is not based on reason, but to a much greater extent on the fact that the statement is unusual, and also somewhat strange. For, if a statement were abstracted from accepted physical theory which read "no material bodies exist in space other than the celestial bodies and the stars or some other thin elastic fluid," it would be considered just as preposterous as the statement: "other material bodies exist there." Neither one can be proved on *a priori* grounds; observations must decide which is correct.[20]

Chladni certainly went to the heart of the problem, but he underestimated the tenacity of the idea. In both Aristotelian and Newtonian theory, it was impossible for there to be small bodies wandering in interplanetary space which might occasionally invade the Earth's atmosphere. The stones that were reported to have fallen must of necessity have a different origin.

To understand the eighteenth-century explanation of meteorites, we must again return to ancient authors. Legends and chronicles in both the Orient and the Occident record numerous instances of the fall of stones and other matter from the sky. Some of these occurrences and their implications for religion and mythology are treated in chapter 7. One fall, that at Aegospotami about 467 B.C., is important because several references to it were well known in early modern times.

According to Pliny, the Greeks honored Anaxagoras of Clazomene because on the basis of his astronomical knowledge he had predicted that on a certain day a stone would fall from the Sun. A stone, Pliny continued, as big as a wagon and of a burned color, did fall in broad daylight in Thrace near Aegospotami. Pliny acknowledged that stones occasionally fell from the sky, but ridiculed both the claim that Anaxagoras had predicted the event and the notion that stones might fall from the Sun.[21] Plutarch, in his life of Lysander, mentioned the same event, but wrote that Anaxagoras foretold that the stone would come from one of the heavenly bodies on which a landslide or earthquake had occurred. Plutarch also cited another commentary, which stated that "before the stone fell, for seventy-five days continually, there was seen in the heavens a vast fiery body. . . . But when it afterwards came down to the ground . . . there was no fire to be seen . . . only a stone lying big indeed, but which had no proportion, to speak of, to that fiery compass. . . . [This] proves those to be wrong who say that a rock broke off the top of a mountain, by winds and tempests . . . and fell to the ground."[22]

In the last sentence, Plutarch undoubtedly referred to Aristotle and his followers. The Aegospotami fall is the only one mentioned in the *Meteorologica* in Aristotle's discussion of comets. He wrote that when comets appeared frequently, the weather was dry and windy; when less frequent, there were excessive winds: "For instance when the stone fell from the air at Aegospotami, it had been lifted by the wind and fell during the daytime: and its fall coincided with the appearance of a comet in the west."[23] Aristotle must have been at his wit's end when he proposed this explanation. Evidently, he was certain that the fall of the stone had occurred. The stone could not have come from the aetherial regions as Anaxagoras asserted, because such an event would have compromised the perfection of the heavens, a cornerstone of Aristotle's natural philosophy. Nor could it have come from the sublunary region, because Aristotle taught that stones were generated in the earth. The dry exhalation by the action of heat produced "fossiles" or infusible stones, whereas fusible stones including metallic ores were formed from moist exhalations.[24] His recourse was to have the wind lift the stone into the atmosphere from which it fell.

Although Plutarch found Aristotle's suggestion ludicrous, Agricola in the sixteenth century accepted the idea. In *De ortu et causis subterraneorum* (1558), Agricola was concerned with the problem of how stones and metallic ores were formed in the earth. Along with the views of Theophrastus, Albertus Magnus, and Avicenna, Agricola presented the teachings of Aristotle, and he questioned the idea that infusible stones were produced from dry exhalations, because he thought they would be generated much more abundantly in the upper regions of fire than in the earth: "Every time that comets, torches, flames, and meteors would be formed, stones or earths would fall; but as we see, this does not happen. Though, of course, authors of miraculous histories report that it has rained stones, yet Aristotle has made no mention of these; indeed, he himself writes that the stone which fell from the sky had first been raised high in the air by the force of the wind."[25] Agricola supported Aristotle's explanation by reference to a nonferrous ore smelting process, in which lighter particles were driven into the furnace flue while heavier matter settled on the hearth. By analogy, he declared, fire in ore bodies in mountainous regions could result in scattering quantities of pumice and ash far and wide, into the fields below the mountains and even out to sea. But he maintained that if stones did fall, they must have a similar composition to those formed in the earth. The reason for Agricola's equivocation was that he was acquainted with the portion of Avicenna's *Kitâb Al-Shitâ* that reported the fall of several meteorites.

Avicenna, or Ibn Sīnā, the encyclopedic Islamic philosopher and physician, lived in what is now Iran from 980 to 1037. The portion of his works which concerns us was translated into Latin about 1300 as *De congelatione et conglutinatione lapidum*. In the few passages in which Avicenna treats meteorites, the first statement that catches the modern eye is his assertion that both iron and stony bodies fall—a clear indication that Avicenna was familiar with meteorites and knew there were different types of them. To be sure, Avicenna followed Aristotle in attributing their origin to "the accidental qualities of coldness and dryness which fieriness acquires when it is extinguished," and he also reported that coppery bodies in the shape of arrowheads occasionally fell during thunderstorms. Most compelling, however, is his account of the fall of a large iron meteorite, which, he wrote, occurred in his own lifetime, and which he regarded, on unimpeachable evidence, to be true. Since the iron mass was too heavy to move, on the orders of the sultan a piece was cut off, but only after the breakage of many tools. This fragment was sent to the sultan, who ordered a sword to be made from it, but the task proved too difficult because of the character of the material.[26]

Avicenna's account clearly puzzled Agricola. He respected Avicenna as an authority, and dutifully informed his readers that Avicenna had documented the fall not only of stones but also of iron. But he wondered whether Avicenna might have been deceived by the "authors of miraculous histories." Agricola was convinced that there were secreted within the earth various kinds of "lapidific juices," each of which concreted into a specific type of stone, mineral, or metallic ore. For stones or iron to form in the air, the particular type of lapidific juice would have to be present, as well as some means by which the matter would be compacted. Unable to imagine how such conditions might occur, Agricola concluded his discussion by reiterating that it was much more acceptable to believe that stones had their origin in the earth.[27]

For a century and a half, from Agricola's time to the early eighteenth century, authorities differed as to whether stones could form in the sky. Ulisse Aldrovandi, an influential professor of natural sciences at the University of Bologna, supported Agricola's reasoning and conclusion. He reported an account of the recent fall in the kingdom of Valencia of a "stone mass filled with metallic veins," but he insisted that metals were only generated with considerable difficulty. Oily vapors were required together with a proper mixture of the mineral quality, as well as a place where the metallic material could remain for an appropriate period while it was being perfected. Such conditions, he emphasized, were found least of all in the air or in the clouds.[28]

Others, however, maintained that stones could be generated in the air. Anselm Boece de Boodt, physician to Emperor Rudolph II, explained that stones sometimes formed in the atmosphere, when "a very great exhalation composed of many terrestrial particles is hardened and confined in a small volume by the cold of the clouds that envelop it."[29] René Descartes also declared that very hard stones could be produced in clouds, but that heat was the agency in the form of lightning. The sediment at the bottom of a jar of rainwater gave evidence that there was earthy material in the clouds, and experience proved that if a mixture of this earth, sulfur, and saltpeter was heated, it would form a stone.[30] Other Cartesians supported this hypothesis, although Jacques Rohault was puzzled that a thunderstone had not been seen to fall on some occasion "in one of the streets of this great city [Paris], or in some courtyard or on the roof of some house."[31]

After 1700, however, there was general agreement that stones did not form in the atmosphere. Nicolas Lemery was apparently the leader in the opposition to the notion, because his dictum was cited often in the following eight decades. It was not absolutely impossible, Lemery wrote, that violent winds rising into the clouds might at times carry along stony or mineral matter, which was softened and compacted by heat to produce what were termed thunderstones. But, he added, "there would be much more reason to believe that it came from mineral matter in the earth itself and formed by the ignited sulfur of lightning, than to suppose that the stone was generated in the air or clouds and ejected along with the thunder."[32]

As already noted, Franklin's electrical theory of lightning superseded Lemery's notion that "ignited sulphur" was the cause. However, the idea that meteorites originated *in situ* by the action of lightning became one of the leading explanations of meteorites falls and finds for the rest of the eighteenth century. Diderot's *Encyclopédie* stated that thunderstones were merely mineral matter that had been fused or scorified by the action of lightning.[33] Johann G. Wallerius, probably the most influential mid-eighteenth-century mineralogist, declared that thunderstones were neither raised into the clouds by winds nor generated there, but produced on Earth by the action of lightning.[34] In a letter written in 1767 to Benjamin Franklin, Giambatista Beccaria attributed the Albareto fall, reported by Domenico Troili, to the agency of lightning: "The soil at Modena is everywhere full of the nearby water. Wherefore the bolt, driving through the stone, which is metallic, into the water beneath, should scatter the water and hurl the stone into the air while covered in its own flash, so that it be not seen until afterwards, when it falls back down."[35] It is hardly surprising, then, that the committee of the Académie Royale des Sciences—composed of Fougeroux de Bonderoy, Cadet de Gassicourt, and Lavoisier, and charged with the investigation of the Lucé fall of 1768—concluded that the stone must have been struck by lightning, and that the peasants who testified to its fall had been deceived.[36]

The reality of thunderstones, supposedly formed in the clouds and cast to earth, was attacked from another quarter—that of archaeologists and natural historians. Some of these so-called thunderstones or *pierres de foudre* were, in fact, unusually shaped crystals of pyrite or marcasite; others called "cerauniae" were Stone Age axs, hammers, or spearheads; and still others, termed "brontia" or "ombria," were fossil echinoids or sea urchins. The first detailed treatise on such objects was written in 1565 by Conrad Gesner of Zurich, a botanist and natural historian, and included sketches of ombria and cerauniae. Four of the latter, either pyramidal or oval-shaped with holes, are obviously hammers or axs. Because all cerauniae had recognizably geometric shapes, Gesner was puzzled about "a stone, which fell from heaven in the year of our Lord 1492, which may be seen suspended in the church at Ensisheim. . . . It had (I think) no definite shape (unless by chance it was diminished by many fragments having been taken away)."[37] Here Gesner described a genuine meteorite, but was so convinced that thunderstones should have figured shapes that he speculated that fragments detached from the stone probably destroyed its geometric form.

A few decades later, Michele Mercati, director of the Vatican botanical garden, voiced the suspicion that cerauniae were stone implements made by humans who were unacquainted with bronze or iron. His view was not widely known in the seventeenth century, because his work on minerals and fossils was not published until 1717. However, others had the same idea. The English antiquarian, Sir William Dugdale, stated his belief in 1656 that cerauniae had been formed by human hands. About the same time, Olaus Wormius, a Danish physician and archaeologist, wondered about some Danish flint blades: "If flint could be easily pierced, you might swear this object had been fashioned rather by art than by nature." And about flint daggers, he wrote: "Considering which I am uncertain whether they are works of art or of nature. Are they what may be referred to as cerauniae, or are they what may be considered to have been old swords?"[38]

Wormius's comments reveal what puzzled him and other archaeologists. The civilized peoples of the early modern period could not conceive that humans had ever expended the patience and care or possessed the dexterity and skill to fashion such tools from hard stone. But when stone implements were sent back from the Americas to Europe, understanding rapidly followed. In 1723, Antoine de Jussieu, referring to stone specimens of an ax, wedge, and arrowheads obtained from Indians of the Caribbean and of Canada, explained in a memoir to the Paris Academy of Sciences how such implements were made without the use of any metal tools. Throughout his memoir, Jussieu stressed that the so-called thunderstones found in Europe were also the tools and weapons of the early inhabitants and did not have a celestial origin.[39]

Further evidence discounting the atmospheric origin of thunderstones rapidly accumulated. Nicolas Mahudel in 1734 gave a detailed description of chisels, hammers, hatchets, picks, digging tools, and lances, all presumed to be cerauniae, and added that the usage of such implements continued well after the discovery of bronze and iron. Brontia, he stated, were petrified echinites, and other alleged thunderstones were figured marcasites and pyrites.[40] Both J. B. L. Romé de l'Isle and René Just Haüy gave good descriptions of the latter *pierres de foudre*. Romé de l'Isle in 1767 wrote that they were large globular specimens of pyrite with spear-shaped protuberances projecting from their surfaces, forming thereby stellated cavities. In 1801 Haüy named this particular type of pyrite *fer sulfuré radié*. This variety, he wrote, was usually found in globular or ovoid crystalline masses. The lower portions consisted of pinnacle-shaped projections pressed against one another and directed toward a common center. The upper portions, jutting out from the surface, were usually the halves of an octahedron, sometimes cubic-octahedrons, and occasionally cubes bearing striations in three directions.[41] It is apparent that both Romé de l'Isle and Haüy were describing massive but not unusual specimens of iron pyrite or marcasite belonging to the isometric system.

In the early eighteenth century, then, there was a convergence of two trends. The first was the growing conviction of chemists and mineralogists that it was impossible for stones to be generated in the atmosphere or in the clouds, whether or not they were thunderheads. The second trend emerged from the studies of early paleontologists, archaeologists, and mineralogists, which tended to prove that the stones alleged to have fallen during thunderstorms were either fossils, ancient stone implements, or crystal masses of a common mineral. One can begin to understand the increasingly militant skepticism among scientists concerning the fall of stones from the sky. But belief was also undermined by reports, both ancient and modern, of the fall of other objects, which eighteenth-century scientists could only regard as fabulous.

Pliny, for example, wrote that rains of milk and of blood had been recorded in the consulate of Manius Acilius and Gaius Porcius (114 B.C.) and on many other occasions; that there had been a rain of flesh in the consulate of P. Volumnius and Servius Sulpicius (461 B.C.), which did not putrefy and which carrion did not eat; and that in the consulate of L. Paulus and C. Marcellus (50 B.C.) there was a rain of wool and another of kiln-dried bricks. Interspersed among these reports were accounts of other falls that may well have been meteorites, such as a fall of iron in Lucania in 54 B.C. and the fall of the stone at Aegospotami, mentioned previously.[42]

Reports of such events continued to accumulate through the centuries. Rains of blood were announced to have fallen in the ninth, fifteenth, and on several occasions in the sixteenth century. A burning object was said to have fallen in 1110 into a lake in Armenia, making its waters bloodred, while at the same time the earth nearby was cleft in several places. In 1618 a shower of blood occurred in Styria, and a red rain fell in 1638 at Tournay, in 1640 at Brussels, and in 1645 at Bois-le-duc. Gelatinous matter was reported to have fallen in 1718 in India.[43] And on 31 January 1686 at a village in the province of Courland [Latvia], a large quantity of a paperlike substance fell from the sky during a snow flurry.[44]

In a memoir read to the French Academy of Inscriptions and Belles Lettres in 1717, Nicolas Fréret attempted to distinguish between fact and fancy in these reports. He dismissed the rains of flesh, asserting that it could not have been flesh, and attributed the red blots seen on the ground, which seemed to give evidence of showers of blood, to the chrysalises of caterpillars. A fall of money reported by Dio Cassius in his history of the reign of Severus, Fréret said, was undoubtedly mercury, which had been elevated as a vapor and fell when it was congealed by the cold. As for the fall of stones, his belief was that they did occasionally fall, and that they had been ejected from volcanoes and might be transported in the air for hundreds of miles and for long periods of time before they fell to earth. To support his position, Fréret referred to a report by Dio that in the reign of Vespasian an eruption of Vesuvius carried ash as far as Egypt, and cited Cardinal Bembo, who, in describing an eruption of Etna in 1537, wrote that its ash was transported more than 200 leagues from Sicily.[45]

Fréret's explanation of meteorites as volcanic ejecta was the last resort for those who believed reports of the fall of stones, and it appears occasionally in eighteenth-century scientific literature. It was used to explain the presence of the Pallas iron [Krasnojarsk] on a high ridge of a remote Siberian mountain. Don Rubin de Celis, who in 1783 rediscovered a portion of the Campo del Cielo iron meteorite, weighing by his estimate about fifteen tons, concluded in his report that it was the product of a volcanic eruption.[46] Domenico Troili employed a variation—ejecta from a nearby vent in the earth—to account for the Albareto fall near Modena in 1766, a theory rejected by Beccaria. The idea of a terrestrial volcanic origin of meteorites also played a central role in the controversy surrounding the Siena fall in 1794, which occurred, coincidentally, only 18 hours after an eruption of Vesuvius.

Eighteenth-century physical theory dictated that pieces of matter could not fall to earth from interplanetary space, because only the great celestial bodies and the aether existed there. The theory of mineral formation eliminated the possibility that stones formed in the atmosphere. There was more than sufficient proof that stones alleged to have fallen from the clouds were Stone Age tools, fossils, or unusual mineral crystals. That rains of blood, milk, or money had occurred was too much for Enlightenment scientists to stomach, and the tales of such miraculous events weakened the credibility of reports that stones had fallen. Nevertheless, it was scientific theory that played the leading role in the disbelief and skepticism that scientists displayed.

DILEMMAS ABOUT FIREBALLS AND GENUINE METEORITES

During the eighteenth century, there were 189 reports of fireballs, and two-thirds of this number appeared after 1750 (fig. 1). In the first decade of the nineteenth century alone, after meteorite falls were accepted, the number of reports increased to 65. Such reports depended not only upon the occurrence of the phenomena but upon population density and a good level of education and information as well.[47] Thus there was an increasing awareness of the phenomenon as the century progressed. The sizable number of fireballs seen

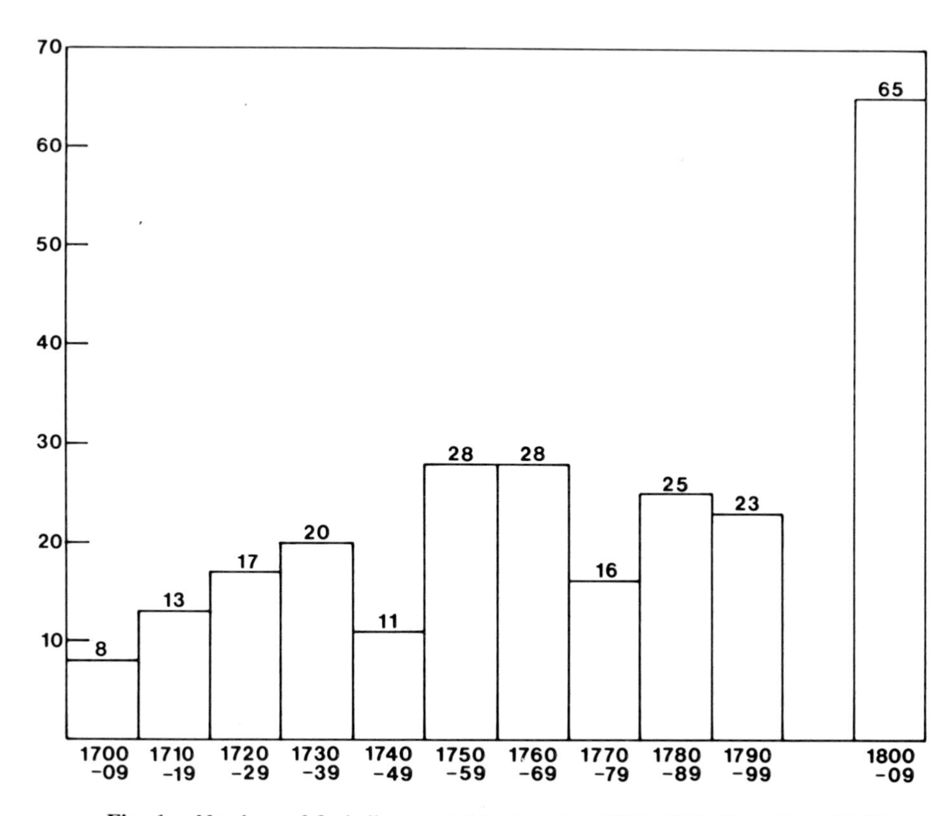

Fig. 1. Numbers of fireballs reported by decades, 1700–1809. From Greg (1860).

in the 1750s and 1760s elicited scientific interest in them. John Pringle published a comprehensive report in the *Philosophical Transactions*, and the French Academy of Sciences instructed Jean-Baptiste Le Roy to study the phenomenon of fireballs in general and the bolide of July 1771 in particular. In Pringle and Le Roy's reports, some of the more important impediments to understanding the nature of fireballs and the reasons why fireballs were not connected with the fall of meteorites become clear.

John Pringle was a physician, a fellow of the Royal Society, and its president from 1772 until 1776. The fireball about which he reported appeared after nightfall on 26 November 1758, and his account was based on the testimony of more than a dozen witnesses. Observers first saw it over Cambridge at an estimated height of 90 to 100 miles. It moved in a west by northwest direction and suddenly disappeared over Fort William in the shire of Inverness, where witnesses judged its altitude to be 26 to 32 miles. Over Auchenleck, part of its tail seemed to break off, and it rose and dipped in its course, which Pringle suggested was due to its impingement on more dense air at the lower elevations, "in the same manner as a cannon ball by water, when it strikes in a very oblique direction."

Pringle thought that the velocity of the fireball was "almost incredible," since it traversed 324 miles in 13 seconds, or 25 miles per second. Estimates of the diameter ranged up to 2 miles, but Pringle, considering that the observers' imaginations had exaggerated its size, reckoned that its diameter was only 0.5 mile. Some witnesses testified that they heard a hissing noise as the fireball passed, but Pringle dismissed these accounts, stating that it was "a deception of that kind which frequently connects sound with motion." Charles Blagden, two decades later, claimed that the noise was physical not psychological, though he confessed he was unable to explain it. Twentieth-century physicists have studied the hissing noise, which sometimes accompanies the passage of a fireball, and a recent explanation is that the fireball may, because of the highly energetic wave turbulence, radiate very low-frequency electromagnetic energy, which excites surface acoustic waves in surrounding objects.[48]

Witnesses at a distance of 75 miles heard the sound when the tail exploded, and compared it to thunder or to volleys of heavy artillery. Pringle thought that the sound gave evidence that the body or at least its surface was solid, because "sounds are either produced by the quick and violent percussion of hard bodies upon the air; or by the sudden expansion of an elastic fluid, after being condensed within some solid substance." Also, after the explosion the body retained its form, so presumably the burning matter was vented through apertures in a hard crust. The velocity of the fireball and the intensity of its light, Pringle wrote, also favored the idea that the body was solid.

Pringle abruptly dismissed the notion that the phenomenon was lightning, stating that it was based wholly "upon the velocity of those balls of fire." Further, he wrote, all observations argued against the "prevailing hypothesis," advanced by Halley in 1719, that such bodies "consist of sulphureous vapours arising from the earth." There was no experimental evidence that such vapors possessed innate levity, as Halley had postulated, which would permit them to rise into the absolute vacuum above the atmosphere. Because of the extreme cold of the higher regions of the atmosphere, the volatility of the vapors would be decreased, and how ignition occurred had not been explained. Finally, attraction should cause the vapors to assume a spherical rather than the lenticular form exhibited in the current instance, extending in a straight line of equal breadth for a distance of some 400 miles.

Pringle, however, did think that another notion of Halley's, expressed in 1714, had some merit—namely, that a meteor "must be some Collection of Matter form'd in the Aether."[49] Because of the great velocity of the fireball in the atmosphere, though, he thought Halley was wrong in supposing that a "fortuitous Concourse of Atoms" met with the orbiting Earth before being attracted to the Sun. He also rejected Halley's suggestion that "such meteors actually came to the ground." The descent of fireballs below the horizon, Pringle wrote, gave the impression that they fell to earth, and the explosion, heard only much later, was mistaken for the noise of the fall. He continued: "Hitherto we have had no certain proof of their fall; and it is to be hoped, that their motions, like those of the comets, have been so regulated at first by a governing Power, that we have nothing to apprehend from their aberration. . . . If it is then probable, that these balls of fire come from regions far beyond the reach of our vapours; . . . surely we are not to consider them as indifferent to us . . . but rather as bodies of a nobler origin, possibly revolving about some center."[50]

In Pringle's account, then, there is an expression of perplexity concerning the incredible velocity and tremendous apparent size of fireballs; a rejection of hypotheses attributing the phenomenon either to the ignition of terrestrial vapors or to the action of lightning; a denial that fireballs fall to earth; and suggestions that fireballs might originate beyond the atmosphere and might be related to comets.

The fireball that Le Roy described in his report first appeared at about 10:30 P.M. on 17 July 1771 over Sussex County, England; traveled in a south-southeast direction over the English Channel, lower Normandy, and the Ile de France directly above Paris; and exploded and disappeared near Melun, about 30 miles away (fig. 2). The estimated height when first seen was about 50 miles, and at the moment of explosion it was 25 miles. Most observers thought that the duration of the phenomenon was 4 seconds, during which time the fireball traversed 180 miles, but Le Roy arbitrarily increased the elapsed time to 10 seconds to arrive at a more credible velocity of 18 miles per second. "From the direction and height of the globe," Le Roy wrote, "there is almost no doubt that it was formed [*formé*] above the shores of England . . . it is there that it originated [*il a pris naissance*]." Here Le Roy's language seems to indicate a bias toward the atmospheric origin of fireballs.

It was the size of the fireball that most astounded Le Roy; his estimate of the diameter, which again was far less than those given by eyewitnesses, was 0.6 miles. "What city," he exclaimed, "could escape a general conflagration and total ruin, if a similar globe fell within its walls! But it appears that this disaster need not be feared by the very nature of things." If the fireballs descended into denser air, they would be broken up and never fall as a single globe of fire. "Some of these enflamed parts," he wrote, might "descend to earth."

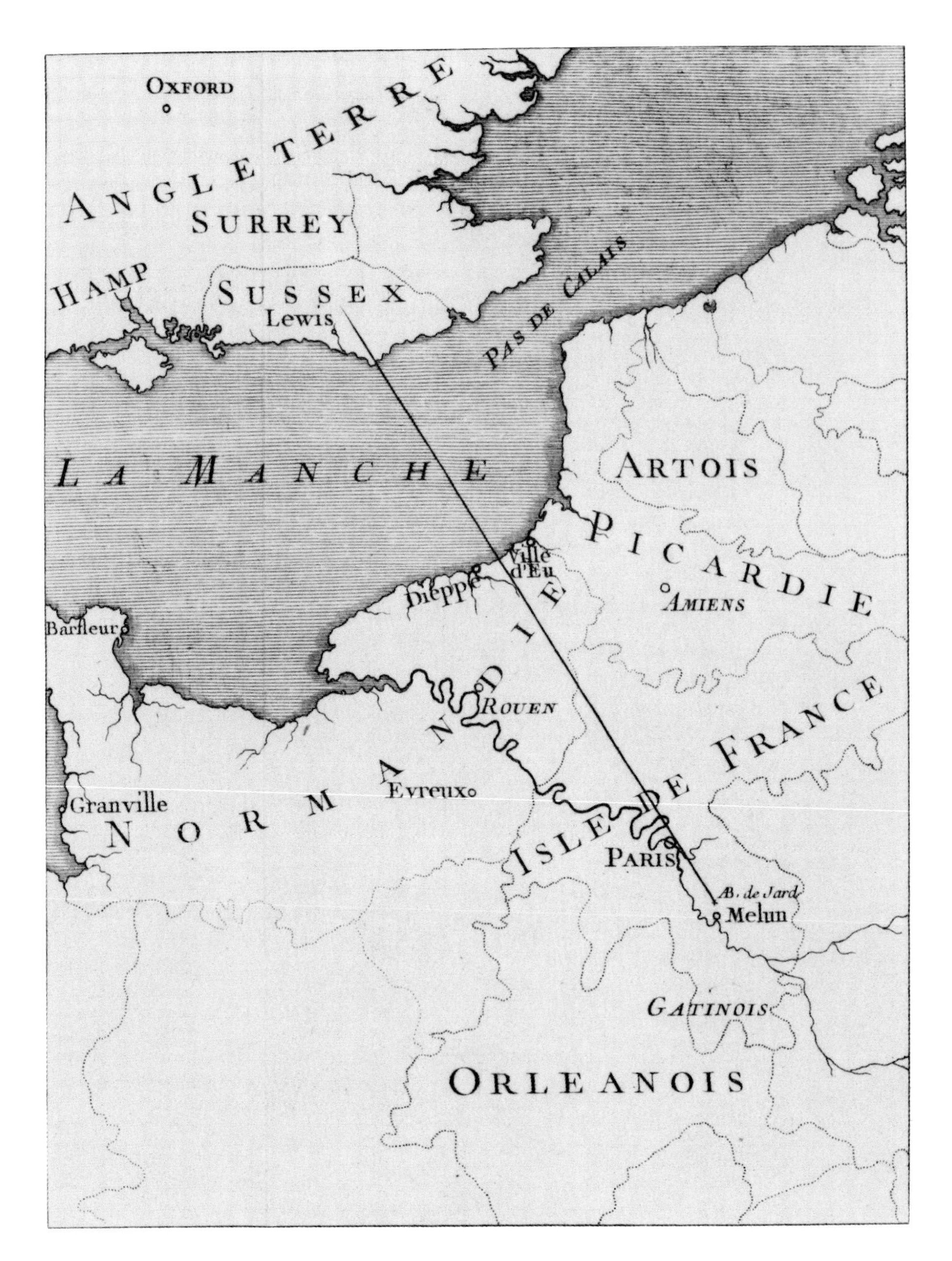

Fig. 2. The route of the fireball of 17 July 1771, described by Le Roy. From *AS* (1771), Plate XIII.

Le Roy worried about this possibility. One of the witnesses, a "highly qualified jurist, a very trustworthy man," had testified that at the instant of the explosion a number of fire particles were close to the ground. But, it was difficult to believe, he wrote, that fire particles could survive a fall to earth from the tremendous height at which the explosion occurred. He suggested that the fire particles might ignite some intermediate substance, or perhaps fell rapidly because they had rarefied the surrounding air. It seems certain that Le Roy did not conceive that the fire particles might be fragments of iron or stone (fig. 3).

As to the cause of the phenomenon, Le Roy rejected the idea that it was a comet. This notion was a reversion to the days when comets were thought to be sublunary phenomena. He also dismissed the hypothesis, which he

Fig. 3. Drawings of three observations of the fireball of 17 July 1771, described by Le Roy. From *AS* (1771), Plate XIV.

attributed to the "illustrious Halley," that fireballs were masses of inflammable terrestrial matter ignited in the higher regions of the atmosphere. Le Roy's own research centered on electrical phenomena, so it was not unusual that he favored this explanation. He could demonstrate, he wrote, that there was a great deal of analogy between fireballs and certain electrical manifestations, but he did not wish to press his conjectures too far in favor of a hypothesis that still presented many difficulties. To show that he had an open mind on the subject, he concluded by pointing out that "the atmosphere is a vast chemical laboratory, in which there are a thousand different chemical combinations, of which we recognize yet only a very small number." He advised that the phenomenon required much more study.[51]

The reports and remarks by Pringle and Le Roy demonstrate why eighteenth-century scientists were puzzled about fireballs and reveal the bases of the various explanations given for their origin. A fireball first appeared high in the atmosphere as a brilliant point of light or shooting star. It advanced rapidly toward earth, increasing enormously in size until it approached that of the full moon, but far surpassing the latter in brilliance with its dazzling light. It emitted flames, smoke, and sparks; occasionally a hissing noise was reported; and it was often accompanied by a loud rumbling noise, which was compared to that of a distant cannonade or to carriages rolling over cobblestones. Sometimes, as Pringle reported, an explosion caused part of the fireball to break off; sometimes, as with Le Roy, explosion resulted in complete disintegration; and occasionally the fireball disappeared over the horizon without an explosion being heard.

The brilliant light that accompanied the fireball throughout its course posed the first problem. Scientists surmised that the air became less dense at higher altitudes until at some point a vacuum existed. It was also known that the resistance of the air could raise the temperature of an object moving through it, as evidenced by a cannonball. But it was inconceivable that at very high altitudes, where the air was very rare or nonexistent, there could be atmospheric friction sufficient to heat an object to the high temperature necessary for it to emit light. This aspect of the physics of a meteorite fall began to be understood only in the 1840s, when James Prescott Joule published the first estimates of the energy exchange resulting from the entry of a meteoroid into the atmosphere.[52] Eighteenth- and early nineteenth-century scientists consequently attributed the light either wholly or partially to electricity, as John Perkins did when offering Franklin his notions on shooting stars.

The relatively short duration of the fireball as a band or swath of fire across the sky, which in reports from Pliny's period forward was compared to a huge beam, did lend support to the notion that a roughly lenticular section of the upper atmosphere extending in length for tens of miles had been progressively ignited. Oily inflammable vapors of terrestrial origin had accumulated and caught fire; after the introduction of the new chemical nomenclature, the vapor was usually considered to be a thick layer of hydrogen gas. The stream of smoke left in the wake of the fireball and the sudden disappearance of the mass of flame after the explosion offered additional evidence for this hypothesis. Alternatively, the swath of flame was thought of as produced by electricity, similar in appearance to lightning. One impediment to this notion was the velocity, since experiments on the velocity of electricity as early as the 1740s showed it to be almost instantaneous, and thus far in excess of the speed of fireballs. One nimble theorist, Pierre Bertholon, overcame this problem by explaining that the electric fluid passed progressively through pockets of aqueous vapors in the atmosphere, the interruptions in the flow being masked by the velocity of the movement as a whole. Bertholon also testified that experiments in which large static electric machines produced appearances resembling fireballs with attendant explosions proved beyond doubt that shooting stars and fireballs were electrical phenomena.[53] It was probably evidence of this kind that caused Le Roy to favor a hypothesis that connected fireballs with electricity.

The velocities of fireballs as calculated from observations were almost beyond comprehension; hence Le Roy arbitrarily extended the time of duration estimated by observers in order to report a velocity that could be considered more credible. At the time, scientists knew that the Earth's orbital velocity was about 18 miles per second, and that the terminal velocity of free fall in air could be no more than about 0.5 mile per second. The velocity of fireballs far exceeded the latter figure, and in some instances was greater than the former. The fact that the calculated velocities of fireballs were somewhat comparable to the Earth's orbital velocity gave rise to the notion that fireballs were Earth-orbiting comets. Thomas Clap, who was president of Yale College from 1739 to 1766, proposed this theory in a treatise published posthumously in 1781. From lack of evidence that fireballs had ever fallen to earth, and from the fact that other celestial bodies were so far distant, Clap concluded that the Earth must be the attractive central body around which fireballs moved. On the basis of observations of three such terrestrial comets, Clap calculated that their closest approach to Earth was about 25 miles and their greatest distance from it 4,025 miles.[54] Clap's idea continued to receive favorable attention well into the 1800s.

A point stressed by Le Roy—namely, that fireballs came from all directions and appeared at all seasons of the year—was intended to forestall any speculations connecting them with the northern lights, but this idea continued to be advanced well into the nineteenth century. Sir Charles Blagden, for example, in his report of the great fireball of 18 August 1783, which traversed most of England and eastern France, rejected the hypotheses of Edmund Halley and Thomas Clap and opted in favor of electricity. Blagden stated that the evidence that swayed him in favor of this theory was that streams of electricity seemed to issue from the fireball in its passage and settle on men, animals, and other objects, producing static electricity on them. But he also thought that there was a connection with auroras, because according to the reports he had read, the paths of most fireballs were almost always to or from the north or northwest quarter of the sky, which corresponded very nearly to the magnetic meridian. Those coming from the north, Blagden concluded, were "masses of electric fluid repelling or bursting from the great collected body of it in the north"; those proceeding toward the north were masses attracted toward the accumulation.[55]

Probably the most confusing aspect of fireballs was their estimated size, and this feature continued to be a major problem until the mid-nineteenth century. A diameter approaching 0.5 mile, as reported by Pringle, Le Roy, Clap, and Blagden, was a usual estimate, although in some instances this figure was increased to 2 miles. Estimates were based both on the assumption that the perceived size of an object in the sky was directly proportional to its distance from the observer, and on a comparison between the perceived size of the fireball and that of the full moon. The Moon's mean distance from the Earth was known to be about 240,000 miles, and its diameter to be somewhat less than 2,200 miles. By applying a simple proportion, an object appearing to be as large as the full moon at a distance of 55 miles was considered to have a diameter of 0.5 mile.[56] As Le Roy emphasized, if such a huge fiery object crashed into a city or town, the destruction would be immense. There was no historical evidence, however, that such a disastrous event had ever occurred. Thus, misjudgment of the true sizes hindered scientists from comprehending the nexus between fireballs and meteorites.

Prior to the publication of Chladni's treatise in 1794, there were only a few fleeting conjectures that shooting stars and fireballs might be extraterrestrial matter. In 1714 Halley, as noted above, surmised that a meteor "must be some Collection of Matter form'd in the *Aether,* as it were by some fortuitous Concourse of Atoms, and that the Earth met with it as it past along in its Orb." In 1719 Halley abandoned this notion, but it did prompt Pringle to postulate that fireballs had a "nobler origin." Another such speculation appeared in an exchange of letters between John Page of Williamsburg, Virginia, and David Rittenhouse of Philadelphia in 1779 and 1780. Page described a spectacular fireball visible over Virginia at 6:10 P.M. on 31 October 1779, which he wrote "was as bright as shining silver, and as broad as the enlightened part of the new moon," and inquired whether Rittenhouse had seen it. Rittenhouse, a respected astronomer and instrument maker, replied that he had indeed observed it, though in its last stages, and declared that "it was certainly a grand appearance near the place where it fell, if any human eye was there." He continued:

> May not these shooting stars be bodies altogether foreign to the earth and its atmosphere, accidently meeting with it as they are swiftly traversing the great void of space? And may they not, either electrically or by some other means, excite a luminous appearance on entering our atmosphere?

He held this opinion, Rittenhouse wrote, because it was improbable that meteors would be generated 50 or more miles high in the air, because the velocity of such bodies was much greater than the force of gravity would induce, and because they only rarely fell perpendicular to the horizon. The only difficulty in the way of such an explanation, he concluded, was "why do they not frequently strike the earth, buildings, etc."[57] With this remark, Rittenhouse joined Pringle and Le Roy in the belief that fireballs never fell to earth. These suggestions by Halley, Pringle, and Rittenhouse that meteors and fireballs might have a cosmic origin are the only ones I have found in eighteenth-century scientific literature before Chladni's fully developed proposal. The lack of contemporary comment testifies to the tenacity of the prevailing view of the contents of the cosmos.

There were, in fact, several reports that appeared in scientific journals of fireballs falling into the sea or striking the earth. Only one, however, included a visit to the site by the observer. While in Jamaica in 1700, naturalist Henry Barham viewed the fall of a fireball having the apparent size of a bomb. At the location of the fall he found several holes, one the size of a man's head and the others the size of a man's fist. All were so deep that he was unable to probe them with a stick, and thereupon discontinued his investigation. Were Pringle, Le Roy, and Rittenhouse ignorant of Barham's report? This possibility, of course, exists. However, the report appeared in *Philosophical Transactions,* so it is more probable that they were aware of it and discounted it. Huge fireballs striking the earth at tremendous velocities would be expected to cause much more devastation and destruction than the several small cavities seen by Barham. Chladni, who mentioned the event, regretted that Barham had not made an excavation; yet even he felt that the cause could have been a bolt of lightning.[58]

Finally, pure chance played an important role in preventing scientists from establishing the connection between fireballs and meteorites. Table 1 shows that of the thirty-eight actual and doubtful eighteenth-century falls, the time of day when the fall occurred is recorded for thirty-two. Twenty-six of the thirty-two, or 81 percent, fell in daylight or twilight. None of these reports mentions any accompanying fireball, although several do record lightning flashes. Witnesses reported fireballs prior to the falls of the other six: Dorpat (1704); Hraschina (1751); Beeston (1780); Barbotan (1790); Salles (1798); and Benares (1798). Of these, there was no record of the doubtful Dorpat fall in the eighteenth and nineteenth centuries; a report of it only recently turned up.[59] Scientists outside of Austria were generally unaware of the Hraschina fall before 1790, when Abbé Stütz published the documentary evidence of its occurrence.[60] Chladni in 1819 first recorded the doubtful Beeston fall.[61] Thus it was only in the 1790s, just a few years before the fall of meteorites was accepted by scientists, that evidence began to accumulate that connected fireballs and meteorites. However, acceptance of the fall of stones did not imply recognition of the fact that fireballs were likely to drop meteorites. Edward Howard, whose chemical analysis of several meteorites influenced their acceptance, had this comment on Chladni's assertion that meteorites were the substance of fireballs: "No luminous appearance having been perceived during the day on which the stones fell in Yorkshire [Wold Cottage], it must be admitted, rather militates against the idea, that these stones are the substance which produce or convey the light of a meteor, or that a meteor must necessarily accompany them."[62] Decades passed before the nexus between fireballs and meteorites was fully recognized and accepted.

Assertions that meteorites were volcanic products or pyrite struck by lightning testify to the primitive state of mineral chemistry in the eighteenth century. Given this condition, it is doubtful that chemists could have discriminated meteorites from terrestrial stones and irons more than a few years before 1802, when Howard did. They could make successful qualitative analyses on simple minerals, such as the oxides, sulfides, and carbonates, in the mid-eighteenth century, and blowpipe analysis, particularly in the hands of the

TABLE 1

ACTUAL AND DOUBTFUL METEORITE FALLS IN THE EIGHTEENTH CENTURY

Meteorite name	Date and time of fall			Location of fall	No.
Dorpat[a]	1704	20 July	evening	Estonia, USSR	1
Barcelona (stone)	1704	25 Dec.	1700 hrs.	Catalonia, Spain	2
Larissa[a]	1706	7 June	1430 hrs.	Greece	3
Schellin[b]	1715	11 Apr.	1600 hrs.	Pomerania, Poland	4
Kloster Schefftlar[a]	1722	5 June	1530 hrs.	Bavaria, Germany	5
Ploschkovitz[b]	1723	22 June	1400 hrs.	Bohemia, Czechoslovakia	6
Mixbury[a]	1725	3 July	—	Oxfordshire, England	7
Carpentras[a]	1738	18 Oct.	1630 hrs.	Vaucluse, France	8
Rasgrad	1740	25 Oct.	1200 hrs.	Bulgaria	9
Nicorps	1750	11 Oct.	1200 hrs.	Normandy, France	10
Hraschina[b]	1751	26 May	1800 hrs.	Croatia, Yugoslavia	11
Tabor[b]	1753	3 July	2000 hrs.	Bohemia, Czechoslovakia	12
Luponnas[b]	1753	7 Sept.	1300 hrs.	Ain, France	13
Terranova da Sibari[a]	1755	July	twilight	Calabria, Italy	14
Brianza[a]	1760	—	—	Milan, Italy	15
Novellara[a]	1766	15 Aug.	—	Emilia, Italy	16
Albareto[b]	1766	July	1700 hrs.	Modena, Italy	17
Lucé[b]	1768	13 Sept.	1630 hrs.	Sarthe, France	18
Mauerkirchen[b]	1768	20 Nov.	1600 hrs.	Braunau am Inn, Austria	19
Aire-sur-la-Lys	1769	—	—	Pas de Calais, France	20
Sena[b]	1773	17 Nov.	1200 hrs.	Huesca, Spain	21
Rodach	1775	19 Sept.	900 hrs.	Coburg, Germany	22
Owrucz	1775 or 1776		1000 hrs.	Ukraine, USSR	23
Fabriano[a]	1776 or 1777		afternoon	Sanatoglia, Italy	24
Pettiswood	1779	—	daylight	Westmeath, Ireland	25
Beeston[a]	1780	11 Apr.	2100 hrs.	Nottinghamshire, England	26
Turin[a]	1782	—	—	Italy	27
Eichstädt[b]	1785	19 Feb.	1200 hrs.	Bavaria, Germany	28
Kharkov[b]	1787	12 Oct.	1500 hrs.	Ukraine, USSR	29
Barbotan[b]	1790	24 July	2100 hrs.	Gers, France	30
Castel Berardenga	1791	17 May	1700 hrs.	Tuscany, Italy	31
Siena[b]	1794	16 June	1900 hrs.	Tuscany, Italy	32
Mulletiwu[b]	1795	13 Apr.	800 hrs.	No. Province, Ceylon	33
Wold Cottage[b]	1795	15 Dec.	1530 hrs.	Yorkshire, England	34
Bjelaja Zerkov[b]	1796	15 Jan.	—	Ukraine, USSR	35
Portugal	1796	19 Feb.	daylight	Portugal	36
Salles[b]	1798	12 Mar.	1800 hrs.	Rhone, France	37
Benares[b]	1798	19 Dec.	2000 hrs.	United Provinces, India	38

[a] Considered doubtful.

[b] Specimen(s) exist in collections.

capable Swedish chemists Axel Cronstedt, Torbern Bergman, and Johann Gahn, increased the ability to detect minor constituents. But the analysis of silicates began only in the late 1780s and 1790s, after Bergman in his *Opuscula physica et chimica* (Uppsala, 1779–1790) described the apparatus, reagents, and techniques that were required. The new chemistry, delineated in 1789 in Lavoisier's *Traité élémentaire de Chimie,* increased the understanding of reactions and combinations, and this was turned to good account by such able mineral chemists as Martin Klaproth in Germany and Louis Vauquelin in France. Crystallography was only employed systematically in mineralogy after René Just Haüy established the science in the 1780s. The confusion in mineralogy before Haüy's studies in the 1790s is well illustrated by the fact that the word "*schorl,*" qualified by an adjective denoting color or configuration, was employed to describe such diverse silicates as tourmaline, feldspar, staurolite, actinolite, and titanite, as well as the mineral anatase. But neither Romé de l'Isle in 1783 nor Haüy recognized that the substance embedded in the Krasnojarsk meteorite was olivine.[63]

One of Howard's major accomplishments was that he detected nickel in all of the meteorites that he analyzed. Cronstedt first isolated nickel in an impure state in 1751, and his countryman Henrik T. Schaffer immediately confirmed the finding, as did Bergman in 1775. But, as Howard emphasized, analysts can easily overlook small amounts of nickel. In 1797 the Berlin chemist, Sigismund F. Hermbstaedt, was the first to publish a suitable method for its determination, and Howard used his technique.[64] J. L. Proust reported the presence of nickel in the Campo del Cielo meteoritic iron in 1799, but since at the time he analyzed only this meteorite, he drew no conclusion from his finding.[65]

It is against this background that one should assess the chemical analysis of the Lucé meteorite, and also the statements made by eighteenth-century mineralogists about the so-called native irons. Abbé Charles Bacheley, an amateur mineralogist and corresponding member of the French Academy of Sciences, sent a report of the Lucé fall, which occurred 13 September 1768, together with a specimen to Paris. This stone and another, which resembled it closely and was reported to have fallen near Coutances (possibly the Nicorps meteorite), were given for investigation to a committee of the academy consisting of Fougeroux de Bonderoy, Cadet de Gassicourt, and Lavoisier. The academy also received a third meteorite, Aire-sur-la-Lys, which fell toward the end of 1769, but since the committee's report did not mention this stone, it is likely that they made their analysis in mid-1769, not in 1772, which is the date of the publication.[66] Because of Lavoisier's subsequent scientific achievements and fame, most accounts of the Lucé analysis cite him as the responsible chemist and the author of the report,[67] although he was the junior member of the committee both in terms of age and academy rank. In 1766 Cadet de Gassicourt was preferred over Lavoisier for a vacant position in the chemistry section of the academy. Lavoisier gained admittance on 20 May 1769 as a supernumary adjunct—an unusual appointment—just four months before the Lucé fall. It is even possible that he did not participate actively in the investigation, since one of Lavoisier's biographers reports that for three years after May 1769 Lavoisier was away from Paris for months at a time, being occupied in his business as a tax farmer.[68] Even given the self-confidence of the young Lavoisier, it is unlikely that, at this early stage of his career as an academician, his would have been the dominant voice in the committee.

The report's preface emphasized that various substances had been called thunderstones. Citing Lemery's 1700 memoir, it stated that true physicists had always doubted the existence of thunderstones, and that the discovery that lightning was identical to electricity was all the more reason for doubt. On this note, the committee's report described the fall, as related by Bacheley. A storm cloud was overhead; there was a brief clap of thunder heard over a wide area and a hissing noise similar to the lowing of oxen, but no flash or flame. Several harvesters looked up and saw an opaque body that fell in an arc and landed in the grass. They ran to the place and found a stone half-buried in the earth and too hot to handle. At first, they were frightened and left the spot, but later they returned to take away the triangularly-shaped stone, which weighed over 7 pounds. Bacheley managed to obtain a piece, which he sent to the academy.

The report described the stone as composed of a pale ash-gray substance, in which an infinite number of small, brilliant, pale yellow metallic points could be discerned through a magnifying lens. A very thin blackish layer, partially blistered and apparently having suffered fusion, covered the exposed exterior surface. The specific gravity of the stone was 3.535, which, the report stated, was higher than that of a siliceous stone, indicating the presence of a considerable amount of metal.

The chemical analysis conducted was standard for that period: initially by the "dry way," and then by the "wet way." The analysts reduced a portion of the stone to powder and calcined it—that is, they heated it in air, gradually increasing the temperature. Just below red heat, the powder emitted an abundant quantity of sulfurous fumes. Intense heat produced a black mass, which they judged was iron combined with an alkali. Unable to separate the constituents further, the chemists turned to wet techniques. Decomposition in sulfuric acid and subsequent evaporation of the decanted liquid yielded diamond-shaped crystals of *vitriol martial* (iron sulfate). They calcined the undissolved sediment, during which sulfur fumes were emitted, leaving a residue of "vitrifiable earth." This step completed the analysis.

The committee concluded that the stone had not been exposed to intense heat, for if it had been, it would have lost all of its sulfur. Therefore it had experienced only momentary heat. It must thus have been struck by lightning, which made it conspicuous to the eyewitnesses, and the heat was sufficient to melt just that portion of the stone that had been hit by the bolt of lightning. As for the other stone, alleged to have fallen near Coutances, it differed only in emitting a less sulfurous odor when attacked by an acid. One could only conclude from its resemblance to the Lucé stone that lightning strikes pyritous materials preferentially.[69]

The Lucé meteorite is now classed as an ordinary L-type or olivine-hypersthene chondrite. This type of meteorite, apart from an iron-rich crust, contains particles of nickel-iron, particles of iron sulfide called *troilite,* and spheroids termed *chondrules,* all of which make up part of an aggregate groundmass consisting of the minerals olivine $(Mg, Fe)_2 SiO_4$ and hypersthene

(Mg, Fe) SiO_3. The analysis, of course, failed to detect nickel. Less understandable was the lack of any attempt to isolate magnesia, which had previously been the subject of analysis by both Joseph Black and Andreas Marggraf, the latter having recognized magnesia in serpentine.[70] In commenting on the Lucé analysis, Howard stated that the results might well yield the conclusion that the stone was a common pyrite, but he was critical of the fact that "it was unfortunately made of an aggregate portion of the stone, and not of each distinct substance irregularly disseminated through it. The proportions obtained were, consequently, as accidental as the arrangement of every substance in the mass."[71] Howard, of course, had the advantage of thirty years of hindsight, and compared to Howard's methods, the analysis was primitive. The simple fact is that in 1769 and for years thereafter, the chemical analysis of meteorites was beyond the capabilities of the most talented analysts.

The crucial factor in the committee's conclusion was that the interior of the meteorite had not experienced excessive temperature, which they believed it would have, had it been fused in the clouds. This inference was not unreasonable. Moreover, the blistered crust did provide evidence of momentary heating, as might be expected from the action of lightning. Many years passed before scientists realized that the conduction of heat in meteoritic stones is so slow that the interior remains cold despite the fact that the surface is raised to a very high temperature during the stone's brief atmospheric passage. Also, although in the 1700s scientists had seen fulgurites—those peculiar tubes often formed in sand after intense local heating due to a stroke of lightning—their origin was not suspected until 1790.[72] Thus the Lucé meteorite analysts were unaware of what a bolt of lightning might actually do to siliceous material.

Several meteoritic irons and stony iron finds were in mineralogical collections in the eighteenth century, but investigations of them did not provide a definite answer to the question of whether matter did or did not fall from heaven. The only references by mineralogists were to the meteoritic irons Senegal [Siratik] and Campo del Cielo and to the stony irons Steinbach and Krasnojarsk, and they focused principally on the latter. The 1600-pound Krasnojarsk mass was found in 1749 on a remote wooded ridge of Mount Emir, Siberia, about 145 miles south of Krasnojarsk. Peter Simon Pallas, a German-born scientist who spent most of his productive years studying the natural history of Russia and Siberia, investigated the location of the meteorite in 1772, took samples of it, and described its appearance and properties in detail in the account of his expedition, which was published in 1772. Although Pallas reported that the native Tartars believed that the iron was a holy relic that had fallen from heaven, he did not state whether he accepted or rejected this belief. He wrote that while there were veins of iron ore nearby, they had never been worked; that there were no iron smelting operations in the region that could have refined such a huge mass; that it would have been next to impossible to transport such a heavy piece to such a remote location; and that there was no trace of any volcanoes in the vicinity. Thus, though Pallas did not openly support an extraterrestrial origin, his description left no alternative. Nevertheless, controversy about the samples, which Pallas or the Russian government distributed to the principal scientific centers of Europe, was not primarily about whether the stony iron had fallen from the sky but about whether "native" iron—a pure, ductile, and malleable material—existed in nature, or whether the mass had been previously smelted, refined, and forged, and was hence a "product of art."[73]

To my knowledge, only one scientist prior to Chladni openly voiced the belief that the Krasnojarsk stony iron fell from the sky. This was Franz Güssman, a native of Lwow, then under Austrian domination. Güssman, a Jesuit who became professor of natural history at Vienna, wrote a mineralogical work entitled *Lithophylacium Mitisianum,* which was published in Vienna in 1785. In treating the subject of "native" iron, Güssman described the Krasnojarsk meteorite in some detail, explaining that some mineralogists believed that it had been produced within the earth just as other native metals were, while others, including the distinguished Swedish scientist Torbern Bergman, considered it to be a product of artificial fusion. The former opinion, Güssman wrote, was insufficient because it did not account for the hardness of the iron, and the latter view was faulty because it ignored Pallas's meticulous report of the nature of the terrain where the iron was found and of the absence of smelting facilities in the neighborhood. Then Güssman abruptly called attention to and provided some information about the observed fall of the Hraschina iron, noting that it had been preceded by a spectacular fireball, and stating that there was no reason to doubt the testimony of witnesses. It was very probable, Güssman concluded, that the Pallas iron had been "flung down from above," and that possibly terrestrial material had been carried into the atmosphere, made flammable by lightning, and then projected to the ground. Apparently few scientists, if any, took note of Güssman's hypothesis. Chladni in 1794 did not refer to the treatise, nor did Haüy in 1801, when he presented the various theories treating the origin of native iron. In 1803, when Güssman wrote a tract opposing the hypothesis that meteorites were ejecta from lunar volcanoes, he complained that his earlier treatise had been completely ignored.[74]

The controversy about the origin of native irons was fruitless, because scientists had no knowledge about the alloys of iron. Although they recognized the different properties of wrought iron, cast iron, and steel, they only began to understand the central role of carbon in developing different structures and hence different properties in iron-carbon alloys in the late 1780s, when Vandermonde, Berthollet, and Monge detected the presence of carbon in varying amounts in these materials.[75] The fact that the presence of nickel might have something to do with the existence of "native iron" was not even suspected. The lack of knowledge of the physics and chemistry of metals in the eighteenth century, then, precluded recognition of the distinct differences between terrestrial and meteoritic irons.

METEORITE FALLS AND SCIENTIFIC ATTITUDES

In 1812 Pierre Bigot de Morogues wrote a history of meteorites that became a standard and influential reference work for later commentaries about eighteenth-century scientific attitudes toward meteorite falls.[76] He wrote:

> I believe it necessary to state how often it is difficult for savants themselves to acknowledge the most evident and the best supported truths. The Academy of Sciences had in vain appointed committees to ascertain the reality of the fall of stones. Many parts of France, Germany, Italy, and England were almost simultaneously the theater [of these events]. Sworn testimony and a multitude of witnesses confirmed the existence [of falls] in all of Europe, without having shaken the opinion of the majority of learned people. Almost all looked with disdain at the pieces of evidence offered to them, and only listened with boredom to the recital of the factual circumstances, without even deigning to compare them in order to deduce some probability [of their reality].

The author's background may explain the severity of this indictment. The son of a naval officer, Bigot de Morogues grew to manhood during the French Revolution. As a student at the École des mines in 1794, he certainly knew about the demise in the previous year of the Academy of Sciences, judged by advocates of the new order to have been the bastion of scientific elitism in the ancien régime. Later, having married well, he gave up mineralogy for the life of a gentleman farmer, managing his wife's estates. Prosperous himself, he turned his attention to the economic problems of society, and wrote several tracts decrying the accumulation of wealth by those who were already rich. Elitism, whether scientific or social, was apparently his bête noire. Chladni noted that Bigot de Morogues's history was derivative and in places erroneous; nevertheless, it has been considered authoritative.[77]

In his assessment, Bigot de Morogues included the frequency of meteorite falls in Europe, the role of the Academy of Sciences, the sworn testimony of witnesses, and the attitudes of scientists. All of these matters require closer scrutiny. We shall begin with the frequency of falls.

In studying meteorite falls for which authentic records exist, Buchwald in 1975 concluded that a population density of over one inhabitant per square kilometer is the minimum requirement for reporting meteorite falls.[78] He also determined that in the century between 1871 and 1970, the total of 549 recorded falls corresponded to a frequency of 0.08 falls per year per 10^6 square kilometers of the Earth's inhabited surface. This figure is a minimum estimate, because some areas, owing to population density and the level of education, record much higher frequencies. Taking these factors into consideration, Buchwald concluded that 0.43 falls per year per 10^6 km^2 is a reasonable frequency of meteorite falls.[79]

Tables 1 and 2 list authentic and doubtful falls for the period 1700–1799 and 1800–1809, respectively, and figure 4 depicts these falls by decade. In table 3 we have extracted the authentic falls and determined their frequency for selected areas of southern and western Europe, and for the total of these areas, in the periods 1700–1789 and 1790–1809. We have also noted the estimated populations in the selected areas in 1700 and 1800; these figures may be inaccurate, but they do demonstrate that the densities are far above the minimum requirement indicated by Buchwald. The calculated frequencies are both informative and surprising. In the period 1700–1789 for the total area, the frequency of 0.07 falls per year per 10^6 km^2 approaches Buchwald's minimum. There is, however, a sharp rise to a frequency of 0.41 falls per year per 10^6 km^2 in the two decades after 1790, the period when scientists accepted

TABLE 2
ACTUAL AND DOUBTFUL METEORITE FALLS, 1800–1809

Meteorite name	Date and time of fall			Location of fall	No.
Loch Tay[a]	1802	Sept.	—	Scotland	1
L'Aigle[b]	1803	26 Apr.	1300 hrs.	Orne, France	2
East Norton[a]	1803	4 July	—	Leicestershire, England	3
Apt[b]	1803	8 Oct.	1030 hrs.	Vaucluse, France	4
Mässing[b]	1803	13 Dec.	1030 hrs.	Bavaria, Germany	5
High Possil[b]	1804	5 Apr.	morning	Glasgow, Scotland	6
Doroninsk[b]	1805	6 Apr.	1700 hrs.	Irkutsk, USSR	7
Constantinople[a]	1805	June	—	Turkey	8
Asco[b]	1805	Nov.	—	Corsica, France	9
Alais[b]	1806	15 Mar.	1700 hrs.	Gard, France	10
Basingstoke[a]	1806	17 May	—	Hampshire, England	11
Timochin[b]	1807	25 Mar.	1500 hrs.	Smolensk, USSR	12
Weston[b]	1807	14 Dec.	0630 hrs.	Connecticut, USA	13
Borgo San Donino[b]	1808	19 Apr.	1200 hrs.	Parma, Italy	14
Stannern[b]	1808	22 May	0600 hrs.	Moravia, Czechoslovakia	15
Lissa[b]	1808	3 Sept.	1530 hrs.	Bohemia, Czechoslovakia	16
Moradabad[b]	1808	—	—	United Provinces, India	17
Atlantic Ocean[a]	1809	19 June	2300 hrs.	off Rhode Island, USA	18
Kikino[b]	1809	—	—	Smolensk, USSR	19

[a]Considered doubtful.
[b]Specimen(s) exist in collections.

falls, and this figure remains constant for the entire period 1800–1966, corresponding almost exactly to Buchwald's estimate. If one assumes that a frequency of 0.41 falls per year per 10^6 km^2 is a more or less constant figure for the total selected areas in Europe, then the number of authentic falls in the period 1700–1789 should have been 70 instead of 12. Even if all doubtful falls are included, the total number in the period 1700–1789 amounts only to 21. How, then, does one account for the other 49 falls that were unrecorded? Either there was a peculiar drought in the number of falls in southern and western Europe at the time, which is possible but not probable, or there was a substantial number of unreported falls. Several plausible explanations for the paucity of reports come to mind. One is disinterest owing to lack of information or educational level. An early nineteenth-century commentator believed that this was the cause. He wrote: "The peasants had not been away from their own districts where aerolites were never mentioned; this leads to the conclusion that the observation, having been neglected in the majority of cases, the phenomenon was thought to be less frequent than it actually is."[80] Moreover, certain folk beliefs may have caused peasants to hide and treasure meteorites (see chap. 7). Another reason is that fear by peasants of ridicule from authorities, and by educated people from their peers, caused silence; we document several instances of this shortly. In any case, it is clear that Bigot de Morogues's account was exaggerated.

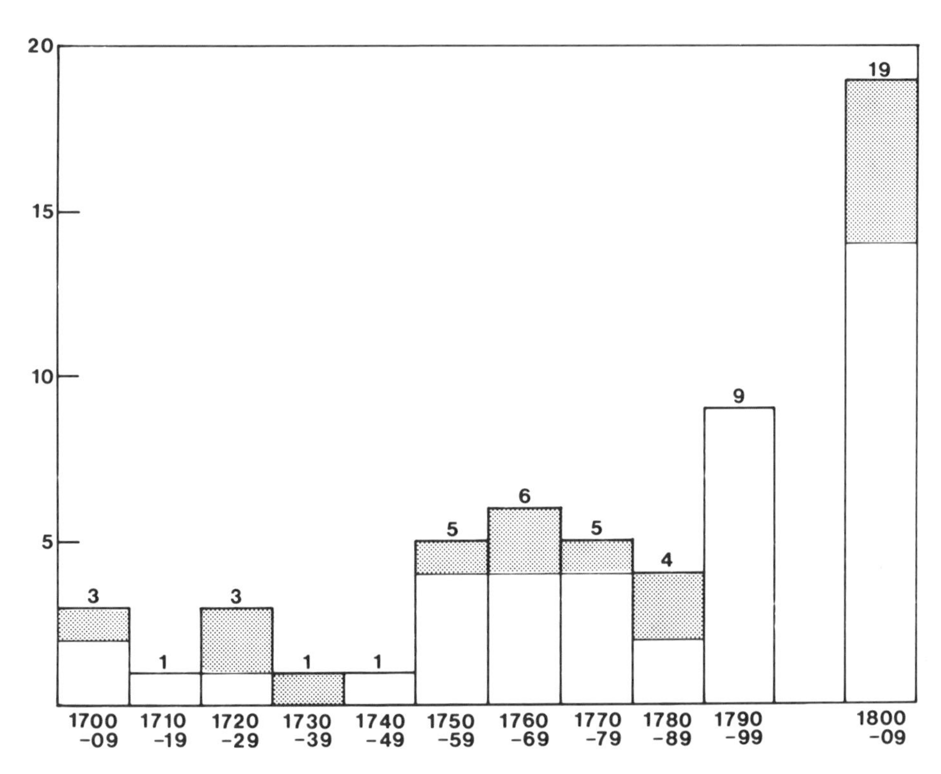

Fig. 4. Numbers of meteorite falls reported by decades, 1700–1809. Shaded areas are those now considered doubtful.

TABLE 3

Frequency of Meteorite Falls in Selected European Areas

	1700–1789		1790–1809		1800–1966		1700	1800
Area	A	F	A	F	A	F	D	D
British Isles (312,700 km²)	1	.04	2	.32	19	.36	25	51
France (551,670 km²)	4	.08	6	.54	50	.54	36	48
Bavaria-Bohemia-Moravia (144,766 km²)	4	.31	3	1.04	14	.58	30?	46?
Spain-Portugal (580,870 km²)	2	.04	1	.09	23	.24	13	22
Italy (continental) (251,406 km²)	1	.04	3	.60	19	.45	45	72
Total (1,841,412 km²)	12	.07	15	.41	125	.41		

Note: A = Actual falls; F = Frequency: falls per year per 10^6 km²; D = Estimated population density: inhabitants per km²; from Beloch (1900), 765–786.

To his credit, Bigot de Morogues did make a distinction between the Academy of Sciences as an institution, its committees, and the scientific community—a point frequently overlooked.[81] The Academy of Sciences never made an official judgment about meteorites, as it did in 1775, when it decided that "henceforth it [the Academy] will not receive nor examine any paper concerned with squaring the circle, trisecting the angle, duplicating the cube, and perpetual motion."[82] But Bigot de Morogues did give the impression that academy committees constantly investigated alleged falls, while there were in fact in the eighteenth century only two inquiries instituted by the academy: the investigation of the 1771 fireball by Le Roy, and that of the Lucé fall by Fougeroux, Cadet, and Lavoisier. The latter report was not published in the *Mémoires* of the academy; it appeared in the 1772 volume of *Observations sur la Physique,* published belatedly in 1777. However, a brief summary of the Fougeroux committee report, probably composed by the academy's perpetual secretary, Grandjean de Fouchy, did appear in the *Histoire* section of the academy's 1769 volume (published in 1772). Entitled "Three curious events of the same kind," it told of the three reported falls at Lucé, Aire-sur-la-Lys, and Cotentin, and stated that a comparison of the stones showed no visible difference. After a short description of the results of the chemical analysis, the note concluded:

> The Academy is certainly far from concluding, on the basis of the similarity of these three stones, that they have been produced by thunder; however, the similarity of events, which occurred in widely separated places, the perfect resemblance between the stones, and the characteristics which differentiate them from other stones, have appeared to it to be sufficient grounds for publishing this note, and to invite physicists to submit anything new on this subject; perhaps these can shed new light on the electric fluid and on its action on thunderstones.[83]

While this semiofficial notice is generally supportive of the committee's conclusions, it does not dismiss out of hand the resemblance among the stones nor gloss over the fact that stone falls were reported to have occurred in localities remote from one another. It gives adequate evidence that within the academy hierarchy there was puzzlement and concern about the three reported falls and the resulting meteorite fragments—enough, at least, to invite the submission of any new information. Instead of the militant disbelief alleged by Bigot de Morogues, this notice indicates a cautious skepticism.

As with the Lucé meteorite, most recorded falls in the eighteenth century were witnessed by farmhands or shepherds who reported the incident to a local authority. In three instances, for different reasons, the reports evoked a request for sworn testimony. The first was Hraschina, and its verification, ordered by Empress Maria Theresa, was probably prompted because of the excitement and fright that the fireball and its explosion caused in an area of many square miles. The Bishop of Zagreb named a commission, which took depositions from seven eyewitnesses, and its report, together with a large piece of the meteorite, weighing 87 pounds, was sent by the Bishop to Vienna. These were stored in the archives for thirty-five years, after which the report was translated and published in 1790 by Abbé Andreas Xaver Stütz, the curator of the Imperial Cabinet. Stütz published the report when he was reminded of its existence by the receipt of a specimen and an account of the Eichstädt fall, which occurred in 1785. Accepting a variation of Beccaria's electrical explanation of meteorites, Stütz called the Hraschina report a fairy tale.[84]

Sworn testimony by five shepherds, who in 1755 witnessed the Terranova di Sibari fall, was furnished to Abbé Domenico Tata by the prince of Tarsia, onto whose estates the stone fell. It was Tata's intention to publish the details of the fall, but he was dissuaded by friends, who advised that he would be ridiculed by "savants and half-savants, who are more to be feared." Tata's account was finally published in 1794, just after the Siena fall.[85]

The Barbotan fall in 1790, in which a shower of stones spread over several communes a few miles southwest of Agen in Gers, France, also elicited a procès-verbal, this time as a joke. J. F. B. Saint-Amans, at the time commissioned to establish the department of Lot-et-Garonne, told his friend Pierre Bertholon, editor of the *Journal des Sciences Utiles,* about the stories of the populace, which both found amusing. To match or top the tall story, to which he thought all the citizens were a party, Saint-Amans requested sworn depositions, which, to his surprise, were furnished to him by the local mayors and three hundred inhabitants. Bertholon published the account, but was outraged that people could have given sworn testimony to an event which he considered to be "physically impossible."[86] It is noteworthy that all three depositions were published in the early 1790s, just a few years before the majority of scientists accepted meteorite falls.

We have seen that Nicolas Lemery in 1700 expressed skepticism about the fall of thunderstones. Nicolas Frérét's memoir of February 1717 gives ample evidence that many scientists at that time doubted that stones fell. The memoirs of Jussieu in 1723 and of Mahudel in 1745, proving that many alleged thunderstones were prehistoric tools or fossil petrifactions, must certainly have contributed to scientific incredulity. When Father Joseph Stepling's account of the Tabor fall was read on 26 February 1756 to the Royal Society of London, it provoked no comment; only "thanks were ordered for this communication."[87] The Fougeroux-Cadet-Lavoisier report is adequate testimony of the disbelief, as are the assertions by Stütz and Bertholon. The incredulity was not limited to scientists. In the 1770s, Edward Gibbon, when describing the reign of Elagabalus (A.D. 218–222) in his *Decline and Fall of the Roman Empire,* referred scornfully to the superstitious reverence that the emperor compelled the citizens of Rome to offer to a black conical stone alleged to have fallen from heaven.[88]

There is evidence, however, that throughout the century there was an undercurrent of belief in the scientific community that stones did fall. Dr. Rost published a report of the Ploschkovitz fall in 1723.[89] Stepling, a fine physicist and mathematician, did not hesitate to publish a monograph on the Tabor fall, nor did Troili waver in his account of the Albareto fall.[90] In contrast to Bertholon, the young scientists who established the Société Philomatique included a report of the Barbotan fall without comment in the first issue of their *Bulletin.*[91] These published reports were sporadic, however, owing to the low frequency of falls. More than twenty years separated the falls at Lucé and Barbotan, so that the young French scientists at the time of the Revolution may have been unaware that a problem existed with respect to the fall of stones.

Indecision and fear of ridicule undoubtedly influenced some scientists not to take a definite stand as to the reality of the phenomenon. We have seen that Pallas did not insist that Krasnojarsk had an extraterrestrial origin. Joseph-Jerome Lalande, the prominent French astronomer, was well acquainted with the reports of the Nicorps and Luponnas falls but preferred "to say nothing" about them.[92] Tata withheld information about the Terranova di Sibari fall for forty years, and the Pettiswood fall in 1779 was not reported until 1796 by the eyewitness, W. Bingley, for fear of ridicule.[93]

There was, then, a wide spectrum of attitudes among eighteenth-century scientists about the reality of meteorite falls. Some were militant disbelievers; some were skeptics; some were noncommittal; some were silent about their belief for fear of ridicule; and some were outspoken believers. Yet the evidence does not support the generalization that all eighteenth-century scientists—or the eighteenth-century scientific community, if it is admitted that there was a scientific community in the modern sense of the phrase—disbelieved in the fall of meteorites. Moreover, the assertion that scientists refused to credit reports of stone falls because they came from nonscientific sources is contrary to the facts.[94] Scientists denied the validity of the accounts because they thought they had sound theoretical reasons for disbelieving that stones fell—namely, the impossibility of their formation in the Earth's atmosphere, and the inconceivability of the existence of solid matter, apart from the planets, comets, and stars, in cosmic space. Even those who credited reports of the fall of meteorites adhered to these hypotheses, postulating that the stones were volcanic ejecta and so forth.

It is reasonable to conclude that in the eighteenth century there was just not enough evidence to shake the faith of scientists in the validity of their assumptions about falling stones. In the 1790s, when well-attested reports of falls began to multiply, the change from disbelief to acceptance followed rapidly.

2

Acceptance

Scientific acceptance of meteorites, both falls and finds, occurred in the decade from 1794 to 1803. Commentators have attributed the relatively rapid shift from disbelief to acceptance to one or a combination of three events: (1) the publication in 1794 of a treatise by Ernst F. F. Chladni, which postulated that material from cosmic space entered the Earth's atmosphere, produced fireball phenomena, and dropped meteorites; (2) the publication in 1802 of a report in *Philosophical Transactions* by Edward Howard, showing that several meteoritic stones had similar chemical compositions, and that nickel was a constituent of these stones and also of "native irons"; and (3) the account presented in July 1803 to the Institut National de France by Jean-Baptiste Biot, giving detailed evidence that a fireball had dropped many stones at l'Aigle in Normandy the previous April. The first two publications undoubtedly influenced acceptance, while Biot's report was more in the nature of a coup de grâce that silenced the few remaining French skeptics. The chronicle of acceptance is, however, far more complex than these brief explanations imply.

There was a fourth influence—namely, the number of meteorite falls that came to the attention of scientists during this period. In his book, Chladni listed a dozen, beginning with the fall of iron in Lucania mentioned by Pliny. He devoted several pages to the fall of the Hraschina iron, and quoted nearly the entire text of the investigating commission's report that Stütz had published in 1790. In isolation, Chladni's tabulation may not have overly impressed his scientific colleagues. But with the occurrence of falls at Siena in 1794; at Wold Cottage in Yorkshire in 1795; near Evora, Portugal in 1796; at Salles in the department of the Rhône in 1798; and at Krakhut near Benares, India in 1798—and with the publication of delayed reports of the Pettiswood, England and Barbotan falls, which appeared in the late 1790s—Chladni's catalog and his assertion that stones did fall acquired new significance. Thus the marked increase in reported falls was an important factor in the acceptance of their reality.

Acceptance or continued skepticism was a matter of conviction on the part of individual scientists. No scientific organization, it should be emphasized, took a position on the matter one way or another. The attitudes of editors of scientific journals varied. Alexander Tilloch, who founded the *Philosophical Magazine* in June 1798, must have been intrigued, because he published notices and commentaries about falls beginning with his second issue. Marc-Auguste Pictet was a cofounder of the Geneva-based *Bibliothèque Britannique,* which was established in 1796 to inform continental readers about current British scientific and literary activities during the prolonged hostilities following the French Revolution. Pictet accepted not only falls but also Chladni's theory of their origin. He was fair, however, and also published opposing views by his fellow Genevans Louis Bertrand and the brothers Jean and Guillaume Deluc. In contrast Jean-Claude de Lamétherie, editor of the *Journal de Physique* at Paris, though publishing reports of falls beginning in August 1798, did not reveal his own acceptance of the phenomenon until January 1803.

Italian scientists led the way in acceptance because of the Siena fall in 1794. Writing in 1804, the expatriate English mineralogist resident in Naples, William Thomson, remarked that the debate over fallen stones had ended in Italy before there was even any question about them in France.[1] In Germany several prominent scientists announced their belief in meteorites before the end of 1797. Several British scientists were convinced about 1800, and Howard's report erased any lingering skepticism. Laplace, Biot, and Poisson were among the first French scientists to acknowledge in autumn 1802 that meteorites fell, and almost all of their French colleagues joined them prior to the l'Aigle fall. In the United States skepticism existed even after the Weston, Connecticut fall in December 1807. European scientific publications took months or even years to reach America, which may explain the lag. There is also some evidence that, by interrupting communication, the Revolution and the subsequent internal political crises and war played a role in the belated acceptance by French scientists.

The acceptance of falls did not provoke a revolution in scientific theory. For more than twenty years many scientists held the view that meteorites were generated in the atmosphere, a reversion to seventeenth-century thought. Another idea, popular in the United States, was that fireballs were terrestrial comets that occasionally dropped meteorites. The most prominent scientists postulated that meteorites were ejecta from lunar volcanoes, an idea that received support until mid-century. But most scientists did not accept Chladni's revolutionary theory of a cosmic origin until the 1840s and later. Thus, scientists accepted meteorite falls long before there was a consensus about their probable origin. They were in fact mystified, and most admitted that their speculations were plausible hypotheses at best.

FROM DISBELIEF TO ACCEPTANCE

The Siena Fall

Until massive urban growth occurred in the late nineteenth century, the probability that a fireball would explode over a city and drop meteorites was very small. This is the answer to the question posed in 1672 by Jacques Rohault as to why stones had never fallen in the great city of Paris. Siena in 1794 had a population of less than 30,000. But then as now, the beauty of its setting and the charm of its medieval art and architecture attracted tourists, especially during the summer months, when the traditional horse races, the Palio delle Contrada, are staged. About 7 P.M. on 16 June, a cloud at very high altitude appeared from the north, discharging sparks and emitting smoke. Flashes of lightning, very red and slower than normal, issued from the cloud, and suddenly there was a loud explosion. A shower of stones then rained down east and south of the city, several of which, witnessed by English ladies, fell in a meadow at Cosona some twenty miles away. Some fragments fell into a pond, which was subsequently drained, and those recovered were sold at high prices to English travelers. Hearing of this, the natives attempted to simulate the originals by artificial means, selling them to the unwary. How many stones fell is not known. One report estimated there were about 200, several of which weighed as much as two to three pounds.[2]

One visitor to Siena at the time was Frederick Augustus Hervey, bishop of Derry and fourth earl of Bristol. Having learned of the eruption of Vesuvius on 15 June, eighteen hours before the Siena fall, Hervey wrote on 12 July to Sir William Hamilton telling him about the event and enclosing a specimen. Hamilton, a British diplomat and husband of Lord Horatio Nelson's mistress, was a respected volcanologist and particularly knowledgeable about Mount Vesuvius. "My first objection," Hervey wrote, "was to the fact itself; but of this there are so many eye witnesses, it seems impossible to withstand their evidence, and now I am reduced to perfect scepticism." The stones, he said, were unlike any that were common in the Siena area. Local philosophers inclined to the belief that the stones were generated in "this igneous mass of clouds," rather than that they were ejected from Vesuvius, which was at least 250 miles away. Hamilton included Hervey's letter in a memoir, which dealt principally with the eruption of Vesuvius and was published in 1795 in the *Philosophical Transactions*. The Siena stones, Hamilton concluded, closely resembled those found on Vesuvius. Thus Hamilton gave the impression that he considered them to be volcanic ejecta, while adding a speculation that the volcanic ashes, which had risen high into the atmosphere, had perhaps been compacted into stones.[3]

The Siena fall, in fact, provoked a lively controversy in Italy involving over a half-dozen scientists, among whom were Giorgio Santi, professor of botany at the University of Pisa; Ambrogio Soldani, professor of mathematics at Siena; Domenico Tata, professor of physics and mathematics at Naples; Lazzaro Spallanzani, professor of natural history at Pavia; and the expatriate English mineralogist, William Thomson. Within three weeks of the fall, Soldani queried witnesses and gathered specimens. It was Soldani who had given Hervey a complete report of the fall and a fragment of the stone, which the latter forwarded to Hamilton. Soldani was also "the local philosopher" who expressed the view that because of their unusual mineralogical appearance the stones must have been generated in the "igneous clouds" by the condensation of atmospheric vapors. Chief among those who derided Soldani's hypothesis was Giorgio Santi, who at first proposed that the stones had been ejected from

Vesuvius and later sharply modified his position by arguing that they came from a submarine volcanic eruption near Siena, the sea subsequently having covered the location. Both Santi and Soldani turned for support to Lazzaro Spallanzani, then the foremost scientist in Italy. He rejected both opinions, declaring that the stones had been carried into the air by a local hurricane and singed by atmospheric electricity before dropping back to earth.[4]

Santi also wrote about the event to Thomson in Naples who in turn alerted Tata. It was the news of the Siena fall that stimulated Tata to publish the report of the Terranova di Sibari fall in 1755, as described in chapter 1. In addition to this report, Tata's brief volume is significant in two other respects. First, the Italian scientists learned both about the Hraschina and Eichstädt falls, because Tata cited Stütz's article in *Bergbaukunde*, and about the Tabor fall, because Tata stated that Thomson had heard of it in a letter from a Captain Tihausky. Italian scientists, then, knew that reports of the falls of stones must be taken seriously. Second, Tata requested Thomson to provide him with a detailed description of the Siena stone, which Tata incorporated into his book. Thomson described the stone as being composed of a black portion and a gray portion, separated by irregular curved lines such as would be produced by a semiliquid or pasty body united with another to which it would adhere. The black part enclosed lamellar, reddish pieces of pyrite, which contained globules of pure iron. These were perfectly malleable and attracted to a magnet. The gray portion enclosed many small particles of pyrite mixed with sand, and also contained many particles of pure iron, with the remainder composed of round quartzous grains and pieces of green glass. One can visualize Thomson's struggles in attempting to identify the minute minerals present in the Siena stone. Thomson named the entire meteorite substance "soldanite," and concluded by stating: "It appears to me that these stones have undoubtedly been pasty in the cloud from which they fell, or that they issued with it from some volcanic explosion, or better yet, that they acquired their friability in the fusion process which gathered and consolidated the substances suspended in the air." Thus Thomson lent support to Soldani's theory, as did Tata, who added the suggestion that the cloud within which the stones formed had been heated by electricity. Thomson saw to it that Tata's treatise was translated into French and published in 1804 in the *Bibliothèque Britannique*. In an addendum, he remarked that had communications in 1794 been better, "there would have been more familiarity with the important phenomenon of meteoric stones than there was not long ago in France, and the time taken laughing at it would have been more usefully employed examining it."[5]

Despite the controversy in Italy concerning the origin of the stones, there was universal agreement there that they had fallen. The event was also significant because it became well known in England following the return of travelers carrying their mementos, spurious or legitimate, and because of Hamilton's publication. Through the latter, the Siena fall also came to the attention of German scientists.

Ernst Chladni

Ernst Florens Friedrich Chladni was born in Wittenberg, Germany (fig. 5). His grandfather was a prominent Lutheran theologian and his father a

Fig. 5. Ernst Florens Friedrich Chladni (1756–1827). Reprinted from Hoppe (1979) with the permission of Akademische Verlagsgesellschaft.

distinguished jurist. Abiding by his father's wishes, Chladni studied law at the universities of Leipzig and Wittenberg, receiving his doctorate at the latter institution in 1782, the year of his father's death. Chladni immediately abandoned law and remained at Wittenberg a few years to study mathematics and science. Thereafter, failing to obtain a professorial position in any university, he began experimental work in acoustics, and in 1787 published his *Entdeckgungen über die Theorie des Klanges*. A lifelong bachelor, he spent most of his time traveling the length and breadth of western Europe giving public lectures on his acoustical discoveries and, in his later years, on meteorites. In June 1824, Wilhelm Olbers wrote to Carl Gauss that Chladni was in Bremen to give lectures: "It is really sad that this man, so worthy in many respects, does not hold a position or an appointment in any institution, and that he must still now in his 67th year seek his meager livelihood in such a way."[6] Very probably one major reason for Chladni's inability to secure a professorial position was that both of his specialties, acoustics and meteoritics, as important as they would become, were at the time only nascent sciences.

One of Chladni's early acoustical discoveries was that sand, spread on circular or rectangular copper or glass plates, formed distinct patterns when a violin bow was drawn across the edges, thus displaying the nodal curves where no vibrational motion occurred. Chladni's demonstrations were strikingly similar to the dust patterns that Georg C. Lichtenberg, professor of physics at the University of Göttingen, had produced in 1784 on an excited electrophorus. In late 1792 and early 1793, Chladni visited Göttingen to demonstrate his newly invented keyboard instrument, the euphonium, and had the opportunity to talk for many hours with Lichtenberg. Physically crippled, Lichtenberg possessed a towering intellect, a genial personality, and a keen sense of humor, which made him respected and admired by philosophers, literati, and scientists throughout Europe. The conversations between Lichtenberg and Chladni can be partially reconstructed from several of Chladni's later references: "I received the initial idea for my work on such masses [meteorites] from Lichtenberg, whose main talent truly was to throw out a few thoughts that gave new perspectives and could lead to further investigations. He had previously been the midwife of my ideas . . . concerning the various vibrations." Chladni asked Lichtenberg why in his lectures he referred to fireballs as electric meteors, when their appearance was so high in the atmosphere that, due to the air's rarity, the electricity must be feeble, enough to cause the aurora but not sufficient to display a burning and smoking mass. Lichtenberg replied that he and other physicists talked in terms of electric meteors because of the similarity between electrical flashes and meteors, but in truth they did not know what to make of them. When he pressed Lichtenberg further, Chladni recalled, "he told me that if all circumstances about fireballs were considered they could be thought of not as atmospheric but as cosmic phenomena, or as bodies which came from outside [of the atmosphere], which gave me the stimulus to study the observations made of fireballs." Lichtenberg, Chladni added, compared this idea with the fact that comets had previously been considered to be atmospheric meteors, although Seneca had been correct in his understanding of them.[7] After spending several weeks in the Göttingen library searching for references to fireballs, Chladni was prepared to propose his theory.

Chladni's book, published in German in early 1794 at Riga and Leipzig, bore the cumbersome and unrevealing title, *Concerning the Origin of the Mass of Iron Discovered by Pallas and Others similar to it, and Concerning a few Natural Phenomena Connected therewith.* However, in the first paragraph Chladni came straight to the point and asserted unequivocally that fireballs consisted of compact and heavy matter, which came from cosmic space, and that the iron found by Pallas in Siberia was a meteorite similar to others that had fallen.[8]

To refute previous explanations and to argue that fireballs were heavy, solid masses of cosmic origin, Chladni first presented a composite description of the characteristics of fireballs compiled from about two dozen reports. They could not be connected with the northern lights, because they came from all directions at every season of the year and, moreover, had a completely different appearance. They could not be electrical in nature, because at the high altitudes where they first appeared there could be no conducting vapors. Further, they were unlike lightning in having a determinate shape and in following a parabolic path, and they displayed actual combustion and were subject to repeated explosions. They could not be exhalations of oily vapors or hydrogen gas from the earth, because it would be impossible for gases to accumulate in long narrow pockets in the atmosphere. Also, their velocities indicated they were solid masses, not gaseous matter, and their trajectories demonstrated the action of gravity. Although admitting that fireballs might be cometlike bodies moving around the Sun or the Earth, Chladni believed that their indeterminate directions favored a movement peculiar to themselves and independent of the Sun and the Earth. All other explanations failing, Chladni argued that there were small masses of matter distributed and traveling through space which came under the influence of the Earth's gravitational force and entered the atmosphere. There, severe friction created excessive heat and electricity, which caused them to become incandescent and molten, to produce gases in their interiors, and to expand to enormous sizes until they were ruptured by explosions. Here, in explaining the visible appearance of a fireball, Chladni was far off the mark. Meteorite interiors do not give evidence of fusion, as chemists and mineralogists were to determine, and a valid explanation of the incandescent aura surrounding a fireball was to be made only fifty years later.

Shooting stars, Chladni thought, differed from fireballs only in that they displayed light briefly as they made their way in straight lines across the sky. Possibly they were at such high altitudes that they evaded capture by the Earth's gravity, and thus displayed only momentarily an electrical appearance or were ignited for just a few instants. Chladni suggested that simultaneous observations of shooting stars taken at places distant from one another should be made to determine their altitudes and paths. And, puzzled by reports that gelatinous or viscous matter had been found where shooting stars had been thought to fall, Chladni admitted the possibility of another type of luminous meteor composed of spongy matter that had risen from earth to a low altitude and been ignited by electricity. This section of his book was to cause much difficulty for Chladni. In 1798 simultaneous observations of shooting stars (unfortunately from too short a base line) did begin to be taken, and apparently demonstrated that some shooting stars rose in the atmosphere. This result, coupled with Chladni's talk about the fall of spongy matter, did give credence to the atmospheric origin of at least some meteors.

In his book, Chladni next turned his attention to reports of falls, documenting a total of nineteen. The most impressive, certainly, was Hraschina, which had been preceded by a fireball that had exploded into several parts and attested to by the sworn statements of seven eyewitnesses. If clouds covered the sky, he pointed out, the fireball's light might be mistaken for lightning and the noise of the explosion for thunder. Eight of the nineteen falls that Chladni cited were eighteenth-century falls: Ploschkovitz, Nicorps, Hraschina, Tabor, Albareto, Lucé, Aire-sur-la-Lys, and Eichstädt.

Chladni's discussion of meteoritic iron centered principally on the Pallas iron [Krasnojarsk], although he gave brief references to Don Rubin de Celis's account of Otumpa [Campo del Cielo], to a huge iron mass found at Aken near Magdeburg, and to scoria found on isolated European mountains. Again, Chladni began his argument for a cosmic origin by excluding other possibilities. They could not have been deposited from an aqueous solution, because they showed evidence of fusion. They could not have been produced artificially by man nor by forest fires or the combustion of coal, because there were no iron ore deposits nearby, the masses were too heavy for primitive man to smelt or transport, the stony ingredient (olivine) of the Pallas iron would not be so uniformly distributed, and the iron would not be malleable. But since the irons had obviously been in a state of fusion at some time, Chladni postulated that nature produced them in a fire that was far more intense than was possible on Earth and that thus resulted in the retention of malleability. A century would pass before this dilemma was resolved. A volcanic source of the masses was rejected for the same reasons as artificial fusion, in addition to the fact that there were no volcanoes in close proximity to the locations where the masses had been found. And although lightning could produce intense heat, it could not fuse such large masses of metal, except superficially. Since all of these possible explanations were faulty, the masses, Chladni argued, must have come from space.

It was at this point in his treatise that Chladni introduced the contention that it was just as reasonable to postulate small material bodies in space as to deny their presence, since neither condition could be proved on a priori grounds. He went on to say that the small masses could either be matter that for one reason or another had not become united to the celestial bodies at the time of their formation, or portions of a celestial body produced by collisions from without or explosions from within. Whatever the reason for their existence, these small masses moved as a result of their initial propulsion through endless space until they approached another celestial body so closely that they were attracted and fell onto it. Chladni concluded his short treatise (63 pages) by calling for metallurgical investigations of meteoritic irons, for detailed reports of fireballs, observed falls, and lightning strikes, and for simultaneous observations of shooting stars in order to determine their altitudes and trajectories.

In assessing Chladni's influence, one must distinguish between the three major hypotheses he proposed: (1) that stones fall; (2) that the stones (or irons) are the substance composing fireballs; and (3) that fireballs have a cosmic origin. The third hypothesis enlisted no support for over twenty-five years. By 1796, a number of German scientists—Lichtenberg, Johann F. Blumenbach, Wilhelm Olbers, Franz X. von Zach, and F. C. Fulda, all connected with the University of Göttingen either as professors or former students, and also Alexander von Humboldt and Abraham G. Werner—had accepted the first two hypotheses on the basis of Chladni's evidence.[9] In England, Chladni's first hypothesis was cited favorably beginning in 1796, and a brief summary of his book appeared in 1798.[10] In France, J. C. de Lamétherie mentioned Chladni's treatise in January 1799, but remarked that the facts had to be authenticated before an explanation was attempted.[11] Eugène Coquebert published a fairly complete translation of Chladni's treatise in 1803.[12] The *Bibliothèque Britannique* in Geneva carried a very favorable review of Chladni's work in early 1801, and later in the year Pictet announced his belief in Chladni's assertion that stones fall, "whatever their origin might be."[13] Without question, then, Chladni's work influenced acceptance of the fall of meteorites, but without the occurrence of other falls in the late 1790s, the impact of his arguments might have been far less.

From Wold Cottage to Howard's Report

In a letter to the *Bibliothèque Britannique* dated 10 May 1803, die-hard skeptic Guillaume Deluc placed the blame for all of the talk about meteorites on the Wold Cottage fall. He wrote: "The York County stone weighing 56 pounds, which was the initial source of all the hypotheses and all of the discussions that have appeared on this matter, had as its basis the report of a laborer, who being in his field, says that he saw it *when it was seven or eight yards from the ground*."[14] To Deluc, it was folly to accept the word of a laborer, but Major Edward Topham, who owned the Wold Cottage property in the East Riding of Yorkshire where the meteorite fell on 13 December 1795, satisfied himself in that respect. He queried the three eyewitnesses who had seen the mass emerge from the clouds after an explosion on that misty afternoon, and as a magistrate, he gathered sworn accounts from all who had heard the stone's passage through the air or seen it fall. Later, he wrote that he had been besieged for information about the fall, "and have answered innumerable Enquiries from the Ingenious and those who had no Ingenuity whatever." To commemorate the event, Topham erected a monument on the exact spot where the stone penetrated the soil.

Topham had permitted the exhibition of the stone together with testimonials of its fall at the Gloucester coffeehouse, Picadilly, in 1796, where the admission price was a shilling. In 1804 he sold the mass to the artistically talented naturalist, James Sowerby, for 10 guineas and a copy of Sowerby's *British Mineralogy,* which featured in the second volume an illustration of the meteorite and Topham's account.[15]

One important result of the Wold Cottage fall was the publication in 1796 of a treatise by the wealthy dilettante, Edward King. He paraphrased Soldani's report of the Siena fall, gave full details of the Wold Cottage fall, and also referred to Chladni's book, repeating the account of the Hraschina fall and of the examination of the Lucé meteorite. Commenting on the Fougeroux committee's conclusion that the latter had been struck by lightning, King wrote, "but since so many corresponding facts in other places have now come to light; such sort of *concurrent* evidence, according to the rule so wisely adopted by the learned Grotius, in his treatise *De Veritate,* ought, surely, to be duly weighed; and may justly lead us to a different conclusion."[16] From later references to King's tract, both in England and Europe, it appears to have reached a sizable audience. Two immediate responses appeared in 1796 in *Gentlemen's Magazine*. One, by a W. Bingley, described the fall of two small stones at Pettiswood, Westmeath, Ireland, in 1779, which Bingley said he had not reported for fear of ridicule. The accounts of other falls by Soldani, King, and Topham persuaded him to write, and he added: "I am not without hope, that, upon a farther investigation by the learned, my cake and Captain [sic] Topham's loaf will be found to have both been baken in the same stupendous oven *according to the due course of Nature.*"[17] A second, anonymous letter listed falls of stones at Hatford, Berkshire, in April 1628; at Woodbridge, Suffolk in August 1642; at Bulkeley, Chester, in July 1657; and at Turin, Italy, in April 1679. The first, at least, from the detailed description, appears to be authentic.[18]

Within the next two years the British public received news of additional falls. In 1796 Robert Southey, poet and man of letters, visited his uncle, the British chaplain at Lisbon, and on his return in 1797 published *Letters Written During a Short Residence in Spain and Portugal*. In it, Southey wrote that "a phenomenon has occurred here within these few days which we sometimes find mentioned in history, and always disbelieve." It was the fall from a clear sky, though preceded by explosions, at 2 P.M. on 19 February 1796 near Evora, Portugal, of a 10-pound stone, having the color of lead. Southey included the testimony of the local justice of the peace, to whom the peasant witnessing the fall had brought the meteorite.[19] Another report, which appeared in the *Philosophical Magazine* in 1798, was by Nicholas Baudin, professor of natural philosophy at Pau, who gave details of the Barbotan fall on 24 July 1790. Tilloch, the editor, solicited Chladni's comments, and Chladni seized the opportunity to present a résumé of his theory.[20]

George Sowerby, James's son, wrote in 1835 that the exhibition in London of the Wold Cottage meteorite "furnished to many learned men of the day, a theme for censure on the blind credulity of the Public."[21] However, Sir Joseph Banks, president of the Royal Society, visited the exhibit, and noting the similarity of the meteorite to a fragment of the Siena meteorite, which he owned, acquired a piece from Major Topham. Banks had begun to collect information on meteorite falls when he received a letter from John Lloyd Williams, F.R.S., who reported that on the evening of 19 December 1798 a large fireball had detonated and dropped several stones near the village of Krakhut, India, 14 miles from Benares. Williams concluded his letter: "I shall only observe, that it is well known there are no volcanos on the continent of India; and, as far as I can learn, no stones have been met with in the earth, in that part of the world, which bear the smallest resemblance to those above described."[22] Banks immediately requested Edward Charles Howard to make a chemical analysis of the stones.

Howard's Report

Edward Charles Howard was educated in France and was already a proficient chemist when he was elected to the Royal Society in 1799.[23] While attempting to determine the composition of hydrochloric acid, he discovered mercury fulminate; for these experiments he was awarded the Copley Medal of the Royal Society for 1800. In presenting the medal on 1 December 1800, Sir Joseph Banks revealed that Howard was "now employed in the analysis of certain stones, generations in the air by fiery meteors, the component parts of which will probably open a new field of speculation and discussion to mineralogists as well as meteorologists."[24]

Howard was still busy six months later, when Marc Pictet paid him a visit, viewed the specimens, and reported Howard's progress to his editorial colleagues of the *Bibliothèque Britannique*.[25] Howard's determination to make a comprehensive study undoubtedly contributed to the delay. Banks gave him fragments of Siena and Wold Cottage, and Charles Greville, who had purchased the mineral collection of Baron Ignaz von Born, located therein a specimen of the Tabor meteorite. From India, Williams sent a fragment of Benares. Howard was also able to examine two irons and two stony irons. He obtained a sample of the Campo del Cielo iron, which Don Rubin de Celis had presented to the Royal Society, and Charles Hatchett, a respected chemist, loaned a specimen of the Senegal iron [Siratik]. Greville made available two pieces of the Pallas stony iron [Krasnojarsk], which had been in the Born collection, and also a fragment called the "Bohemian iron," which one may legitimately infer was a piece of the Steinbach stony iron.[26]

Whether of his own volition or urged to do so by Charles Greville or Sir Joseph Banks, Howard sought the collaboration of Count Jacques-Louis Bournon, a French émigré whose detailed mineralogical descriptions of the meteorites comprised a considerable portion of the report and added substantially to its impact. Numerous references in the report show that Howard and Bournon were familiar with most of the extant literature on meteoritic stones and irons.

Meteoriticists now classify the four meteoritic stones that Bournon and Howard examined as ordinary chondrites—ordinary because they greatly outnumber other types of meteoritic stones in our collections. Tabor is a H-type or olivine-bronzite chondrite; Wold Cottage is an L-type or olivine-hypersthene chondrite; and Siena and Benares are LL-types or amphoterites, which are very similar to but distinct from the L-type. Bournon distinguished, apart from the crust, the four principal constituents of the stones. They were, first, the millimeter-sized chondrules (fig. 6), which he described as gray-brown, mainly opaque, elongated or elliptical globules. Particles of troilite (stoichiometric iron sulfide) he identified as reddish-yellow martial pyrites. Third, Bournon saw small particles of iron, which could be separated by a magnet. These are, in fact, nickel-iron inclusions, as Howard learned in his analysis. The fourth constituent was the silicate groundmass, mostly olivine and pyroxene (either bronzite or hypersthene), which Bournon described as a whitish-gray earthy substance that "serves as a kind of cement to unite the others." There were slight perceptible differences among the stones, Bournon noted; for example, the Tabor stone contained more iron particles, giving it a higher specific gravity, and the "martial pyrites" in it were visible only with the aid of a magnifying lens. "These stones," he concluded, "although they have not the smallest analogy with any of the mineral substances already known, either of a volcanic or any other nature, have a very peculiar and striking analogy with each other."

Following Bournon's description, Howard presented the methods used and the results obtained in his chemical analysis of the stones. He found that there was iron and nickel in the crust, along with other matter, but admitted his inability to make a satisfactory quantitative analysis. He and Bournon attempted meticulously to separate the four constituents that Bournon had identified—a procedure, Howard noted, that had not been used by the French academicians in their examination of the Lucé meteorite. Obviously this

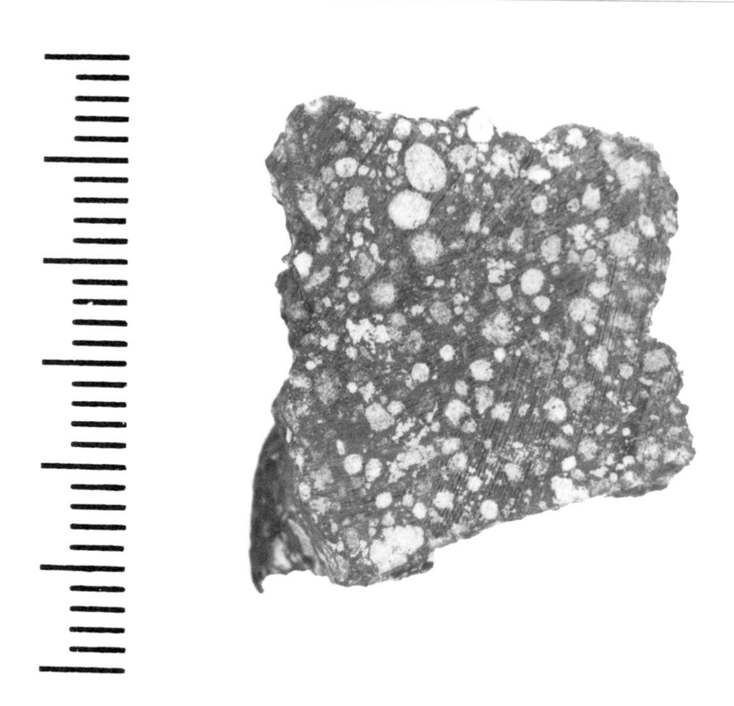

Fig. 6. Chondrules in the Bali carbonaceous chondrite. The scale is in mm. U.S.N.M. no. 4839. Courtesy of the Division of Meteorites, NMNH, Smithsonian Institution.

separation process took an inordinate length of time, and probably accounts for the long delay before the report's publication. To isolate 10 grains of troilite and 100 grains of chondrules—the initial amounts Howard used for his analyses of these constituents—must have taken weeks and exhausted the patience of both scientists. And unfortunately, the analytical results show that the separation was incomplete. The sulfur content of the troilite is too low, while the magnesia content of the chondrules appears too high and the silica and iron content too low. But the most striking result of Howard's analysis was that the iron particles contained nickel, ranging from 9 percent in Tabor to 28 percent in Benares. He reported the contents of silica, magnesia, iron oxide, and nickel oxide in the groundmass. These percentages were fairly uniform in the four stones, except that Tabor had a higher percentage of iron oxide, a result predicted in Bournon's examination.

The report next considered the irons. Bournon postulated that the stony irons represented a possible transition stage between the stones and the irons (fig. 7). He suggested that the iron particles in the stones might become interlaced, forming cavities that were originally filled with the substance found in the Pallas iron. The irons, in contrast, were compact masses of iron particles. Bournon described in considerable detail the vitrified substance found in the cells of the Pallas iron, and insisted correctly that it had a great analogy with peridot. Howard's analytical results of the substance, he stated, were "very nearly the same" as those of Klaproth in his analysis of peridot. Later confirmation of Howard's analysis did establish the fact that the material was olivine (peridot being gem olivine). Howard's analysis of the iron portions of the stony

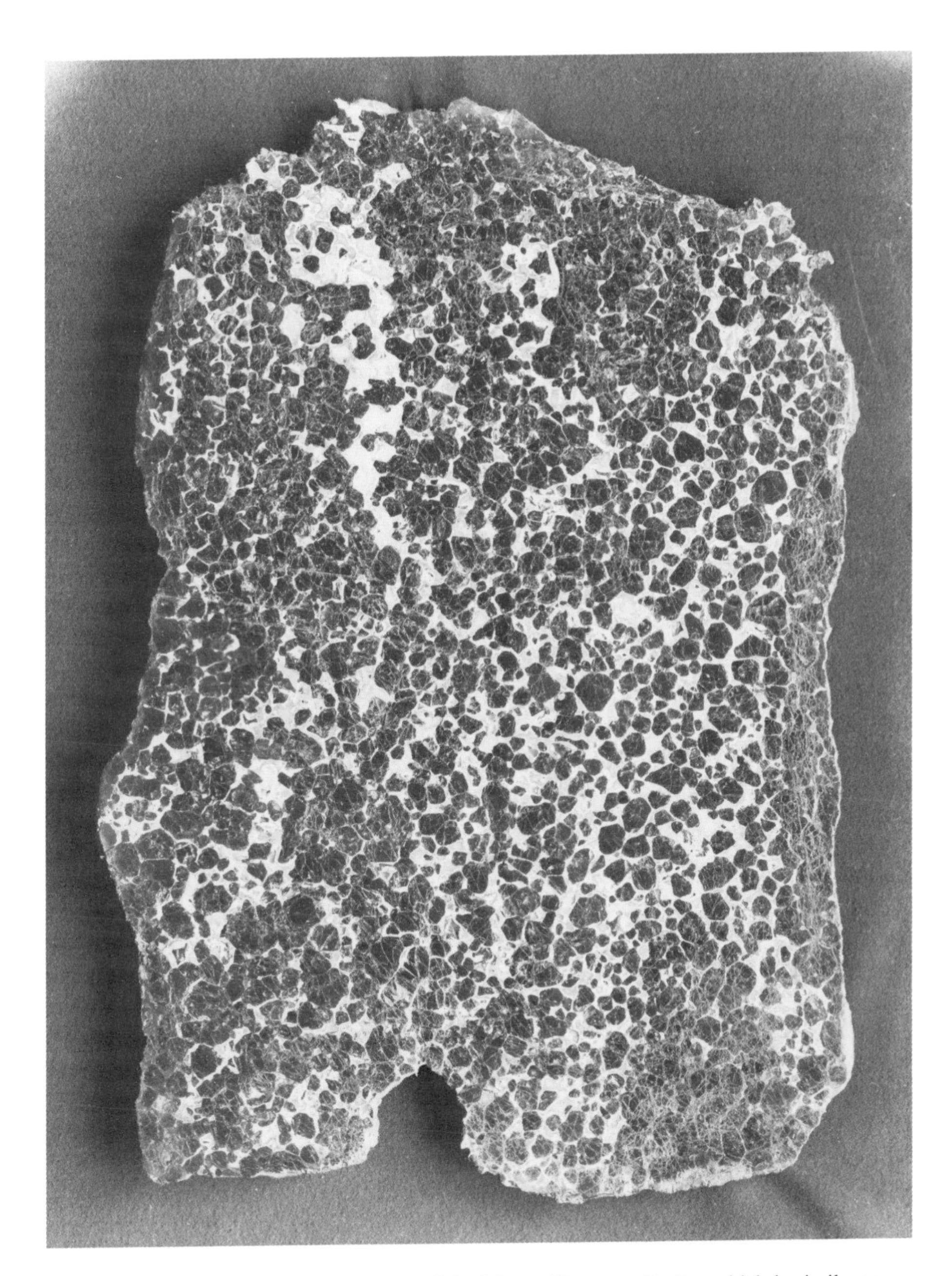

Fig. 7. Polished and etched section of the Mount Vernon pallasite, which is similar to the Krasnojarsk meteorite found by Peter Simon Pallas. The dominant olivine is dark, and the metallic nickel-iron is light. U.S.N.M. no. 300. Courtesy of the Division of Meteorites, NMNH, Smithsonian Institution.

irons and of the irons determined that they all contained nickel: Campo del Cielo, 10 percent; Steinbach, 4 percent; Krasnojarsk, 17 percent; and Siratik, 5 percent. These findings established that nickel was a constituent of the irons alleged to have fallen as well as of the stones.

Howard, as previously noted, was uncertain whether meteorites comprised the substance of fireballs. Witnesses had not seen fireballs prior to the Wold Cottage and Portugal falls, but they did report one prior to the Benares fall, and described lightning at Siena that might have been a fireball. To test the theory that electricity might be involved, Howard attempted to produce a crust on the interior surface of one of the stones by means of a substantial electrical charge, measured in England at the time in terms of equivalent areas (panes or squares) of glass. When he applied an "electrical charge of about 37 square feet of glass, it was observed to become luminous in the dark, for nearly a quarter of an hour; and that the tract of the electrical fluid was rendered black." Howard did, then, produce a sort of a crust, which gave credence to the notion that electricity was responsible for the crust's formation. In addition, he did note the thermoluminescence of meteorites, which has recently received more thorough scientific attention.[27]

In his conclusion, Howard emphasized that all of the stones had black iron-oxide crusts; they all had pyrites of a peculiar nature; they all contained an alloy of nickel and iron; and their groundmasses were similar in chemical composition. These coincidences and the authoritative reports of the falls, he declared, must "remove all doubt as to the descent of these stony substances." As for the irons and stony irons, they all contained nickel, and he ended by asking whether the stones and native irons did not have the same origin, and whether they might be the "bodies of meteors."[28]

Howard's report presented powerful arguments supporting the view that stones closely resembling one another and unlike any terrestrial rocks fell from the heavens, and also that "native irons" had a meteoritic origin. His comprehensive report commanded attention, and many scientists, particularly in France, accepted his conclusions after confirming the accuracy of his results and Bournon's descriptions. Howard, however, was fortunate. All of the stones he examined were ordinary chondrites. Had one or two been achondrites, which have no chondrules and many of which contain little or no nickel-iron, Howard's report would have lacked conviction.[29]

Controversy in France

Just what effect the French Revolution and its aftermath had on the attitudes of French scientists toward the fall of meteorites is a matter of speculation. Events certainly prevented or delayed the flow of information for about a decade. The Barbotan fall of 24 July 1790, though briefly noted in the Philomatic Society's bulletin, occurred during the period when the members of the Academy of Sciences were preoccupied with the mandated reform of its constitution. The Siena fall on 16 June 1794 coincided with the height of the Reign of Terror. The Salles fall near Villefranche occurred on 12 March 1798. According to Marquis Étienne de Drée, who reported it to the Institut five years later, a fragment was sent at the time to a member of the National Assembly. Drée added: "I do not know what attention was paid to it, and what effect it produced at a time when every mind was absorbed in politics." This was the period of the Second Directory, when France was a virtual police state. Drée only learned about the fall in February 1802. Moreover, many of the best scientists in France were actively working for the government, aiding in the production of iron and steel, gunpowder, and weapons. Others were teaching in the newly established institutions of higher learning, the École polytechnique and the École des mines. Thus, interruption of normal communication or preoccupation with more important activities may have caused French scientists to be unaware of the rising interest in falling stones that was occurring in Italy, Germany, and Britain. Beginning about 1800, however, with the publication of articles concerning meteorites in French and foreign journals, French scientists could hardly have been unaware that the fall of stones had become a topic of serious scientific discussion.

There is also a problem with respect to the number of French scientists who, after Pictet published a résumé of Howard's article in the May 1802 issue of the *Bibliothèque Britannique,* continued to oppose the idea that stones fell from the atmosphere. One can infer from a statement by Lamétherie that there was considerable opposition, for he reported that when Pictet read a memoir to the Institut on the results of Howard's report "he met with such disfavor that it required a great deal of fortitude for him to finish his reading."[31] However, Biot's description of the session varies a good deal. Pictet's account, he wrote, was received with a "cautious eagerness," though the "desire to explain everything" caused the phenomenon to be rejected for a long time.[32] There were, in fact, only three scientists who publicly expressed their opposition: the brothers Jean-André and Guillaume-Antoine Deluc of Geneva, and Eugène Patrin, an associate member of the mineralogy section of the Institut and librarian at the École des mines.

When Pictet early in 1801 published a favorable review of Chladni's treatise, it drew immediate fire from the Deluc brothers. Jean, a strict Calvinist, employed the same explanation of a fall that the Fougeroux committee had used thirty years before: stones did not fall; the event was imagined when lightning struck close to the observer. Just as no fragment of our globe could separate and become lost in space, he wrote, fragments could not be detached from another planet. It was also very unlikely that solid masses had been wandering in space since the creation, because they would have long since fallen into the sphere of attraction of some planet. And even if they did fall, they would penetrate the earth to a great depth and shatter into a thousand pieces.[33]

Guillaume, the younger and somewhat more conservative brother of Jean, rejected Chladni's hypothesis that fireballs consisted of matter from cosmic space because it was contrary to the laws of gravitation, "which maintain in their place and in their entirety all of the globes that move." He had seen a piece of the Pallas iron, he continued, and there were many indications that it had been artificially smelted and that the crust was due to rust. According to Pallas, the mountain had at its summit a very rich lode of iron, so that the Pallas iron was merely the product of a smelter used long ago to exploit the iron deposit. "I am obliged to believe," Guillaume concluded, "that the author of an hypothesis which demolishes planets in order to make this mass one of the fragments should at least make certain that nothing similar to iron existed in the place where it was found, nor in the vicinity."[34]

Patrin also responded to the review. The best explanation for the Pallas iron, he wrote, was the action of lightning. Lightning was responsible for the equal distribution of the vitreous globules in it, and also caused the mass to roll down the mountain. He proposed to place an iron-rich mineral with an iron point attached thereto on glass supports on top of a mountain. Undoubtedly, he said, it would be struck by lightning, and he presumed that a mass similar to the Pallas iron would result.[35] There is no evidence that Patrin ever conducted the experiment.

In the May 1802 issue of the *Bibliothèque Britannique,* Pictet published a thorough summary of Howard's report.[36] This sparked a brief but telling interchange between Patrin and Bournon. Patrin published a caustic critique of Howard's report, asserting that all accounts of falling stones were attested to only by unreliable witnesses and supported only by rumors. Valid physical principles explained the existence of such stones: they had been struck by lightning. "The love of the marvellous," Patrin wrote, "is the most dangerous enemy of natural science."[37] Since the Treaty of Amiens had at this time opened up normal lines of communication between Britain and France, Bournon, an émigré since 1792, in May 1803 addressed a scathing reply in the form of a letter to the *Journal de Physique.* Bournon applauded Patrin's zeal, but regretted that Patrin, in his eagerness to accuse Howard and himself, did not possess the facts necessary to conduct a judicial proceeding. Patrin had not seen any of the stones that formed the basis of their memoir, and apparently had not examined any others. Howard's analyses, Bournon wrote, were at least as decisive as his own descriptions. Patrin ought to explain why nickel had been found in all of the stones, which according to him had been struck by lightning, whereas it had not been detected in any other pyrite. Patrin believed that the action of lightning on a vein of iron ore produced the 1600-pound Krasnojarsk mass and the 30,000-pound South American iron [Campo del Cielo]. Bournon countered: "What a bolt of lightning would produce such an effect! . . . If this explanation were adopted, it could as easily give a reason for the formation of those isolated buttes of basalt, which until now we have considered as owing their origin to the expulsion of lava from a subterranean current." Patrin, Bournon continued, must also confirm that lightning either changes part of the iron into nickel or introduces nickel into the iron masses. Becoming more serious, Bournon reported that since their initial memoir, he and Howard had examined other stones, Salles and Barbotan, which were entirely similar to Benares and Tabor, respectively: "It is not within the laws of chance to find time after time the same type [of stone] where people say they have seen them fall, whatever might be the weakness of the degree of authenticity which their rank in society credits their testimony." Patrin, Bournon concluded, would be well advised to await the reports of further facts before offering in refutation a notion that was not only improbable but also bordered on the marvelous.[38]

In May 1803 Patrin published a contrite, even abject, response. He regretted that he disputed the evidence given for the fall of stones because, apart from his respect for the scientists who offered it, he was greatly interested in establishing the certainty of a fact that was relevant to his theory of volcanoes. Atmospheric fluids, he wrote, escape from volcanoes in the form of inflammable metallic and stony matter; these combine in the atmosphere and their fall is accompanied by fiery meteors. He had not wished to dispute the fact that stones fell; on the contrary, he wished to see it confirmed. Patrin concluded: "But the new proofs that M. de Bournon has furnished for it leaves nothing to be desired; I can only be gratified to have given to this respected scientist the opportunity to authenticate more and more facts which are of such great importance in the physical history of our terrestrial sphere."[39] This was the last communication by Eugène Patrin on the subject of meteorites, and it was written two months before Biot's report.

Pictet's summary of Howard's report also elicited a response from Biot. In August 1802 Biot published an article that evidenced his growing conviction that stones did fall. Citing the reported falls at Hraschina, Barbotan, Siena, Wold Cottage, and Benares, he stated that the agreement of facts in all of these cases gave the hypothesis a high degree of probability. Referring to Howard's analyses he wrote: "But what is striking . . . is that all of these stones contain nickel, a substance that is found rarely on earth, and iron in a metallic state, which is never a volcanic product."[40] The following October, Pierre-Simon Laplace, referring to Howard's report, suggested that the meteorites might be ejected from lunar volcanoes, and Biot and Siméon-Denis Poisson joined him in this opinion. About the same time, Antoine-François Fourcroy made a new analysis of the Ensisheim stone, and found that it contained 2.4 percent nickel.[41] Fourcroy; Louis Vauquelin; Balthazar Sage, founder of the Ecole des mines; Louis-Auguste Tonnelier-Breteuil, former president of the academy; and the mineralogist Marquis Etienne de Drée—all announced their acceptance of the fall of meteorites within the next few months.[42]

At least one staunch opponent remained. On 10 May 1803 Guillaume Deluc wrote yet another memoir. He attacked the hypothesis of lunar origin and also the renovated theory of the formation of stones in the atmosphere, which was then becoming prominent and gaining acceptance. Stones simply had not fallen. To say that the composition of these stones was unusual, he declared, proved nothing, since the composition of all terrestrial minerals was not known. Besides, the composition of the Benares stone had little in common with that of the Pallas iron. Deluc's most poignant argument, however, had to do with Chladni's idea of a cosmic origin. This hypothesis, he wrote, meant that the universe owed its origin to blind nature rather than to an intelligent being. It was a sad idea, leaving one with neither consolation nor hope.[43]

Biot's l'Aigle Memoir

The l'Aigle meteorite shower of about three thousand stones occurred about 1 P.M. on 26 April 1803. News of the event reached Paris on 3 May, and soon afterward fragments were sold on the Paris streets, as they had been in Siena a decade earlier.[44] On 9 May Fourcroy read a letter to the Institut, sent to Vauquelin by J. B. Leblond of l'Aigle, a corresponding member, which described the fireball, its explosion, and the shower. Vauquelin and Fourcroy analyzed several specimens, and on 19 June Fourcroy presented their results to the Institut. The stones were very similar to those previously examined by

Vauquelin and Howard. L'Aigle is, in fact, an ordinary L-type chondrite, but with a brecciated texture, a feature recognized and well described by Fourcroy.[45]

In view of the Fourcroy-Vauquelin report giving the details of the fall and the analysis of the meteorite, Biot's presentation of another memoir, less than a month later, on the same event is puzzling, and his investigation almost redundant. The recital of facts and the analyses of two of the most prestigious chemists in France could hardly be put into question. Nevertheless, Biot left Paris on 26 June, spent nine days conducting his investigation, returned to Paris on 4 July, and on 17 July read his report to the Institut. In the preface Biot stated that he had been requested to make the investigation by Minister of the Interior, Jean-Antoine Chaptal, who was a respected scientist and who, in 1803, happened to be president of the first class of the Institut. To be sure, Chaptal appears not to have received any information through official channels on the l'Aigle fall. The prefect of the Orne department at Alencon had not even heard of the fall when Biot questioned him, because the subprefect, located at Sees, had not forwarded to him the report submitted by the mayor of l'Aigle. The subprefect had wished to obtain further information, but because of other pressing duties in the intervening two months, had failed to do so. Yet had Chaptal wished to have a report for official reasons, it seems more reasonable that as minister of the interior he would have demanded an immediate, comprehensive investigation from the local administration and reprimanded the officials for their delinquency. Had he wished a scientist to participate, he could have called on the local corresponding member of the Institut, Leblond. Instead, he selected Biot, a young physicist whose convictions concerning the fall of stones were public knowledge, rather than a mature academician known to be fair-minded but unconvinced of the reality of the phenomenon. And if there were indeed "many" disbelievers among the members of the Institut, this would have been an easy task.

The episode becomes more comprehensible if we consider the activities and status of Laplace and his relationship with Biot. Jean-Baptiste Biot was to become one of the most brilliant scientists of the early nineteenth century, publishing over three hundred memoirs and more than a dozen books on magnetism, sound, geodesy, heat, and light. Gifted in mathematics and educated thoroughly in the classics, he studied at the École polytechnique from 1794 to 1797. Thereafter, while holding the position of professor of mathematics at the École centrale at Beauvais, Biot offered to read the proof of the first two volumes of Laplace's *Traité de la mécanique céleste,* then in the process of publication. Laplace immediately perceived Biot's talents, and in 1800 he was successful in persuading Chaptal, just named by Napoleon to take over Laplace's former position as minister of the interior, to appoint Biot a professor of mathematics at the Collège de France.[46] At the time, Laplace was very influential. He became chancellor of the Senate in 1803, and as the foremost figure in the scientific world, he had a dominant voice in the first class of the Institut. In 1800 Biot became a nonresident member of the Institut, and was elected to full membership in the first class in April 1803. Biot, in collaboration with Laplace, published his initial memoir on the velocity of sound in 1802, and in that year also, Laplace conjectured that meteorites might be ejecta from lunar volcanoes, an idea that Biot immediately endorsed. Then in early 1803, the hypothesis of the atmospheric origin of falling stones reappeared in a new guise, apparently supported by contemporary chemical theory. It is likely that Laplace and Biot, seeking to confirm the hypothesis of lunar origin, wished to gather additional scientific facts, which the Fourcroy report either omitted altogether or mentioned only superficially. These were the velocity of the fireball, its direction, and its angle of approach to Earth, from which they hoped it would be possible to compute its orbit. Confirmation of the fall would be an additional bonus, convincing the last diehards of the reality of meteorites. Laplace probably requested Chaptal, in his capacity as minister of the interior, to give Biot an official charge; hence any accusation by Institut members that an investigation initiated by Chaptal, as president of the first class, was biased could be avoided. To underscore the idea that the Institut was not involved, Louis-Jacques Thenard, a young chemist, analyzed the fragments collected by Biot, and Biot presented the report to Chaptal before reading it to the Institut.

This reconstruction of events may be pure conjecture, but Biot's report lends it verisimilitude. He described, of course, in overwhelming detail his interviews with scores of eyewitnesses: children, aged men and women, peasants, laborers, war veterans, estate managers, and ecclesiastics. He told of the broken tree branches, the holes still visible in the soil, and the large number of fragments still in the hands of the inhabitants of the l'Aigle area. All these pieces, Biot emphasized, were exactly similar to one another, and none resembled any of the minerals common to the countryside nor any of the products or slag from nearby foundries. This physical and moral evidence of the fall, he wrote, placed "beyond doubt one of the most astonishing phenomena which men have ever witnessed."

Biot's investigation nevertheless went minutely into other aspects as well. He did not go directly to l'Aigle, but instead took a circuitous route, making inquiries along the way in an attempt to determine the greatest distance at which the fireball had been seen and at which the sound of its detonation heard. In visits to more than a dozen farms, hamlets, and villages in the neighborhood of l'Aigle, he questioned people as to the path and height of the fireball before its detonation and the precise location of the cloud from which the shower of stones had come. At the same time, Biot was able to locate the perimeter beyond which no stones fell, finding that it formed an ellipse 2.5 leagues in length and almost 1 league in width (10 × 4 km), with the major axis lying in a southeast to northwest direction. The calculated declination of 22° agreed with the estimates of the angle of approach made by those eyewitnesses who had directly faced the approaching fireball and, surprisingly, with the actual direction of the magnetic meridian at l'Aigle. The problem of the fireball's velocity was, however, too difficult for Biot to solve. The high-altitude cloud from which the stones had fallen was above the southeast end of the ellipse, but it was rectangular, with the longest dimension lying in an east-west direction. Further, the largest stones fell at the southeast end of the ellipse and were reported to have been the first to fall. Biot could only conclude that the horizontal component of the velocity of the fireball at the time of the detonation was "not very considerable." Biot, then, noted the "dispersion ellipse," a feature of most meteorite showers. His detailed observations gave no support to the theory of lunar origin, but they did provide crushing evidence that stones fell from the sky.[47]

The United States and Jefferson

Very few articles about meteorites appeared in scientific journals in the United States prior to the fall of the Weston, Connecticut, stone of 14 December 1807. This is not to say that scientists were unaware of the controversy in Europe. Both Benjamin Silliman at Yale and James Woodhouse in Philadelphia, when they analyzed samples of the Weston stone, knew that they should search for nickel, and had at hand the relevant articles by Howard, Klaproth, Vauquelin, and Fourcroy, which gave details of the analytical methods. They also knew that some European scientists had postulated that the meteorites were ejecta from lunar volcanoes, and that others believed they were formed in the atmosphere.[48] But the subject received little attention, probably because there had been no reports of falling stones in America.

We get only a glimmer from manuscript sources as to what the attitudes of American scientists were, and this extant literature focuses almost entirely on the incumbent president, Thomas Jefferson, and his circle of friends. On 30 September 1803, Robert Livingston, the U.S. minister to France, wrote to Andrew Ellicott, a surveyor, mathematician, and friend of Jefferson: "The fact of the fall of stones from the sky, having been put, by some late enquiries almost beyond a doubt the Philosophers are now disputing whether they are generated in the atmosphere or whether we owe them to volcanic eruptions in the moon, as much remains to be said on both sides; prudent men have not yet thought it proper to pronounce judgment."[49] Ellicott informed Jefferson of Livingston's news, to which Jefferson replied on 23 December 1803:

> I find nothing surprising in the raining of stones in France, nor yet had they been mill-stones. There are in France more real philosophers than in any country on earth; but there are also a greater proportion of pseudo-philosophers there. The reason is that the exuberant imagination of a Frenchman gives him a greater facility of writing, & runs away with his judgment unless he has a good stock of it. It even creates facts for him which never happened, and he tells them with good faith.

Jefferson went on to say that during his seven years in France, new discoveries were made "which overset the whole Newtonian philosophy." In such instances, "the evidence of nature, derived from experience, must be put into one scale, and in the other the testimony of man, his ignorance, the deception of his senses, his lying disposition."[50]

On 20 January 1804 Ellicott wrote to Livingston:

> I confess my scepticism with respect to the fall of stones from the sky, of course, the arguments concerning their generation in our atmosphere, or being the effects of volcanic eruptions in the moon, appear useless, unless the facts of their fall be placed beyond all doubt. . . . Notwithstanding my present scepticism, which arises from the very dubious statement of all of the cases which have come to my knowledge, my mind is open to conviction on such evidence as ought to satisfy a mathematician.[51]

It was not until October 1805 that Ellicott received published material from France, which convinced him that stones did fall, that they had an unusual composition and texture, and that they were generated in the atmosphere. He advised Jefferson of his conversion, and Jefferson responded on 25 October 1805. He wrote that he had not seen the documents to which Ellicott referred, but that he had read Izarn's *Lithologie atmosphérique,* which was "an industrious collection" of facts of the same kind:

> I do not say that I disbelieve the testimony but neither can I say I believe it. Chemistry is too much in its infancy to satisfy us that the lapidific elements exist in the atmosphere and that the process can be completed there. I do not know that this would be against the laws of nature and therefore I do not say it is impossible; but as it is so much unlike any operation of nature we have ever seen it requires testimony proportionately strong.[52]

This passage indicates that Jefferson's skepticism was not about the fall of meteorites, but about their generation in the atmosphere. It is in this light that we should attempt to judge whether or not the remark so often attributed to him following the fall of the Weston meteorite two years later is apocryphal—namely, "It is easier to believe that two Yankee professors would lie than that stones would fall from heaven." In his *Discourse* on Jefferson, Samuel Latham Mitchill reported that soon after the Weston fall, he received an account and a specimen from friends. A senator who was to dine with Jefferson that evening asked to borrow the report and sample to show to the President and request his comments. When presented with the evidence, Jefferson, according to Mitchill's friend, said that "it is all a lie."[53] Later, on 15 February 1808, in a reply to a letter from a citizen offering to send a fragment of the Weston stone for an official examination by the Congress, Jefferson suggested that the members of a scientific society would be better qualified to examine the stone, "supposed meteoric," than those of the national legislature. He continued:

> We certainly are not to deny whatever we cannot account for. A thousand phenomena present themselves daily which we cannot explain, but where facts are suggested, bearing no analogy with the laws of nature as yet known to us, their verity needs proof proportioned to their difficulty. A cautious mind will weigh the opposition of the phenomenon to everything hitherto observed, the strength of the testimony by which it is supported, and the error and misconceptions to which even our senses are liable. It may be very difficult to explain how the stone you possess came into the position in which it was found. But is it easier to explain how it got into the clouds from whence it is supposed to have fallen? The actual fact however is the thing to be established.[54]

The tenor and even the wording of this letter is quite similar as that in Jefferson's December 1803 reply to Ellicott. It is possible that, upon reflection, he dismissed the notion of the atmospheric generation of stones and reverted to his original ambivalence about their fall. One other point is relevant. At the time of the Weston fall, the New England states were in an uproar about the economic effects of the Jeffersonian-sponsored Embargo Act of November 1806, and there was even talk of secession. Jefferson was antagonistic to the New Englanders, because they sought to circumvent the embargo by smuggling goods into Canada. It is therefore possible that soon after the fall and before the American Philosophical Society in March 1808 heard Silliman's report and accepted his memoir for publication, Jefferson, in a fit of temper, made the remark. But scholars have not yet located the source, so that at this time it must remain conjectural.

The episode does give evidence that in the early 1800s news of events in Europe relative to meteorites reached America only belatedly, and that a few American scientists continued to be skeptical about falls long after they were accepted abroad. The complete dearth of reported falls in America may be significant. But a generalization about the attitudes of American scientists toward meteorites requires more testimony.

EARLY NINETEENTH-CENTURY THEORIES OF METEORITE ORIGIN

When scientists accepted the fall of meteorites, there was no theory of their origin that was satisfactory to a majority. The hypotheses proposed demonstrate that there was certainly no lack of "exuberant imagination," as Jefferson put it, among either French or non-French scientists. They also reveal the tenacity of Aristotelian meteorological ideas, the rudimentary knowledge about the constitution and physics of the atmosphere, and the confusion about the nature of heat and light and about electricity and magnetism prior to the development of thermodynamics and electromagnetic theory.

While it is true that most hypotheses postulated an extraterrestrial origin of meteorites, it is incorrect to say that most scientists accepted a cosmic source. The ingrained notion that the Earth-Moon system was a closed entity still reigned, so that theories proposing an origin within this system enjoyed the most support. These theories held that meteorites were formed in the atmosphere, that they were ejecta from lunar volcanoes, or that they were pieces dropped in passage by comets orbiting the earth. Scientists made a clear distinction between such theories and those postulating that meteorites were remnants of a destroyed planet or debris from when the Solar System formed. A few scientists speculated that meteorites were ejecta from active terrestrial volcanoes or from massive primitive volcanoes in the Earth's early history.[55] Here, however, we describe only the leading theories.

Atmospheric Origin

If Lavoisier is to receive any criticism for impeding meteoritics, he should be blamed for the several passages in his *Traité élémentaire de Chimie* (Paris, 1789) that furnished sustenance for the hypothesis of atmospheric origin. A vital component of Lavoisier's chemical theory was that a gas—for example, oxygen—was a compound of the element oxygen and of a weightless elastic fluid, which was the matter of heat, or what he termed "caloric." When combustion occurred, say with carbon, the element oxygen combined with the carbon to form oxides and simultaneously released the combined caloric, which produced the heat released during combustion. Different gases, he taught, are formed daily on Earth, and if lighter than the atmosphere, they rise into the higher regions and form strata above the common air. "The phenomena which accompany igneous meteors," Lavoisier wrote, "induce me to believe that there exists in the upper part of our atmosphere a stratum of inflammable fluid in contact with those strata of air which produce the phenomena of the aurora borealis and of other fiery meteors."[56] But gases were not the only substances that passed into the atmosphere. "It is not impossible," he said, that "we may discover in our atmosphere, certain substances naturally very compact, even metals themselves."[57] Lavoisier added that he intended to pursue the topic of fiery meteors in a separate treatise, but his tragic death intervened.

The most popular version of the atmospheric theory incorporated Lavoisier's ideas. Gases—principally hydrogen, marsh gas, and oxygen—were said to rise high in the atmosphere where they formed layers or spherical envelopes. Metallic and earthy elements—either in the gaseous state or as particulates, the products of smelters, whirlwinds, or volcanic dust—also rose into the air and became entrained in the inflammable gases. Electricity caused the gases to ignite, which consolidated the metallic and earthy matter into solid masses. The release of heat and light produced the phenomenon of a fireball and gave it a horizontal propulsion, so that in falling the meteorite assumed a curved path. The atmosphere, then, was a vast storehouse of chemicals, upon which the forces of nature acted to produce shooting stars and meteorites.

Soldani, Thomson, and Tata, as mentioned previously, proposed this hypothesis in 1794, and Alexander von Humboldt referred to it in 1798.[58] In early 1803, Joseph Izarn, physician, physicist, and later inspector general of the University of France, expounded the theory in detail in his *Des Pierres tombées du Ciel, ou Lithologie atmosphérique,* the work mentioned by Jefferson. Lamétherie gave Izarn's book an extensive but noncommittal review, which gave it widespread visibility.[59] But Ludwig W. Gilbert, editor of the *Annalen der Physik,* wrote very critically that: "It seems to me from all that is exposed that the author is quite a stranger to most areas of physics and has a very lively imagination." Throughout his abstract, Gilbert made devastating comments on Izarn's ideas, showing them to be illogical and unscientific, and stated that Izarn had either misread or not understood Dalton's theory of atmospheric gases and Fourcroy's and Berthollet's teachings about chemical combination and affinity.[60] Later, David Brewster charged that Izarn had "dragged his readers into a tedious and somewhat obscure explanation."[61]

Izarn's work initiated a stream of publications favoring atmospheric origin. These included a new edition of Soldani's book and treatises or articles by G. A. Marichal, Armand Séguin, and Serres de Mesplès in France; Ernst Wrede in Germany; and Thomas Forster, William Beauford, Eusebius Salverte, and John Murray in England.[62] In the United States in 1810, Lyman Spalding, a chemistry lecturer at Dartmouth Medical School, submitted an essay to the American Philosophical Society in which he proposed the hypothesis and cited Lavoisier's *Traité* as supporting evidence. A committee of referees, however, considered that it was "founded on mere hypothesis, unsupported by experiment or facts," and recommended against publication.[63]

There were differences in details. Wrede, a professor of mathematics at Berlin's Friedrich-Wilhelm Gymnasium, maintained that the explosion of the gases created a vacuum that caused compaction of the meteorite, an idea supported by Séguin, who was Lavoisier's assistant from 1789 to 1794. Salverte proposed that atmospheric water vapor dissociated, thus producing the necessary oxygen and hydrogen. Murray, a chemist, supposed that the atmospheric masses of oxygen and hydrogen were differently electrified, so that upon collision they exploded and fused the meteoritic stone or iron. Murray's article received a scathing critique from Henry Atkinson, an astronomer and schoolmaster who pronounced Murray's reasoning "vague," "inconclusive," and "strange."[64]

The effects of magnetism also entered the theory. Johann W. Ritter, a German Naturphilosoph, was struck by Biot's observation that the direction of the l'Aigle fall was precisely that of the magnetic meridian. He researched reports of the appearances of auroras and of fireballs and meteorite falls back to 1492, and announced that he had found a definite correlation. Fireballs must be connected with magnetism, he wrote, and in addition, the meteorites that they dropped contained iron and nickel, both of which were attracted to the magnet. Electricity, Ritter thought, probably played a role in the explosion of the fireball, and "what was electrical in the explosion became heat and light, and what was magnetic became corporeal."[65]

Ritter's ideas favorably impressed P. N. C. Egen, professor of physics at Doetinchen Gymnasium, and in 1829 John (?) Butler ascribed the gradual compaction in the atmosphere of volatilized metals and earths to the action of magnetism.[66] Karl E. von Hoff, a geologist and mineralogist, thought that the idea was reasonable, given the presence of iron and nickel, but added that "how the force is exercised, which gathers the volatilized and dissociated substances in the highest regions of the atmosphere, and there presses them into solid masses with the known appearance of meteorites remains to be explained."[67] In 1848, Charles U. Shepard attempted an explanation, at least in part. Ashes from active volcanoes, he thought, rose high in the atmosphere into regions influenced by polar magnetism. There, the highly magnetic elements—iron, nickel, cobalt, and chromium—separated and, as suspended metallic particles, formed the columnar clouds visible in auroras. The other substances, among them diamagnetic elements, collected into masses at right angles to the columns, thus producing the transverse arches of auroras. Disturbances of the magnetic forces caused precipitation of some of the suspended matter. If the disturbance was confined to the metallic matter, then iron meteorites fell; if confined to the nonmagnetic matter, then ordinary stones; and if both types were affected, then stony meteorites dropped.[68] Shepard's hypothesis received no scientific support as far as can be determined, although at least two articles in popular journals mentioned it favorably.[69] Shepard himself abandoned the hypothesis almost immediately.[70]

It was the tenacious belief that the Earth and its moon constituted a closed system which kept the theory of atmospheric origin alive for such a long time. In 1822 Egen wrote: "As long as men have observed nature, the Earth has been a closed entity of a type that nothing ponderable has manifestly or demonstrably come to it or left it. There is, thus, great probability that everything ponderable, which seems new on the Earth and about which we know nothing further, belongs exclusively to the Earth. The law of property of the Earth is, as it were the law of probability."[71] Almost forty years later, Mrs. G. S. Silliman advanced the same argument in support of atmospheric origin:

> It seems not in accordance with ascertained science to ascribe mysterious appearances on the earth, or in its atmosphere, to causes proceeding from the planets, or spheres, moving in space, *independent* of the earth and its system. . . . Is it not more in harmony with the integrity and perfection of His works that this phenomenon [meteorites] should originate in a meteorological process, than that the symmetry of the creation . . . should be violated by a visit to the earth of a lone, foreign intruder from the depths of space?[72]

There was recognition that the theory of the atmospheric origin of meteorites was fundamentally Aristotelian. Thomas Forster pointed this out in 1823, and a decade later Julius L. Ideler stressed the fact. Ideler translated and edited an authoritative text of Aristotle's *Meteorologica,* and in the process wrote a book on the origin of meteorites and auroras. Ideler's work included a comprehensive summary of all hypotheses relative to the atmospheric origin of meteorites, and almost triumphantly he declared that not all of Aristotle's ideas were outmoded. With a few exceptions, he thought, Aristotle's meteorology had stood the test of time.[73]

Lunar Origin

Although Robert Hooke in 1665 had concluded that the craters on the Moon had been caused by volcanoes, sustained observations of the craters and of any volcanic activity began only in the 1780s.[74] Franz U. Aepinus in 1781 published an article on the theory of lunar volcanoes, and Lichtenberg discussed the craters in articles published in 1779 and 1781. Antonio de Ulloa, a natural historian, noted a red spot on the Moon during a total eclipse in 1778, which Beccaria explained as a lunar volcanic eruption. In England William Herschel reported observing an eruption in 1783 and three more in 1787. Lichtenberg sent Herschel his congratulations, stating that: "Your last three volcanoes delighted me. . . . Perhaps in the end we shall find out from the Moon the most reasonable theory of the Earth. It would be curious, if the Moon is now busy forming its own atmosphere, and perhaps after a few centuries we would see with greater intensity more indistinct twilights and clouds on it."[75] Within a few years, apparent volcanic activity on the Moon was reported in England by Nevil Maskelyne; in France by Lalande and Jean-Dominique Cassini; and in Germany by Johann Bode, Franz Xaver von Zach, and Johann Schröter.[76]

It was Heinrich Wilhelm Olbers, however, who apparently was the first to conceive that meteorites were lunar volcanic ejecta. Olbers, a Bremen physician and astronomer, had been one of Lichtenberg's students at Göttingen. They corresponded frequently, and when Chladni left Göttingen in February 1793, he carried a letter of introduction from Lichtenberg to Olbers. It is very likely that Olbers knew about Chladni's book in 1794, when it was published, but it was the Siena fall that provoked Olbers's interest in the origin of meteorites.[77] In a letter to Gauss in 1802, Olbers wrote:

> What do you think about the stones that have fallen from the sky, and about Laplace's idea that they are probably the products of lunar volcanoes? Seven years ago in a lecture held at the museum here concerning the shower of stones in Siena, I advanced the *possibility* of a lunar origin of these stones. *At that time,* to be sure, I thought that the stone shower had been propelled from Vesuvius, but the famous stone fall at Aegospotamos induced me to investigate how great the velocity of a ponderable body on the Moon's surface would have to be to permit its fall.[78]

Lichtenberg had argued that the Siena stone could not have been propelled such a great distance from Vesuvius, but Olbers, misled by Hamilton's remark about the similarity of the Siena meteorite and Vesuvian stones, found his reasons unconvincing. When Howard's report appeared, Olbers decided to publish his calculations on the lunar escape velocity.[79] Lichtenberg had not accepted Chladni's arguments; he remarked to friends that when he read the book he felt as if a stone had hit him on the head, and wished Chladni had not written it. But in 1796, after Olbers made his tentative proposal, Lichtenberg coined the aphorism: "The Moon is such an unfriendly neighbor that he welcomes the Earth with stones."[80]

Lamétherie reported that Laplace made his calculations on the lunar escape velocity after learning that "some English scientists" supposed that meteorites had been ejected from lunar volcanoes."[81] Laplace proposed the hypothesis in a letter to von Zach in summer 1802, and emphasized then that it was merely a conjecture that required careful investigation.[82] It appears that, not only to Laplace but to other early proponents as well—namely, Olbers, Biot, Poisson, Heinrich W. Brandes, and Charles Hutton—the lunar hypothesis was just a speculation. It was based in part on calculations of the lunar escape velocity, determined by all to be about 7,900 feet per second, and in part on estimates of the maximum muzzle velocity of guns and the velocity of ejecta from terrestrial volcanoes.[83] In fact, in his 1803 article, Olbers listed most of the major objections that were subsequently raised against the hypothesis.

Gauss mentioned one difficulty even before Olbers published his paper. In his reply to Olbers's letter containing the lunar escape velocity calculations, Gauss wrote:

> Thank you for sending your results on the selenitic runaways. It must be an interesting problem to calculate the *probability* that a body projected from the Moon with a given appropriate velocity reaches the Earth. I believe it must be very small, and the *direction* of the propulsion confined within narrow limits. Perhaps one time I'll do it myself.[84]

In his paper, therefore, Olbers emphasized that, for a body projected from the Moon to fall to earth, the body would have to be in an elliptical orbit whose perigee fell within the Earth's atmosphere. Lunar volcanoes would have to eject a great number of bodies to satisfy this condition, with the result that the Moon's mass would suffer constant diminution. If there were a large number of lunar ejecta in space, Olbers said, they ought to be seen through the best telescopes. The difficulties in the way of the lunar hypothesis were numerous and important, he concluded, and he hoped that Chladni would publish a new edition of his book to shed more light on the hypothesis of cosmic origin.[85]

At first, Chladni was not disturbed by the lunar hypothesis. In 1803, he noted that a shower of beans had allegedly fallen in Leon, Spain, and that 12 bushels were collected, cooked, and found to be delicious, and he jokingly suggested that a lunar eruption had perhaps catapulted a food warehouse into space. More seriously, he wrote, meteorites might come from the Moon, but there were the difficulties against the hypothesis mentioned by Olbers.[86] However, his confidence in his own theory of cosmic origin had already been shaken by the results of the simultaneous observations of meteors, a development to be described shortly. Further, the examinations and analyses of the stony meteorites by Howard, Bournon, Drée, Vauquelin, Fourcroy, and Klaproth indicated that they all had the same texture and approximately the same chemical composition, which indicated a common origin and argued against the notion that they came from all parts of cosmic space. In addition, the fact that the iron and nickel were not oxidized pointed to an origin of meteorites in an environment that was deficient in oxygen, which agreed with the knowledge that the Moon had little or no atmosphere. Finally, the average specific gravity of meteorites corresponded to the specific gravity of the Moon.[87] Confronted with this evidence, Chladni, in 1805, endorsed the idea of lunar origin. He wrote:

> At the present time, I completely agree, *that the stone and iron masses, which often fall with a fireball, are nothing other than ejecta from volcanoes on the Moon,* and it is enough for me to have been the first in modern times to demonstrate in my treatise . . . 1. *that the extant reports of such falls were not fabrications but actual observations;* 2. *that these masses come to us from outside* [of the atmosphere].[88]

The iron and stone masses, he thought, might come from adjacent volcanoes.

Despite this frank admission, however, Chladni never gave his full allegiance to the lunar hypothesis, nor did he abandon his cosmic hypothesis. In 1808 he wrote that "in the case these masses are ejecta from volcanoes on the Moon," the stones could be considered as altered species of rock and the irons and stony irons as lava.[89] In 1816 he stated that it had not yet been determined whether meteorites were material refuse moving in cosmic space that had never united with a celestial body, as he had first thought, or ejecta from lunar volcanoes, debris from a destroyed planet, or pieces torn from a celestial body.[90] Then in 1818 Chladni reverted to his original idea of a cosmic origin, maintaining that there was a possibility that lunar ejecta could fall to earth, but that the majority of meteorites originated on the Moon was highly improbable. The velocity with which meteors encountered the Earth's atmosphere was much too great for them to have come from the Moon, and it corresponded to the orbital velocity of celestial bodies.[91]

The theory of lunar origin was termed "Laplace's hypothesis" in the early nineteenth century, although what little Laplace wrote about meteorites appears contradictory. Even in the sixth and last edition of his *Exposition du système du monde* (1835), he described how lunar volcanoes could provide the requisite escape velocity, and how in time the orbits of ejecta could change such that meteorites would penetrate the Earth's atmosphere. The similar composition of meteorites, he wrote, gave *vraisemblance* to this idea of their origin.[92] But a few pages later he said that meteorites stood in relation to the Earth as comets to the Solar System: "they would appear strangers."[93] And years before, in 1820, he had questioned whether planetary motions would be altered by the impact of meteorites, which "appear to come from the depths of celestial space."[94] Yet the theory of lunar origin remained "Laplace's hypothesis."[95]

Although Olbers, as early as 1805, remarked that he preferred Chladni's cosmic hypothesis, he did not publicly relinquish the idea of lunar origin until 1836.[96] From observations made by Brandes of the velocities of meteors entering the atmosphere, Olbers calculated that the required velocity of propulsion from the Moon would be too large to be credible (i.e., 22.6 mi/sec). "I consider," he wrote, "the Moon in its present condition to be a very quiet neighbor."[97] Poisson, in 1837, indicated that he had abandoned the lunar hypothesis by remarking that meteorites were the result of the fall through the atmosphere of bodies, apparently inexhaustible in number, which "move in the heavens, either around the Sun, or around the planets, or perhaps around their satellites."[98]

Despite these defections, there were still longtime adherents of the lunar hypothesis, as well as new converts. Benzenberg, about whom more will be said later, wrote a curious book in 1834, in which he postulated that there were 1,200 volcanoes on each side of the Moon, and that over 4 million moonstones had fallen to earth since the creation 6,000 years before. Benzenberg, for some unknown reason, asserted that Chladni had held from 1794 the view that all meteors were moonstones, some of which fell to earth. He himself, Benzenberg said, only became converted to Chladni's view after the meteor shower of November 1833, which he attributed to a lunar volcanic eruption.[99]

In 1834, also, the great chemist Jöns Jacob Berzelius came out in favor of the lunar hypothesis, although he did add that there was a possibility that meteorites might be fragments of a shattered planet. Berzelius offered many of the same arguments: the volcanic activity observed on the Moon; the relatively small velocity required for ejecta to escape the Moon's gravity; the texture testifying to a common origin; the chemical composition evidencing a source deficient in oxygen and water. Berzelius, however, advanced a new idea. Those meteorites rich in unoxidized iron and nickel, he thought, must come from mountain ranges at the center of the Moon's disc that faced the Earth. It was the magnetic attraction between the Earth and this area of the Moon that caused the Moon always to present this face to the Earth. Meteorites deficient in native iron—for example, Stannern, Jonzac, and Juvinas—came from other volcanoes, whose location only occasionally permitted ejecta to attain orbits that would intersect with the Earth's atmosphere; hence such meteorites were much rarer.[100]

American chemist J. Lawrence Smith was among the last scientists in the period under consideration who attempted to sustain the hypothesis of lunar origin. The mainstays of his refined theory, presented in 1855, were the testimony of a common origin of meteorites, a prolonged igneous action, a source deficient in oxygen, and an average specific gravity corresponding to that of the Moon. However, Smith held that meteorites had either been ejected long ago from currently extinct volcanoes, the craters of which were now visible, or they had been scraped and propelled from the Moon's surface by some unknown force.[101] In what was probably a response to Smith's theory, Benjamin Apthorp Gould published in 1859 what Gauss had thought about doing years before—a calculation of the probability of a body with the velocity necessary to escape from the Moon actually reaching the Earth. He estimated that of 5 million such bodies ejected, only three would fall to earth. Given the masses of known meteorites, Gould concluded that the Moon would suffer such a loss of mass that its precession and nutation would be affected, and these consequences had not been observed.[102] Thereafter, the vast majority of scientists agreed with Brewster's comment that the hypothesis of the lunar origin of meteorites was "a romantic notion."[103]

Though J. Lawrence Smith's reasons for thinking that meteorites came from the Moon were inadequate, his speculation that they had been propelled from the surface of the Moon turned out to be correct. Not all were, but just a few, as scientists discovered about 1980, when two meteorites were found in Antarctica that had almost certainly been flung into an eventual Earth-crossing orbit by the impact of a giant meteorite on the lunar surface.

Origin from Comets Orbiting the Earth

The idea that terrestrial comets were in elliptical orbits around the Earth and occasionally dropped meteorites had few adherents. A few European scientists lent support, but the hypothesis was kept alive until the 1840s primarily by professors at Yale College. This notion is important, however, because it reveals how some scientists explained the apparently huge size of the fireball and its sudden disappearance.

The hypothesis rested on the fireball theory of Thomas Clap, noted in chapter 1. Clap's idea was not widely known; Chladni did not mention it. Soon after the fall of the Weston meteorite, Jeremiah Day, Benjamin Silliman, and James Kingsley, all professors at Yale, evidently discussed the various theories of meteorite origin and concluded that Clap's theory was more tenable than the others. Silliman and Kingsley's memoir pointed out the defects in the hypotheses that meteorites were ejecta from lunar or terrestrial volcanoes or formed in the atmosphere, and offered Clap's theory "with much deference."[104] In 1810, Day gave a more extensive description of the various theories of origin and applauded Clap's idea, remarking that "the first step towards an important discovery, is often, an ingenious conjecture."[105] In 1813, John Farey and Benjamin Bevan in England endorsed the theory in an article that elicited Chladni's criticism.[106] Given the frequent meteorite falls, Chladni wrote, the numbers of the terrestrial satellites would soon be diminished and their activity would finally come to an end. Yet Chladni had no objection to an alternative advanced by Zach that the masses might orbit the Earth several times until they fell.[107]

The most vocal supporter of the terrestrial comet theory was Benjamin Silliman. During the 1830s and 1840s, Silliman gave many public lectures on meteorites to large audiences in New York and New England, displaying a large chunk of the Weston meteorite. In this period he favored the idea that the comets were portions of a destroyed planet. Toward the end of the lecture, he undoubtedly electrified his audience by saying: "May they not one day come down entirely? Shall we desire it! They might sweep away cities and mountains—deeply scar the earth and rear from their own ruins colossal monuments of the great catastrophe."[108]

The mainstay of the terrestrial comet theory was the belief that the fireball was of enormous size and passed rapidly out of sight after a few pieces had split off. Referring to the Weston fireball, Day wrote:

> The great body of the meteor must be a solid compact substance, capable of sustaining the resistance of the air. Its magnitude is such, as to illuminate, at once, a region of one or two hundred miles in extent. It is inconceivable, that a body only two or three feet in diameter, however luminous, should attract, at the same moment, the gaze of a whole country; and appear, to the distant spectators, one third, one half, or three fourths as large as the moon. The real diameter of the meteor, according to the lowest computation, must be some hundreds of feet. No such body has ever come to the ground.[109]

Nathaniel Bowditch, in fact, calculated that if the diameter of the Weston fireball was 491 feet—a minimum estimate from all of the reports—and if it had the same specific gravity as the meteorite that fell, then it must have weighed 6 million tons.[110] One European journal reported Bowditch's estimate as fact, an error that Chladni hastened to correct, explaining that one could conceive of a fireball in miniature as an inflated soap bubble, from which, after it burst, only a few little droplets fell.[111]

The fact that fireballs are perceived to be much larger than they actually are emerged gradually. In 1837 Olbers wrote: "The illusion which occurs naturally with bright luminous objects, that we believe we see them disproportionately larger than they really are, makes it almost impossible to determine the true diameter of meteors with any certainty at all."[112] In 1854 J. Lawrence Smith, doubting reports of the huge size of the Weston fireball, experimented to assess the extent of the illusion. Viewed from a distance of 10 inches, the light emitted by carbon points, across which an electric arc was passing, appeared to be 0.3 inch in diameter. From a distance of 0.25 mile, however, the apparent size of the light source was three times the diameter of the Moon's disc. A small piece of lime and a sliver of steel heated to incandescence yielded similar but less spectacular results. The large estimated sizes of fireballs, Smith wrote, were based on comparisons with the size of the Moon's disc. His experiments demonstrated conclusively that an incandescent body at a distance of a half-dozen miles from an observer might exhibit an apparent diameter equal to half the size of the Moon's disc and yet not be more than a few inches or a foot or two in diameter, depending upon the intensity of the incandescence. The Weston fireball, he continued, disappeared at the time of the explosion. Numerous luminous fragments were seen to descend, and "the natural conclusion appears to be, that the whole meteorite was contained in the fragments that fell."[113]

Smith's experiments were convincing. Although the reason for the luminosity of a fireball remained a problem, its apparent size was no longer stressed. And without the existence of huge fireballs, the terrestrial comet hypothesis passed into oblivion.

Cosmic Origin

When Chladni proposed his hypothesis of the cosmic origin of meteorites in 1794, he preferred the explanation that they were fragments of matter that had not united with any of the planets when those were formed. As an alternative, he suggested that meteorites might be small pieces of a previously existing planet that had been destroyed either because of an interior explosion or a collision with another celestial body. Early on, every scientist who wrote on the subject of meteorites considered Chladni's preferred explanation to have little probability. Drée wrote that it adhered only slightly to accepted physical and astronomical principles,[114] and Brewster felt that it was stated in terms that were "too vague and gratuitious."[115] Chladni attributed the rejection to a deep-rooted and arbitrary belief that each celestial body was considered to be a self-contained entity into which nothing extraneous arrived and from which nothing departed.[116]

Chladni soon encountered difficulties with his initial theories concerning meteorites and meteors, on two counts. As noted earlier, the similar appearance of meteorites examined thus far seemed to indicate a common origin, and their average specific gravity corresponded to that of the Moon. This evidence—and, probably, the scientific eminence of its proponents—led Chladni to accept the hypothesis of lunar origin. There was yet another difficulty, which resulted from early simultaneous observations of the paths and heights of meteors entering the atmosphere.

In his 1794 treatise, Chladni proposed that meteors were probably the same as fireballs, differing only in that they might be too far distant for the Earth's gravitational attraction to act on them. The momentary light they displayed was a result either of electrical action or actual combustion for a brief period. But Chladni was worried, as he continued to be throughout his life, by reports that meteors or fireballs had dropped viscous, spongy, and gelatinous matter. The loose texture and bituminous odor of the Alais carbonaceous chondrite, which fell on 15 March 1806, gave later credence to such reports. Chladni left open the possibility that there could be a second type of meteor, formed from terrestrial residues and ignited at much lower altitudes, from which this peculiar matter fell. It was this admission of two kinds of meteors that created the problem.

Chladni's suggestion that the altitudes and paths of meteors should be determined by simultaneous observations excited Lichtenberg, who in 1798 enlisted two of his students, Heinrich W. Brandes and Johann F. Benzenberg, to make the observations. In September and October 1798, Brandes and Benzenberg, positioned at opposite ends of a baseline 15.25 km in length, observed 400 meteors, of which 22 were deemed to have been seen simultaneously. Of these, they calculated that the onset of luminosity occurred at a maximum altitude of 226 km and at a minimum altitude of 10.5 km. What was surprising was that two of the meteors seemed to rise rather than to descend. One first appeared at an altitude of 37 km and disappeared at an altitude of 81 km; the second became luminous at an altitude of 40 km and vanished at an altitude of 97 km. Moreover, the calculated path of the second, No. 12, was such that it had either traversed the Earth or had originated in the

Earth's lower atmosphere.[117] Lichtenberg could hardly contain his enthusiasm. On 3 November 1798, he wrote to Benzenberg that "in such a short time you have accomplished more in this science [meteorology] than all other physicists since the creation of the world or certainly since the Flood and the time of Aristotle." And, he added, "If your observation of No. 12 is correct, it also seems to me that the cosmic origin of the phenomenon is improbable."[118] Lichtenberg died three months later, but Brandes and Benzenberg continued to make observations. In the spring of 1802, they noted simultaneously three more meteors, one of which rose in its path.

At first Chladni appeared unconcerned by these findings. In his 1794 treatise, he said, he had only suggested the identical origin of meteors and fireballs. From the observations of Brandes and Benzenberg that some meteors rise almost perpendicularly, he was now satisfied that meteors were atmospheric events whose nature was unknown and merited further investigation. But fireballs, he maintained, have a cosmic origin.[119] However, as the theory of the atmospheric origin of fireballs and meteorites gained ground, Chladni undoubtedy realized that his concession that meteors were atmospheric phenomena substantially weakened his argument that fireballs had a cosmic origin. In 1817 Chladni responded in two articles. In one he attempted to explain that the viscous, spongy, and gelatinous matter reported to have fallen from meteors and fireballs—with the exception of Tremella, bird droppings, pollen, and the like—could have come from cosmic space. The dust, he wrote, may have been a type of chaotic matter in space or may have arrived as a cometlike mass, and there was no reason why matter from space could not be moist or bituminous as well as in the form of compact stones.[120] In the second article Chladni asserted that the upward motion of meteors observed by Brandes and Benzenberg was the result of a ricochet that occurred when they met the denser layers of the Earth's atmosphere at lower altitudes. As evidence he cited Pringle's report of the fireball seen over England in 1758, and several others that mentioned an apparent ricochet. He was returning, Chladni concluded, to his original contention that meteors and fireballs were identical phenomena and had the same cosmic origin.[121]

In 1818 Chladni again attacked the observations, but with new arguments. One could not legitimately draw conclusions from only 22 observations, he wrote; there must be at least 200. If there were two kinds of meteors, as Benzenberg maintained, then one kind had a cosmic origin and the other a lunar origin. Citing the meteor shower observed in South America, Greenland, and western Europe on 12 November 1799, Chladni postulated that the Earth could have traversed a region of cosmic space where there was at the time an accumulation of cosmic matter, or that there could have been an unusually strong eruption of lunar volcanoes a few days earlier.[122] Later in 1818, Chladni rejected the lunar hypothesis altogether, and never wavered thereafter from his conviction that meteors and meteorites had a cosmic origin.

Brandes and Benzenberg's conclusion that some meteors ascended in the atmosphere was wrong. However, Chladni's explanation of the phenomena was also incorrect: once fireballs have penetrated deeply into the atmosphere, they only very rarely can ricochet out of it. Not until 1837 was there a suspicion, mentioned first by Olbers, that observations of meteor altitudes might be in error, owing to the small parallax. Two years later, Friedrich Bessel at Königsberg, in a complete analysis, demonstrated how the accuracy of observations of meteors could be seriously compromised by using a baseline that was too short. The most damaging case, No. 12 of the original observations, he stated, was undoubtedly due to an observational error amounting to just one degree.[123]

Meanwhile, Chladni's alternative explanation—namely, that meteorites might be fragments of a destroyed planet—began to receive attention. It gained initial support from the discovery of the asteroids: Ceres on 1 January 1801 by Giuseppi Piazzi, and Pallas on 28 March 1802 by Olbers. At that time Olbers postulated that they were portions of a planet that had been destroyed either from an internal upheaval or as the result of a collision with a comet, and predicted that more pieces would be found.[124] Chladni, of course, viewed the discoveries as vindication of his theory of cosmic origin. When he was eleven or twelve years old, he wrote, he had been impressed by the vast space between Mars and Jupiter and had fully expected that another planet would be discovered there. No doubt other parts of the destroyed planet would be found as time went on.[125] Subsequent to Carl F. Harding's discovery of Juno in 1805 and Olbers's discovery of Vesta in 1807, Lagrange and Wildt made independent orbital calculations and concluded that the four asteroids were once fragments of a larger planet.[126] By 1832, further analysis of the orbits had convinced David Brewster that

> the striking coincidences between theory and observation in the eccentricity of their orbits, in their inclination to the ecliptic, in the position of their nodes, and in the places of their perihelia, are phenomena which could not possibly result from chance, and which concur to prove, with an evidence amounting to demonstration, that the four new planets have diverged from one common node, and have therefore composed a single planet.

Smaller fragments, which composed the central portion of the destroyed planet, Brewster continued, were propelled beyond the gravitational attraction of the larger pieces. If they entered the sphere of the Earth's attraction, they might revolve about it for a time, and then plunge to earth as meteorites.[127]

Surprisingly, it was Olbers, the discoverer of two of the asteroids, who was not convinced. In 1837 he wrote that he was far from accepting the destroyed planet theory as more than a hypothesis, and had considered the lunar origin of meteorites as more plausible than the idea that they were small fragments of a former planet. Though abandoning the theory of lunar origin, Olbers at the same time accepted the cosmic origin of meteorites. The analysis of the meteor shower of 1833, he wrote, proved conclusively that meteors did not participate in the Earth's rotation, and that they had entered the Earth's atmosphere from cosmic space.[128] To illuminate Olbers's conclusion, we turn to the subject of meteor showers.

3

Mathematical Astronomy and Statistics

Prominent meteor showers are associated with comets. Thus the Leonid meteor shower, which at present reaches a maximum about mid-November each year, derives from the Periodic Comet 1866 I—that is, the first comet which passed its perihelion in the year 1866 and which was discovered almost simultaneously in December 1865 by the French astronomer Tempel in Marseilles, and the American astronomer Tuttle at the Washington Naval Observatory. Similarly, the August Perseids are associated with Periodic Comet Swift-Tuttle (1862 III), the November Andromedes with Periodic Comet Biela (1852 III), and the April Lyrids with Comet Thatcher (1861 I). Solid particles, ranging in mass from fractions of a gram to several grams and composing the cometary debris, are eventually distributed in a more or less continuous stream around the comet's orbit. If the orbits of the Earth and the comet around the Sun intersect, then each year at the point of intersection the particles enter the Earth's atmosphere and produce a meteor shower. The shower is particularly spectacular when the Earth meets the main concentration of the cometary debris. In the case of the Leonids, such an event occurs every 33 to 34 years, since the orbital period of Comet Tempel-Tuttle is 33.25 years. The last major Leonid shower occurred on 17 November 1966, when observers at Kitt Peak, Arizona, during a period of 40 minutes, estimated that meteors fell at a rate of 60,000 per hour.

The November Leonid showers had attracted attention and comment before 1833. However, the shower that occurred in the early morning hours of 13 November 1833, particularly in the eastern United States, was such a fantastic display that it commanded the attention of astronomers and other scientists both in America and Europe. The event had two rather immediate results. First, astronomers proved to their satisfaction that the shower was not an atmospheric phenomenon, but that the Earth had encountered a stream or a swarm of objects or a "nebulous body" that was moving in cosmic space. And as observations of other meteor showers at various times of the year increased, the ingrained idea of *le vide planétaire* dissipated. Interplanetary and cosmic space was now visualized as being filled with bits and pieces of matter, just as Chladni had predicted.

This new view of the contents of the cosmos presented several problems. What was the origin or the source of this matter? Why did it appear to be concentrated in clouds or swarms? A definitive answer to the latter question came during the 1860s, when astronomers proved that the orbits of several comets and meteor streams were identical. This chapter describes the process of observation and analysis which, after three decades, led to the solution. There was, of course, speculation, both before and after the orbits of comets and shower meteors were found to correspond, that meteorites were larger fragments within these swarms that survived their fiery passage through the atmosphere. The nineteenth-century cometary theory and the evidence offered in its support are treated in chapter 5.

During the period when mathematical astronomers were attempting to determine the orbital characteristics of meteor streams, meteorites began to receive statistical attention. By the early 1850s, over three hundred actual or alleged meteorites had been reported to have either fallen or been found. By analyzing past records, scientists attempted to determine the number and total weight of meteorites that fell to earth each year, whether meteorites fell or were found more frequently in certain geographical locations, whether more fell during certain months of the year, and whether correlations might be established between the average specific gravity of meteorites and those of the planets. As time went on, such statistical analyses were refined such that the frequency of falls during each hour of the day and night was calculated.

Scientists also began to look more closely into the phenomena accompanying the fall of meteorites in the atmosphere. The cause of luminosity received much attention, as did the reason for the explosion of a fireball. There were also attempts to determine more exactly the paths and velocities of fireballs that dropped meteorites, to acquire the data from which their cosmic orbits might be calculated. Mathematical and statistical analysis entered almost every field of science in the nineteenth century, and the area of meteors and meteorites was no exception.

THE 1833 LEONIDS AND AFTERMATH

The early morning hours of November 13, 1833, Professor Denison Olmsted of Yale reported, were

> rendered memorable by an exhibition of the phenomenon called SHOOTING STARS, which was probably more extensive and magnificent than any similar one hitherto recorded. . . . The firmament was unclouded; the air was still and mild; the stars seemed to shine with more than their wonted brilliancy. . . . Probably no celestial phenomenon has ever occurred in this country . . . which was viewed with so much admiration and delight by one class of spectators, or with so much astonishment and fear by another class. For some time after the occurrence, the "Meteoric Phenomenon" was the principal topic of conversation in every circle.[1]

Few preachers could resist the opportunity presented by the spectacular shower to indulge in homilies. Quoting from Revelation, "And the stars of heaven fell unto the earth. . . . For the great day of wrath is come" (fig. 8), one continued:

> We do not call that solemn, awful, majestic, and glorious exhibition of the Works of the Almighty . . . by the cold name of METEORIC PHENOMENON. . . . We . . . pro-

Fig. 8. Woodcut by Albrecht Dürer illustrating Revelation vi:12–17.

> nounce the Raining Fire . . . a sure *Forerunner* . . . of the great and dreadful Day . . . when the SIXTH SEAL SHALL BE OPENED![2]

Given the estimates that well over two hundred thousand meteors fell in a period of six to seven hours, the consternation and fear they aroused is fully understandable.[3]

As professor of mathematics and natural philosophy, Olmsted felt a responsibility to record the event for posterity, and immediately wrote an account of his observations, which was published the same day in the New Haven *Daily Herald*. There was, he wrote, "a constant succession of fire balls, resembling sky rockets, radiating in all directions from a point in the heavens, a few degrees south-east of the zenith, and following the arch of the sky towards the horizon." About 5:45 A.M., he continued, it occurred to him to mark the point of apparent radiation accurately among the fixed stars.

> The point was then seen to be in the constellation of Leo, within the bend of the sickle, a little to the westward of Gamma Leonis. During the hour following, the radiating point remained stationary in the same part of Leo, although the constellation in the mean time, by the diurnal revolution, moved westward to the meridian, nearly 15 degrees.

Upon reflection, Olmsted realized that the apparent radiant point was due to perspective and that the meteors were actually moving in straight lines that appeared to converge toward a vanishing point. Since the radiant point moved but remained fixed in Leo, he concluded that the meteors were entering the atmosphere from cosmic space. Consequently, the next day he published a request for communications from other observers, asking particularly whether any had also noted a radiant point.

Olmsted's account and request were reprinted in numerous other papers in the United States, and within several weeks he received replies from as far west as the Mississippi River, Buffalo, New York in the north, and Mobile, Alabama in the south. Further, one intensely interested witness, Alexander C. Twining, a former Yale student, journeyed to the New York docks, where he interviewed the masters of over a dozen ships that had been at sea the night of the shower, the majority sailing westward on the Atlantic and two in the Gulf of Mexico. From all of these correspondents, Olmsted learned that the area in which the meteor shower was most heavily concentrated was bounded by the 18th and 43rd parallels of latitude north and the 61st and 91st meridians of longitude west.

The *American Journal of Science,* whose editor was Benjamin Silliman, published in consecutive issues two articles by Olmsted on the meteor shower. The first included the verbatim replies he had received from a dozen correspondents, and his résumé of the various observations of the numbers and sizes of the meteors, the lengths of their trails, the prevailing weather, whether sounds had been heard, whether any had been seen to fall to earth, and whether the position of an apparent radiating point had been recorded. In the second article, Olmsted provided the texts of a few belated replies of importance, and then presented his explanation of the phenomenon. Before submitting this article, Olmsted recalled, he went to Jeremiah Day, who had become president of

Yale, and also to Professors Silliman and Kingsley, all of whom advised him to be sure that he was right before he published.[4] We recall that all three were adherents of the terrestrial comet theory, and although Olmsted's hypothesis diverged significantly from this idea, a sufficient number of its features remained to indicate that Olmsted was influenced by the elder scientists.

Olmsted described the source of the meteors variously as "a cloud (so to speak)," a "cloud, which . . . consisted of nebulous matter, analogous to that which composes the tails of comets," and "in order to avoid circumlocation . . . a comet." The origin of the meteors, he wrote, was beyond the atmosphere—first because the radiant point moved westward with the stars, and second because the phenomenon occurred at about the same time in places differing in longitude. This first correct conclusion was followed by the assertion that the source was at a height of 2,238 miles, a figure that, two years later, Olmsted admitted was erroneous.[5] What led him astray was that a number of his correspondents placed the radiant point in different parts of the constellation Leo, and in one case even in Leo Minor. Determining the distances of each of four other observers from New Haven, Olmsted used triangulation to calculate the height resulting from pairs of observations and then took the average of these figures. The observations were, of course, too crude, a weakness that Olmsted only later recognized.

The cometary body, Olmsted continued, could not have been a terrestrial satellite, because it remained stationary with respect to the Earth over a considerable period of time. It could not have been "a Nebula, which was either stationary, or wandering lawless through space," because it would soon have been separated from the Earth, which during the shower moved 550,000 miles through space. Since the radiant point remained in the constellation Leo, he wrote, the Earth "was moving almost directly towards the comet," and the comet must have been moving in the same direction as the Earth around the Sun, with nearly the same angular velocity. Noting that many shooting stars had been observed in 1832—on 13 November over the Red Sea, on 14 November in the Tyrol, and on 19 November in England—Olmsted concluded that the cometary body had a period of nearly six months. While the Earth made its annual revolution, the comet orbited the Sun twice, at an angle slightly inclined to the ecliptic. It reached aphelion about 13 November each year, intersecting the Earth's orbit and producing the shower (fig. 9). Since the perihelion of the comet was very close to the orbit of Mercury, he wrote, it might suffer perturbations from that planet or from the Earth, which might shorten or lengthen its period slightly. Here, Olmsted was misled by the reports of annual showers that he believed were as brilliant as the one he had observed.

Other conclusions flowed from Olmsted's main hypothesis. Since the Earth and the comet, he thought, were moving with the same angular velocity, the velocity of the meteors was due only to the Earth's gravitational attraction. Therefore, he calculated that their velocity upon entry into the atmosphere was the same as that of a body falling freely from a height of 2,238 miles to a height of 50 miles above the Earth—that is, approximately 4 miles per second. The matter of the meteors, he stated, was combustible, because it burned in the atmosphere. It was transparent, because the cometary body did not shine by reflected light. It must have had a low specific gravity, because it did not

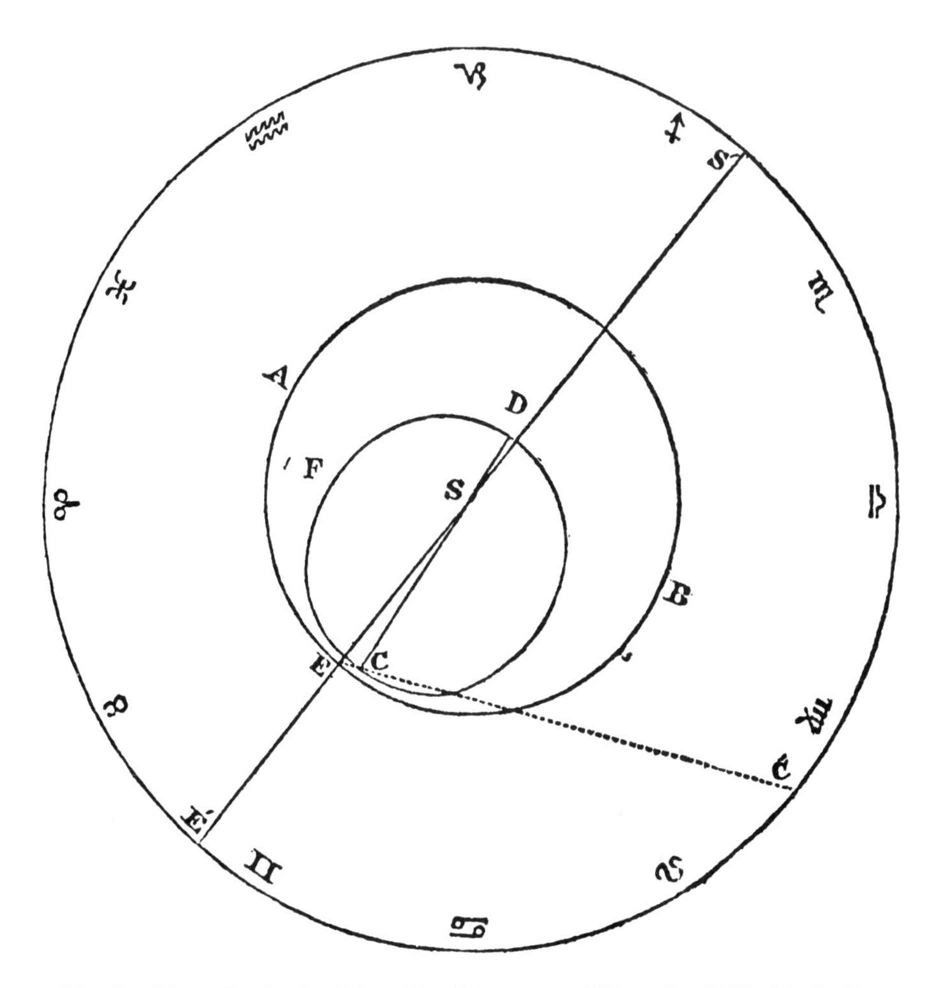

Fig. 9. Olmsted's sketch of the orbit of the comet of November 1833. S is the Sun; AEB is the Earth's orbit; and CDF is the "comet's" orbit. At the time of the meteor shower, the Earth is at E, and the "comet" is at C. The outer circle is the ecliptic with the signs of the zodiacal constellations noted thereon. From Olmsted, *AJS* (1834), 26:164.

have enough momentum to enable it to survive the passage through the atmosphere. Its substance, therefore, was different from that which composed meteorites, Olmsted wrote. Nevertheless, some of the meteors must have been of great size, he said, since they appeared as large as the full moon, which indicated that at a distance of 11 miles they had a diameter of 528 feet. Here Olmsted shared Day and Silliman's confusion relative to the actual sizes of meteors and fireballs.[6]

Despite his many misconceptions, Olmsted correctly concluded that the source of the November meteors was cosmic, since the radiant point remained in the constellation Leo during several hours of the shower, an observation he and a half-dozen others had made. The validity of this inference and of his hypothesis concerning the period of the comet had to be judged on the basis of observations made in subsequent Novembers. Meanwhile, as Olmsted wrote much later, he received a "cold reply" from Nathaniel Bowditch when he sent the latter a copy of his articles, and was "ridiculed by T. Paine, editor of the astronomical part of the American Almanac." Olmsted also noted that Elias Loomis, in 1833 a tutor at Yale and later one of America's prominent meteorologists, "had little confidence" in his views, but with the reappearance of the shower in November 1834 was "favorably inclined" to the theory.[7]

If Loomis was impressed by the November 1834 display, others were not. Twining, who had estimated the velocity of the meteors to be 14 miles per second while luminous, beginning at 80 miles and ending at 30 miles above the Earth's surface, reported that although the radiant of the meteors of 13 November 1834 was again in Leo, their maximum rate of appearance was about four per minute.[8] Alexander Dallas Bache, professor of natural philosophy and chemistry at the University of Pennsylvania, noted in the *American Journal of Science* that the number of meteors seen on 13 November 1834 was not unusual, nor was there an apparent radiant.[9] When Olmsted persisted in emphasizing the size of the 1834 shower, Bache, an outstanding West Point graduate and an intimate of politicians both then and later, prevailed upon Secretary of War Lewis Cass to send a circular to all military posts in the states and territories requesting information about the shower. Replies from the commanders of twenty-two posts, Bache reported, agreed with one exception that there had been no unusual display on either 12 or 13 November. Bache added that the keen-eyed sentries on their nightly rounds would have been immediately aware of the appearance of such a phenomenon.[10] He persisted for some time in denying that there was any recurrence of the Leonid showers, but failed to shake Olmsted's conviction.[11]

In Europe there had been sporadic scientific interest in the observations of meteors following the initial work of Brandes and Benzenberg at the turn of the nineteenth century. In the early 1820s, while a professor of mathematics at Breslau, Brandes organized a fairly widespread network of amateur observers, but interest lagged when intermittent periods of inclement weather interfered with their observations. There had also been reports of unusual meteor showers. One widely publicized shower was seen on the night of 12 November 1799 at Cumana [Venezuela] by Alexander von Humboldt and the French natural historian Goujaud Aimé Bonpland, during their exploration expedition of Central and South America. For four hours, at this latitude of 10.5° north, they watched thousands of meteors come from above the horizon east-northeast and spread in an arc over the sky from 30° north and south of east. The shower was observed as far north as Greenland, and Andrew Ellicott, en route by ship from New Orleans to Philadelphia, gave a vivid account of the display at latitude 25° north.[12] Also, Chladni reported that an unusually large number of shooting stars were seen on 10 August 1815, and Brandes noticed an extraordinary display of meteors on 10 and 11 August 1823.

Correlation of these random observations by scientists began only in the 1830s. One of the most intensely interested was Lambert A. J. Quetelet, a Brussels astronomer who later achieved lasting fame for his basic work in statistics. Another was that dynamic and inspiring scientist, Dominique François Arago, director of the Paris observatory and perpetual secretary of the Academy of Sciences, In Breslau, there was the former Prussian army officer and senior astronomer at the observatory, Palm H. L. von Boguslawski. Polymath Georg Adolph Erman, who married Marie Bessel, daughter of the noted Königsberg astronomer, worked in Berlin, and the indefatigable Olbers was busy at Bremen.

The news of the great November 1833 shower in North America reached Europe many months before the 1834 publications of Olmsted, and thus alerted the European astronomers. A prominent display of meteors on 12 and 13 November 1832 over England, the Netherlands, eastern France, Germany, and western Russia reminded Arago that Captain Bérard, master of the brig *Loiret,* had reported the appearance off Cartagena, Spain, of a huge number of shooting stars on the night of 13 November 1831. Olmsted's articles confirmed in Arago's mind the periodicity of the November meteors, and the observation that the radiant remained in the constellation Leo led to his immediate acceptance of the cosmic origin of the meteors. He wrote:

> Chance, one can say, rather than scientific ingenuity, led to the establishment of a fact which guarantees henceforth to shooting stars a very important rank in our planetary system. More and more there is confirmed the existence of a zone composed of millions of small bodies, whose orbits meet the plane of the ecliptic at a point which the earth occupies every year from the 11 to the 13 of November. It is a new planetary world which is beginning to be revealed to us.[13]

Olbers agreed. In Europe on 13 and 14 November 1834 there had been a recurrence of the meteor shower, but on a much smaller scale. The meteors' orbits of the sun, he wrote, probably take several years, so that in 1834 they were not the same little bodies that were seen in 1832 and 1833. Whether they gather in other similar wide streams, Olbers continued, awaits further investigation. Arago had mentioned a possible stream about April 22 each year, and Olbers thought that there might also be recurrent showers on August 10 and 11.[14]

In both 1834 and 1835, Quetelet had observed an unusual number of shooting stars on August 10, and on looking at earlier nineteenth-century records, he found that there had been nine occasions on which prominent showers were observed between 6 and 14 August. On 8 August 1836 both he and Arago noticed that the radiant of the meteor shower was in the constellation Perseus. Both claimed priority for the discovery, though neither realized that Professor John Locke of Cincinnati, Ohio, had noticed this fact on 8 August 1834.[15]

Jean-Baptiste Biot, meanwhile, proposed that the origin of the meteor streams was the zodiacal light. This phenomenon had been recognized as cosmic by Kepler, and in 1683 Jean-Dominique Cassini, following his investigation of Saturn's rings, proposed that the zodiacal light consisted of a great number of tiny corpuscles that orbited the Sun almost in the plane of the Sun's equator. This theory found few adherents; most eighteenth-century astronomers favored the idea that the zodiacal light was actually the Sun's atmosphere. However, Laplace proved that the Sun's atmosphere could not assume a flattened, lenticular shape, nor extend even to the orbit of Mercury. Laplace reverted to Cassini's theory, and Biot, adhering to Laplace's view, postulated that on November 1 each year the Earth was in the neighborhood of the ascending node of this nebulous background. Since the zodiacal light was then above the Earth's orbit, Biot stated, the particles composing the light could encounter the Earth's atmosphere and form meteors.

Olbers made short work of Biot's hypothesis. The relative velocities of the meteors made the idea quite impossible, he wrote, and besides, at the time, the node of the Sun's equator was not in the 20th degree of Taurus, as Biot said, but in the 20th degree of Gemini.[16] Biot soon discarded the idea of the origin of meteors as being the zodiacal light, but still maintained that the Earth's collision with a solar nebula produced the annual phenomenon.[17] Whether influenced by Biot or not, Olmsted in 1837 discarded the notion of a cometary body and accepted the zodiacal light as the source of meteors.[18] Thereafter, Olmsted published nothing further on meteors; his lecture notes indicate, however, that he adhered to the zodiacal light hypothesis at least into the 1840s, and also thought that meteors were connected with auroras.[19]

In the late 1830s observations of periodic meteor showers became almost a sacred duty for astronomers. Almost all pressed friends and relatives and, when possible, students into service during the nights when showers were expected to appear. The Breslau observatory was probably the best organized. The octagonal observation room there had windows on each side, except at the north and south, where clocks faced the center of the room. Two or three of the most knowledgeable students (as far as astronomy was concerned) were posted at each window. When one announced the fall of a shooting star, the time of the observation and his window designation were recorded by an adjacent student facing a clock. This memo was returned to the window, where the magnitude, duration, and path of the meteor were noted on it, and where one of the observational partners entered the information on a star chart. Only when one of the window posts become vacant did Captain von Boguslawski, now professor of astronomy at the University, have an opportunity to make observations, since the direction of the operation required his constant attention. Olbers was enthusiastic about the data gathered at Breslau on the August 1837 meteor shower.[20] A similar situation prevailed in the United States. Following his account of the 1838 November Leonids, Professor Joseph Lovering of Harvard College commended seven students, by name, who had aided his observations.[21] One can only speculate as to how many dozens or hundreds of young people learned the rudiments of astronomy or came to appreciate the majesty and vastness of the universe from such experiences. This also explains in part how astronomy became a popular science in the nineteenth century.

By the early 1840s, astronomers almost without exception believed that the source of shower meteors was cosmic. The basis of this belief was the fact that the radiant point remained fixed with respect to the fixed stars while the Earth rotated.[22] The radiants of four apparently annual showers had been fairly well determined: the shower occurring about 13 or 14 November had its radiant in the constellation Leo; that of about 10 December in Gemini; that of about 20 April in Lyra; and that of about 10 August in Perseus. Some puzzling aspects of these showers, however, were that the number of meteors in the August epoch was about the same every year, that the showers could be observed on several nights before and after the maximum, and that the radiant appeared to shift ever so slightly during the period of the showers. In contrast, during the late 1830s the November shower never produced the number or brilliance of the displays that had been observed in Europe in 1832 and America in 1833. These observations gave rise to the assumption that the "asteroids" producing the August showers comprised millions of particles strewn along an elliptical orbit, whereas those causing the November shower might be concentrated in a small segment of the orbit. Olbers suggested that there might not be another November meteor shower comparable to those of 1799 and 1833 until 1867, thereby inferring that the denser aggregation had a period of about thirty-four years.[23] This question of annual periodicity, or lack of it, caused scientists to search annals and ancient chronicles for records of star showers. Quetelet was the first to make this effort; another was Edward C. Herrick, a former Yale student and New Haven bookman who found the radiant of the 19 April 1839 shower in the constellation Lyra.[24] Herrick, who became librarian at Yale in 1843 and its treasurer in 1852, pointed out that the date of the August maximum gave substance to the legend of the "burning tears" of St. Lawrence, which fell as shooting stars each year on the saint's feast day, August 10.[25] He was searching through hundreds of volumes, he wrote, in order to compile a record of all meteor showers reported in ancient times.[26] Though such research did pay handsome dividends later on, at the time other astronomers did not consider it worthwhile. American astronomer Sears Walker wrote: "It seems more natural . . . in the absence of proof of identity with the groups of recent appearance, to regard the ancient displays of Quetelet's catalogue as isolated clusters, whose meeting with the earth was purely *accidental*."[27]

Several astronomers, including Boguslawski, Erman, and Walker, attempted to determine the orbits of the particles producing the meteor showers. They considered that a meteor stream was necessarily within a few seconds of its node—the intersection of the orbit with the ecliptic—so that its heliocentric radius vector differed from that of an observer by only a negligible quantity. Similarly, the heliocentric longitude of the stream was the same as that of an observer and, of course, the time was known. In addition to these known elements, two others were required to determine the orbit. The first was the direction of the stream with reference to the plane of the ecliptic and to another plane at a known angle to the plane of the ecliptic. This direction, it was thought, could be determined by calculation from the position of the radiant at the given time. The second unknown element was the cosmic velocity of the stream, and it was on this point that all efforts at the time to determine the orbit of a meteor stream failed. Some astronomers, Olbers and Walker among them, recognized that the November meteors were in a retrograde orbit—that is, that they moved in a direction around the Sun opposite to that of the Earth.[28] However, the values of the velocity of the August or November meteors, which were calculated from hundreds of observations of the duration of luminosity and of the length of the paths of individual meteors, as viewed against the background of the fixed stars, were uniformly too low.[29] It was not just that the observations were imprecise, which in many cases they were. Rather, the possibility of a diminution of the cosmic velocity upon entrance into the upper regions of the atmosphere was not even considered. The mean velocity derived by dividing the estimated length of the path of a meteor by the duration of its visibility was thought to be the cosmic velocity, whereas the former is actually less than the latter owing the effect of atmospheric resistance. It was only some twenty years later, in the mid-1860s, that this discrepancy began to be addressed and that the orbits of meteor streams were calculated.

A DIGRESSION ON THE FALL OF PECULIAR MATTER

Reports of peculiar types of matter found after the passage of fireballs and the appearance of meteors presented another puzzle. Chladni, as already noted, had always been concerned both about accounts of the fall of red or black dust and about finds of viscous, slimy, or gelatinous matter, and finally conceded that various types of matter could come from cosmic space and need not be in compact masses. In the letters to Olmsted about the 1833 Leonid shower, there were four reports of the fall of peculiar matter. In Virginia, a small amount of substance resembling an egg white was observed to fall; in Rahway, New Jersey, witnesses discovered lumps of jelly where particles of "fiery rain" struck the ground. Gelatinous matter resembling soft soap, which evaporated immediately upon the application of heat, fell to earth with "prodigious force" in Newark, New Jersey. And a woman at West Point, New York, who was milking at sunrise, heard something come down "with a sposh" before her. She described it as a flattened, transparent mass looking like boiled starch, but when she wished to show it to some people a few hours later it had vanished.

Olmsted took these accounts at face value. The residuum from meteors, he wrote, was uniformly described as gelatinous, from which one could infer that the "matter of which the meteors were composed was both highly volatile and transparent" and that it possessed a high degree of combustibility.[30] However, Edward Hitchcock, professor of chemistry and natural history at Amherst College, who thought that the 1833 shower was a modification of the aurora, had a quite different opinion about the gelatinous substances. Some years before, he had closely examined a gelatinous substance said to have fallen from a meteor: "I recognized it in a moment as a species of gelatinous fungus, which I had sometimes met with on rotten wood in damp places, during dog days." Hitchcock had predicted that more would spring up within a day, which happened, so that there was "an entire mistake."[31] Learning of Hitchcock's study, Olmsted concluded in his second article on the November 1833 shower that "it is even somewhat doubtful whether any palpable substance reached the ground, which could fairly be considered as a deposit from the meteors."[32]

In 1837, Olbers reviewed the evidence for the fall of gelatinous and slimy substances, but stopped short of terming it erroneous. He regretted that none of the material had been chemically analyzed, and concluded that most of this type of matter attributed to meteors belonged to the plant kingdom, various species which grew rapidly in damp and rainy weather.[33] The phenomenon of "red snow" fell prey to scientific explanation in 1840. The so-called *protococcus nivalis,* Louis Agassiz stated, was the ova of a species of rotiferous animal called *Philodina roseola*. These animalcules died in the snow, most abundantly in ditches, at the bottom of which the ova produced a red deposit.[34] However, in the 1840s and 1850s, scientists continued to list reports of the fall of viscous matter along with meteorite falls; examples are A. Boisse in 1850 and R. P. Greg in 1854.

An unpublished manuscript by Charles Upham Shepard yields an approximate date when reports of the fall of nonmeteoritic matter were no longer seriously considered by scientists. Dated 19 March 1867, the document lists instances in which nonmeteoritic substances were reported to have fallen or been found under circumstances that indicated a fall. In one case, an Alabama resident stated that for a period of two weeks during violent storms solid matter—clay, mortar, brick fragments, beef bones, and wood, all hot—had fallen on his property. Whether Shepard submitted the manuscript for publication and it was rejected, or whether he decided against submission is not known. In any event, above the title there is written: "I do not attach any weight to the following record. CUS Nov 12, 1881."[35]

Were scientists wrong in banishing from serious scientific consideration the reports of such "anomalous events"? Charles Fort, an anti-science writer of the early twentieth century, thought so. The reports of showers of blood or of red snow, or of the fall of gelatinous matter, he thought, were "falsely and arbitrarily excluded" from science. He wrote—facetiously, one hopes:

> I shall have to accept, myself, that gelatinous substance has often fallen from the sky—Or that, far-up, or far away, the whole sky is gelatinous? That meteors tear through and detach fragments? . . . I think, myself, that it would be absurd to say that the whole sky is gelatinous: it seems more acceptable that only certain areas are.[36]

It appears that Fort's general attitude toward science finds sympathy at the present time. Perhaps this is justifiable, given the pretentiousness of some contemporary scientists. Or perhaps the critics of science fail or do not wish to understand the interplay between evidence and theory. In any case, in the 1860s two areas of research yielded results that caused scientists to ignore reports of the fall of peculiar matter in the wake of meteors. The first gave evidence that the matter causing the appearance of meteors was at least in part associated with comets. The second, which is treated later in this chapter, produced a much better understanding of the phenomena accompanying the entry and passage of meteors and fireballs in the atmosphere.

COMETS AND METEOR STREAMS

Proof of the correspondence of the orbits of some comets and meteor streams was established following several concomitant developments. Mathematical astronomy became an important scientific discipline in mid-nineteenth-century universities, and this stimulated the construction of new observatories and the installation of more modern equipment in the older ones. Telescopes achieved greater magnification, more light-gathering power, and increased accuracy through such innovations as the capability of glass manufacturers to cast large discs of homogeneous flint and crown glass for the lenses of refractors, the introduction of silver coating of the glass mirrors of reflectors, the employment of means to reduce the flexure of lenses, and the adoption of the Fraunhofer equatorial mounting. A new type of telescope, the comet-seeker, also appeared, which had a short focal length and large aperture designed especially for searching for comets. Also, astronomers prepared new and more accurate charts of the fixed stars. In this area, Friedrich Wilhelm Argelander was an indefatiguable worker. His efforts culminated in the 1850s and 1860s with the publication of successive volumes of the *Bonner Durchmusterung,* which gave the positions and magnitudes of over 324,000 stars. By reference to such charts, astronomers could determine with some confidence whether an object observed in the heavens was a cataloged star or a stranger. It is small wonder, then, that beginning in 1845 asteroids were discovered at the rate of three to four each year, that Neptune was first observed in 1846, and that comets were identified in increasing numbers.

Scores of well-educated and mathematically adept astronomers emerged from the universities and served their apprenticeships in the observatories. Foremost among those who made a direct contribution to the demonstration of the correspondence of the orbits of comets and meteor streams was Giovanni Schiaparelli. Educated as an engineer in Milan, Schiaparelli was sent by his government to study mathematical astronomy at Berlin, and returned to work in the Brera observatory in Milan, of which he became the director in 1862. Three Germans were prominently involved. Johann Gottfried Galle, while serving under Johann Encke at Berlin, first observed Neptune following receipt of the computed orbit from Le Verrier, and became director of the Breslau observatory after Boguslawski's death. Heinrich Louis d'Arrest, who discovered the asteroid Freia and comets 1845 I, 1851 II, and 1857 I, received his Ph.D. at Leipzig, and in 1858 was named professor of astronomy and director of the observatory at Copenhagen. Carl F. W. Peters, the son of the astronomer C. A. F. Peters, studied at Kiel, Berlin, and Göttingen, and then became his father's assistant at the Altona observatory near Hamburg. In Vienna, there was Edmund Weiss, who, following the receipt of his Ph.D. at Vienna, taught at the university and worked at the observatory, of which he became director in 1878. Theodor Ritter von Oppolzer also worked in Vienna in his private observatory, which was equipped with a 7-inch refracting telescope, the largest in the Austro-Hungarian Empire. Three other scientists, noted more for their contributions to celestial mechanics than to observational astronomy, were also involved. Urbain Jean Joseph Le Verrier held the chair of celestial mechanics at the Sorbonne and was named director of the Paris observatory in 1854, after Arago's death. John Couch Adams taught at Cambridge, where he was director of the observatory, and shared with Le Verrier the honor of having made the computations which led to the discovery of Neptune. In the United States, there was Hubert Anson Newton, who was appointed to the chair of mathematics at Yale College at the tender age of twenty-five, and who studied in Paris with the gifted geometer Michel Chasles.

Historical research by scientists also contributed to finding the correspondence between the orbits of some comets and meteor streams. As noted earlier, in 1839 Quetelet published the first major catalog of the dates of meteor showers recorded in medieval and modern times, and in 1860 he gave a more complete list of 315 showers. The results of Edward Herrick's research—39 annotated dates—appeared in 1841, and in the same year Michel Chasles published an additional 46 dates of showers. Though not published in its entirety until 1845, the work of Eduoard Constant Biot, presented to the Paris Academy of Sciences beginning in 1841, yielded a vast store of information.[37] Eduoard, the son of Jean-Baptiste Biot, was an engineer by profession and a distinguished historian of Chinese science by avocation. In the 1830s he published several memoirs on early Chinese mathematics, and then collaborated with his father in producing a number of studies of ancient Chinese astronomy, a subject concerning which the elder Biot became an authority in his later years. In his astronomical research, Eduoard Biot came upon the work of a Chinese historian, Ma Tuan-Lin (ca. 1240–1280), entitled *Wên Hsien Thung Khao* (Historical Investigation of Public Affairs). This work listed many recorded extraordinary celestial phenomena dating from ca. 687 B.C. in the Chou period to the beginning of the Sung dynasty (A.D. 960), and included some reports from the latter. Because the Chinese believed that any unusual heavenly phenomenon was an infallible precursor of a momentous event that would occur in their empire, imperial astronomers kept detailed records of such incidents, which were preserved together with other records in the annals of successive dynasties. Ma Tuan-Lin had compiled all of these records, and Biot extracted from them 149 accounts of meteor showers, to which he added 272 more for the Sung period (960–1275), and a further 74 reports for the Yuan and Ming dynasties (1276–1744). Biot's translation and statistical survey of these Chinese records was a significant achievement that substantially aided the historical dating of the showers occurring in mid-November.

A great deal of calculation had to be accomplished before the ancient dates were of any use. Islamic and Chinese dates had to be correlated with the Julian calendar, and the advance in the number of days occasioned by the change from the Julian to the Gregorian calendar had to be considered. But even before fairly reliable dates were established, it became evident that both in the Orient and the Occident the greatest shower activity in ancient and medieval times occurred in July and October, moving gradually thereafter to early August and early November. The October-November showers had been more spectacular through time, and it occurred to astronomers who were attempting to determine the orbit and period of this meteor stream that the gradual advance of dates could be explained by the precession of the nodes of the orbit—that is, the positions at which the orbit intersected the ecliptic changed as time went on. In the 1840s and 1850s several astronomers attempted to calculate orbits from the old records. Boguslawski did determine that the November meteor stream had a positive annual node variation, but he was not successful in determining the orbit.

H. A. Newton attacked the problem of the November meteors in the early 1860s. Teaching at Yale, he knew Herrick and his work, and having studied at Paris with Chasles, he was also familiar with the latter's research. He was assisted in his investigation, he wrote, by "Mr. J. W. Gibbs." The renowned theoretical physicist, Josiah Willard Gibbs, was then earning his Ph.D. at Yale, and was recognized at the time as a brilliant student of both the classics and mathematics. In the first of a series of articles published in 1863 and 1864,

Newton at first used the dates of the showers provided by Herrick, Chasles, Quetelet, and Biot. If the meteors were cosmic, he stated, then they should follow the sidereal year instead of the tropical year. Since the sidereal year is longer than the tropical year by 20 minutes 23.5 seconds (0.01416 days), the date of the shower should move forward in the calendar one day every seventy years. However, after taking this factor into account, Newton noted an additional discrepancy of about the same magnitude: "By giving to the nodes of the November ring a procession of one day in 70 years," he wrote, "most of these [dates of showers] would be brought into the November period."[38] Most but not all, and for this reason Newton was compelled to go back to the original documents in order to be absolutely certain of the dates—and times, if possible—of the showers.

An example illustrates Newton's problem. Seven chronicles pointed to the occurrence of a tremendous star shower in October of A.D. 902. One, however, reported the date as 899, and another as 903. Important recorded concurrent events were the capture of Taormina on the east coast of Sicily by the Saracens and the death of the cruel Aghlabite prince, Ibrahim Ibn Ahmad, who ruled over a fair portion of western North Africa and most of Sicily from his capital city of Qairwan [Tunisia]. From Arab chronicles Newton ascertained that there could be no doubt that the year was 902. As to the actual date of the shower, only two appeared in the records—12 and 13 October. Here, the Arab annalists were of no help, since the date assigned to the prince's death varied from October 25 to November 3. In this predicament Newton selected the morning of Wednesday, 13 October 902 as the day of the shower, since it was so stated in the *Chronicon Salernitanum* and since subsequent showers pointed back to that date.[39]

In addition to the year 902 and to the years 1799, 1832, and 1833, in which the dates and times of the showers were known, Newton determined the dates in nine intervening years. Four from Biot's Chinese records were quite straightforward, and another in 1366 occurred on the morning following the Feast of Saint Ursala and the Eleven Thousand Virgins, hence 22 October. It was clear from Newton's data that there had been 28 cycles of major showers in the 931 years between 902 and 1833, so that, by division, Newton found the length of each cycle to be 33.25 years. Moreover, during the 931 years, the date when the maximum occurred moved forward in the calendar by 19 days. Taking into consideration the change from the Julian to the Gregorian calendar and the difference in longitude between New Haven and Italy (where the shower of 902 occurred), Newton calculated that the length of an annual period was 365.271 days, equivalent to an annual increase in the movement of the node of the stream along the ecliptic of 1.711 minutes. (This value included the increase due to the precession of the equinoxes and also the increase due to perturbations of the stream by any planets.)

Newton's data also showed that noteworthy showers occurred just prior to the end and after the beginning of each 33.25-year cycle; chroniclers reported showers both in 931 and 934, and there were also showers in Europe in 1832 and America in 1833. Newton concluded, therefore, that the particles in the stream were not strewn out uniformly along the entire orbit, but instead formed a "group" that took two or three years to pass through the node. The number of complete revolutions of the group in a year, Newton figured, could not exceed two, because otherwise the orbit would not extend far enough from the Sun to intersect that of the Earth. Hence, the number of revolutions per year that the group described must be limited to five values: 2 ± 1/33.25, 1 ± 1/33.25, or 1/33.25. Not all of these values, he continued, were equally probable. With the first two—2 ± 1/33.25—only a very small portion of the orbit near the aphelion would fulfill the conditions. Similarly, if the group made only one revolution in 33.25 years, only a small portion of the orbit near the perihelion would be suitable. Of the other two values, Newton favored the period 1 + 1/33.25 revolutions per year, equivalent to 354.621 days. Given this value, he then calculated the elements of the orbit, from which it followed that the velocity of the particles when they entered the Earth's atmosphere was 20.17 miles per second. The total precession of the node during the 931 years, he calculated, was 26 degrees 33 minutes, of which 12 degrees 58 minutes (equivalent to 27.8 minutes in a 33.25 year period) was due to the precession of the equinoxes; the rest, 13 degrees 35 minutes (29.1 minutes in a 33.25 year period), was due to perturbation. Since the actual motion of the node was now known, he concluded that if the motion of the node due to planetary perturbations was computed for each of the five periods, the correct period could be ascertained.[40] It was this calculation that Adams completed three years later.

It will be noted that Newton calculated the cosmic velocity of the meteors when they entered the Earth's atmosphere to be 20.17 miles per second. This value was close to that which Walker had found twenty years earlier by averaging many observations and arriving at a velocity of 21.07 miles per second. In assessing Newton's work, Charles P. Olivier thought that Newton was misled, or perhaps reassured, by the fact that the calculated and observed velocities were about the same, and that if he had known the true velocity, "the theory of a short period would at once have been disproved."[41] There is, however, a problem with this explanation. In 1862 Newton published a report concerning two fireballs observed over the United States on 2 August and 6 August 1860. From estimates of the lengths of the visible paths and the duration of their visibilities, Newton concluded that both were in hyperbolic orbits—that is, the velocities indicated they would pass out of the Solar System unless perturbed by a planet.[42] It is clear from the data presented that Newton's estimates of the elapsed times were inaccurate. Nevertheless, the episode demonstrates that Newton had no difficulty in conceiving of very high cosmic velocities. Had he then calculated the velocity of the November meteors in an orbit having a period of 33.25 years, he would not have been surprised at the value obtained, particularly since he knew that the orbit of the meteors was retrograde. In favoring the shorter period, Newton was in all likelihood motivated by the reason he mentioned—namely, that close to perihelion the meteor stream having an orbital period of 33.25 years would apparently just graze the orbit of the Earth, making the appearance of spectacular meteor displays less probable.

While Newton in New Haven was reasoning from historical data, Schiaparelli in Milan was occupied with observations and calculations of the orbits of comets and meteor streams. In particular, he investigated Comet 1862

III, first observed in April 1862, and also the orbit of the meteor stream producing the August showers. He sent his results, which incorporated his theory, in a series of four letters to Father Angelo Secchi, director of the observatory of the Collegio Romano in Rome. These letters appeared in 1866 in the last four monthly issues of the *Bullettino Meteorologico dell-'Osservatorio del Collegio romano,* although Secchi included part of the fourth letter—the most important—in his account of the 1866 November shower, published in *Les Mondes* on 20 December 1866.[43] Astronomers in other countries, then, were not aware of Schiaparelli's results until early in 1867, a fact that may explain his later complaint that his accomplishments were almost unrecognized.[44]

The basis of Schiaparelli's theory and results was his consideration of cosmic velocities of comets and meteor streams and of the effect on the velocity of atmospheric resistance when the particles entered the atmosphere at different angles of inclination in either direct or retrograde orbits. From Coulvier-Gravier's data on the horary increases in the numbers of meteors during showers, and from the known directions of radiants, Schiaparelli's developed a formula from which he calculated cosmic velocities. As a first approximation, he found that the velocity was about 1.447 times the Earth's orbital velocity. Comets in parabolic orbits at the same distance from the Sun as the Earth would have a velocity 1.41 times that of the Earth. Given the fact that these velocities were of the same order of magnitude and the knowledge that both comets and meteor streams move not only in direct but also retrograde orbits, which are inclined to the ecliptic at various angles, Schiaparelli concluded that meteor streams closely resembled comets.

The matter of the universe, Schiaparelli then postulated, exists in every degree of subdivision, from the huge stellar masses to minute particles forming "cosmical clouds." When a cosmical cloud enters the sphere of the sun's attraction, he wrote, it is transformed into a parabolic "current" that may take years or centuries to complete its perihelion passage. When the current passes the perihelion, its composing particles, which were originally thousands of miles apart, become a densely populated stream both in depth and breadth, and in the following years they are dispersed miles apart in a closed ring around the Sun. If the cosmical cloud has a short period of revolution around the Sun, Schiaparelli continued, it will disintegrate, and planetary perturbations may divert some of the particles from the central track, which then become sporadic meteors if they meet the Earth's atmosphere. He believed that a very large number of meteoric currents were traversing the Solar System in all directions, and that such currents might intersect and pass through one another without disturbance. Matter that fell into the Earth's atmosphere and dropped meteorites differed only in size from the smaller particles seen as shooting stars.[45]

Certain facets of Schiaparelli's theory were not new. Olmsted and others had postulated that meteors were particles belonging to a cometary cloud or nebula, and Olbers and Erman had proposed that meteor swarms eventually formed an orbital ring around the Sun. But what made Schiaparelli's theory convincing was the demonstration in his fourth letter to Secchi that the elements of the orbits of the August meteors, which Schiaparelli called Perseids, and of Comet 1862 III were almost identical. The longitude of the perihelion, the longitude of the ascending node, the inclination of the orbit, the perihelion distance, the direction of motion around the Sun, and the estimated period of revolution of the orbit of the Perseids and that of the comet were so nearly the same that the fact of their correspondence was indisputable. In the same letter, Schiaparelli gave the elements of the orbit of the November meteors, assuming that their period of revolution was 33.3 years. Subsequently, following the lead of Schiaparelli in naming the showers according to the constellation containing their radiant, astronomers agreed in naming the November showers the Leonids, those of April the Lyrids, and so forth. Although the word "uranatmi," meaning sidereal exhalations, was proposed to designate the particles producing meteors, Newton's term, "meteoroids," became generally accepted.

In 1866, thirty-three years after the great November shower of 1833, astronomers eagerly awaited the return of the Leonids. They were not disappointed, and another fine shower occurred in November 1867, which scientists in Paris attempted to observe above the clouds in one of Henri Giffard's "free" balloons. Early in January 1867, Le Verrier announced in the *Comptes Rendus* the results of his computation of the orbit of the Leonids, assuming a period of 33.25 years. Almost simultaneously, Oppolzer's report to the Vienna Academy, giving the elements of Comet 1866 I, became available. C. F. W. Peters, at Altona observatory, immediately recognized that the orbits calculated by Le Verrier and Oppolzer were almost identical, and inserted a note to this effect in the *Astronomische Nachrichten* on 29 January 1867. Peters, however, did not know of Schiaparelli's earlier publication of the Leonids' orbit, which the latter meanwhile had recalculated with greater accuracy and had identified, in a fifth letter to Father Secchi, as the same as Comet 1866 I. Schiaparelli's result, included in Oppolzer's announcement of the identity of the orbits, appeared in the *Astronomische Nachrichten* on 6 February 1867.[46] In this instance, simultaneous discovery was the result of the intense interest of astronomers in the phenomena.

Both Le Verrier and Schiaparelli had assumed that the orbital period of the Leonids was about 33.25 years, which was one of the five values assigned to it by Newton. After making more precise observations of the radiant of the November 1866 shower, John Couch Adams undertook the task of calculating the effects on all five possible orbits of planetary perturbations, and published his results in March 1867. He determined that the perturbations affecting the smaller orbits would be insufficient to account for the precession of the node. But Adams found that if the periodic time was 33.25 years, in one revolution Jupiter increased the longitude of the node by 20 minutes, Saturn by 7 minutes, and Uranus by 1 minute, for a total of 28 minutes, the other planets having no appreciable effect. Newton, it will be recalled, calculated from historical data that there was an increase of 29 minutes in 33.25 years, and the close correspondence of Adams and Newton's results firmly established the period of the Leonids. There was also the conclusion that the Leonids were not yet in a continuous ring, as were the Perseids, but instead composed a limited stream, extending for some distance along the orbit and requiring two or three years for the denser portion to pass the node.[47]

In the spring of 1867, Edmund Weiss, assuming that meteor streams were normal concomitants of comets, published a list of thirty-three comets whose orbits closely approach the orbit of the Earth. Almost immediately thereafter, Arrest determined that the orbit of Biela's comet (1852 III) corresponded with that of the late November Andromedes, and Johann Galle found that the April 20 Lyrids shared the orbit of Comet 1861 I.[48] By 1868, astronomers had determined the radiants of over a hundred meteor streams in both the Northern and Southern hemispheres, and were firmly convinced that in time the comets associated with these streams would be identified. Twenty years later, in praising the discoveries of Schiaparelli and the other astronomers, Norman Lockyer, himself an astronomer and the editor of *Nature,* wrote: "The demonstration of the identity of meteor swarms and comets is one of the greatest physical triumphs of the century."[49] However, though meteoroids had been demonstrated to be cosmic particles producing shooting stars and associated with comets, there was in the 1860s no evidence that fireballs which dropped meteorites also shared cometary orbits. Later in the nineteenth century, both Newton and Lockyer became proponents of the view that meteorites were merely larger meteoroids that survived their passage through the atmosphere. We describe these developments in chapter 5.

METEORITE FALLS AND FINDS: THE START OF STATISTICAL ANALYSIS

Between 1800 and 1850 there were slightly over 100 attested meteorite falls, and in the same period about 60 irons were found or brought to the attention of scientists. While the origin of a few of these irons was doubtful, because of their high nickel content the great majority was known to be meteoritic. Scarcely any attention had been given to the geographical location of these falls and finds. The fall at Benares, India, in 1798, far distant from Europe, had aroused interest and contributed to the acceptance of meteorite falls. When reporting the fall of a shower of stones on the surface of ice near Luotolax, Finland, on 13 December 1813, Nils Nordenskiöld expressed surprise. "I am not aware," he wrote, "that any meteoric bodies of this nature have hitherto been observed to fall in so high a latitude."[50]

In the late 1840s, Charles Upham Shepard became aware of this oversight. About 35 of the irons had been found in the United States, and of these no less than 23 were discovered in the rather limited area of the contiguous states of North and South Carolina, Georgia, Alabama, Kentucky, and Tennessee. This geographic concentration struck Shepard as odd. In compiling the locations of meteorite falls on the American continents from 1800 to 1849, he found that of the 14 that had occurred, 13 were in the eastern United States of which 7 fell in the aforementioned states, and that the only other fall was at Macao, Brazil. Thus he determined that the "meteoric region" in America extended about 25° in longitude and was between the parallels of 33° and 44° north latitude, with "the line of most frequent deposit" cutting "obliquely across the 37th parallel." Turning his attention to the "old world," Shepard found that 50 of 55 falls occurred between 41° and 56° north latitude, 45 of them between the 43rd and 54th parallels, "a zone of the same breadth as that found to be the American region." The longitudinal extent, however, was greater than in America, although the greatest number of falls, Shepard stated, was spread over 30° of longitude from the Atlantic seacoast in the west to western Russia in the east. Moreover, 11 of the 14 meteoritic irons found in Europe were located between the 46th and 52nd parallels.

In his analysis of these statistics, Shepard wrote:

> If it then appears that these aerial strangers alight upon our earth in such great preponderance over limited areas, can we help admitting that there presides over their descent some great law, or in other words, that these falls take place in accordance with some fixed plan?

He admitted that current knowledge was inadequate to determine just what the plan was, but pointed out that the meteoritic zones on both sides of the Atlantic were roughly bounded by isothermal parallels and isodynamic lines—that is, lines of equal magnetic intensity. He concluded that

> we are strongly tempted to refer the forces of greatest activity concerned in the phenomenon, to an union of thermal and magnetic action, although it is, at the same time, possible that more powerful local attractions in the surfaces concerned, than exist elsewhere, may also exert some influences over the deposition of these singular bodies.[51]

Shepard's fertile imagination had, of course, prevailed over his analytical abilities. In Europe, the "meteoric zone" corresponded with the area of greatest population density and of highest educational level. In the United States, settlers responding to the cotton boom had begun to cultivate virgin land in the mid-South and were curious about iron masses that stopped or broke their plows. However, Shepard's report did have one significant result. It provoked the interest of scientists in gathering data and making statistical analyses.

In 1854 Robert P. Greg, a first-rate English mineralogist with an active interest in meteorites (and meteors), made an emphatic response to Shepard's ideas. The delay in his answer was due to the fact that he assembled a list of every reported meteorite fall, both authentic and doubtful, from 1478 B.C. on, together with the locations of meteorites that had been found. Of the 338 falls and finds, 11 were in the Southern Hemisphere and 327 in the Northern Hemisphere, between 5° and 70° north latitude. Less than half of the latter, Greg showed, were located in Shepard's meteoric zone. Going one step further, Greg made an analysis, taking France as the unit of comparison, of the number of falls that had actually occurred since 1790 and of the number that might be expected to occur in most western European countries and in the United States, given the areas and population. Only in Spain and European Russia was the number of actual falls substantially less than the expected number. The statistics, Greg wrote, demonstrated a

> pretty equable occurrence of meteorite falls on the surface of our earth. . . . Due allowance must, of course, be made for various counteracting influences, such as preponderance of sea and uninhabited countries in certain latitudes, and want of historical or scientific records among particular nations, etc.

In disposing of Shepard's theory, Greg declared that it would indeed be strange if these bodies, varying in size and in weight from half an ounce to 30,000 pounds, and traveling at velocities of up to 30 miles per second, should "be attracted by particular countries more than others, or arrange themselves in zones parallel to the isothermal or isodynamical lines."[52]

Greg's analysis did not settle the matter at the time nor, as we shall see, much later. P. A. Kesselmeyer, in 1860, noted Greg's argument that there was no relation between meteorite falls and geography, but thought there might be some substance to Shepard's ideas. Although not completely convinced by the evidence, Kesselmeyer felt that there was some connection between meteorite falls and earthquakes and that possibly meteorites were terrestrial volcanic ejecta. To substantiate this belief, he prepared three maps showing the locations of meteorite falls and finds and active volcanoes in Europe, Asia, and North and South America. Kesselmeyer's maps proved nothing, but his method of denoting the place of a fall or find by degrees and minutes of both latitude and longitude became a standard practice. It has proved difficult in some cases and impossible in others to determine the precise location of a find or fall. For example, according to contemporary reports the Lenarto iron was found in 1814 on a forest-covered slope on one of the highest summits of the Carpathians. In 1819 Chladni added the information that the site was "three hours from Bartfeld" (now Bardejov, Czechoslovakia). Kesselmeyer gave the coordinates as 49°18′ N, 21°4′ E. In 1896 Brezina stated that Lenarto was found west of Bartfeld, but gave the coordinates as 49°18′ N, 21°41′ E, which would place the locality east of Bartfeld. It was probably this confusion that caused G. T. Prior in his *Catalogue of Meteorites* (1923) not to list the coordinates of Lenarto, but M. H. Hey, in the third edition of the *Catalogue,* gave the location as "approx. 49° N, 21° E." After examining all of the evidence, including old detailed maps, V. F. Buchwald recently concluded that the most probable coordinates were 49°20′ N, 21°1′ E, which places the site a few kilometers from that denoted by Kesselmeyer.[53]

The reported location of the find or fall, the classification, and the name of the meteorite serve as means of identification. But the names of meteorites have been changed, occasionally by the whim of classifiers, and more often because of alterations of political boundaries. Also, scientists have modified classification systems, resulting in the transfer of some meteorites from one category to another. Moreover, the reported location may have been imprecise, as exemplified by the Lenarto meteorite, or a meteorite may have been moved from the actual site of fall, either by human hands or by such natural forces as glaciers. Yet even eliminating doubtful locations, the growing total of falls and finds and the spread of population over the planet are giving scientists the opportunity to assess whether Greg's conclusion that meteorites fall randomly to earth is valid. Greg, for example, could list only four falls and finds for South Africa and none for Australia, whereas at present there are about 50 authentic localities in South Africa and some 140 in Australia. Recently, the recovery of over 7,000 meteorite fragments in Antarctica, some probably from the same fall, is increasing the probability that Greg was correct.

The annual number of falls and the total weight of fallen material also intrigued investigators. Early in the nineteenth century, Carl von Schreibers estimated that about 700 meteorites fell each year, a number that Olbers thought was reasonable.[54] Greg, taking the area of Europe and the number of falls there in sixty-four years as a standard, calculated that 219 meteorites were likely to be recorded as falling annually on the Earth's surface, provided it had the population density and degree of civilization prevalent in Europe. Considering, however, that many falls at night would probably escape observation, and that the population density and awareness in other parts of the world were less than in Europe, Greg concluded that the total number of recorded and unrecorded falls would exceed 400 by one-third.[55] In 1858, Karl von Reichenbach in Vienna challenged Greg's estimate. Though using Greg's data, Reichenbach asserted that Greg had grossly underestimated the total. By an extrapolation that was not free of error, Reichenbach estimated that meteorites fell at an annual rate of 4,500. Assuming that each fall averaged over a hundredweight and that the age of the earth was a million years, Reichenbach concluded that about 5,000 million hundredweight of meteoritic matter had fallen. Always ready to speculate, Reichenbach concluded that the Earth and other planets had originated from the consolidation of meteorites, small particles which circled the Sun.[56] His theory is described in detail in chapter 5.

In an unpublished manuscript, possibly written as early as 1851 or as late as 1858, C. U. Shepard also speculated about the total weight of meteorite falls, coming to a far different conclusion than Reichenbach did. The total weight of iron meteorites that had fallen in the past century, Shepard calculated, was 177 pounds (Hraschina, Charlotte, and Braunau). This figure was only 3.7 percent of the 4,784 pounds of stones that had fallen. However, the total weight of iron meteorites that had been found just exceeded 59 tons. If, prior to a hundred years before, he wrote, stones fell in the same ratio of weight to irons as they had in the past century, then 1,599 tons of stones would have fallen, and the aggregate weight of meteorites would amount to 1,661 tons. Shepard then saw a flaw in his analysis. "But the ratio of the undiscovered unobserved meteoric matter," he wrote, "to that here inferred can only be conjectured. The former may have exceeded the latter 20, 50, 100, or even a greater number of times." Nevertheless, Shepard carried on,

> There is strong reason to believe that they did not begin to fall anterior to the diluvial [Quaternary] period. No traces of bodies analogous, in the slightest degree to them, have been met with below the diluvium & in but one instance much imbedded in that. They are generally found at or near the surface of the ground.

Shepard was apparently the first to speculate that meteorite falls occurred only in recent geological periods, a hypothesis defended until the 1950s. Shepard now jettisoned his 1848 theory that meteorites were connected with auroras, and conceived of a new possibility:

> Could all this closely resembling matter—have once existed in a single mass—a satellite perhaps to our earth which approaching at length so near to its surface was unequally acted upon by its attraction (when within 50 or 100,000 miles), & rent into pieces—the fragments for a time revolving in a ring about the planet in the former orbit of the satellite (like one of Saturn's present rings), but these fragments from time to time being precipitated, the largest and the heaviest first upon the earth—the deposition commencing in the early diluvial period? . . . If such an hypothesis can be entertained, it would explain the

> fact of the local deposition of the meteorites, inasmuch as 90 p.c. of all the falls occur in a stream of 10 or 15° of latitude crossing America obliquely from Mexico through the southern and middle States of the Amer. Union, France, Germany & southern Russia.[57]

It is a pity that Shepard did not publish this hypothesis, since it was superior to his abandoned theory. Possibly, Maxwell's 1857 analysis showing that Saturn's rings were particulate influenced Shepard. It is significant, however, that from analyses of the statistics on falls and finds available in the 1850s, Reichenbach and Shepard developed completely different theories.

Greg went further in his analysis than either Shepard or Reichenbach, and arrived at yet another hypothesis. Plotting 197 falls according to the month in which they fell, Greg determined that substantially fewer meteorites had fallen in December and January than in other months, and that there was a marked increase in the number of falls in June and July. He compared his findings with those of Baden Powell, who had made a similar analysis of the frequency of the appearance of luminous meteors by months. There was no apparent correlation between the two tables, since Powell's data showed very large numbers of meteors observed in August and November, months in which meteorite falls were about the average. Further, Powell's table showed that far fewer shooting stars appeared in June, a month in which a greater than average number of meteorites fell.

From these data and from the results of astronomers' studies of the two dozen asteroids discovered before 1854, Greg drew several conclusions. First, because of the lack of correlation between the monthly variations of meteorite falls and luminous meteors, Greg felt justified in stating that there was a definite distinction between meteors and meteorites. The latter, he said, were composed of solid matter, whereas the former were either fluid or gaseous. Second, he thought that there were periodic epochs for the fall of meteorites, just as it had been observed that there were periodic epochs in the appearance of shooting stars. Third, Greg wrote, if meteorites were associated with "the system of asteroids," their orbits would be, as astronomers had determined for the asteroids' orbits, principally outside of the Earth's orbit. However, there would be a greater probability of the orbits of meteorites and that of the Earth intersecting when the Earth was at "the period of aphelion or greatest distance from the sun," that is, in June and July. According to Le Verrier, Greg continued,

> there is a probable predominance of the perihelia of the asteroids in the mean direction of our summer solstice; a circumstance, if true, quite confirmatory of the opinion I have expressed, that the increase observable in the number of falls recorded for the months of June and July is not quite the result of chance.

Chladni, Greg wrote, was the first to propose that meteorites were the fragments of a destroyed planet, and now there was increasing evidence in support of the hypothesis that the asteroids were the larger pieces:

> Although the strange coincidences attending this group may be *accidental,* in general phrase, yet their phaenomena cannot but be considered as evidence tantamount to demonstration, of their having once composed a single planet, and having diverged by

the explosive force of a tremendous cataclysm: and in addition to their orbital vagaries, the bodies themselves are not round, as is said to be indicated by the instantaneous diminution of their light on presenting their *angular* faces.

The observations that the asteroids circled the Sun in a counterclockwise direction, as did the other planets; that they had highly elliptical orbits variously inclined to the plane of the ecliptic; and that they were angular rather than spherical masses, all appeared to Greg as more than adequate testimony to their common origin from a prior single planet. Meteorites, Greg concluded, were the smallest surviving pieces, "the minute outriders" of the asteroid group. With the assembly of more data, scientists could probably determine the periods of epochs of meteorite falls.[58]

Greg continued to gather data on meteorite falls, and in 1860 he went one step further in his analysis. Selecting 72 meteorites whose "precise hour of fall" had been recorded, Greg tabulated the horary distribution of meteorite falls. He determined that only 13 falls had occurred before noon, while 58 had fallen between noon and 9 P.M. Greg believed that this horary distribution was not the result of chance. The singularity, Greg thought, might be explained by "the greater tendency to encounter those bodies in their orbits, as they recede from the sun; that side of the earth most directly opposite to the sun being naturally most likely to come into actual contact with them."[59] Greg was on the right track. He assumed that all meteorites, as "minute outriders" of the asteroids, were in direct orbits around the sun. Meteorites that fall between noon and midnight must be in direct orbits, having either overtaken the Earth or been overtaken by the Earth.[60]

Wilhelm Haidinger, a highly respected Austrian mineralogist who had studied meteorites for two decades, was dubious about Greg's conclusions. In a series of articles published in 1867, Haidinger reported the results of his own survey of horary distributions of falls, which was more precise and more complete than Greg's. Whereas Greg had given just the average coordinates of his selected falls as 48° north latitude, 10° west longitude, Haidinger assembled data on falls in both the Northern and Southern hemispheres, and in Asia and America as well as in Europe. Moreover, since local times were, in the period before there was general agreement on international time zones, selected sometimes haphazardly and sometimes whimsically, Haidinger attempted to make certain that the reported local time of a fall was actually Greenwich time plus or minus the necessary horary correction, as determined by the longitude of the fall west or east of Greenwich. His findings of the horary distribution of 178 falls were:

Midnight to 6 A.M.	17 falls
6 A.M. to noon	57 falls
	74 total
Noon to 6 P.M.	77 falls
6 P.M. to Midnight	27 falls
	104 total

Haidinger's results, then, were generally supportive of Greg's. He pointed out, however, that it was less likely that meteorite falls would be observed during the late night and early morning hours, a condition that would bias the results. Greg's hypothesis would carry much more conviction, Haidinger concluded, if one could be sure that all meteorite falls during the night had been reported, and that all of the statements that cited the precise time of fall were correct.[61]

Greg's speculation that meteors were either fluid or gases was soon proven to be erroneous. His conclusion that all meteorites were "minute outriders" of the asteroids now appears to have been too sweeping. However, his and Haidinger's findings relative to the monthly and hourly incidence of meteorite falls are not too inaccurate in light of current research. Figure 10 shows a recent analysis of the monthly variation, which indicates a broad maximum of falls from April through July and a minimum from October to March. Figure 11 indicates that the horary variation of falls differs according to latitude, but it does show a pronounced minimum between about 3 A.M. and 9 A.M. As for the annual number of falls to earth, Reichenbach's estimate was the closest to modern predictions, although his presumed weights were not. Extrapolations of data from the Canadian camera network indicate that 19,000 meteorites weighing more than 0.1 kg fall on the entire planet annually; of these, 4,100 have a minimum total mass of 1 kg, and 830 a total mass of 10 kg.[62]

THE PHENOMENA OF FALL

The light, heat, and sound accompanying the fall of a meteorite in the atmosphere became subjects of scientific investigation in the early 1800s. In the mid-twentieth century, with the aid of newly developed photographic and radar techniques, scientists have arrived at general solutions to the physics of a meteorite's fall. However, there are still some problems about which scientists do not agree, such as the cause of the hissing noise reported to accompany the fall of numerous meteorites, as mentioned in chapter 1. There are also other problems that defy exact solution; for example, the partition of the energy released by a falling meteorite between the body itself and the surrounding atmosphere depends on many unknown factors and cannot be calculated accurately. For the most part, nineteenth-century scientists attempted to explain the phenomena of fall qualitatively, but from mid-century onward several began to adopt a quantitative approach. These early efforts illustrate the slow diffusion within the nineteenth-century scientific community of the understanding of newly discovered physical principles, in particular the laws of energy conservation and conversion.

Calculations from observational data and from experiments with artificial meteorites discharged from satellites indicate that the cosmic velocity of meteorites decreases from 95 percent to 5 percent of its initial value in a stratum of atmosphere 28 km in thickness.[63] However, the location of this stratum depends crucially on the cosmic velocity of the meteorite and its angle of entry into the atmosphere. If a very large iron meteorite with a high velocity should

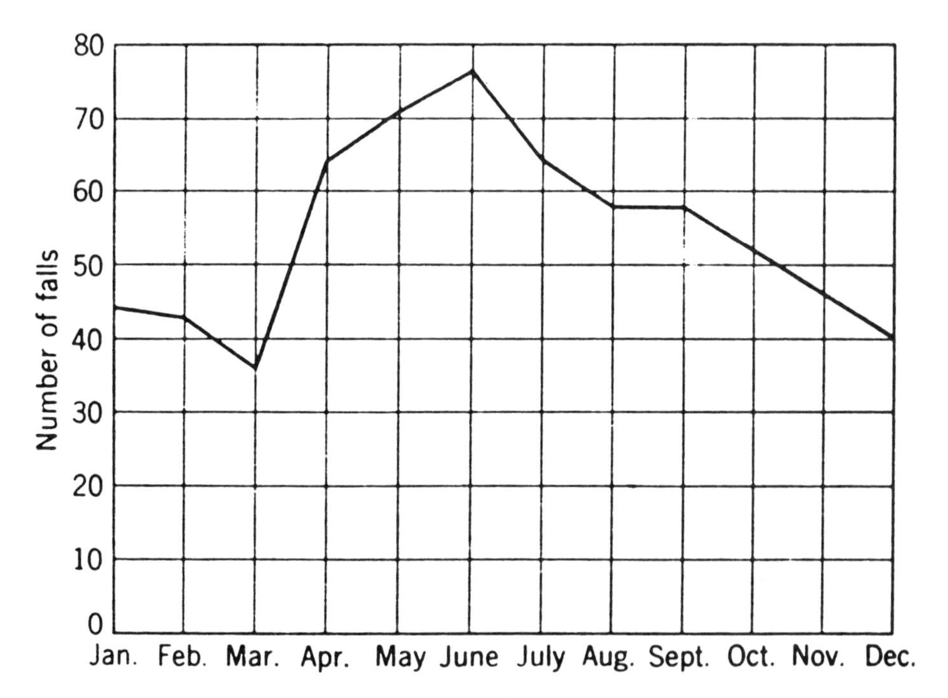

Fig. 10. Monthly variation in the incidence of meteorite falls, 1800–1960. From Mason (1962*b*). Reprinted with permission of the author.

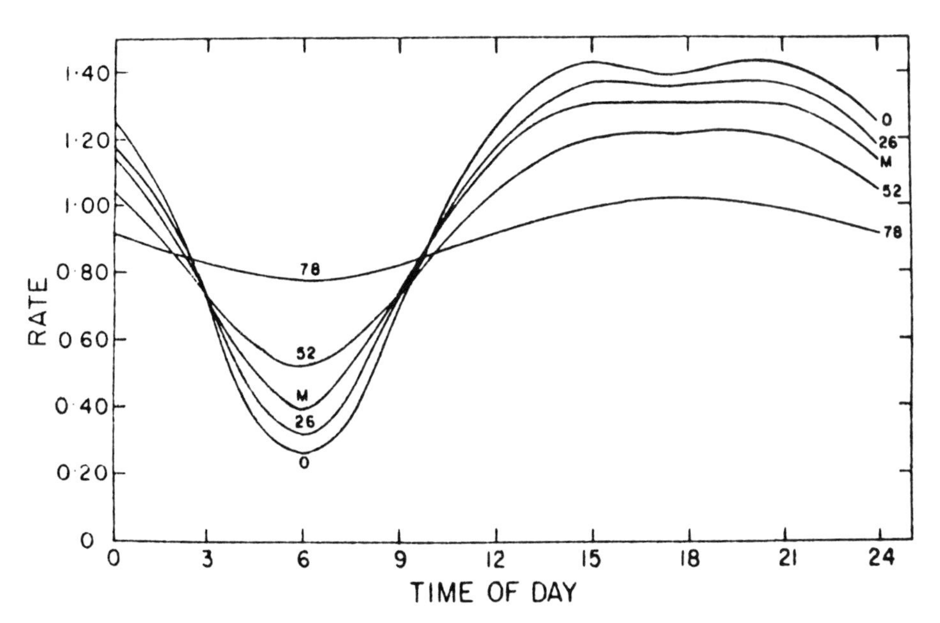

Fig. 11. Mean values for the year, at each hour of the day for latitudes of 0°, 26°, 52°, 78°, and the weighted mean of the whole Earth, labeled M. From Halliday and Griffin (1982). Reprinted with permission.

enter the atmosphere in a vertical direction, it would strike the earth with about 90 percent of its cosmic velocity. If the angle of entry is small, the maximum deceleration occurs high in the atmosphere, and thereafter the velocity of the meteorite fall depends only on gravity and air resistance—that is, the meteorite falls at a terminal velocity. The majority of meteorites probably begin this stage of free-fall at altitudes ranging from 5 to 25 km.

At the moment of entry into the atmosphere, the meteorite possesses both potential and kinetic energy: the former is a product of its mass and its height above earth, and the latter a product of its mass and its cosmic velocity. In the formula that converts kinetic energy to other forms of energy per unit of time, the velocity is raised to the third power. Hence the entry velocity, which may be as high as 72,000 meters per second, results in an enormous amount of energy release and conversion in each moment after atmospheric entry.

Throughout the fall, only a small fraction of the kinetic energy converted in the process develops heat in the meteorite, although this quantity results in continuous melting of the exposed surface. The major portion of the transformed energy heats and ionizes the atmospheric atoms to produce the fireball, displaces the air and forms shock waves to create the sound and explosions, and causes the meteorite to rotate and fragment. The bursting of the meteorite produces little noise; the sound results from sonic booms generated by the rapid displacement of the air along the meteorite's path. With the onset of free-fall, the energy converted is insufficient to cause atmospheric ionization, and luminosity ceases. Usually, the meteorite is cooled in its fall through the lower atmosphere and strikes the ground barely warm.

This brief explanation of the physical causes of the phenomena accompanying a meteorite's fall serves to illustrate the problems that confronted early nineteenth-century scientists. Their calculations of the velocities of meteors and fireballs, which were based on the estimated durations of luminosity and lengths of trajectories, were inaccurate, as were the reported heights of the beginning and end of luminescence, determined from purportedly simultaneous observations. The height of the atmosphere as well as its physical constitution were matters of speculation. The behavior of gases at high temperatures and pressures was unknown. Only in the 1830s did scientists begin to offer proof that the various forces of nature were subject to laws of conservation and conversion. Given this state of affairs, explanations flowed mainly from comparisons with phenomena in the more familiar sciences of electricity, chemistry, and exterior ballistics; at times, scientists cleverly combined principles drawn from all three disciplines.

In 1803 Jean-Baptiste Biot and Pierre Prevost, both scholars of the classics as well as scientists, noted passages in the writings of Roman poets that appeared to offer hints of the underlying causes for the appearance of heat and light during a meteorite's fall. Lucretius had mentioned, Prevost wrote:

> . . . how motion will o'erheat
> And set ablaze all objects—verily
> A leaden ball, hurtling through length of space,
> Even melts . . .[64]

Biot recited the same passage from Lucretius, and cited others to the same effect from Ovid's *Metamorphoses* and Virgil's *Aeneid,* adding that it would be profitable to determine by experiment whether a lead ball projected at high velocity would actually liquefy.[65] Stimulated possibly by Biot's challenge or by another event to be described shortly, Marc Auguste Pictet did design an experiment, but with the air in motion and the body at rest. He permitted compressed air to escape from an orifice that was positioned a short distance away from the bulb of a thermometer. With an estimated air velocity of 327 feet per second, the temperature rose from 18 to 33 degrees on a scale with 80 parts. The experiment convinced Pictet that a projectile having a much higher velocity could, as a result of the friction of the air, easily attain a temperature at which combustion occurred.[66] He postulated that the matter of heat, which many scientists at that time thought was a weightless subtle fluid (caloric), rose along with other evaporated gases to the top of the atmosphere, occupying a region from 300 to 450 miles high. This gaseous mixture was not only combustible but also offered resistance to a meteorite, and light and heat resulted from ignition of the inflammable mixture effected by friction.[67]

At the time, however, Prevost remained unconvinced (as other scientists were later) that a body would encounter sufficient resistance in the upper atmosphere to cause it and the surrounding air to be heated to an incandescent temperature. But after Humphry Davy's discovery in 1806 that soda and potash were metallic oxides, the idea that actual combustion of a meteor occurred in the upper atmosphere gained support. Commending it, Jeremiah Day wrote that "the bases of magnesia and silex, may originally exist in the meteor, in the state of *pure metal*: . . . when the body . . . comes into our atmosphere, a sudden and violent combustion is produced, by the very strong affinity of these substances to oxygen."[68] However, the suspected insufficiency of oxygen in the upper reaches of the atmosphere caused scientists to postulate two or more simultaneous processes. In 1818, William Higgins, an Irish chemist and staunch adherent of Chladni's theory of the cosmic origin of meteorites, proposed that a meteorite became increasingly electrified as it passed through regions high above the earth, where uncombined electricity supposedly existed. As the meteorite became electrified, it simultaneously lost its specific heat (caloric), which emitted light at the meteorite's surface. At the same time, the iron and sulfur in the meteorite combined with oxygen, producing heat. The released heat, Higgins wrote, melted the surface, and the dense electrical atmosphere surrounding the falling body kept the contiguous air in an incandescent state. If the meteorite passed into a cloud that possessed an opposite electrical charge, the meteorite's electrical charge would be lost, giving rise to the appearance of thunder and lightning. At this point, the mass would almost always break into pieces, and the specific heat would return to the latent state, resulting in the cessation of luminosity.[69] Poisson also favored electricity as the principal cause of a meteorite's luminosity, although his explanation incorporated the presence of two electrical fluids.[70]

During this period, attention also became focused on the heat developed by air compression. In 1802, Joseph Mollet, professor of physics at the École Centrale in Lyons, learned from a workman in a nearby armory that a linen shred accidentally lodged in the exit tube of the condensing pump of an air gun had caught fire, and also that there was a momentary flash of light whenever an air gun was discharged in the dark.[71] Mollet communicated these facts to the Institut in Paris, where unsuccessful experiments were made to duplicate the phenomena. Pictet, in Geneva, was aware of these events, and in attempting to simulate the process arrived at the results already mentioned. The final product of Mollet's further research was the fire piston, in which the ignition of an inflammable material placed at the bottom of a cylinder occurred when a tightly fitted piston compressed the air column within it.

We shall not trace here the many difficulties encountered by Mollet in attempting to explain the principle of the fire piston to his fellow scientists. Its importance in this context is that an analogy was soon made between its action and the fall of a meteorite. Johann Benzenberg, in 1810, was apparently the first to draw the comparison. He wrote:

> The incandescence perceived around fireballs in a state of ignition may be the result either of *combustion,* although with difficulty admissable in air so very rarefied, or of *friction,* as generally believed. I think it results still more from the *compression* of air, as in our newly invented tinder-boxes air produces fire by mere compression. Could not *electricity* become free in the same way?[72]

Benzenberg went on to speculate that the massive compression of a large volume of air would release electricity. This would accumulate around the meteorite, producing a fireball, and would finally discharge with an explosion, causing dispersion of the fireball and fragmentation of the meteorite.

The hypothesis of air compression received further support in 1821 from Theodor Grotthus, the eminent Lithuanian scientist. The intense and rapid compression of the air in front of the meteorite upon entry into the atmosphere, Grotthus wrote, ignited the air, melted the surface of the mass, and caused the explosion. The rumbling noise accompanying the passage of the fireball was owing to the rapid flow of air into the vacuum that the fireball created behind it. The bursting of the meteorite resulted from the fact that the cohesion of the mass was insufficient to resist the unusually high air pressure.[73]

During the next few decades communications on the subject of the phenomena of light, heat, and sound during a meteorite's fall were merely repetitions of the rival hypotheses of combustion, friction, electrical action, and air compression. In 1848, however, James Joule made the first attempt to describe the phenomena in terms of energy conversion. The ignition of meteoric stones in collision with the atmosphere, he reported, was a remarkable illustration of the equivalence of heat to mechanical power, or *vis viva* (kinetic energy). Suppose, he said, that a 6-inch cube entered the atmosphere with a velocity of 18 miles per second, the atmosphere at that height having one-hundredth the density of air at the Earth's surface. The resistance offered by the air would be at least 51,600 pounds, and if the stone traversed 20 miles, the heat developed would raise 6,967,980 pounds of water one degree Fahrenheit. The largest portion of the heat would be given to the displaced air, every particle of which would sustain shock: "the stone may be considered as placed in a blast of intensely heated air." Even if the stone received only a hundredth portion of the heat, it would still be a quantity "quite equal to the melting and dissipation of any materials of which it may be composed." It was for this reason, Joule wrote, that many meteors appeared in the sky but few meteorites were found on Earth. In conclusion, he voiced his gratitude to the "Author of nature" for thus protecting his creatures.[74]

In analyzing Joule's figures, it is apparent that in calculating the resisting force, he employed *vis viva* = mv^2, instead of kinetic energy = $mv^2/2$. And in arriving at the total heat developed, Joule assumed that the force of 51,600 pounds acted constantly through the entire distance of 20 miles, and employed the value 782 foot-pounds as the mechanical equivalent of heat. He neglected loss of mass or of velocity. Nevertheless, his was the first mathematical approach to the problem.

Though Joule had shown that the *vis viva* of a meteorite upon entering the atmosphere was converted into an enormous quantity of heat, he was unable to demonstrate conclusively that most of the energy was expended in displacing and heating the air and that only a small portion caused heating of the meteorite. Together with William Thomson, Joule in 1857 conducted experiments to determine the temperatures at various distances from a body placed in a current of rapidly moving air. The results, however, were inconclusive; Joule and Thomson were able to determine only that the stratum of air around the body had a higher average temperature than the air farther off.[75] Consequently, some scientists, such as John Tyndall, though admitting the fact of energy conversion, thought that the heat was primarily expended in raising the meteorite to the temperature of incandescence.[76] And Gabriel-Auguste Daubrée, the leading meteorite scientist in France, wrote that the survival of the carbonaceous material in the Orgueil meteorite, which fell in 1864, contradicted the hypothesis that the heat of meteorites was owing to the loss of *vis viva*.[77]

In 1861 Wilhelm Haidinger attempted to give a comprehensive description of the phenomena.[78] The compression of the air, Haidinger wrote, caused heating of the air around the meteorite, some heating of the meteorite, and a "rotatory movement around an axis corresponding with the meteorite's orbit." The resistance at first forced a portion of the air, made luminous by compression, outward and perpendicular to the meteorite's trajectory, and then backward, developing an "igneous globe, either round, or as frequently happens oviform, occasionally extended so far back as to form even an actual tail" (fig. 12). To prove that the air became luminous, Haidinger referred to experiments conducted in 1859 by Julius Plücker, professor of mathematics and physics at the University of Bonn. When an electrical discharge occurred in a highly evacuated Geissler tube, Plücker determined that light was produced and the surface of the tube became warm. According to Plücker, the electricity acted as an exciter of heat, causing the gaseous particles to become incandescent and a transfer of some heat to the tube walls. Plücker thus discovered the ionization of gases, although a satisfactory physical explanation of the phenomenon was only made three decades later by J. J. Thomson. Haidinger, however, surmised that the same process occurred in the rare atmosphere high above the Earth.

The igneous globe or fireball, Haidinger thought, contained a vacuum, and when the meteorite lost its cosmic velocity, the vacuum collapsed with a

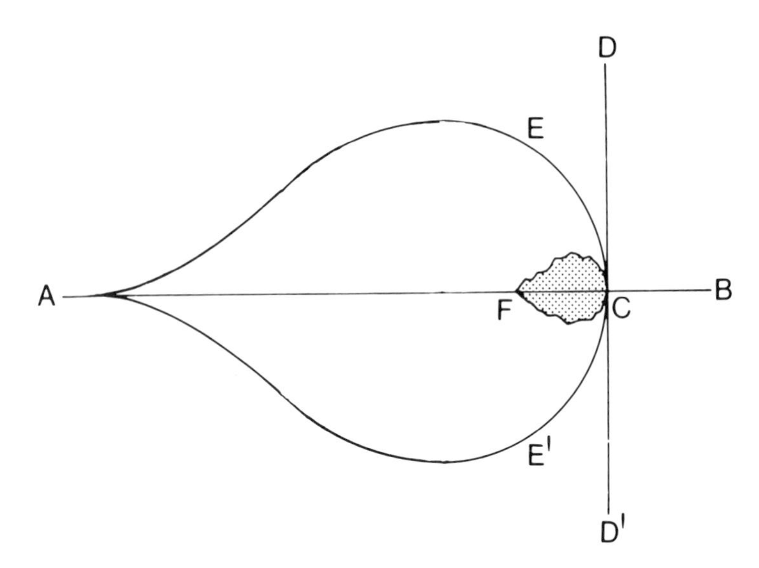

Fig. 12. Haidinger's concept of the formation of a fireball. The air, made luminous by compression, is forced outward along the line DD′, which is perpendicular to the meteorite's trajectory AB. It is then forced backward along the curve ECE′ to form a spherical or oviform globe that sometimes has a tail. The interior of the globe is a vacuum. From Haidinger (1861*a*).

detonation and the luminosity ceased. If the meteorite broke into fragments, there was a less intense noise. With the expenditure of its cosmic velocity, the meteorite began to fall freely due to the action of gravity. During this stage, the surface of the meteorite cooled, and there was a conduction of heat from the hot exterior to the intensely cold interior, such that stones were scarcely warm to the touch when they struck the ground. As proof of this point, Haidinger cited the case of the Dhurmsala meteorite, which fell in July 1860, and which had frost on its surface when it was recovered.

Also in 1861, J. Lawrence Smith published an article explaining the phenomena of light and sound accompanying the meteorite fall at New Concord, Ohio, on 1 May 1860, which agreed in most respects with Haidinger's views. Smith, however, thought that the sound emanating from a fireball resulted from "concussion of the atmosphere arising from the rapid motion of the body through it or in part due to electrical discharge."[79] Smith also insisted that the black crust of a meteorite was not formed in the atmosphere but was already present in space. Although he offered no explanation for this assertion, it is likely that Smith, who still adhered to the hypothesis that meteorites were ejecta from lunar volcanoes, thought that the crust was a by-product of the eruption.

One conclusion shared by Haidinger and Smith brought a negative response. Both maintained that such meteorite showers as occurred at l'Aigle and New Concord were the result of a swarm of meteorites entering the atmosphere and were not produced by the disintegration of a single body into many fragments. In 1863 Nevil Story-Maskelyne published a report on the fall of five stones at Butsura, India, on 12 May 1861. Aided by the presence of a vein of nickel-iron that traversed all of the pieces, Maskelyne was successful in fitting together the pieces. He postulated that the breakage was caused by the temperature differential between the interior and the exterior, since parting occurred in the directions of planes of weakness. Further, detailed examination showed that several layers of crust had formed on the single mass during flight, successively spalling off and reforming.[80] Nonetheless, Haidinger stood firm. In 1869, in an article answering Maskelyne and rebuking Daubrée for stating that the phenomena of fall were still mysterious, Haidinger reiterated his position that swarms were responsible for showers, citing the fall of a reported hundred thousand stones at Pultusk in January 1868.[81] Smith remained silent on the subject of showers, but later agreed that the crust was formed in the atmosphere.

Haidinger, in 1861, had also given attention qualitatively to the deceleration of meteorites in the atmosphere, although, surprisingly, he did not cite Joule's work.[82] In a sketch (fig. 13), he proposed that the cosmic velocity continually decreased to a point where it was totally lost, after which the meteorite entered a stage of free-fall due to gravity. In 1868 Schiaparelli attempted to treat the deceleration quantitatively, using as a basis a formula developed primarily by Guillaume Piobert to determine the effect of air resistance on the velocity of artillery shells.[83] Schiaparelli recognized that the angle of a meteorite's entry into the atmosphere would affect the deceleration, but used the extreme case of vertical entry in his calculation. He computed the decrease in initial velocity as a function of the increase in atmospheric density, and constructed tables for two initial velocities. These showed the percent decrease of the initial velocities at various values of atmospheric pressure, the latter obtained by a conversion of density into pressure. Taking an initial velocity of 16,000 meters per second, which Schiaparelli considered to be the usual velocity of a meteorite approaching the Earth in a direct orbit, he calculated that three-fourths of the cosmic velocity would be lost at a point

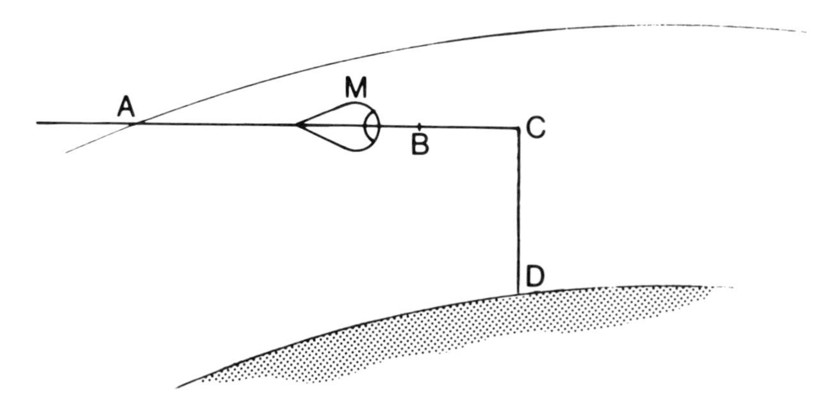

Fig. 13. Haidinger's concept of a meteorite's trajectory. During the period when it still retains a part of its cosmic velocity, the meteorite traverses the path ABC, forming the fireball. When, at point C, the cosmic velocity is completely expended, the fireball collapses, luminosity ends, and the meteorite falls freely from C to D. From Haidinger (1861*a*).

where the atmospheric pressure was 12.41 mm of mercury, and fifteen-sixteenths of the meteorite's *vis viva* would have been converted into heat. A meteorite in retrograde orbit entering the atmosphere at 72,000 meters per second, upon reaching a point where the pressure was only 7.58 mm of mercury, would have lost eight-ninths of its cosmic velocity, and eighty-eighty-firsts of its *vis viva* would have been converted to heat. In approximately the same amount of time, Schiaparelli wrote, the higher velocity meteorite would have lost twenty-one times more *vis viva* than the one moving at lower velocity. From this fact, it followed that meteoroids entering the atmosphere in retrograde orbits appeared to be both more numerous, because smaller ones became visible, and more brilliant, because a larger quantity of *vis viva* was converted into light and heat. It was no wonder, Schiaparelli concluded, that meteoroids in retrograde orbits were almost always totally consumed, whereas those in direct orbits would have a much greater probability of survival in the atmosphere, and would fall more frequently in the afternoon and early evening hours, when they overtook or were overtaken by the Earth.

Schiaparelli's work raised two questions. First, would meteoroids that were not members of the Solar System and that moved in hyperbolic orbits at velocities much higher than those of comets survive entry into the Earth's atmosphere? Second, did a meteorite, as Schiaparelli maintained, lose practically all of its cosmic velocity at a very high altitude? An apparently valid answer to the first question came almost immediately. On 30 January 1868 a shower of stones, estimated to have numbered a hundred thousand pieces, occurred at Pultusk, Poland, a few miles north of Warsaw. Many persons witnessed the fireball, and from estimates by twenty-six observers, including two amateur astronomers, of the point of its appearance in the sky, its trajectory, and its duration, Johann Galle calculated that its heliocentric velocity was at least 56 km per second, and that therefore its orbit must have been hyperbolic.[84] For decades, Galle's calculations were accepted as valid, thereby casting a cloud on Schiaparelli's conclusions. In the 1930s and 1940s, however, there was considerable analysis of and discussion about Galle's assumptions and results. Most scientists today believe that his conclusion was erroneous.[85]

A tentative answer to the second question—the final velocity at which a meteorite strikes the earth—came in the 1880s, after the fall of a meteorite near Middlesbrough, Yorkshire, on 14 March 1881. Alexander Herschel, who for decades had been calculating the heights at which the luminosity of meteors began and ceased and determining the radiant points of meteor showers, was teaching at Newcastle, a scant 40 miles away. He was beside himself with excitement, and immediately went to the site to ensure that the recovery of the stone would be careful and thorough. On April 5, Herschel informed Lazarus Fletcher at the British Museum (Natural History) that

> operations are still proceeding at the actual place of fall to photograph the site and scene of the discovery, to exhume and preserve the hole bodily in a proper earth-box built round it, (and to mechanically measure with a pile-driving machine the quantity of motion required to drive an iron imitation of the meteorite to a depth of 12<u>in</u> into the soil which it penetrated, at which depth it was actually found).*
>
> *This experiment I am only at present contemplating, having not yet had time to contrive, and forward to the R<u>y</u> Engineers at Darlington & *Middlesbro'* the necessary Apparatus.[86]

Herschel did see to the completion of the experiment, and his published article reported that the measured quantity of work required to duplicate the hole 11 inches deep (the average depth) was 9,307 foot-pounds. Inasmuch as the stone weighed 3 lb 8.5 oz, he calculated that the velocity of fall at impact was 412 feet per second. In free-fall in the atmosphere, a stone would acquire this velocity in a height of 0.5 mile.[87] Herschel's investigation did not substantiate Schiaparelli's contention that all meteorites reached earth at velocities of free-fall, but it demonstrated that at least some did.

Schiaparelli's use of equations that were applicable to the relatively slow velocities of artillery projectiles was inappropriate for calculating the resistance encountered by meteorites entering the atmosphere. Further, he neglected to take into account the loss of mass suffered by a meteorite in its path, and also assumed that all of the kinetic energy was converted into light and heat. Nevertheless, he was on the right track, and he refined his conclusions in further studies during the next thirty years. In the 1880s Ernst Mach's investigations of the behavior and the effects of projectiles moving at velocities greater than the speed of sound shed further light on the subject, as did studies in the same period of the ionization of gases. By the 1880s, then, scientists understood in broad outline the causes of the phenomena of light, heat, and sound that accompanied the fall of a meteorite, even though they were still unable to express the details with any degree of mathematical precision. The advance in seventy years had been remarkable.

4

Nineteenth-Century Foundations of Meteorite Analysis

About 550 meteorites fell or were found during the nineteenth century, so that chemists and mineralogists received a continually increasing supply for analysis and investigation. As a rough estimate, about 250 scientists published articles relating to the chemistry, mineralogy, or petrography of meteorites during the century. For most, a description or analysis of a meteorite fragment that happened to come into their hands was a matter of passing interest. For others, such as Gustav Rose in Berlin, Karl Reichenbach in Vienna, Gabriel-Auguste Daubrée in Paris, Lazarus Fletcher in London, or J. Lawrence Smith in Louisville, Kentucky, meteorites provided an irresistible attraction, and despite other duties or interests, they pursued meteoritic studies until old age or death intervened.[1]

These scientists had a number of goals, some explicit and some implicit. One was to assess reported falls or finds to determine whether the stones or irons were indeed meteorites. In the first few years of the nineteenth century, after Edward Howard's analyses revealed that all of the meteorites he had examined contained unoxidized nickel-iron, this determination appeared to be a simple task. However, scientists' confidence in the criterion of the presence of nickel was badly shaken after the analysis of the meteoritic stone that fell at Stannern in Moravia on 22 May 1808. It contained no nickel-iron, and Vauquelin's analysis disclosed only a trace of nickel. Klaproth, in 1812, found no nickel in a supposedly meteoritic iron mass unearthed at Collina di Brianza near Milan, but he hesitated to label it nonmeteoritic. The analysis of the Stannern meteorite, he wrote, indicated that the absence of nickel could not be considered the sole criterion as to whether or not a stone or iron was a meteorite.[2] A few years later, Gillet de Laumont was worried when Carl von Schreibers sent him a sizable stone from the Stannern fall. He had received another fragment earlier, he reported, and neither of the pieces was attracted by the magnet, nor did they contain metallic iron, leading many people to believe they were just pieces of a clay crucible. However, the black crust of the piece presented by Schreibers was almost intact, and Schreibers "had himself verified that this stone had really fallen," so that Gillet de Laumont somewhat reluctantly accepted them as meteoritic.[3] The certification of a reported fall or find as meteoritic was a continuing problem in the nineteenth century, and it remains so at the present time, though to a far lesser degree.[4]

Another goal of chemists and mineralogists was to identify the chemical elements and compounds and minerals in meteorites to learn whether and in what ways meteorites differed from terrestrial rocks and cast and smelted irons. Nineteen elements had been found in meteorites by the 1850s, so scientists were confident that the same laws of chemical combination, affinity, and crystallization applied to both terrestrial and meteoritic matter.[5] However, there were minerals in meteorites that had never been found on Earth. In this chapter we describe the identification by nineteenth-century scientists of some common meteoritic minerals among the more than one hundred now known.

Classification of meteorites was a purposeful activity. By the 1840s the differing chemical composition and mineralogical constitution of meteoritic stones and the various structures exhibited by meteoritic irons caused scientists to classify them and to assign names or terms to those that appeared to be similar from the point of view of chemical composition, texture, or structure. The increasing number of meteorites and the acquisition of new knowledge about them ensured that their reexamination and reclassification were continual goals.

Scientists were also intensely interested in the origins of meteorites, and by the 1860s interpretations of chemical composition, mineralogical constituents, texture, and structure resulted in hypotheses of origin based on scientific considerations, rather than sheer speculation. We outline these theories in chapter 5, although we give some attention here to different interpretations that were made relative to composition, constitution, and structure.

There were practical and theoretical problems as well. Meteoritic stones are composed mainly of complex silicate minerals, and meteoritic irons are primarily alloys of iron and nickel of varying composition. When chemical analyses of samples taken from the same meteorite yielded different results, corroboration was impossible, and controversies about techniques and methods inevitably followed. It took decades to work out analytical methods that chemists agreed were fairly reliable. Almost all of the mineral crystals in meteorites are tiny and imperfect, and very few could be measured and identified by the techniques available to nineteenth-century crystallographers. The development of improved microscopes in the 1860s greatly enhanced mineral and textural analysis, and the introduction of the technique of studying and later photographing carefully prepared thin sections of meteorites was of major importance. The first qualitative spectroscopic studies of meteorites also occurred about 1860.

Moreover, the rudimentary state of chemical and mineralogical knowledge in the early decades of the nineteenth century was as much an obstruction to the understanding of the composition and constitution of meteorites as were the imperfect methods. In 1801 Haüy, in referring to the silicates, wrote: "Not only has the chemical art, been slow in its progress relative to the analysis of the substances of this class . . . but it has not even taken the first step toward their synthesis."[6] The few analyses that Haüy did report in his text were understandably erroneous, because the atomic weights of elements were at that time unknown, and the combining weights were based on limited experiments. Even in the 1850s, there was still dispute about the atomic weights of several common elements. Just as important, however, were certain erroneous underlying assumptions held by mineralogists and chemists. For example, despite the discovery in 1819 by Eilhard Mitscherlich of the phenomenon of isomorphism and its widespread acceptance, most scientists throughout the entire century continued to regard the replacement of one atom [ion] for another in a compound as a vicarious substitution. The majority held that all chemical combinations were in definite proportions by weight, and this belief created misunderstanding of the constitution of meteoritic irons as well as of the chemistry of stones. Another unfortunate assumption was that there were various silicic acids that formed salts just as sulfuric acid does, leading to misjudgments of the compositions of the silicate minerals in meteoritic stones.

During the course of the nineteenth century scientists gradually overcame many of these problems, although the most difficult were not resolved until the twentieth century, after X-ray diffraction analysis placed increased emphasis on the intimate structure of matter. Throughout the century there was an increasing interaction among the sciences of chemistry, mineralogy, crystallography, and geology, culminating in the appearance of scientists such as Gustav Tschermak of Vienna, who, because of his comprehensive knowledge of all of these areas, was able to make significant contributions to the understanding of meteorites. In addition, new techniques and improved instruments played a prominent role in nineteenth-century advances. Both features—namely, broadly trained scientists and more sophisticated analytical equipment—would become prerequisites for further progress in the twentieth century.

METHODS

Specific Gravity

It is almost as certain as death and taxes that no meteorite fragment ever left the hands of a nineteenth-century mineralogist before its specific gravity was measured. The determination of specific gravity extends back to the noted bathtub incident of Archimedes, which, according to Vitruvius, caused him to invent a method for ascertaining whether King Hiero's crown was pure gold or an alloy. We have noted that the first step taken by the Fougeroux committee in its investigation of the Lucé meteorite was to measure its specific gravity. It was 3.535, higher than that of a siliceous stone, from which the committee concluded that it contained a considerable amount of metal, later found to be iron.

There is little information as to what instruments were used in the nineteenth century to measure the specific gravity of meteorites, or what degree of care was employed, although values were usually reported to three and sometimes to four decimal places. It seems likely that early measurements were made with improved versions of a hydrometer invented by William

Nicholson in 1785. Haüy in 1801 described one of these, and recommended it as being easy to dismantle and carry, and sufficiently precise.[7] During the century more sophisticated instruments appeared; Lazarus Fletcher in 1901 used a pycnometer, or specific gravity bottle, when he analyzed the Zomba meteorite.[8] Haüy cautioned that corrections should be made for water temperature and that representative samples should be taken, and it appears that these admonitions were taken seriously. In 1825 Stromeyer, when comparing the chemical compositions and properties of several terrestrial olivines and the minerals found in stony iron meteorites, recorded water temperatures and barometric pressures attendant on his specific gravity measurements.[9] And in 1855 J. Lawrence Smith noted that the three fragments of the Tazewell iron he had used for the determination of its specific gravity were "selected for their compactness and purity."[10]

More important than the question of accuracy, however, is why scientists measured the specific gravity of the mixed meteoritic matter in the first place. Determining the specific gravity of particular minerals, as Stromeyer did with the olivines in stony irons or as Fletcher did with enstatite in the Zomba meteorite, is entirely comprehensible. But scientists were aware very early on that meteoritic stones were conglomerates of minerals, that stony irons contained substantial quantities of silicates, and that the irons had differing compositions of iron and nickel and also various types and variable amounts of inclusions. The specific gravity (density) of certain classes of metal-free carbonaceous meteorites varies from 2.2 to 2.9; of other stones from 3.2 to 3.9; of the stony irons from 4.6 to 4.9; and of the irons from 7.7 to 8.0. Thus specific gravity reflects the quantity of contained iron. It is possible that nineteenth-century scientists made the measurements from force of habit, since specific gravity was and still is an important character used to identify minerals, or that they wished to obtain one verifiable quantitative datum. It is more likely, however, that they thought that specific gravities might have some possible future utility in the classification of meteorites. Although this was never the case, specific gravities did play a role in the development of theories of the origin of meteorites, an aspect we consider later on.

Chemical Analysis

Criticism of results and the exposure of errors are part and parcel of experimental science and have contributed greatly to its development. Nevertheless, the tone of comments by mid-nineteenth-century scientists concerning the results of chemical analyses of meteorites reveals a sense of impatience and even frustration, rather than constructive criticism. In 1859, for example, Reichenbach scoffed at Philip von Holger's analysis of the Wessely meteorite, deeming it completely erroneous. He doubted the accuracy of John W. Webster's analysis of Nobleborough and of Alexander Scherer's analysis of Doroninsk. He thought that Shepard's analyses of the Hartford [Marion, Iowa] and Forsyth [Mincy] meteorites had obviously been completed with too much haste, and he considered that the older analyses of Laugier, Vauquelin, and Klaproth needed revision.[11] Story-Maskelyne in 1863 wrote that "the chemical methods adopted for the analysis of a meteorite are probably unsatisfactory to every chemist who has employed them."[12] And J. Lawrence Smith in 1870 questioned all analyses that did not report the presence of cobalt, tin, and copper in meteoritic irons. He had analyzed, he said, over a hundred irons, and had invariably found cobalt and minute quantities of copper, so he considered the presence of these elements to be important.[13] During the century chemists learned that the analysis of several samples, extreme carefulness, and infinite patience solved part of the problem. Lazarus Fletcher, for example, filled four laboratory notebooks in analyzing the Zomba meteorite, and spent the best part of a year completing it.[14] He gave the Youndegin meteorite the same careful treatment (fig. 14).

Yet nineteenth-century meteorite analyses, even when made by experienced chemists, were usually defective in three respects: (1) analysts failed to obtain complete separation of the nickel (and cobalt); (2) they also had difficulty in determining the quantities of the alkalis (sodium, potassium) present; and (3) the presence of iron in a number of different combinations was the most important problem. Chemists made advances in the first two areas during the century, but the third remained a problem well into the twentieth century. Hence it is useful to describe the nature of these difficulties and the steps taken to overcome them.

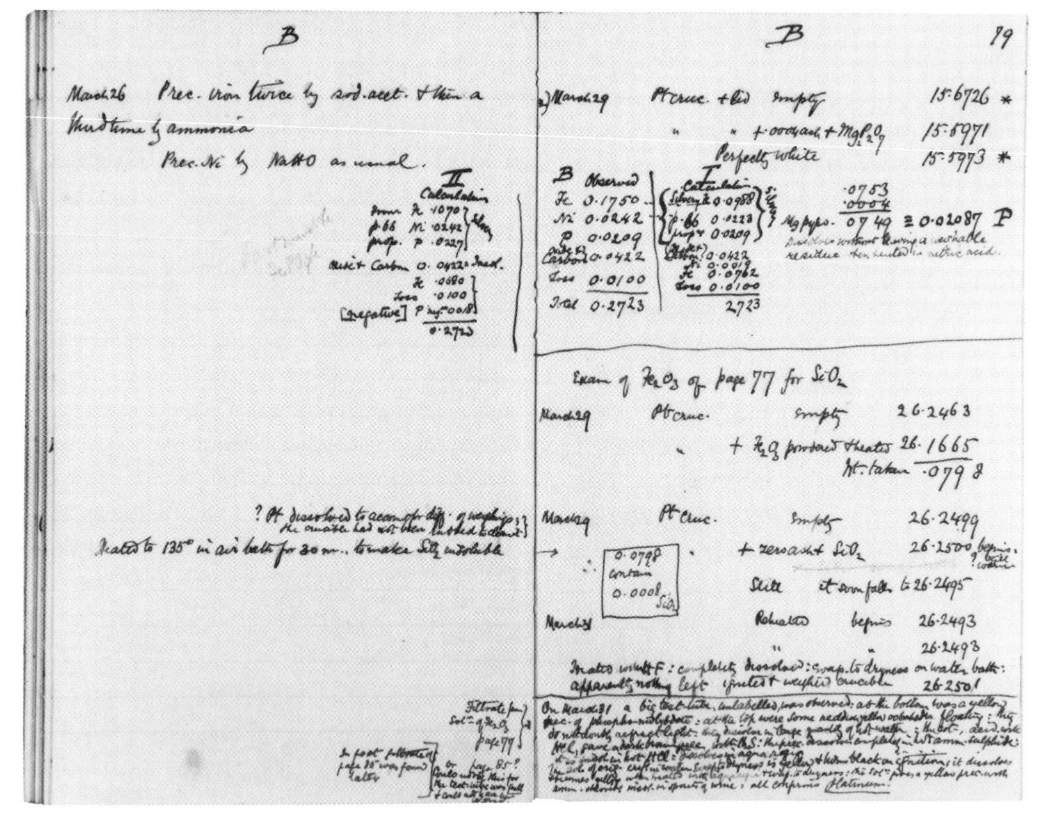

Fig. 14. A page from one of the four notebooks Lazarus Fletcher used in his analysis of the Youndegin meteorite, 1886–87. Courtesy of the British Museum (Natural History), London.

Nickel-iron is normally present as two distinct minerals having different and variable chemical compositions. In addition, the nickel-iron may contain cobalt, chromium, phosphorus, copper, and carbon. Early on, chemists recognized that the complete separation of the nickel from the other constituents was a difficult task. Berzelius, in his 1834 analysis of the Krasnojarsk stony iron, noted the great discrepancy in previous reports of the nickel content, Howard having found 17 percent and Klaproth only 2 percent. Berzelius determined that the amount of nickel was 10.78 percent, not too far removed from a modern analysis of 8.9 percent.[15] Reported nickel analyses of meteoritic irons during most of the nineteenth century are, by and large, uniformly lower than modern analyses show. Various reagents were used to separate the iron from the nickel by precipitation, but treatment with sodium acetate became preferred. Chemists gradually realized that a single precipitation was insufficient to separate all of the nickel. Fletcher, for example, employed four precipitations and satisfied himself by a further test that there was no residual nickel. Others, however, made as many as seven precipitations.

The alkalis are normally present in the acid-insoluble portion of meteoritic stones, and their complete separation defied the efforts of chemists until the late nineteenth century. In 1853 J. Lawrence Smith clarified the matter in two lengthy articles.[16] He described: (1) the various means to render the silicates soluble; (2) the technique of separating the other ingredients, in particular magnesium, from the alkalis; (3) the method of removing sal-ammoniac that unavoidably accumulated in the process; and, finally, (4) the technique of separating the alkalis from one another. Smith's detailed description was a definite aid in the analysis of meteorites, and he returned to the subject in 1871, outlining improvements he had made in the intervening period.[17] At the time, Benjamin Silliman, Jr. hailed Smith's accounts as an important contribution to analytical chemistry, and the fact that Smith's 1871 article was still being cited a century later provides distinct evidence of its considerable value.[18]

Iron is present in several different combinations in meteorites: as an alloy with nickel and cobalt; as a sulfide and possibly a phosphide or carbide; and as a component of the silicates. Chemists, of course, sought to determine each of these combinations in the individual percentage amounts, and they gradually developed methods for the analysis of ordinary meteoritic stones, which were, however, not completely satisfactory. The first step was to attempt, as Howard did in his 1802 analyses, to isolate the nickel-iron alloy from the remainder of the stone by the use of a magnet. Howard noted at the time that some of the earthy matter could not be separated by mechanical means, and chemists soon learned that Howard had been correct. In 1834 Berzelius commented: "This mechanical separation by means of a magnet appears to be very easy, but it proves as good as impossible to accomplish completely."[19] Berzelius and others thought the separation might be performed more effectively under water or alcohol. Edouard Baumhauer, when analyzing the Utrecht meteorite in 1845, used an electromagnet, carefully transferring the attracted portion of the powder, which was spread on a plate under alcohol, to a dish filled with alcohol, at which time he interrupted the current of the magnet.[20] Others tried the method of flotation, thinking that the higher density of the nickel-iron particles would cause them to sink while the lighter constituents would float on the surface. This technique was of no avail, since the shape of the particles in this process is just as important as the density. Friedrich Wöhler, in particular, attempted to achieve chemical separation. He initially proposed that the proportion of the nickel-iron could be calculated by measuring the volume of hydrogen emitted when the powdered stone was decomposed by an acid, the sulfur content being determined at the same time. The objection to this method was the uncertainty of the calculation, so Wöhler suggested separation by iodine and water, then by bromine water, and later by a moderately concentrated solution of neutral cupric chloride. All of these methods, and another, the dissolution of the nickel-iron by mercuric chloride, were flawed because part of the iron sulfide and a small amount of the iron silicate in the stone were dissolved at the same time.[21] As a result, such chemists as Lazarus Fletcher reverted to the method of magnetic separation. Fletcher extracted the nickel-iron from the Zomba meteorite "by means of a seven-toothed iron comb . . . which had been magnetised: the powder being worked through in small quantities at a time."[22] He assumed that any silicate adhering to the metal, which was dissolved in the subsequent acid treatment, had the same composition as the bulk of the acid-soluble silicate in the stone, and merely combined the two quantities in his final calculation of the bulk analysis. This compromise was also elected by many other chemists.

Having separated the nickel-iron as much as possible and having found the respective quantities of the elements therein, the chemist next determined the quantities of the constituents of the nonmagnetic part. Here, the total amount of iron was an important figure. The chemist deducted from the total iron an amount sufficient to combine with the sulfur (or phosphorus and carbon), and reported the balance as iron oxide, which was considered to be in combination as a silicate. The accuracy of the amount of iron oxide thus depended on how completely the chemist had separated the nickel-iron and the precision of his sulfur determination.

When attempting to analyze the nonmagnetic portion of most meteoritic stones, chemists prior to 1834 found that part of it was soluble when the powder was digested with hydrochloric acid and that an insoluble residue remained which required different treatment. In his 1834 analysis of several meteoritic stones, Berzelius reported that the soluble portion constituted about one-half of the nonmagnetic portion and that it corresponded closely with terrestrial olivines, in which the relative proportions of magnesium and iron vary. He was uncertain whether certain other constituents that he found in small quantities in the soluble portion really belonged to the olivine or whether they might be traces of the beginning dissolution of the insoluble part.[23] Later, chemists determined that some of the insoluble silicate did in fact go into solution, the quantity depending upon the temperature, the acid concentration, the fineness of the powder, and the duration of the process. Berzelius's result, however, did simplify the analytical process, and his techniques were adopted immediately by other chemists.

The total amount of oxygen in a meteorite was not ever determined by analysis, but was instead a calculated figure. It comprised the amount required to form the normal oxides of those elements present that are more electropositive than iron, plus the amount necessary to form iron oxide from the iron not considered present as a sulfide. This circumstance, as we shall see shortly, affected the identification of the silicates present in the meteorite.

Mineralogical-Petrographical Analysis

Bournon, in his examination of the stony meteorites analyzed chemically by Howard in 1802, used a magnifying glass to distinguish the most prominent constituents, which were the globules, "martial pyrites," small particles of iron, and an earthy groundmass. Careful descriptions of the constituents visible under a magnifying lens and comparisons of the textures with those of other meteorites soon became commonplace, providing early classifiers with invaluable information.

In 1825 Gustav Rose attempted to measure with a reflecting goniometer the minute crystals that he found in the Juvinas meteorite. In his account one reads the complaint echoed by crystallographers throughout the century—namely, that the crystals were either too minute, too broken, the surface curvature too great, or the edges too rounded to give images distinct enough to permit accurate measurements.[24] Despite these difficulties, Rose was able to identify crystallographically one major constituent as augite[25] and another tentatively as labradorite. He also gave a complete description of the forms of a few iron sulfide crystals, but he was surprised because they displayed no magnetism until heated. They were in fact stoichiometric iron sulfide (later called troilite). Rose, however, thought that they were pyrrhotite, and knowing that the magnetism of pyrrhotite varied and was occasionally masked by the presence of nickel sulfide, he tested for this but found no trace. There he left the subject, but continued on to give a complete description of the forms of a large olivine crystal that he had found in a section of the Krasnojarsk stony iron, and to demonstrate that the values of the interfacial angles were almost exactly the same as several terrestrial olivines. Rose, at this time, did not refer any of the crystals he measured to crystal systems, although he did use the notations introduced by his colleague at Berlin, Christian Samuel Weiss, to describe the forms. In 1815, Weiss had designated the four crystal systems now known as the isometric or cubic, the tetragonal, the hexagonal, and the orthorhombic systems, but had failed to distinguish the monoclinic and triclinic systems, whose axes are not mutually perpendicular. Friedrich Mohs had in 1822 recognized the latter possibility, and in 1824 Mohs and Karl F. Naumann independently announced the existence of these systems, although they used completely different names for them. Until 1839, when William Hallowes Miller published his *Treatise on Crystallography,* the nomenclature of the crystal systems and the notation employed for the description of crystal forms were hopelessly confused owing to the idiosyncracies of the numerous classifiers. Thereafter, Miller's notation was generally accepted, although his system nomenclature was subsequently modified.

Although in the next few decades crystallographic analysis became an important tool for mineralogists, only a few working with meteorites reported angular measurements of mineral crystals they found. Shepard was one; in 1846 he drew sketches and reported some values of the interfacial angles of a half-dozen crystals he had observed.[26] Undoubtedly, the difficulty of finding large enough crystals and of measuring them were the reasons for the lack of reports.

Reichenbach in 1857 appears to have been the first to study the minerals and textures of meteorites with a microscope. He did not report what kind of microscope he used, but he did state that the magnification was 200X and that he polished the surfaces of the irons that he investigated.[27] During the decade of the 1850s, however, Henry Clifton Sorby had been studying thin sections of rocks with a polarizing microscope,[28] and both Henri de Sénarmont and Alfred Des Cloizeaux in France had been investigating the optical properties of mineral crystals. Nevil Story-Maskelyne, professor of mineralogy at Oxford, who was appointed keeper of minerals at the British Museum in 1857, was not only familiar with but also intensely interested in these developments. In 1861 he supervised the construction of a polarizing microscope equipped with a revolving graduated stage and an eyepiece goniometer for the study of mineral crystals and the thin sections of meteorites (fig. 15). Maskelyne hired Victor von Lang, later professor of physics at Graz and Vienna, as an assistant from 1862 to 1864, and von Lang's diary and notebook show that he was continuously engaged from September 1862 until March 1863 measuring crystals in the Gorruckpore [Butsura], Parnallee, and Juvinas meteorites.[29] He and Maskelyne published preliminary results of these investigations in 1863, and during the next seven years Maskelyne examined over 140 meteorites by this technique. In an article published in 1870, though, it is clear that he was disappointed and frustrated by the results. He wrote:

> The crystallography . . . of the numerous crystals seen in such a microscopic section is almost hopelessly difficult. . . . A long series of measurements and determinations of the directions parallel to the principal sections in the crystals met with in these microscopic slides has convinced me that . . . the microscope is only partially of use in determining the mineralogical character of the constituents.[30]

The microscope was useful, Maskelyne continued, in picking out distinct particles of minerals, which, in a very few cases, could be analyzed crystallographically. More often, however, if a sample of several minute crystals amounting to 0.2 gram could be collected, a chemical analysis could be carried out successfully.

Maskelyne's pessimism as to the value of the polarizing microscope in studying thin sections of meteorites was premature. Mineralogists in Germany and Austria had begun to use this technique, and when Gustav Tschermak was appointed curator of the mineral collection at the Vienna museum in 1869, he immediately proceeded to assemble and study a sizable collection of meteorite thin sections.[31] In April 1876 Tschermak began to photograph the minerals and textures revealed by the microscope, and his numerous petrographic studies of meteorites culminated in 1886 in the publication of his classic work, *Die mikroskopische Beschaffenheit der Meteoriten.*[32] Thereafter, the microscopic study of thin sections of meteoritic stones became a necessity.

The maximum magnification used by Tschermak in 1886 appears to have been about 100X. Meanwhile, Sorby in 1876 described that he had been able to achieve a magnifying power of 650X with the use of vertical illumination—

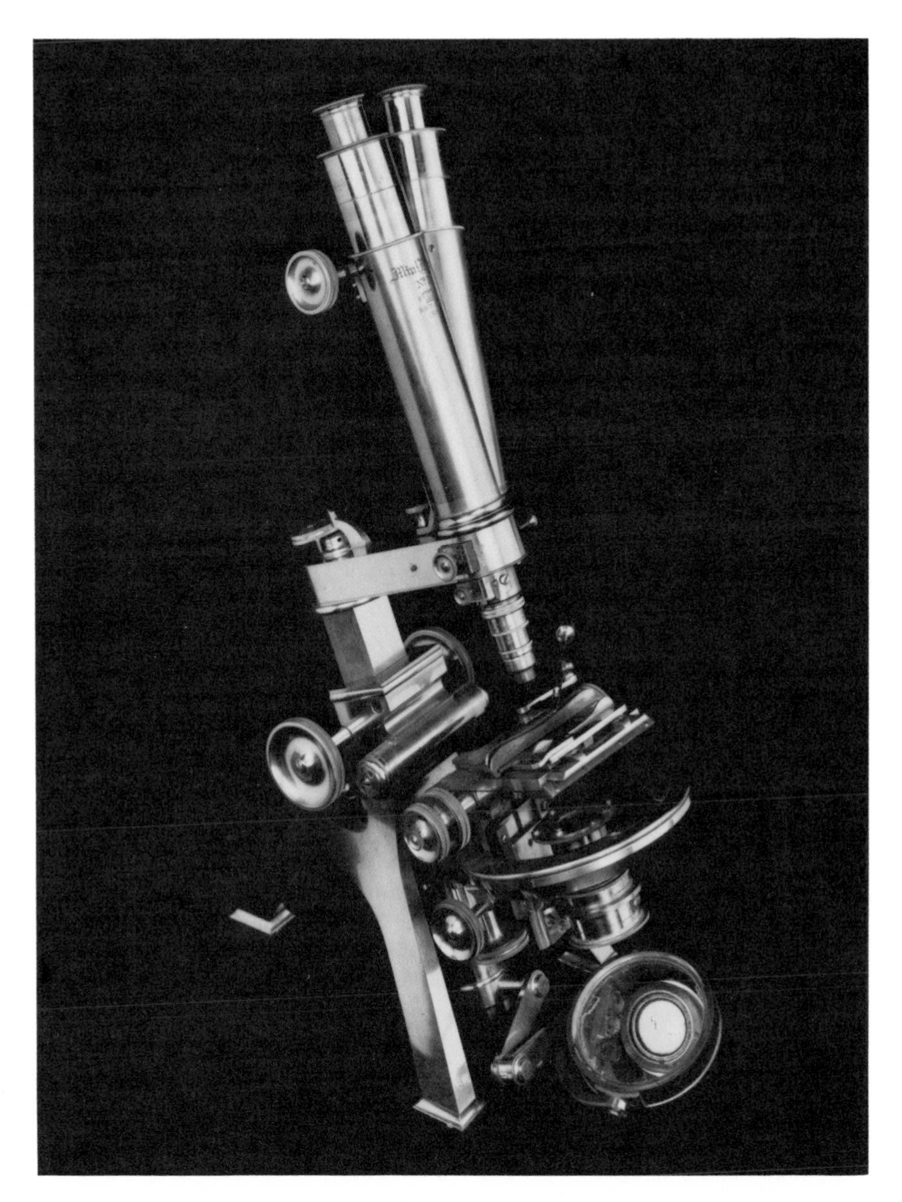

Fig. 15. Petrographic microscope devised by Nevil Story-Maskelyne and fabricated by Powell and Lealand in 1863. It is probably the earliest polarizing microscope with a rotating stage. Accessories include a direct vision spectroscope built by Browning. Courtesy of the British Museum (Natural History), London.

a technique, he wrote, "contrived many years ago by Richard Beck."[33] He explained that

> it was not until last autumn, that I employed what may be called *high powers*. This was partly because I did not see how this could be satisfactorily done, and partly because it seemed to me unnecessary. I had found that in almost every case a power of 50 linear

> showed on a smaller scale as much as one of 200, and this led me to conclude that I had seen the ultimate structure. Now that the result is known, it is easy to see that my reasoning was false, since a power of 650 linear enables us to see a structure of an almost entirely new order.

The next year, however, Sorby reported his observations of a number of polished and etched meteorite thin sections, some of which he had borrowed from Maskelyne and some of which he had prepared himself. There, he stated: "When properly prepared, the surface [of meteoritic iron] may be satisfactorily examined with a magnifying power of 200 linear, which is required to show the full detail."[34] The apparent limit set by Sorby probably retarded the use of vertical illumination microscopes in the study of meteorites. Emil Cohen in 1894 wrote that 30 to 40X magnification was satisfactory in general for the study of the structure of meteoritic irons, 100X for the finer details, and added that Sorby had recommended for strong magnifications (300–400X) the use of the Beck illuminator.[35] Whereas scientists investigating the structure of metals began using high magnifications in the 1890s, those studying meteorites did not do so until the second decade of the twentieth century.

Spectrum Analysis

Few scientists in the nineteenth century employed the spectroscope for meteorite analysis, due partly to the fact that it was then only a qualitative technique, and partly to disagreements concerning the results and their interpretations.[36] Soon after 1859, when Gustav Kirchhoff proposed his theory that each element had a uniquely characteristic spectrum, Robert Bunsen, according to Wöhler, used a spectroscope to analyze the Juvinas and Parnallee meteorites. He found that both contained lithium, which had not been previously discovered in meteorites.[37] In 1862 Theophilus Engelbach, professor of chemistry at Bonn, analyzed the Cold Bokkeveld carbonaceous meteorite. He reported a perfect calcium spectrum, an intense sodium line, three less prominent strontium lines, and very weak potassium and lithium lines—results that now appear suspect.[38] Several scientists in the following decades investigated the spectra of occluded gases in meteorites. Only Norman Lockyer, however, pursued the spectroscopic studies of meteorites vigorously over a period of years. Initially, he investigated meteorites to aid his studies of the solar spectrum, and later to support his theory of the origin of meteorites. His work is described in chapter 5.

CONSTITUENTS: OLIVINE, PLAGIOCLASE, PYROXENE

Apart from nickel-iron, which is considered in the next section, various silicate minerals are the most common minerals in meteoritic stones. The basic unit of the silicate minerals is the silicate tetrahedron (fig. 16) in which four oxygen ions surround and are bonded to a central silicon ion, giving the combination

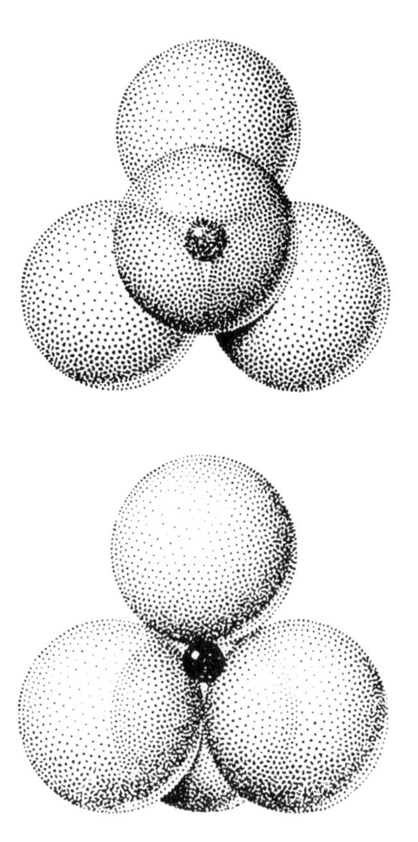

Fig. 16. SiO_4 tetrahedra. The four oxygen ions are at the apexes of a regular tetrahedron, surrounding the silicon ion, which has a valence of four. From *Dana's Manual of Mineralogy* (17th ed. © 1959). Reprinted with permission of John Wiley & Sons, Inc.

a net charge of -4. In some silicates an aluminum ion may replace the silicon ion, but the degree of such substitution is limited. The various types of silicate minerals depend upon the manner in which the silicate tetrahedra are bonded together and also on the kind of elements that are located in the interstices between adjacent tetrahedra. Olivine, pyroxene, and plagioclase feldspar are the most common silicate minerals found in meteoritic stones. Only rarely, however, were grains of these minerals large enough to permit identification by either chemical or crystallographic methods employed in the nineteenth century. Furthermore, the minerals are usually so intermingled and mixed with other nonsiliceous minerals that it was difficult to find a pure sample (pl. 1). For these reasons, several decades passed before scientists identified all three minerals in meteorites.

The olivines have a structure in which positively charged divalent metal ions (cations) occupy the interstices between neighboring silicate tetrahedra. Each cation is bonded to six oxygen ions of adjacent tetrahedra, thus producing a regular stacking of tetrahedra and octahedra, and giving a structural symmetry belonging to the orthorhombic crystal system. If the interstitial ions are all magnesium, the resulting mineral is forsterite (Mg_2SiO_4); if all are ferrous ions, then it is fayalite (Fe_2SiO_4). Because either a magnesium or ferrous ion may occupy an interstitial site, most olivines have a chemical composition intermediate between these two pure minerals. The product of such random ionic substitutions is called a solid solution, whose composition varies between those of the end members—in the case of olivine, between pure forsterite and pure fayalite—and whose general chemical formula is $(Mg,Fe)_2SiO_4$. The olivine in most stony and stony iron meteorites ranges in composition from 60 to 85 percent forsterite and 15 to 40 percent fayalite, although some contain pure forsterite, and in a very few the fayalite predominates. Olivine also occurs as inclusions in iron meteorites. Manganese, calcium rarely, and chromium very rarely may substitute for magnesium and/or ferrous ions in some olivines. Further, if olivine is subjected to high temperature and pressure, its crystal structure may be altered, but this phenomenon appears to be unusual in meteorites.

The silicate tetrahedra may also be joined by the sharing of oxygen ions of adjacent tetrahedra. When two of the four oxygen ions in each silicate tetrahedron are shared with neighboring tetrahedra, a simple chain structure is formed. This structure yields a ratio of one silicon to three oxygen ions, and offers interstitial sites for cations. If magnesium and/or ferrous ions occupy the sites, the crystal symmetry is orthorhombic and the resulting mineral is an orthopyroxene, $(Mg,Fe)SiO_3$, or alternatively $(Mg,Fe)_2(Si_2O_6)$. Again, a solid solution exists between pure $MgSiO_3$, enstatite, and pure ferrosilite, $FeSiO_3$. In meteorites, however, the iron component is rarely above 30 percent; below 10 percent the mineral is termed enstatite, between 10 and 20 percent bronzite, and between 20 and 30 percent hypersthene. Under certain conditions of temperature and pressure, these pyroxenes may crystallize in the monoclinic system, and are then called clinoenstatite, clinobronzite, and clinohypersthene.

Calcium, a larger cation than magnesium or iron, may occupy some sites in the simple chain structure, substituting for ferrous and magnesium ions. When this occurs to the extent of about 10 mole percent calcium silicate, the resulting mineral crystallizes in the monoclinic system and is called pigeonite, $(Ca,Mg,Fe)SiO_3$, which is a pyroxenic constituent of numerous meteorites. Other pyroxenes—diopside, augite, and wollastonite—occur in meteorites only infrequently.

In the plagioclase feldspars, aluminum-oxygen tetrahedra almost identical in size and shape to the silicate tetrahedra are present, and they link with the silicate tetrahedra by sharing oxygen ions. However, the total charge of the AlO_4 tetrahedron is –5 instead of the –4 charge of the SiO_4 tetrahedron, which permits the introduction into the structure of a monovalent cation for each aluminum tetrahedron, in order to balance the charge. If a divalent cation enters the structure, then the charge must be balanced by the presence of two aluminum tetrahedra. Thus, in the plagioclase or soda-lime feldspars, where a monovalent sodium ion is replaced by a divalent calcium ion, a greater amount of aluminum is necessarily present. Plagioclase feldspars, then, form a complete solid solution series from pure albite, $NaAlSi_3O_8$, to pure anorthite, $CaAl_2Si_2O_8$. The series is divided arbitrarily into the following minerals:

	Percent albite	Percent anorthite
Albite $NaAlSi_3O_8$	100–90	0–10
Oligoclase	90–70	10–30
Andesine	70–50	30–50
Labradorite	50–30	50–70
Bytownite	30–10	70–90
Anorthite $CaAl_2Si_2O_8$	10–0	90–100

Plagioclase crystallizes in the triclinic system, and the crystals are frequently twinned; that is, there is intergrowth according to a known law. Oligoclase is the most common feldspar in meteorites, although in certain classes the plagioclase is bytownite or anorthite.

An account of the development of modern knowledge of the chemistry and structure of silicates would be a lengthy and fascinating history of a completely neglected area. Here we can only mention that prior to the initial-phase equilibrium investigations by Norman L. Bowen and others in the early 1900s, and the X-ray diffraction studies in the 1920s and 1930s by William L. Bragg, the structures and constitutions of the silicate minerals were largely unknown. Scientists had to depend upon chemical analysis and the frugal results of crystallographic analysis. As Harry Berman wrote in 1937: "Except for the important modifications by Tschermak, . . . the theory is substantially today as it was in 1811, and the classification now used in latest textbooks is the same as that found in texts more than fifty years ago."[39]

In citing 1811, Berman was referring to the year in which Berzelius, Johann W. Döbereiner, and James L. Smithson (whose bequest to the United States provided the foundation of the Smithsonian Institution) proposed almost simultaneously the theory that silicates were salts of a silicic acid. In 1846 Auguste Laurent suggested that several silicic acids existed, and in 1856 Edmond Frémy lent support, pointing out that acids attacked some silicates and not others.[40] Eventually, theory called for four silicic acids: metasilicic acid, H_2SiO_3; orthosilicic acid, H_4SiO_4; disilicic acid, $H_6Si_2O_7$; and trisilicic acid, $H_4Si_3O_8$.[41] Thus, for example, olivine, termed an orthosilicate, was a salt of orthosilicic acid, and albite, a trisilicate, was a salt of trisilicic acid. When heated, these various acids, whose presence was inferred from gels, gradually lost water and eventually formed the anhydride SiO_2. There were a few staunch opponents of the notion that there were several acids. Karl Rammelsberg continued to maintain that only metasilicic acid existed, and that all silicates derived from it.

The silicic acid theory, in one form or another, colors the explanations and the formulas that nineteenth-century mineral chemists gave for the results of their analyses of the silicate minerals in meteorites. Since they were unable to measure directly the amount of oxygen present, they inferred in what ways it was combined with silicon, and the resulting chemical formulas of the minerals were based on these assumptions.

The phenomenon of isomorphism or solid solution complicated matters further. It was first vaguely perceived early in the nineteenth century by four researchers working independently—Nicholas Leblanc, William Hyde Wollaston, Johann Fuchs, and François Sulplice Beudant.[42] But in 1819 Eilhard Mitscherlich clearly expressed the law of isomorphism, which, in its revised version, stated that "the same number of atoms, if combined in the same manner, will produce the same crystalline form, so that the form does not depend upon the nature of the atoms, but upon their number and manner of combination."[43] Scientists accepted isomorphism, but with reservations, since some found it difficult to visualize how a particular mineral could be said to exist. Gustav Rose, for example, held that it was next to impossible to specify definite boundary lines for the various mineral species intermediate between the end members of an isomorphic series, and that in such cases the definition of a mineral species "must be left to individual opinion."[44]

Individual preferences in the interpretation of analyses, noted by Rose, are readily apparent in the scientific literature. Many scientists seem to have taken the position that chemical combinations must have definite proportions by weight within solid solutions. This conviction appears not only with respect to silicate minerals but also with respect to nickel-iron and nickel-phosphorous-iron, as will be described shortly. In some cases isomorphism was also quite difficult to understand. This was particularly true of the plagioclase feldspars, where calcium was somehow substituting for sodium at the same time that aluminum was substituting for silicon. Even after the 1860s, when Tschermak demonstrated that a complete solid-solution series existed between albite and anorthite, a number of scientists still shook their heads.[45]

The silicate in meteorites which scientists distinguished most readily was olivine. Its occasional presence as fairly large and well-defined crystals in stony irons obviously helped. After Bournon described and Howard analyzed the olivine in the Krasnojarsk meteorite in 1802, there was a period of doubt because Howard's analytical results varied from those of several terrestrial olivines reported by Klaproth. But Stromeyer's 1824 assertion of the existence of isomorphism in olivine, coupled with Rose's demonstration that the forms of the crystals in terrestrial olivines corresponded almost exactly with a large crystal he had found in Krasnojarsk, settled the matter. Then in 1834, Berzelius determined that the acid-soluble silicate in several meteoritic stones he analyzed was predominantly olivine, that it composed about half of the stones, and that the relative proportions of magnesium and iron in the olivine varied, thus confirming Stromeyer's judgment of its isomorphism (pl. 1).

Berzelius based these conclusions on his analyses of the Blansko and Chantonnay meteorites, both of which contain olivine and orthopyroxene as major constituents. However, even the great chemist Berzelius could blunder occasionally. Something went awry when he analyzed the Luotolax meteorite, which is composed principally of plagioclase and orthopyroxene. Berzelius reported that over 90 percent of the sample was soluble in acid, and stated that the soluble portion was also olivine.[46] This error persisted for more than three decades, primarily because there were only a few meteorites of the same kind as Luotolax. One was the Mässing meteorite, upon which a bulk analysis was made shortly after its fall in 1803; hence it was impossible from the analysis to identify the minerals it contained. Another was the Bialystok meteorite, which Gustav Rose described in 1837. Rose did not make an analysis, but he did emphasize its striking similarity to the Luotolax meteorite. There were numerous large and small grains of olivine, he wrote, which were intermingled with white grains of feldspar.[47] Orthopyroxene (bronzite-hypersthene) may be mistaken for olivine, and this is undoubtedly what Rose described. This misconception initiated by Berzelius was clarified in 1872 by Tschermak, who determined that what had been called olivine was in fact orthopyroxene.[48]

There was a delay in the identification of plagioclase, partly owing to misinformation, and partly because in the first half of the century only four meteorites were available that contained enough plagioclase to permit a successful chemical analysis (pl. 2). These were the Stannern, Jonzac, Juvinas, and Petersburg meteorites, all containing plagioclase and pyroxene; however, because the Juvinas fall amounted to 91 kilograms, most analyses were made of its fragments. André Laugier first analyzed the Juvinas meteorite in 1821, soon after its fall, and gave particular attention to an abundant whitish constituent which appeared to him to be similar to leucite or white garnet. Small crystals of this substance had noticeable cleavage, and crystallographers whom he consulted concluded that the substance was feldspar. Laugier considered that his finding of over 10 percent alumina and 0.2 percent potassia, apart from the silica, confirmed this opinion.[49] Gustav Rose disagreed, because, as he wrote in 1825, feldspar contained over 16 percent potassia, and in addition the crystals appeared to him to be similar to either albite, labradorite, or anorthite.[50] Here is expressed the early nineteenth-century belief that the potash feldspar, orthoclase, bore no relation whatsoever to the soda-lime feldspars. Rose pursued the possibility that the substance was plagioclase. He eliminated anorthite, which is readily soluble in acids, because Laugier had reported that the meteorite was not attacked by acids. To determine whether it was albite, Rose tested for sodium, but finding that it contained only 0.6 percent, he concluded that the crystals were probably labradorite and discontinued the investigation.

There the matter rested until 1846, when Shepard reported that anorthite was a constituent of meteorites, and that crystals of it were "massive in the Juvenas stone," noting also that Rose had observed them and suggested that they were labradorite.[51] Shepard's report alerted Rammelsberg, who immediately analyzed a fragment of Juvinas.[52] Rose, Rammelsberg noted in his report, had been misled by Laugier's statement that the stone was not attacked by acids. In fact, almost 37 percent of his sample was acid-soluble, and the results showed this portion to be almost pure anorthite.

A few years earlier, Rammelsberg had found plagioclase in stony meteorites, but in quite a different manner.[53] He was concerned about the insoluble portions of meteoritic stones, in particular the alumina content, reported by Berzelius to be almost 6 percent in the Blansko meteorite and by Dufrénoy to be over 10 percent in the Chateau-Renard meteorite. Alumina in such amounts, Rammelsberg thought, could not be a vicarious admixture, and it could not be one of the bases because the analytical results showed otherwise. The only other alternative was that alumina was isomorphous with silicic acid. Pehr Adolph von Bonsdorff, professor of chemistry at Helsingfors, Rammelsberg pointed out, had expressed this view in 1822;[54] however, there was not a complete analogy because only one silicon atom combined with three oxygen atoms in silicic acid, whereas the same number of oxygen atoms combined with two aluminum atoms in alumina. As further evidence of isomorphism, however, Rammelsberg referred to the work of Hermann Kopp at Giessen. Kopp had proposed that isomorphism occurred when an atom having the same volume as another replaced or substituted for it. If the atom volumes were not precisely the same, there would be a slight change in the angles of the resulting crystal. Kopp attempted to calculate atom volumes by a formula using molecular weights and specific gravities, arriving at a value for the atom volume of alumina of about 182 and for that of silicic acid of about 200. The values of the volumes were close enough, Rammelsberg thought, for isomorphism to occur. With this justification for his procedure, Rammelsberg arbitrarily separated the constituents that Dufrénoy had found in the Chateau-Renard stone into two parts. He assigned all of the sodium and potassium and some of the silicic acid and alumina to one part, and all of the iron oxide, magnesia, lime, and the rest of the silicic acid and alumina to the other part. The first part, comprising about 6 percent of the entire stone, Rammelsberg stated, was albite, and the second part amounting to about 32 percent of the stone, was hornblende, since the ratio of the oxygen in the bases to that in the silicic acid and alumina was approximately 4 to 9, the usual ratio found in hornblende. Rammelsberg treated the bulk analyses of the acid-insoluble portions of the Blansko and Chantonnay meteorites in the same fashion, concluding that one portion of each was labradorite and the other hornblende.

Rammelsberg's belief that isomorphous substitution was the reason for the various compositions of the plagioclase feldspars was correct, although his explanation was, of course, erroneous owing to ignorance of structure. And as it turned out, his ambitious and ingenious attempt to identify the mineral constituents of the acid-insoluble portion of meteoritic stones by partitioning the composing chemicals was ultimately unsuccessful. Nevertheless, a number of scientists adopted his method. Wolfgang Sartorius von Waltershausen was of the opinion that the insoluble part of the Kakowa meteorite consisted of 82 percent magnesia-wollastonite and 17 percent anorthite. Wöhler also probably favored the method, since his student, Elijah Harris, reported in 1859 in his doctoral dissertation that the insoluble portion of the Ausson meteorite consisted of about 20 percent labradorite and 77 percent augite.[55] Scientists later found that the feldspar in the pyroxene-plagioclase meteorites was either bytownite or anorthite and that in most other stones containing feldspar the mineral was oligoclase.

Identification of the pyroxenes was a much more formidable task. One major reason is that a number of different ions may occupy the interstitial sites in the silicate simple chain structure, and thus produce a number of minerals with different chemical compositions and diverse properties. Another is that the minerals in the enstatite series may crystallize in the orthorhombic or in the monoclinic system, depending upon the conditions of pressure and temperature in the crystallizing environment. A third reason is that the mineral nomenclature for many decades in the nineteenth century was inexact, to say the least; mineralogists gave different names to the same mineral, and while some insisted that chemistry should be the criterion for classification, others preferred to employ crystallographic results. Thus, the names "augite," "pyroxene," or "feldspar" might be used by different mineralogists to denote completely different minerals.

Again, it was an observation by Charles U. Shepard that paved the way toward the identification of the pyroxenes. In 1846 he described a mineral which, he wrote, "is a ter-silicate of magnesia . . . [and] forms more than two-thirds of the Bishopville stone"[56] (pl. 2). He named the mineral chladnite "in honor of Chladni, the scientific founder of this department of knowledge." Two years later Shepard reported his analytical results: 70 percent silicic acid, 28 percent magnesia, and 1 percent soda, so that the ratio of oxygen in the magnesia to that in the silica was 1 to 3.[57] In 1851 Sartorius von Waltershausen analyzed a fragment of the Bishopville meteorite and arrived at about the same results, but also found 1.5 percent alumina. Though making errors in his calculations, Sartorius did produce the correct formula—MgO,SiO_2; however, he postulated that chladnite was a kind of wollastonite, in which magnesia substituted for lime. The issue was confused further in 1861, when Rammelsberg found by analysis almost 3 percent alumina, 35 percent magnesia, and only 57.5 percent silicic acid. Doubting the existence of a definite mineral, Rammelsberg did not attempt to devise a chemical formula.[58]

Meanwhile, Shepard in 1854 described the Tucson iron meteorite and speculated that certain inclusions were chladnite. J. Lawrence Smith immediately corrected him, pointing out that the inclusions were actually olivine, and added a note that he suspected "chladnite is likely to prove a pyroxene."[59] At about the same time, in 1855, Gustav A. Kenngott, professor of mineralogy at Zurich, published a memoir giving details of the minerals of what he termed the "augite group" of the pyroxenes. One member of the group was enstatite, which, Kenngott wrote, was a bisilicate of magnesia, was "augitic in crystallization," and had the formula $3MgO,2SiO_3$. In 1861, when Kenngott saw Rammelsberg's analysis of chladnite, he insisted that the mineral was identical with enstatite.[60] Smith then made two analyses of the Bishopville meteorite and reported in 1864 that chladnite consisted of 60 percent silica and nearly 40 percent magnesia. He agreed with Kenngott that the mineral was the magnesian pyroxene, enstatite, and accepted Kenngott's formula, in which the oxygen content of the magnesia to that of the silica was 1 to 2.[61] Both Rammelsberg and Maskelyne acted to clarify the formula of enstatite, and through his work on the Breitenbach, Bustee, and Manegaon meteorites, Maskelyne recognized the existence of solid-solution series that included enstatite and bronzite.[62] By the 1870s mineralogists began to report regularly these constituents in meteorites.

We shall not trace later developments that led to the identification of other pyroxenes. Suffice it to say that both Maskelyne and Tschermak were in the forefront of the investigation of the monoclinic pyroxenes, and that scientists found it necessary to employ more precise nomenclature to identify what had been known as "augitic" constituents. However, we should emphasize the significance of the identification of pyroxene, plagioclase, and olivine in meteorites. It was not just that their presence in substantial quantities in some meteorites and their absence in others aided in classification. Rather, from the study of terrestrial rocks, scientists suspected or knew that these minerals would be present only under certain environmental conditions. They applied this knowledge to the formulation of theories of the origins of meteorites, a subject reviewed in chapter 5.

CONSTITUENTS: NICKEL-IRON, TROILITE, SCHREIBERSITE

Nickel-iron containing small amounts of cobalt is present in almost all meteorites. The irons consist almost entirely of nickel-iron; nickel-iron is a major constituent of the stony irons; and most meteoritic stones contain nickel-iron in quantities ranging from 1 to over 20 percent. A few stones, such as the Stannern meteorite mentioned earlier, either lack nickel-iron completely or have only trace amounts.

Twentieth-century advances in understanding the structure of meteoritic irons are described in chapter 8. The basis of current knowledge is the nickel-iron equilibrium or phase diagram, the lower temperature portion of which is shown in figure 17. There are two phases. Kamacite is α-iron with up to 7.5 percent nickel in solid solution. In this phase the nickel atoms substitute for iron atoms in a body-centered cubic structure, and the kamacite displays cubic cleavage. Meteorites known as hexahedrites contain between 5.3 and 5.8 percent nickel, and consist almost entirely of large cubic crystals of kamacite. Apart from inclusions of precipitated particles, the principal feature of most etched hexahedrite surfaces is Neumann lines or bands, which form on twin planes owing to the shock of probably cosmic collisions (fig. 18). Taenite, the

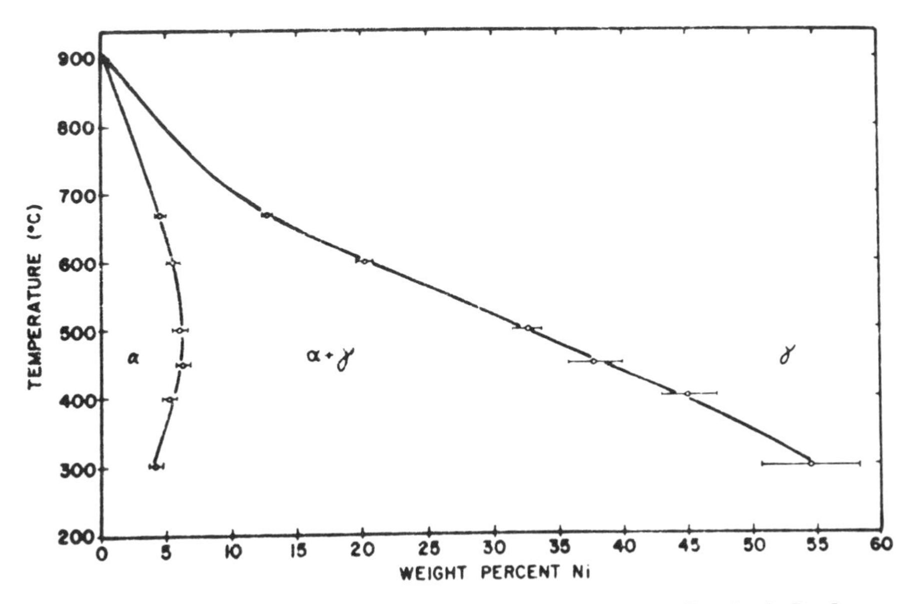

Fig. 17. Nickel-iron phase diagram, low temperature portion. From Romig, A. D., Jr., Goldstein, J. I., *Metallurgical Transactions*, Volume IIA, July 1980, p. 1155, American Society for Metals, Metals Park, OH 44073, USA.

Fig. 18. The Edmonton (Canada) hexahedrite, U.S.N.M. no. 1634. Neumann bands cross the section in at least seven different directions. Courtesy of the Division of Meteorites, NMNH, Smithsonian Institution.

other phase, is γ-iron, which has a variable nickel content of 25 to 50 percent, and which crystallizes in a face-centered cubic structure. During cooling, part of the taenite transforms to kamacite, which envelops such inclusions as troilite and schreibersite, a feature termed swathing kamacite (fig. 19). The intimate mixture of taenite and kamacite is called plessite. This assemblage of the two minerals appears in a number of different forms: as a lamellar intergrowth, as spheroids, as spindles, or as cells. A number of meteorites, having a nickel content of 20 to 30 percent, display at low magnifications only structureless plessite, and for this reason are termed ataxites.

Fig. 19. Swathing kamacite surrounds silicate inclusions in the Brenham pallasite, U.S.N.M. no. 2174. Courtesy of the Division of Meteorites, NMNH, Smithsonian Institution.

The most striking feature of the predominant class of iron meteorites called octahedrites is the Widmanstätten pattern (fig. 20), which appears on a polished and etched surface. This structure develops upon cooling over a period of hundreds of thousands or millions of years at temperatures between 700° and 350° C. Kamacite very gradually forms platelets parallel to the four octahedral planes of the face-centered cubic structure of the original taenite phase. Over long periods of time, the kamacite lamellae thicken, owing to solid-state diffusion, into plates or bands containing up to 7.5 percent nickel. These intersect mutually and produce the characteristic pattern. Some of the original taenite with increased nickel content usually remains as very thin platelets between the kamacite bands. In addition, there are areas of plessite in meteorites containing between 6 and 20 percent nickel. Recent studies using the electron microprobe reveal the nickel content of the kamacite bands, the taenite platelets, and the plessite fields (fig. 21).

Carl von Schreibers, director of the Vienna mineral and zoology cabinet, named the structure after Aloys von Widmanstätten, who, he wrote, observed it in 1808 on a polished and etched surface of the Hraschina meteorite.[63] In fact, William Thomson, who was involved in the Siena fall controversy, published a description of it in 1804.[64] He polished a section of the Krasnojarsk

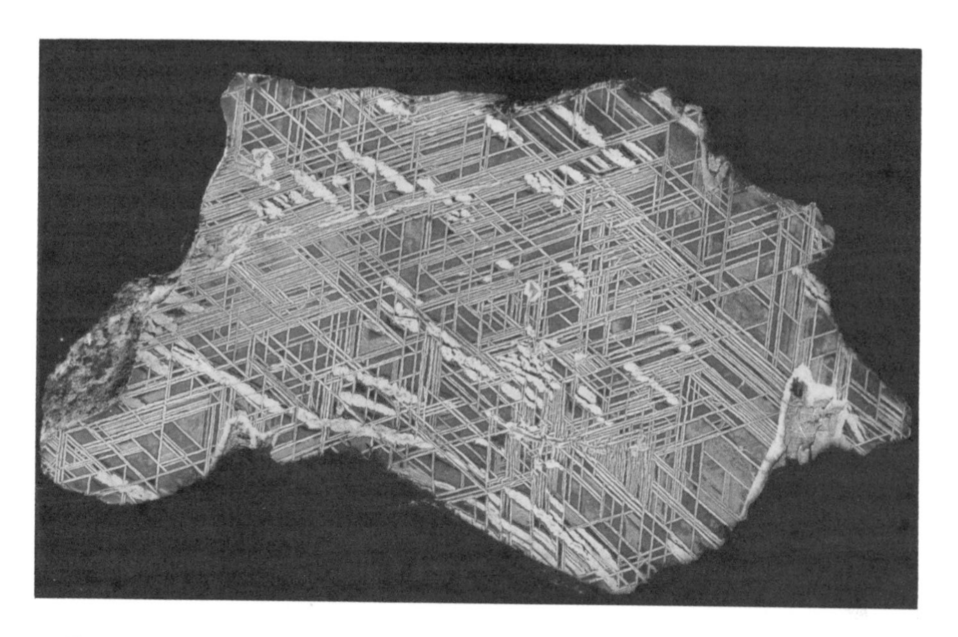

Fig. 20. Widmanstätten structure in a fine octahedrite. Edmonton (Kentucky), U.S.N.M. no. 1413. Courtesy of the Division of Meteorites, NMNH, Smithsonian Institution.

pallasite and treated it with weak nitric acid to prevent rusting. Thomson observed that there were three kinds of iron: the broad lamellae that intersected mutually at angles of 76° and 104°; thin strips adjacent to these; and material in the rhombic or triangular areas enclosed by the lamellae and strips (fig. 22, no. 5). He also found that the three kinds of iron dissolved in nitric acid at different rates: first, the broad lamellae [kamacite]; next, the iron enclosed in the geometric spaces [plessite]; and finally the thin brilliant strips [taenite].[65]

The second part of Thomson's article in the *Bibliothèque Britannique* contains his speculation that the iron had an octahedral structure, although from the values of the angles he measured, he concluded that it was not the structure of a regular octahedron (fig. 22, no. 4). He also reported that he etched a fragment of Rubin de Celis's iron [Campo del Cielo], which produced a light indication of the lamellar texture. As to the malleability of the Pallas iron, he stated that it was difficult for him to believe that the nickel increased the ductility. The malleability of the iron had not been produced by hammer forging, because it was obvious that the iron and the enclosed peridot had been in a state of fusion. He discounted the idea of a "celestial origin" and preferred instead to believe that just because 99 percent of cast iron was not malleable, there might be an exception. If a large mass of cast iron cooled very slowly, he wrote, a cavity would form in the interior, as he had observed in 1772 in a Warrington foundry. In addition, Charles Greville in 1783 had shown him a specimen of iron, which contained tiny malleable dendrites, very similar to the Pallas iron.

Thus Thomson, the first to describe the Widmanstätten pattern and to differentiate the kinds of iron present, did not believe that meteoritic iron had

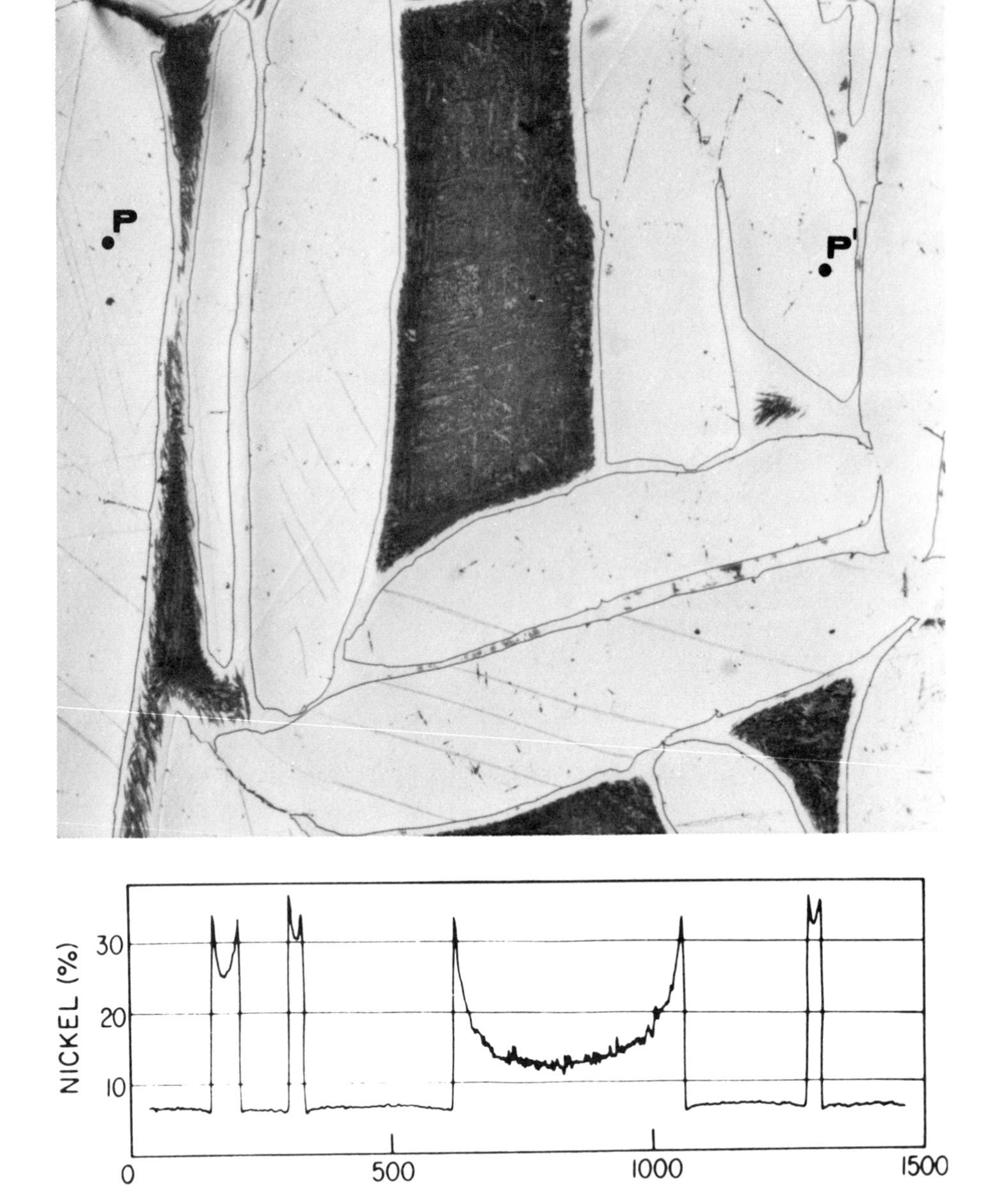

Fig. 21. Above: Polished and etched section of the Anoka iron. Below: Electron microprobe trace of the nickel concentration, which is highest in taenite and lowest in kamacite at the interface between these phases. Nickel in the plessite area (dark) fluctuates, reflecting the presence of microscopic grains of taenite and kamacite. From Wood (1964). Reprinted with permission of the author and Academic Press.

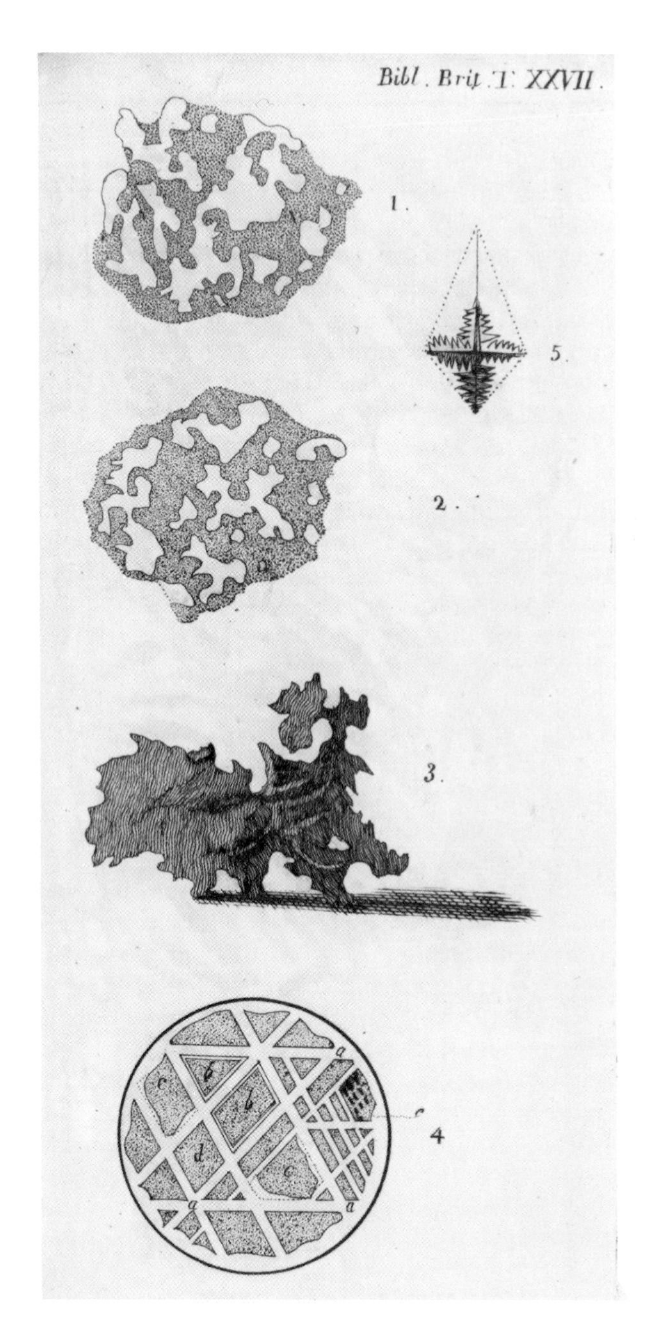

Fig. 22. William Thomson's sketches of sections of the Krasnojarsk pallasite and of the Widmanstätten structure. From Thomson (1804).

a cosmic origin, feeling that it was more likely to be a volcanic product, subject to a very unusual cooling environment. His conclusion is all the more puzzling because his reference to the nickel content indicates his familiarity with Howard's 1802 investigations, and because he mentioned Biot's report of the l'Aigle meteorite, an abbreviated version of which had appeared in 1803 in the *Bibliothèque Britannique*.[66]

By the time François P. Gillet de Laumont published an account and drawings of the Widmanstätten structure in 1815, its extraterrestrial origin (whether from the upper atmosphere or from the Moon) was unquestioned. Gillet de Laumont received his sample of the Elbogen meteorite (as well as a fragment of the Stannern fall) directly from Schreibers, and he immediately confirmed that it was attacked unequally by dilute nitric acid, producing the Widmanstätten pattern.[67] The angles formed in the pattern, he wrote, measured 60° and 120°, indicating the effect of a "law of crystallization," and he added that Schreibers considered the pattern as a characteristic of "all native irons that had fallen from the atmosphere." Comparing it with a damask pattern, Gillet de Laumont believed that the "blackish" portions contained more carbon and had solidified first, while the whiter lamellae were either free of carbon or contained very little. To test Schreibers's hypothesis, he etched several pieces of cast and forged steel, none of which showed the Widmanstätten pattern. These tests, he said, were "favorable to the opinion of M. de Schreibers." However, several pieces of cast iron from the collection of Romé de l'Isle, when etched, were attacked unequally, and revealed striations roughly analogous to the figures displayed on the Elbogen iron, forming angles of 60° and 120°. Tests on many more specimens would have to be conducted, Gillet de Laumont concluded, before Schreibers's view that the pattern was characteristic only of meteoritic irons could be challenged.

Gillet de Laumont's hypothesis that varying carbon content produced the visible pattern of differently colored lamellae did not receive support. In 1820 Joseph Moser dissolved completely with nitric acid—as, evidently, Thomson did—the wide bands [kamacite] of a fragment of the Elbogen meteorite, leaving a skeleton of the finer strips [taenite] and the intermediate area [plessite].[68] Moser, however, made an analysis, and found 4.18 percent nickel in the solution and 9.83 percent nickel in the skeleton. He thus confirmed the earlier view, proposed by Neumann in 1812, that the pattern resulted from a difference in the nickel content of the alternating lamellae.

Berzelius's meteorite publications in 1833 and 1834 complicated this hypothesis. In his investigations of iron meteorites, he found a residue containing iron, nickel, and phosphorus. While not denying that the alternating lamellae contained different quantities of nickel, Berzelius postulated that the iron-nickel-phosphorus compound was "probably the basis of the crystalline figure" of the Bohumilitz iron.[69] Following his analysis in 1834 of the Krasnojarsk and Elbogen meteorites, he made the same conjecture.[70] So great was the authority of Berzelius that later chemists concerned with the structure only rarely failed to cite his opinion. In 1894, Emil Cohen flatly rejected the idea, stating that Berzelius's conclusion was erroneous.[71] Recent studies, however, give strong support to the theory that the nucleation of schreibersite promotes the formation of the Widmanstätten structure.[72]

Several scientists postulated before the mid-1840s that the Widmanstätten pattern reflected an octahedral crystal structure. They also recognized that deviations from the values of 60° and 120° between the intersecting lamellae resulted from lack of correspondence between the octahedral planes and the surface exposed by cutting the meteorite, although it was only much later in the century that Gustav Tschermak and Aristides Brezina made a complete analysis of this feature (fig. 23).[73] Moreover, until the 1840s, very few irons that were considered to be meteoritic failed to show the Widmanstätten figure when etched. Shepard, for example, remarked in 1847 that the Walker County iron, found in 1832, did not display a Widmanstätten figure, nor did the Babb's Mill meteorite, which was ploughed up in 1842.[74] The evidence suggested, then, that Schreibers might be correct in his belief that all meteoritic irons, when etched, displayed the Widmanstätten structure.

Studies of the Braunau meteorite demolished this hypothesis. This meteorite—only the third iron observed to fall—landed in two pieces near the Benedictine monastery at Braunau, Bohemia, in the early morning of 14 July 1847. The abbot, Johannes Rotter, recognized its scientific value, and immediately permitted the larger 22-kilogram mass to be sectioned and its pieces

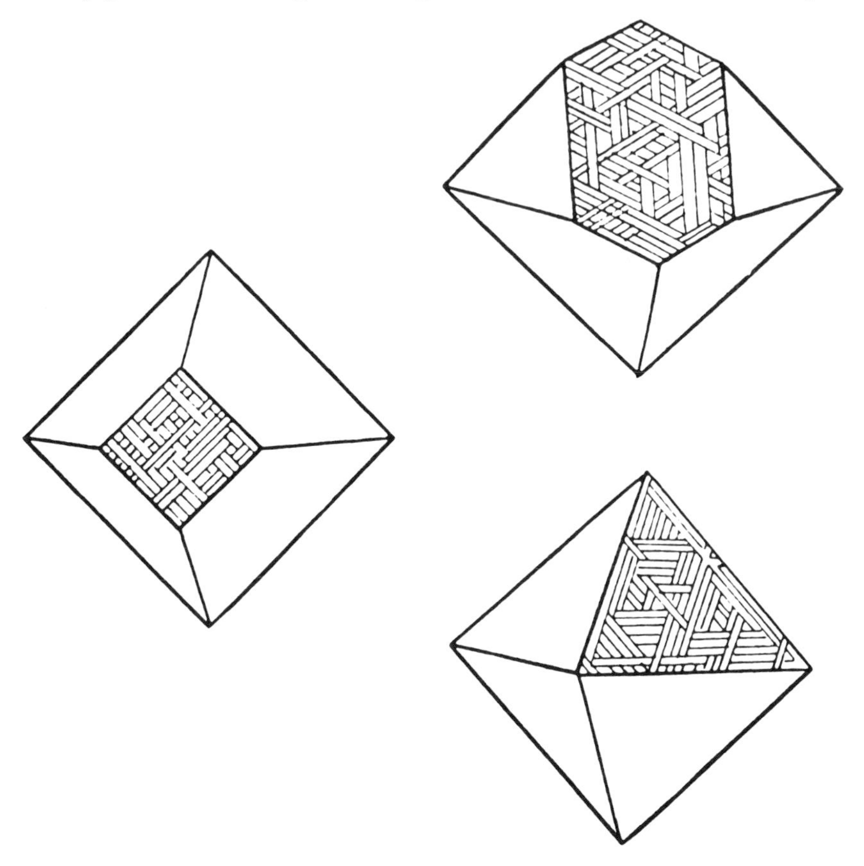

Fig. 23. Sketches by Tschermak of the Widmanstätten structure as it would appear on sections parallel to (100), (110), and (111) of the parent taenite crystal. From Tschermak (1885*b*).

distributed to a number of universities and natural history museums. In return, he received sufficient funds to establish a small hospital, so that the meteorite was to the abbot a "hypothetical gift from heaven."[75] There was an immediate flow of reports concerning the Braunau meteorite, which is a hexahedrite showing no Widmanstätten structure. The iron, Haidinger wrote in 1847, was "completely homogeneous" and displayed "perfect parting surfaces parallel to the three directions of a cube, just as galena did." The piece obtained by the Vienna Natural History Museum was "apparently a *single individual*" [crystal]. "The completely crystalline structure," Haidinger wrote, "distinguishes the Braunau meteorite from all others previously known." There were numerous fine lines visible on the sectioned surface, which Haidinger thought were traces of the Widmanstätten figure, but in 1848 Johann G. Neumann postulated that they were the result of repeated twinning.[76] "No part of the mass," Neumann stated, "is completely free of twins, since the linear design of the etched surfaces is analogous everywhere." While scientists, because of the regular orientations, accepted twinning as the cause of Neumann lines or bands, there was disagreement for several decades about the nature of the twin law, the location of the planes on which twinning occurred, and the ultimate cause of the twinning. Neumann thought that the lines came from twinning on a rhombohedral axis {111}, as in the penetration twins of fluorspar [fluorite]. Tschermak believed that the lines could originate from either of two forms, {211} or {221}, of the isometric system. He also thought that the twins formed at the same time that the nickel-iron crystallized out of a molten solution. But in 1875 Alexander Sadebeck, pointing to the fact that high pressure caused the formation of twins in calcite, postulated that shock caused the twins to form.[77] Scientists now accept that almost all Neumann bands were formed by the shock of a long past cosmic event, though a few might have been the result of entry into the atmosphere. The twin plane, as Tschermak thought, is {211}, so that twelve differently oriented lines are possible.[78]

Studies of the Braunau meteorite also served to clarify the distinction between what became known as hexahedrites and octahedrites. In 1848 Friedrich Glocker made a comparison of the Braunau iron and the Seeläsgen meteorite, unearthed before 1847 during the construction of a drainage ditch near Oder, Germany (now Poland). Glocker found that the Seeläsgen iron, while displaying the Widmanstätten structure after etching, also had the same cubic cleavage as the Braunau iron. A clear distinction, Glocker wrote, should be made between cubic cleavage and octahedral structure: "Crystal structure is a capability, a predisposition or natural disposition of substance, something primitive as an essential property inhering in the substance; cleavage, on the other hand, is something secondary, something merely superficial, which can be associated with the crystal structure or not."[79] Thus there was a dawning understanding of the difference in crystal structure between the two nickel-iron phases.

In the period from the 1820s to the 1860s several chemists used Moser's method and found, as he had, that the alternating lamellae of the Widmanstätten structure had different nickel contents. Most analysts, however, merely reported the bulk percentage compositions of iron, nickel, cobalt, and other elements they found. J. Lawrence Smith in 1855, for example, reported that the composition of the nickel-iron in the Tazewell meteorite, an octahedrite showing the Widmanstätten pattern, corresponded to the formula Fe^5Ni. In the same report, however, he presented without comment an analysis of a specimen of the Coahuila meteorite (and another of a pseudometeorite), whose nickel-iron composition bore no relationship to the formula that he postulated a few pages earlier.[80]

Reichenbach's studies in the early 1860s abruptly changed this situation.[81] There were, he wrote, two different ways to exhibit the structure of meteoritic irons: by etching with acid, or by *Anlauf*—namely, heating to differentiate colors. These two methods demonstrated that there were three types of nickel-iron, which Reichenbach called the TRIAS. The first type, of a light-gray color and actively attacked by acids, Reichenbach called *Balkeneisen,* or kamacite, from the Greek meaning "rod" or "picket." The second, consisting of paper-thick cream-colored or reddish-yellow lamellae, he named *Bandeisen,* or taenite, from the Greek "band" or "strip." The third kind, a dark gray, fine-grained constituent that filled up the spaces between the first two, Reichenbach termed *Fülleisen,* or plessite, from the Greek meaning "to make full." In some meteorites showing no Widmanstätten structure, such as Braunau, Reichenbach wrote, the entire mass was predominantly kamacite, along with a few spots of plessite. He found that the nickel-iron contained in stony meteorites was primarily kamacite, and that kamacite not only formed the wide bands of the Widmanstätten structure but also enveloped foreign ingredients. He pointed to the Zacatecas meteorite as a good example of the presence of this enveloping or swathing (*wulstige*) kamacite. Reichenbach did not make a chemical analysis of kamacite, but he did separate mechanically several taenite lamellae, which his son analyzed. The result showed 85.7 percent iron and 13.2 percent nickel—corresponding, Reichenbach thought, to the formula Fe_6Ni. Plessite, he said, predominated in the Cape of Good Hope and the Babb's Mill irons (both nickel-rich ataxites).

Most scientists readily accepted Reichenbach's nomenclature, but there were differing views concerning the status of his three constituents even after the beginning of the twentieth century. Rammelsberg maintained that nickel-iron was an isomorphous mixture of these elements, and in 1870 wrote the formula as Fe^nNi.[82] Meunier agreed with Reichenbach that taenite was a compound of definite atomic proportions, Fe^6Ni, and stated also that kamacite had the formula $Fe^{14}Ni$, and plessite the formula $Fe^{10}Ni$.[83] Both Sorby and Tschermak thought that there were just two alloys, taenite and kamacite, and that plessite was a mixture of these two. In 1891 John M. Davidson at the University of Rochester meticulously separated by mechanical means the kamacite, taenite, and plessite of the Welland, Ontario, meteorite. He was able to divide the plessite into "kamacite-like" and "taenite-like" portions, and upon analysis determined that their respective percentages of nickel and iron corresponded very closely to the percentages he found in the kamacite and taenite portions of the meteorite.[84]

Cohen, in 1894, summarized the various findings and gave his own views. "Earlier," he wrote, "I thought it probable that kamacite had a constant composition with the formula $Fe_{14}Ni$. . . . After new analyses, the validity of this assumption appears to be questionable." As for taenite, fourteen separate analyses had yielded three different formulae. There was a possibility, Cohen thought, that taenite was an alloy of constant composition that became mixed with varying quantities of iron. Plessite, in contrast, was certainly not a simple substance, as Davidson had definitely demonstrated.[85]

Other scientists, however, persisted in efforts to establish formulas for kamacite and taenite. Lazarus Fletcher in 1908 reported that his analysis of the Youndegin iron yielded the formula Fe_5Ni_3 for taenite and $Fe_{15}Ni$ for kamacite.[86] As we shall see in chapter 8, this belief on the part of many scientists that kamacite and taenite were compounds having definite proportions of iron and nickel unduly influenced early physical chemists in their attempts to construct equilibrium diagrams of the iron-nickel system. Until the 1940s there continued to be controversy as to whether taenite and kamacite were phases with variable compositions.

There were attempts, particularly in the closing decades of the nineteenth century, to duplicate the Widmanstätten structure in artificial irons. One certain conclusion emerged from these studies: although the pattern did not appear on the etched surfaces of all iron meteorites, those which did display the pattern were unquestionably meteoritic, because attempted reproductions failed. Sorby came closest to the truth of the matter. He wrote in 1877 that the Widmanstätten structure was "due to a very regular crystallization, and to the separating out one from another of different compounds of iron and nickel, and their phosphides. When meteoric iron showing this structure is artificially melted, the resulting product does not show the original structure, and it has therefore been contended that meteoric iron was never in a state of igneous fusion." This was not the case, Sorby continued, because a figure corresponding most closely to the Widmanstätten pattern could be seen in the central portions of Swedish bar iron kept for several weeks at a temperature below the melting point but high enough to permit recrystallization. In this instance, the figure resulted from a complete separation of pure iron from a compound containing some carbon, rather than from the separation of compounds containing varying amounts of nickel.[87] After further experiments, Sorby returned to the subject in 1887. From what was known, he wrote, he was led to believe "that the present crystalline structure of normal meteoric iron was developed at a temperature much below that of fusion. . . . That very profound changes can quickly take place in iron merely somewhat softened by heat admits of no sort of doubt, and further research may prove that similar great changes may take place at no very high temperature, when the time of action is indefinitely long."[88] Though Sorby believed that the phases were compounds and could not specify the temperature range in which the Widmanstätten structure formed, he was certain that a process of separation, now termed diffusion, occurred far below the melting point. His ideas came to fruition in the twentieth century.

Troilite is stoichiometric iron sulfide, FeS. It is nonmagnetic, bronze-colored, and crystallizes in the hexagonal system. In meteoritic irons, it appears as rounded or irregular nodules (pl. 3), or as rods or plates arranged symmetrically with the Widmanstätten pattern. Troilite occurs as grains or tiny veins in meteoritic stones. Thus, although common in meteorites, troilite has been found only rarely in terrestrial rocks. Pyrrhotite is nonstoichiometric iron sulfide with the formula, $Fe_{1-x}S$. It is also bronze-colored and crystallizes in the hexagonal system, but it is magnetic. Termed "magnetic pyrite" or *Magnetkies,* it was named "pyrrhotin" by Johann F. A. Breithaupt in 1835. James Dwight Dana, who often disagreed with Breithaupt on questions of nomenclature, renamed it pyrrhotite in 1869. Pyrrhotite exists in only a very few meteoritic stones and irons. Pentlandite is another sulfide mineral, with the formula $(FeNi)_9S_8$. Again, it is a yellowish-bronze color and is nonmagnetic, but it crystallizes in the isometric system. Terrestrially, it usually occurs associated with pyrrhotite, and it is found infrequently, as a weathering product of troilite, in certain iron-poor meteoritic stones. To complicate matters further, there is also common pyrite, FeS_2. It is brass-yellow, displays magnetism when heated, and crystallizes in the isometric system. But it has been found thus far in only one meteorite. Establishing troilite as by far the predominant iron sulfide in meteorites engaged scientists for more than a century. Controversy resulted both from impure samples and because of the lack of adequate crystallographic analysis.

Bournon in 1802 described iron sulfide in the Benares, Tabor, Siena, and Wold Cottage meteorites. It was nonmagnetic and reddish-yellow, and he called it "martial pyrites." Howard reported only the results of his analysis of the "pyrites" in the Benares stone. He had difficulty in extracting the material, and, as Berzelius remarked later, his analysis was flawed by the intermixture of nickel and other ingredients. Nevertheless, Howard did conclude that it contained "pyrites of a peculiar character."[89] In 1805, however, Joesph Proust detected very small particles of iron sulfide in American "native" irons and concluded from tests when they were powdered that they were magnetic.[90] Rose in 1825 was fortunate in being able to extract a crystal of the substance from the Juvinas meteorite. He found that it could not be distinguished either crystallographically or by blowpipe analysis from Magnetkies (magnetic pyrite), but he was surprised that it was nonmagnetic. He conjectured that its magnetism was masked by the presence of nickel sulfide, but having exhausted the Juvinas sample, he analyzed the iron sulfide of the Stannern stone. Finding no trace of nickel in it, Rose faced a blank wall and made no conclusions.[91]

Shepard in 1829 found what he termed a "proto-sulphuret" of iron in the Richmond meteorite. It occurred, he wrote, in distinct crystals whose fundamental form was hexagonal, and it was nonmagnetic.[92] Possibly because Shepard was just embarking on his scientific career at the age of twenty-four, he did not emphasize his belief that the substance was stoichiometric FeS, nor did he, as he would have later, give it a name. However, Berzelius in 1834 was not tentative. The iron-sulfur compound found in meteorites, he wrote, was neither pyrite nor Magnetkies, but a compound containing one atom each of iron and sulfur.[93] Further analyses appeared to settle the question. In 1848 Rammelsberg confirmed that the iron sulfides in the Seeläsgen iron had the theoretical composition of FeS, and Nicolaus W. Fischer found that the iron sulfide in the Braunau meteorite was a simple combination.[94] William S. Clark, in his inaugural dissertation under Wöhler in 1852, wrote that sulphuret of iron was a common ingredient in iron meteorites with the formula FeS.[95]

However, there were opposing views. Benjamin Silliman, Jr., in a note to the translation of Fischer's article in the *American Journal of Science,* wrote curtly: "The occurrence of both white and yellow iron pyrites in the American meteoric irons has been repeatedly mentioned both by myself and Professor Shepard."[96] And Carl Bergemann, professor of chemistry at Bonn, in 1849 noted that he observed in a 2-inch section of the Zacatecas meteorite a half-dozen yellowish-green inclusions that apparently were Magnetkies.[97] Paul Partsch, curator of the Vienna mineral collection, Bergemann continued, believed that his Zacatecas specimen contained both magnetic pyrite and pyrite, and that these ingredients retarded the appearance of the Widmanstätten structure, a conclusion with which Bergemann did not agree. J. Lawrence Smith in 1855 attempted to clarify matters. The common inference, he wrote, was that the composition of "sulphuret of iron usually designated as pyrites is the same as the terrestrial variety." This was certainly not the case, and Smith agreed with Rammelsberg that it was a simple compound of iron and sulfur. Smith's analysis showed that the substance contained 62.38 percent iron, 35.67 percent sulfur, 0.32 percent nickel, and a trace of copper. The formula FeS, he said, requires 63.64 percent iron and 36.36 percent sulfur, in close agreement with the analytical results. The occasional presence of nickel, cobalt, and copper, he added, could understandably make some chemists unsure as to whether they too were constituents of the iron sulfide.[98]

In 1861 Reichenbach entered the picture. He reported observing iron sulfide in the form of paper-thin parallel plates up to 1.5 inches long in the Lenarto iron, and in 1862 found them also in the La Caille and Tazewell meteorites.[99] Tschermak determined in 1871 that these were oriented in directions parallel to the surfaces of a cube, and in 1880 Brezina, after a complete investigation, termed them Reichenbach lamellae.[100] Reichenbach's 1862 article was entirely devoted to iron sulfide.[101] He maintained that there were four types. The first, identified by Rammelsberg as FeS, was bronze-colored, but its crystal form was unknown. The iron sulfide lines he described were of this type; there could be no doubt about its chemical analysis, and it had never been found in terrestrial rocks. Rose had found the second kind, Magnetkies [pyrrhotite], in the Juvinas meteorite. It was also bronze-colored, had the crystal form of a six-sided pyramid, and its chemical combination was unknown, although it dissolved in acid with the release of some sulfur. The third type was whitish-yellow, and both the crystal form and chemical combination were unknown. The chemical combination of the fourth type was also unknown, but it was certainly a variety of pyrite, unattacked by acids, and was readily observed in the Blansko and Lockport [Cambria] meteorites. It was spice-yellow, and although belonging to the cubic [isometric] system, it frequently had in meteorites the form of an octahedron. Thus Reichenbach swept aside Smith's confident assertion.

Haidinger in 1863 suggested that stoichiometric FeS be called troilite in honor of Father Domenico Troili, the Jesuit who reported the fall in 1766 of the Albareto meteorite.[102] Troili had stressed repeatedly, Haidinger wrote, that the fact that a stone fell could not be denied, and he had noticed iron sulfide in it, which he called "marchesita"—that is, pyrite. Rose in 1863 accepted the name troilite, stating that it occurred in irons, but added that he was uncertain as to whether part of it was Magnetkies. Yet he maintained that the iron sulfide in the Juvinas and Richmond meteoritic stones was Magnetkies, which was recognizable because of its crystal form, and considered it probable that the iron sulfide in other meteoritic stones was also Magnetkies.[103]

Rammelsberg in 1870 agreed that there were two types of iron sulfide in meteorites: troilite, FeS, and pyrrhotite, Fe^8S^9.[104] Meunier, however, was skeptical. Several analyses of the iron sulfide in meteoritic irons, he wrote in 1869, convinced him that it had the formula of pyrrhotite, so that troilite did not exist.[105] Smith, after an analysis of the troilite in the Cosby's Creek iron in 1875, charged that Meunier's analyses were erroneous.[106] But Meunier persisted. When in 1881 Brezina found a troilite crystal in the Coahuila iron whose crystal form agreed with that of pyrrhotite, Meunier saw this as additional proof that all meteoritic iron sulfide was pyrrhotite. Brezina, in contrast, interpreted his finding as demonstrating that all iron sulfide in meteorites was troilite, FeS. He supposed that terrestrial pyrrhotite was probably also simple iron sulfide contaminated by admixtures of other elements, or perhaps altered because it had formed in air instead of an atmosphere of hydrogen, as meteoritic irons had.[107]

As Emil Cohen noted in 1903, problems remained at the beginning of the twentieth century. There was good evidence that troilite and pyrrhotite were crystallographically identical members of the hexagonal system, but there were also indications that troilite crystallized in the isometric system. A few scientists maintained that both troilite and pyrrhotite had the formula FeS, while others thought that pyrrhotite contained a surplus of sulfur. Cohen himself believed that one fact was certain: the iron sulfide in both meteoritic irons and stones was identical. Cohen considered that the formula of troilite was (Fe,Ni,Co)S, although he admitted that the contained quantities of nickel and cobalt might be very small.[108] Thus, although troilite had been established as a meteoritic mineral by the end of the nineteenth century, the determination of its actual chemical composition—and of the very uncommon presence of pyrrhotite, pentlandite, and pyrite in some meteorites—awaited the more refined chemical and crystallographic techniques of the twentieth century.

Schreibersite is another meteoritic mineral that caused scientific controversy for many decades of the nineteenth century. It is a phosphide with the formula $(Fe,Ni)_3P$. The lowest nickel content found is 7.0 percent, the highest 65.1 percent. The crystals that are the first to precipitate from the taenite phase of cooling nickel-iron are low in nickel; those precipitating later have a high nickel content.[109] Schreibersite is yellow in reflected light; it is magnetic, and it crystallizes in the tetragonal system. It occurs in many but not all iron meteorites, and it is apparently a very rare terrestrial mineral (pl. 3). Rhabdite is a morphological variety of schreibersite, which appears as tetragonal prisms or as platelets in the interior of kamacite, and which has a medium-to-high nickel content depending upon the meteorite and its location within it (fig. 24).

Berzelius in 1832 was the first to investigate schreibersite from the Bohumilitz iron.[110] He noted that it was a compound containing 66 percent iron, 15 percent nickel, and 14 percent phosphorus, with minor amounts of silicon and carbon. As mentioned earlier, he postulated that it was the cause of the Widmanstätten pattern. He gave no reason for this conjecture, but it is

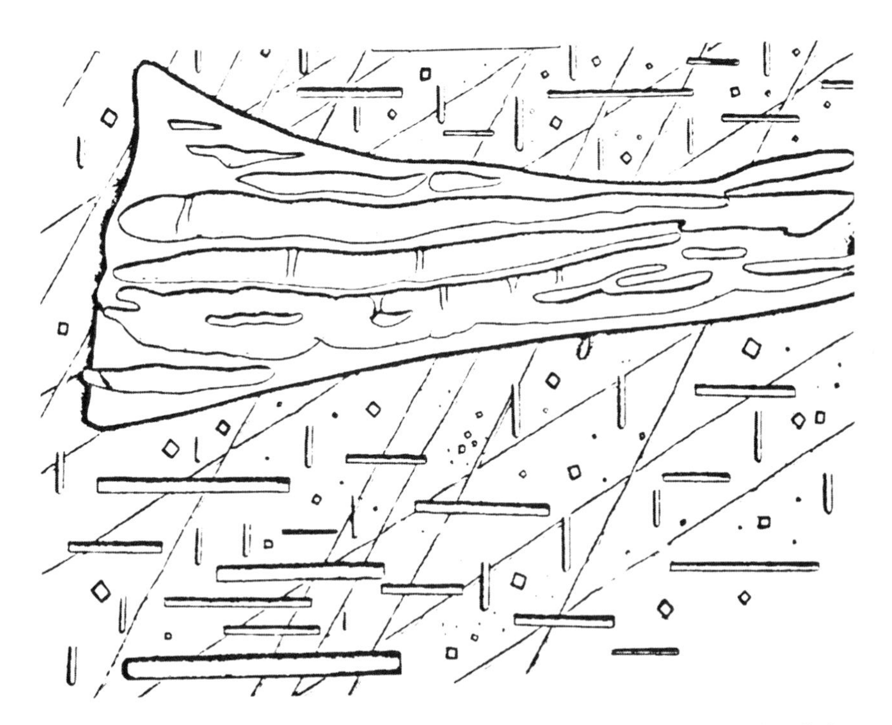

Fig. 24. Rhabdite. One of Rose's original drawings that he used in his description. From Rose (1864*b*).

possible that it stemmed from metallurgists' knowledge that an appreciable phosphorus content substantially affects the quality and properties of iron and steel. "It is worth investigating," Berzelius wrote, "whether in other types of meteoric iron the Widmanstätten figures are formed from an alloy of iron containing more nickel and with phosphorus, as is probable from this study." In 1834 Berzelius concerned himself with the "*Phosphormetalle*" in the Krasnojarsk meteorite. Upon analysis, the small, glittering metallic crystals yielded 68 percent iron, 17.7 percent nickel and magnesium, and 14 percent phosphorus. Berzelius conjectured that the constituents were combined in definite proportions, but lack of sufficient material, he wrote, prevented a more comprehensive study. Again, he postulated that the Widmanstätten pattern might be caused by the compound.[111]

Shepard in 1846 wrote that he had isolated a "blackish brown powder" from the Asheville iron, which corresponded with the "phosphoret of iron, nickel and magnesium" mentioned by Berzelius.[112] Shepard named the mineral *dyslytite,* from the Greek meaning "insoluble." In the same article, a few pages later, he named another mineral *schreibersite,* which he described as a "sesquisulphuret of chromium," consisting of brown and black deeply striated prisms.[113] It is not certain what this mineral was; Emil Cohen conjectured that it was what was later termed *daubréelite,* $FeCr_2S_4$.[114] The nomenclature became confused in the next year when Adolphe Patera and Haidinger proposed the name schreibersite for the residue found in their analysis of the Arva [Magura] iron, consisting of 87 percent iron, 4 percent nickel, and 7.3 percent phosphorus.[115] Their name was accepted in 1848 by Nicolaus Fischer, who analyzed the grayish-white, glittering platelets in the Braunau iron and found 56 percent iron, 25 percent nickel, and 11.7 percent phosphorus.[116] Fischer agreed with Berzelius that schreibersite was the cause of the Widmanstätten figure, and also postulated that the three elements were combined in definite proportions. When Shepard read Fischer's article, he immediately protested that he had already given the name schreibersite to the "sulphuret of chrome" he had found in the Bishopville meteoritic stone, and the name dyslytite to what Fischer and the "Vienna naturalists" (Patera and Haidinger) had called schreibersite.[117]

Some scientists, however, thought that the dyslytite of Shepard and the schreibersite of Patera were probably different forms of the same mineral. Clark, writing in 1852, stated that phosphorus seemed to be as characteristic an element in iron meteorites as was nickel.[118] Clark's doctoral adviser, Wöhler, thought that the phosphorus ingredient indicated a nonterrestrial origin, since it constituted 0.5 to 2.5 percent of most meteoritic irons. The form of the "phosphoret," they thought could be either a black flocky residue (the dyslytite of Shepard), thin, elastic bronze-yellow plates (the schreibersite of Patera), or brilliant, four-sided prismatic crystals. All varieties, however, consisted of iron, nickel, and phosphorus, occasionally containing carbon and magnesium.

In 1853, when reporting his analysis of the Seneca Falls meteorite, Shepard attempted to clarify matters, but only succeeded in further confusing the situation.[119] An initially silver-white component that became a brownish powder when partially attacked by aqua regia appeared, he wrote, to be identical to the insoluble ingredient in other meteorites that he named dyslytite. It contained iron, nickel, phosphorus, chromium, carbon, and silicon, and perhaps some nitrogen gas. In the residue, there was also a silver-white substance, tinged with reddish-gray, which had not previously been described, and for which he proposed the name *partschite*. This substance occurred in four-sided oblique prisms, contained iron, nickel, magnesium, and phosphorus, and "would appear to be confounded with the foliated metallic substance, called Schreibersite by Patera." But Shepard insisted that he had already given another substance that name. The partschite of Shepard and the third variety of the "phosphoret" noted by Clark and Wöhler were what Gustav Rose later termed rhabdite. Shepard never retreated, but, as Emil Cohen noted, Patera was partially to blame for the confusion, because his description of schreibersite left much to be desired.[120]

J. Lawrence Smith in 1855 reported on three analyses of schreibersite, all on samples from the Tazewell iron, and proposed that it had the formula, Ni_2Fe_4P.[121] Bergemann immediately challenged the formula and the notion that schreibersite was a compound having definite proportions by weight. He reported that the phosphorus:nickel:iron ratios of the schreibersite in the Cosby's Creek, Ocatitlon [Toluca], and Cocke County [Cosby's Creek] irons were 1:3:30; in the Arva [Magura] iron 1:3:15, and in the Misteca meteorite 1:3:6.[122] The combination as reported by Smith, then, was not general, and the iron, nickel, and phosphorus in schreibersite could be in various proportions.

Reichenbach, along with the Dutch chemist Edouard Baumhauer, was inclined to doubt the existence of schreibersite. In addition, he objected to the name because of its stridency (*Uebellautes*), and thought in any case that

Schreibers did not deserve the honor.[123] In 1861, however, after etching the surfaces of several irons, Reichenbach detected a sharply delineated, almost tin-white substance, which he called *Glanzeisen* or *Lambrit* (lamprite). It appeared extended along kamacite bands, and formed rows of lustrous, irregular blebs, or little lines or strips.[124] In the next year, he distinguished some reddish-yellow and crystalline needles, which, he wrote, were not lamprite but were possibly crystallized taenite.[125] Gustav Rose in 1863 went far in clarifying matters. He described schreibersite in detail, noting that it had been called lamprite by Reichenbach, but insisted that the name designated by Haidinger was older and had preference. He also described and named rhabdite, although he stated that it was possible that schreibersite and rhabdite were merely different appearances of the same substance. Both, he admitted, must contain phosphorus, and if it was found that they were the same substance, the name rhabdite must, of course, be abandoned.[126]

After studying the Braunau meteorite, Tschermak proposed that schreibersite and rhabdite were identical, that the rhabdite needles over time gradually became platelets.[127] However, Walter Flight at the British Museum of Natural History challenged this hypothesis. In 1882 he analyzed the phosphides in the Cranbourne meteorite and found that rhabdite corresponded to the formula $(Fe_4Ni_3)P$, whereas schreibersite had the formula $(Fe_2Ni)_7P$.[128] Beginning in 1889, however, Ernst Weinschenk made comprehensive studies of schreibersite and rhabdite, both independently and in collaboration with Emil Cohen. Part of the confusion, Weinschenk learned, stemmed from the fact that occasionally another mineral, which he named cohenite (Fe_3C), was intermixed with the schreibersite. Rhabdite and schreibersite were different forms of the same mineral. Further, as Cohen wrote, Smith's idea that the mineral was a combination of iron, nickel, and phosphorus in definite proportions could not be accepted; instead its formula was $(Fe,Ni,Co)_3P$.[129] It remained for twentieth-century scientists to study the role of schreibersite in the development of the Widmanstätten structure.

CLASSIFICATIONS

In attempting to prove the extraterrestrial origin of meteorites, early investigators such as Howard, Klaproth, and Bournon stressed their similarity by referring to the ubiquitous presence of nickel-iron. Their reports, however, acknowledged implicitly that there were three distinct types: stones; spongy irons, or irons with substantial amounts of embedded stone; and pure or "native" irons. Further, they recognized that the specific gravity, texture, and color of the stones varied, and that the tiny spheroids could be seen more distinctly in some than in others. The investigation of the Stannern meteorite in 1808 exploded the idea that all meteorites contained nickel-iron, and also showed that the spheroids were not a characteristic feature of all meteoritic stones. Thereafter, the problem facing scientists who described and attempted to classify meteorites was, just as with terrestrial minerals, whether their chemical compositions or their mineral characters should take precedence, or whether some compromise between the two could be successfully achieved. This ambivalence became apparent in the early 1820s. Nils Nordenskiöld, engineer of mines at Åbo, Finland, proposed one approach in his report on the Luotolax meteoritic stones, which fell in Finland in 1818 and which closely resembled those of the Stannern fall. Nordenskiöld could find no nickel, and reported that the fragments resembled lava, a friable aggregation of volcanic ashes. "Aerolites," he wrote, "ought to be regarded and examined, not as homogeneous masses, but as a kind of compound rocks."[130] But at the same time, André Laugier, in his report on the Juvinas meteorite, insisted that meteoritic stones could be divided into two groups with primarily different chemical compositions. The first contained little or no nickel-iron, were low in sulfur and magnesia but high in lime and alumina, and had large grains and were easily pulverized. The second group had a much higher nickel-iron, sulfur, and magnesia content, but were low in lime and alumina, and had a finer and more compact texture. Laugier also thought that all meteorites contained chromium, which would provide an infallible test of their meteoritic nature.[131] Thus Laugier stressed chemistry and Nordenskiöld mineral petrography.

Attempts to classify meteorites began in the 1840s, when sufficient numbers became available. In 1843 Paul Partsch, the curator of the Vienna imperial collection, separated the stones from the irons and divided the irons into two categories: dense or compact irons, and spongy or knotty irons containing stone.[132] Partsch further subdivided the stones into what he termed "anomalous" and "normal" groups. Placed in the anomalous group were the very unusual Chassigny meteorite, which fell in 1815, and the only two carbonaceous meteorites that were well recognized at the time—Alais (1806) and Cold Bokkeveld (1838).[133] Following Laugier's lead, Partsch separated the normal group into two sections: the magnesium-poor meteorites, such as Stannern and Luotolax; and the magnesium-rich meteorites, with the exception of the carbonaceous variety.

In 1846 and 1847 Charles U. Shepard made a much more ambitious attempt at classification.[134] There were, he wrote, two broad divisions of the science of meteorites. The first he called "astrolithology," which was a tabulation of the chemical constituents and of the mineral species found in meteorites. Shepard noted 37 minerals, and listed, by the supposed order of prevalence, 21 chemical elements claimed to have been detected. The second division was "astropetrology," which embraced the enumeration and arrangement of meteorites. Shepard separated meteorites into two classes—metals and stones—and divided each of the classes into three orders. Those of the metals were "malleable homogeneous," "malleable heterogeneous," and "brittle." He named the three orders of the stony class "trachytic," "trappean," and "pumice-like."

Shepard's classification was idiosyncratic and in several respects erroneous, but it does demonstrate the pitfalls that confronted him and his contemporaries. There were at least four pseudometeorites and several doubtful falls among the meteorites he listed. He was unaware that in two instances he placed different specimens of the same meteorite in different classes. He was unable to make distinctions among the stones and irons that later classifiers would.

And although Shepard attempted to adhere to a petrological scheme, metallurgical properties influenced his division of the irons, and chemical criteria appeared in the classification. His effort was too ambitious and probably too pretentious for his contemporaries to accept.

In the late 1840s, A. Boisse published a meteorite classification worthy of mention, because he based it primarily on petrographical criteria and on density.[135] His three classes were (1) ductile meteorites, (2) stone meteorites, and (3) incoherent or uncompacted meteoritic materials. The ductile meteorites contained only one species, meteoritic iron. He divided the stones into two species: "metal-bearing magnetic" and "deprived of magnetic grains." The two species of the third class were dry powders or colored rain or snow, and viscous or gelatinous matter. His inclusion of this class illustrates the tenacity of the notion that gelatinous matter reported to have fallen must be meteoritic in nature. Boisse made another contribution. He was the first to emphasize that the density of the meteorites—which varied, according to his figures, between 2.58 and 7.76—corresponded to the density of minerals that one would expect to find according to "the laws of affinity and of gravitation" in the successive layers of a spherical planetary body. If the body had been destroyed, Boisse added, then the fragments would have "all the varieties of composition and structure which aeroliths present to us."[136] Boisse's observation appears to have been elicited from his classification, and it lent additional support to the idea that meteorites were remnants of a destroyed planet.

Reichenbach, in 1859, brushed aside Shepard's classification and failed to mention Boisse's. With his characteristic antipathy toward anyone connected with the Vienna collection, he wrote that Partsch's classification lacked any sound basis and that Partsch appeared to be investigating the pathology of meteorites by dividing the stones into normal and anomalous categories.[137] Classification according to the mineral constituents present was the only correct basis, Reichenbach declared, but he could not employ this method because the minerals in meteorites had not been sufficiently studied. Instead, he would have to base his classification on electrochemical relations. Because those meteorites composed almost entirely of nickel-iron were the most electropositive, and those lacking iron the most electronegative, his classification reflected nickel-iron content and, ultimately, specific gravity. Initially, he listed 155 meteorites in the order of their densities, from Alais (1.70) to Asheville (7.90). Then, he separated the meteorites into nine genera, most of which were divided into groups. Although nickel-iron content was supposedly the core of Reichenbach's schema, color, presence or absence of the small spheroids, and development of the Widmanstätten structure on etching also became criteria. Emil Cohen wrote later that Reichenbach's classification was "a motley assemblage" and that it found little favor.[138]

It was probably because of the work of Gustav Rose that Reichenbach did not again pursue classification. Rose published his classification in 1863 as a description and ordering of 153 meteorites in the Mineralogical Museum at the University of Berlin, and revised it slightly the next year, when the total number in the collection reached 181.[139] The most significant features of Rose's system were his introduction of terminology, much of which is still employed, his use of primarily mineralogical criteria in classifying meteoritic stones, his careful description of many meteorites, and his reliance on detailed drawings to illustrate important constituents and features.

Rose's two classes were irons and stones. He divided the irons into three categories: (1) almost pure nickel-iron; (2) pallasites—nickel-iron and olivine; and (3) mesosiderites—nickel-iron with magnetic iron oxide, olivine, and augite. Thus the terms *pallasite* and *mesosiderite* entered the literature. His attempt to subdivide the almost pure irons into five groups differing in structure was by later standards not too successful. There were seven divisions among the meteoritic stones. The first and by far the most predominant were the *chondrites*. Rose named the tiny spheroids, mentioned repeatedly for six decades, chondrules; the stones in which the chondrules were visible, and which consisted of a mixture mainly of a magnesium silicate, olivine, and nickel-iron, were the chondrites. The second division, named in honor of Edward Howard, were the *howardites,* which consisted of an intermixture primarily of olivine and possibly anorthite. Rose, here following Berzelius, thought erroneously that the orthopyroxene in the howardites was olivine. The third division, termed *chassignites,* had only one member, the Chassigny meteorite, which was almost entirely an iron-rich olivine. The fourth division, *chladnites,* also had only one representative, the Bishopville stone. We recall that Shepard had named the magnesium silicate that he found in the stone chladnite. Rose now proposed that the meteorite type be called chladnite and that the mineral be named shepardite. This recommendation, however, was not accepted. The name enstatite took precedence over shepardite, and enstatite achondrites, such as the Bishopville stone, were later called aubrites. The name chladnite is now infrequently used to designate all orthopyroxene achondrites. The Shalka meteorite was the only member of the fifth division, aptly named *shalkites*. Rose described it as a grainy mixture primarily of olivine and shepardite. Here, as with the howardites, Rose believed that the greenish-yellow grains of orthopyroxene were olivine. The sixth division consisted of *carbonaceous chondrites*. Rose listed the Renazzo stone as an ordinary chondrite instead of placing it in this division. He remarked, however, that because of the small size of his specimen he was not sure that his judgment was correct. Rose named the seventh division *eucrites,* from the Greek meaning "clearly determinable," because the constituents of these meteorites were easily recognized. They were, he wrote, a mixture predominantly of augite and anorthite, with a very small amount of nickel-iron.

Rose's intimate knowledge of meteorites had been gained over a span of forty years and thus commanded respect. His classification system won wide acceptance, particularly in Germany and eastern Europe, and became the basis of the later classifications of Gustav Tschermak and Aristides Brezina. Rose's authority also helped to establish in the literature the names troilite, and schreibersite (instead of Shepard's dyslytite and Reichenbach's lamprite), and also the term "Neumann lines." His classification appeared just before there was recognition that enstatite and bronzite were members of the orthopyroxene solid-solution series, a defect that Tschermak corrected.

Tschermak modified Rose's system in 1872 and made further refinements in 1883. His revisions reflected the advances made in the interval in the knowledge of the mineralogy and petrography of meteorites, both by chemical methods and by the microscopic study of thin sections. Consequently, they were not trivial.[140] There were five classes, the first two comprising Rose's irons, and the other three Rose's stones. They were: (1) meteorites consisting chiefly of iron; (2) meteorites having an iron groundmass with enclosed silicates; (3) meteorites consisting mainly of olivine and bronzite with some iron, whose texture was primarily chondritic; (4) meteorites consisting mainly of olivine, bronzite, or pyroxene in varying amounts; and (5) meteorites consisting principally of augite, bronzite, and lime feldspar with shining crusts. Tschermak separated the almost pure irons into octahedrites (0), hexahedrites (H), and ataxites (D, for *dicht* or "compact"—i.e., showing no coherent Widmanstätten structure). He also listed separately two irons (Zacatecas and Cape of Good Hope) which he felt did not fit into these categories. He then subdivided the octahedrites further into groups bearing subscripts on the basis of the approximate width or character of the lamellae in the Widmanstätten figures: O_f (*fein,* fine lamellae); O_m (*mittlere,* medium lamellae); O_g (*grob,* coarse lamellae), and O_k (*krummlinig,* somewhat curvilinear figures).

Tschermak divided his second class into four groups: (1) pallasites—olivine in iron; (2) mesosiderites—bronzite and olivine in iron; (3) siderophyres—bronzite in iron; and (4) grahamites—plagioclase, olivine, and bronzite in iron. He thus added two new but transient terms to the literature—*siderophyre* and *grahamite*—the latter honoring Thomas Graham, the English chemist who had analyzed a number of meteoritic irons. Tschermak's third class corresponded to Rose's chondrites. Tschermak, however, distinguished by subscripts nine groups among the chondrites on the basis of color and/or texture; C_w chondrites, for example, had a whitish groundmass with or without whitish chondrules. He also included in this class the carbonaceous chondrites (K, for *kohlige* or "carbonaceous"), which Rose had ordered separately. The subdivisions of Tschermak's fourth class included several of the groups that Rose had distinguished among the stones: chassignites, chladnites, and shalkites. But Tschermak changed the name shalkites to *diogenites,* to honor Diogenes of Apollonia, who, Tschermak thought, had been the first to voice the cosmic origin of meteorites. He also added to this class the amphoterites, which were mixtures of olivine and bronzite, and the bustites, which were mixtures of diopside and enstatite. The fifth class comprised Rose's howardites and eucrites, and it was here that Tschermak corrected the erroneous opinion, initiated by Berzelius and held by Rose, that the orthopyroxene in the howardites was olivine.

Tschermak's classification was almost entirely petrographic, and he departed from these criteria only in employing textural characteristics among the irons and chondrites. He also followed Reichenbach in using color to differentiate the chondrites; Brezina later remarked that this device was the only useful part of Reichenbach's system.[141] Tschermak's system illustrates further that the members of the orthopyroxene solid-solution series were not yet clearly distinguished, nor were the clinopyroxene constituents. In 1884 Tschermak prepared another classification, almost identical to that of 1883, but with one major difference.[142] He reversed completely the sequence of the classification, beginning with the eucrites and ending with the almost pure irons. Tschermak wrote that classification in this way was more natural, because it corresponded to a geological sequence. According to prevalent geological opinion, the metallic core of the Earth was the oldest structure, and the silicates, as their density decreased, were progressively younger. The identical structural sequence undoubtedly occurred in the formation of other planets, so that the new ordering reflected Tschermak's conviction that meteorites were the fragments of a partially or completely demolished planetary body.

Meanwhile, other scientists were attempting to classify meteorites. In 1863, Maskelyne offered a brief tripartite division into aeroliths—stones; aerosiderites—irons; and aerosideroliths—irons containing appreciable amounts of silicate.[143] Later, he settled on the terms *aeroliths, siderites,* and *mesosiderites,* and these were used in England until the mid-twentieth century.[144] In 1867 Shepard made a final attempt to establish a classification, dividing meteorites into litholites, lithosiderites, and siderites.[145] He consulted his Greek dictionary in naming many of the subclasses and orders within these three classes, and the unfamiliar terminology was probably more responsible for the lack of acceptance than were Shepard's taxonomical principles. Again, the criteria were mixed—mineralogical, metallurgical, chemical, and textural—and again Shepard included four pseudometeorites.

In France, Daubrée also offered in 1867 an abbreviated classification.[146] His two major classes were: *sidérites,* those meteorites containing metallic iron; and *asidérites,* those lacking metallic iron. He divided the sidérites into two groups. The first, consisting only of iron, were *holosidères.* The second, containing iron and silicate, had two divisions: *syssidères,* which were predominantly iron; and *sporadosidères,* iron in isolated grains. The latter were further subdivided into *polysidères,* with substantial iron content; *oligosidères,* with small iron content; and *kryptosidères,* with insignificant iron content. Thus, Daubrée based his classification entirely on iron content, and later added explanatory notes to his subdivisions, indicating how they corresponded to Rose's system.[147] Stanislaus Meunier accepted Daubrée's nomenclature. Indeed, the names of the principal classes of meteorites, which Daubrée introduced, with the exception of asidérites, remained in usage in France during much of the twentieth century.[148] Meunier, however, united those meteorites which he thought had the same constitution and structure into over sixty types, to each of which he gave a name that was usually the place of fall or find (e.g., Erxlebenite, Richmondite).[149] He then placed these types under Daubrée's classes and divisions. Meunier's system was professedly petrographic and based on his comparison of meteorites and terrestrial rocks. His critics faulted him for not differentiating those stones that contained chondrules from those that did not; for including stones with some nickel-iron, such as the carbonaceous meteorites in the iron-free class; and for arbitrarily adding new group names to the already overburdened terminology of meteorites.[150]

Brezina expanded the classification system initiated by Rose and modified by Tschermak in a number of publications spanning the two decades from 1885 to 1904.[151] Brezina was almost too meticulous. He subdivided the chondrites, for example, into 32 groups differentiated by color, texture, and occasionally by chemical composition or mineralogical constituents, and separated the octahedrites into 11 structurally different categories. He did, however, take into account the narrow striations, called veins, that some meteoritic stones display, and introduced the term *achondrite* to designate stones lacking chondrules. He also distinguished among the octahedrites by setting up standards of measurement of the width of the kamacite lamellae in Widmanstätten patterns. Thus the width of the lamellae in a fine octahedrite (O_f) was defined as between 0.5 and 1 mm, and that in a coarse octahedrite (O_g) as between 1.5 and 2.0 mm. Brezina was also the first to use latitude and longitude consistently, and he attempted to determine the correct place-names of meteorite falls and finds, and to change those names he thought were in error. His final classification had more than 70 subgroups, many represented by only one meteorite. Twentieth-century efforts were directed at reducing the number of subdivisions and finding a more satisfactory classification basis.

The increasingly complex nineteenth-century systems of classification reflected growing knowledge of the chemistry and mineralogy of meteorites and of the details of texture and structure revealed by microscopic studies. The growing number of meteorites and the data about them were put into an orderly arrangement. Some simple explanations could immediately be made; for example, the meteorites that fell in greatest numbers were the ordinary chondrites, and there were obviously some very odd or rare meteorites, such as the Chassigny stone. A few classifications, Tschermak's for instance, reflected a preconceived theory. But did the classifications form the basis of theoretical generalizations? It is my view that no, in any direct sense they did not. They did raise a myriad of questions, however. Why do some meteorites contain chondrules and others not? Under what conditions might chondrules form? How does brecciation occur? Why do some meteorites contain little or no nickel-iron? Why do only a relatively few meteorites contain carbonaceous matter? In attempting to answer such questions—and in combining tentative answers with hypotheses that contemporary astronomers were formulating about the Solar System, the galaxy, and the distant nebulae—scientists did propose theories. We turn now to a description of several late nineteenth-century theories that sought to explain the origin of meteorites, and to others that attempted to account for the origin of life on Earth.

5

Late Nineteenth-Century Meteorite Theories

Professor Hubert A. Newton, in 1886, delivered his presidential address to the American Association of the Advancement of Science on the subject of "Meteorites, Meteors, and Shooting Stars." In the introduction he declared that reputable scientists had postulated that meteorites "came from the moon; . . . from the earth's volcanoes; . . . from the sun; . . . from Jupiter and the other planets; . . . from the comets; . . . from the nebulous mass from which the solar system has grown; . . . from the fixed stars; [and] from the depths of space." There were speculations, he continued, that meteorites "supply the sun with his radiant energy; . . . give the moon her accelerated motions; . . . break in pieces heavenly bodies; . . . threw up the mountains on the moon; . . . made large gifts to our geological strata; . . . cause the auroras; [and] give regular and irregular changes to our weather." He went on: "A comparative geology has been built up from the relations of the earth's rocks to meteorites; a large list of animal forms has been named from their concretions; and the possible introduction of life to our planet has been credited to them." And, as for their location in the heavens, Newton remarked, scientists had hypothesized that "they are satellites of the earth; they travel in streams, and in groups, and in isolated orbits around the sun; they travel in groups and singly through stellar spaces; it is they that reflect the zodiacal light; they constitute the tails of comets; the solar corona is due to them; the long coronal rays are meteor-streams seen edgewise."[1]

Newton's allusions to the various theories concerning meteorites, though concise, were sufficiently comprehensive to give his scientifically diverse audience a good idea of the wide-ranging speculations that had appeared in the scientific literature in the previous three decades. Some of those Newton mentioned were obsolete. There were no longer any scientists who held that meteorites were ejecta from active lunar volcanoes or maintained that the fall of meteors or meteorites influenced terrestrial weather. Newton's own belief, which he expounded and defended in his lecture and to which we turn later, was that meteors and meteorites were the fragments of comets. His conviction was understandable, because it was this theory that Newton had played a major role in shaping during the early 1860s.

The various other hypotheses to which Newton referred had, in the interim, been the product of scientific studies in two areas. The first concerned meteorites themselves: comparisons with terrestrial rocks and irons; investigations into those features or chemical combinations that were either absent or rarely found in terrestrial minerals; and considerations of the prevailing conditions that might have produced their varied chemical compositions and peculiar and characteristic structures. These studies also included attempts to synthesize meteorites artificially. The second area was astronomy, and embraced not only telescopic observations of the Sun, the planets and their satellites, asteroids, comets, the stars, and nebulae but also spectroscopic investigations of these objects. It was in this period that astrophysics had its genesis, and that various theories concerning the origin of the Solar System, the planets, novae, and comets began to strike deep roots. Meteoroids (then also called meteorites, for the present distinction had not yet been made) became major evidence in support of several leading cosmological theories. The ideas of Kant and Laplace began to receive close scrutiny as a result of new discoveries. We begin by sketching their views and the findings which indicated that modifications were necessary.

Before the invention of the telescope, nebulae were not objects of particular astronomical interest. The cloudy appearance of the Milky Way had, of course, puzzled natural philosophers in antiquity, and astronomers had noticed "nebulous" stars, particularly in the Andromeda constellation. Portuguese navigators, sailing in the Southern Hemisphere in the fifteenth century, recorded the nebulae that became known as the Magellanic Clouds. Galileo's *Sidereus Nuncius,* published in 1610, changed this attitude. The Milky Way, he wrote, was a collection of "innumerable stars grouped together in clusters." Further, "the stars which have been called 'nebulous' by every astronomer up to this time turn out to be groups of very small stars . . . [and] the mingling of their rays gives rise to that gleam which was formerly believed to be some denser part of the aether that was capable of reflecting light from stars or from the sun."[2] Thereafter, astronomers began to observe and record the positions of nebulae in order to distinguish them from comets. In 1771 Charles Messier in France described 45 nebulae, and ten years later cataloged a total of 103 nebulous objects. Between 1783 and 1802, William Herschel added to this number enormously. He made systematic sweeps of the sky with his increasingly more powerful telescopes, locating over 2,500 nebulae. Edmund Halley in 1715 was unable to resolve the stars in several nebulae, and postulated that they might be extended masses of self-luminous fluid that contained no stars.[3] Most eighteenth-century astronomers, however, held the view that nebulae were very distant star clusters, and that improved telescopes would permit the resolution of the individual stars. Herschel's observations gave strong support to this belief, and it was reinforced in the 1840s when William Parsons, the third earl of Rosse, employing his 6-foot metal reflector, resolved many nebulae into stars and noted that several apparently had a spiral shape. In 1864, however, William Huggins used a spectroscope to observe a nebula in the constellation Draco and detected a few isolated bright lines. From a comparison with laboratory spectroscopic data, Huggins identified hydrogen lines among these, and when he observed that a few other nebulae displayed similar spectra, he concluded that there were two classes of nebulae—those that were star clusters, and those that consisted of luminous gases.[4] Thus Halley's conjecture that some nebulae were expanses of a "lucid medium" was vindicated.

Cosmological theories proceeded hand in hand with the discoveries in observational astronomy. Immanuel Kant, in 1755, postulated that the Sun and the planets had come into existence from the condensation of primordial diffused matter—that is, from a nebulous mass. The attraction and repulsion of the nebulous particles caused the flattening of the entire Solar System into a disclike shape, which explained the fact that the planets were almost on the same plane. It was probably Kant's theory that Chladni had in mind in 1794 when he suggested that meteorites might be pieces of matter that never joined the planetary masses and wandered freely in space in all directions.

Georges-Louis Leclerc, Comte de Buffon, writing at about the same time as Kant, had a different view. He hypothesized that a large comet had struck the Sun, which caused the expulsion of incandescent matter that cooled and coalesced into rapidly rotating planets. Material flung from the still fluid planets, owing to centrifugal force, became their moons or satellites. But in 1770, when Anders Johan Lexell, a Swedish astronomer working at St. Petersburg, determined that comets had small masses and low densities, the foundation of Buffon's theory crumbled. Laplace's theory, expounded in 1796, was a more viable alternative. He proposed that the Sun's atmosphere had contracted, and that the planets and their satellites had formed from the condensation of the atmosphere in the plane of the solar equator. Laplace carefully made a distinction between the Sun's atmosphere and a nebula. Along with contemporary astronomers, he thought that nebulae were clusters of very distant stars.

The theories of both Kant and Laplace could accommodate the discoveries of new planets, new satellites, and asteroids; they were merely previously unobserved celestial masses, which had come into existence at the same time as the known planets. However, the recognition in the 1830s that space contained particulate matter, some of which occasionally penetrated the Earth's atmosphere as meteors, and some which fell to earth as meteorites, posed a problem for cosmologists. How, when, and where did this particulate matter come into existence? Further, the growing body of evidence that some comets were associated with streams of these particles, which became established fact in the 1860s, increased the dilemma. According to Kant, comets formed from portions of the primeval nebulous mass and moved thereafter in their highly eccentric orbits. In Laplace's view, they originated independently of the Solar System; those close to the Sun were captured in its contracting atmosphere, while the more distant comets continued in their orbits.

Both these hypotheses required modification in light of the coexistence of some comets in identical orbits with matter that produced meteor showers. One revision, having to do more with the consequences of the existence of meteoritic matter than with its source, was that the planets were formed by the accretion of material particles instead of by a process of condensation. In 1840, for example, Ernesto Cappocci de Belmonte, a Neopolitan astronomer, postulated that cosmic particles united due to magnetic attraction, the smaller aggregations being those which produced meteors, somewhat larger ones becoming fireballs and meteorites, and the largest forming comets that escaped the Earth's gravity.[5] John J. Waterston, in 1853, was among the first of many scientists to suggest that the Earth formed over a very long period of time by the accretion of meteoritic particles. At the same time, Waterston advanced a theory which, unbeknownst to him, Julius R. Mayer had voiced in 1848—namely, that the Sun's heat was constantly replenished by the influx and combustion of meteoritic particles.[6] In 1890 Lazarus Fletcher wrote that this theory was "generally acknowledged."[7]

The astronomers or physicists who suggested these early theories of the role of meteoritic matter neglected mention of its nature or constitution—which, since its properties were unknown, is understandable. About 1860, however, scientists who had been studying meteorites began to offer theories of their origin that they inferred from their structures and chemical and mineralogical constitutions. Though the conclusions of these scientists differed sharply, the beneficial result was that cosmologists could no longer disregard the evidence elicited from meteorites. Meteorites became, as Edward Anders remarked a century later, a poor man's space probe.

THE THEORIES OF REICHENBACH AND HAIDINGER

Karl L. von Reichenbach first proposed his theory in 1858, and during the next two years he devoted all or portions of a half-dozen articles to its elaboration. These publications are repetitive to the point of tedium, and they contain contradictions. Taken as a whole, however, they represent the first fairly coherent theory of the origin of meteoritic matter and of meteorites that was based on evidence drawn from investigations of the structure and composition of many meteorites. Reichenbach proposed his theory prior to the time when scientists examined comets with the spectroscope. Hence it displays the contemporary belief that the light emanating from comets was completely reflected sunlight.

Reichenbach's theory proceeded directly from his study and descriptions of the Hainholz mesosiderite, found near Minden in Westphalia in 1856.[8] He succeeded in obtaining for his collection "the largest part" of the 16.5-kg meteorite and considered it "the most remarkable" that had ever fallen or been found in Germany. It contained, he wrote in a preliminary note, spheroidal nodules that reached "the size of a hazel nut," and which, when detached from the groundmass, appeared as dark green, dense bodies that took a high polish. "They form obvious meteorites within the meteorite," he wrote, "and they give us valuable clues about the physical processes involved in their formation and about its geology." Shortly thereafter, he described the Hainholz meteorite more completely. At 200 magnifications, he perceived particles of iron sulfide and of metallic iron within the dark green nodules. More important, whereas in other stony irons the iron only formed a network around the silicate, in the Hainholz meteorite metallic iron also appeared as inclusions up to 22 mm in diameter. The inclusions were "necessarily older," Reichenbach concluded, because they became part of the meteorite "after they were completely formed." Similarly, the iron particles in the olivine nodules originated prior to the time they became embedded in the younger mass of the stone. Meteorites did not come into existence rapidly, Reichenbach wrote, but slowly, with each constituent taking its place sequentially. As with the study of the Earth's crust, one must differentiate the various epochs of formation. Further observations were necessary, however, to develop a kind of geology of meteorites.

The idea that there were meteorites within meteorites, giving evidence of a sequential formation, obsessed Reichenbach. His initial preoccupation with the fine structure of meteorites coincided with the appearance in June 1858 of Donati's comet, stimulating the publication of his hypothesis. Reichenbach wrote that he was among the few men in the fortunate position of having at his disposal for careful study a large number of meteorites; therefore he felt a duty to make known the connection between meteorites and "the beautiful comet, whose presence recently delighted us."[9] The known facts about comets, he continued, were that they were transparent; that they shone by reflected sunlight, since their transmitted light was polarized; and that they had no phases. Further, the tail consisted of a swarm of tiny particles that were far apart from one another, whereas the nucleus was a denser mass of the same particulate matter. Also, according to the reports of Piazzi, the contours and the inner configuration of Donati's comet did not remain the same but changed almost every hour.

A meteoritic stone, Reichenbach stated, consisted of spheroids of various sizes and colors, some of which contained smaller spheroids or inclusions of iron. These particles had become spherical by the mechanical action of abrasion against one another, and they served as the building blocks of meteorites. "Each spheroid in itself," Reichenbach declared, "is an embryo of the tangible world," and the spheroids might have existed for millions of years before becoming the constituents of a meteorite. The chemical elements from which the tiny spheroids formed, he continued, must originally have been gaseous. At some time, countless crystallizations occurred in a space having a diameter of many millions of miles, forming billions of particles. This vast expanse of particles was a nebula. Just what force caused the host of particles to move through space was unknown. However, the motion resulted in the unequal distribution of the particles in space, and interaction between the particles created denser aggregations. If the swarm entered the Solar System, it became a comet, the denser portions forming one or more nuclei.

Reichenbach thought that there were probably a hundred thousand comets that orbit the Sun, some large and some small. Meteorites were the products of the consolidation of countless small particles within the nuclei and tails of comets. They were essentially small planets and parts of the universal consolidation process. Continued impact caused the meteorites to fracture, and the resulting rubble reconsolidated. The brecciation visible in many meteorites, the poorly formed crystals, and the presence of fine dust demonstrated that mechanical forces had been at work. The changing appearance of a comet was the reflection of the constant motion, impact, and abrasion of its constituent particles. Heat, Reichenbach emphasized, was not a factor in the consolidation.

The heat that comets received from the Sun, according to authorities, was hardly sufficient to boil water. Consequently, the interior of meteorites showed no indication of the action of intense heat; iron sulfide crystals and graphite, for example, which would melt at high temperature, existed in the midst of iron. While inside a comet, meteorites were only loosely consolidated, the iron acting as a binder to unite the stony matter. When a meteorite entered the Earth's atmosphere, the atmospheric pressure forced the large and small particles together into a compact mass. Meteorites and comets, Reichenbach concluded, were one and the same phenomenon, and "meteorites are the progeny of comets."

In subsequent articles, Reichenbach attempted to clarify and strengthen his hypothesis by assembling additional evidence. He had demonstrated in his classification system that meteorites were chemically and mineralogically dissimilar and could therefore be separated into numerous genera and subgroups. To explain these marked differences, he postulated that, at the time of the initial crystallization from the gaseous state, the atoms of the composing elements were not uniformly distributed in space. In the first stage of the consolidation process, the atoms combined into molecules of silica, magnesia, iron oxide, and so forth, and later these molecules united to form, for example, olivine, augite, and labradorite. However, the nonuniform distribution of the atoms in space resulted in the formation of different minerals in different areas. "There must be and must have been," he wrote, "entire provinces in space, which were occupied only with that atomic matter and in the relative amounts from which *Stannern-type stones* could arise." The Erxleben meteoritic stone was the prototype of others that came from a different province. There were, then, "many workshops" in space, and meteorites did not come from just a single source.[10]

Reichenbach devoted a great deal of attention to the universal consolidation process. On a grand scale, it produced nebulae and swarms, the particles of which united to form meteoritic matter. Some swarms became comets, while others aggregated to form the smaller asteroidal bodies or larger planets. All of the planets and satellites in the Solar System, then, had a meteoritic origin.[11] On a small scale, every meteorite originated as a molecule and grew gradually by the apposition of other particles and dust. At a later period, particles of iron sulfide, graphite, and nickel-iron joined the growing masses, when all or almost all of the ambient oxygen had been depleted by combination with the more active bases.

Reichenbach continued to be concerned about the addition of the iron, however. He insisted that the olivine and the iron in the Krasnojarsk meteorite could never have been in the state of fusion, since their differing densities would have caused their solidification in layers. Instead, the iron formed a network around the olivine, indicating that it was a later addition. At the same time, however, he found it difficult to explain how sizable inclusions of nickel-iron were embedded in the silicate of the Hainholz meteorite, which implied that they were in place prior to the solidification of the silicate. To overcome this contradiction, Reichenbach conjectured that pieces of silicate formed in an earlier stage of consolidation had encountered an atmosphere of gaseous iron, from which particles of iron crystallized within cavities and around the silicate masses. This gradual crystallization deposited layer upon layer of iron, forming the Widmanstätten structure. This process was coeval with the formation of the cometary nuclei.[12]

In postulating how a meteorite originated, Reichenbach added, we must distance ourselves from terrestrial conditions. Everything performed in our laboratories occurred under the influence of gravity, polarity, dynamics, strong light, proportionately greater heat, and atmospheric pressure. Almost all of these agencies were absent during the formation of a meteorite. When a huge number of the mineral crystals of a comet congregated in the nucleus, they rotated around one another and pushed, pressed, and rubbed against one another, producing powder, dust, and spherules that finally aggregated into a coherent mass. Iron played its role by crystallizing in the interstices and acting as a kind of cement.[13]

Reichenbach's final two articles were recapitulations. He mentioned comets only in passing and concentrated on the formation of meteoritic stones and stony irons, citing evidence inferred from the structure of meteorites that certain constituents had been the first to form and had been united with others that came into existence later. He concluded that there were seven distinct periods or ages, beginning with the crystallization from the gaseous state and ending with the formation of meteoritic iron.[14]

Reichenbach's hypothesis incorporated both old and new elements. His nebula, composed of crystalline particles or chondrules, resembled the primeval nebula of Kant. Also, from the mid-1830s on, a number of scientists had speculated that meteor swarms were associated with comets, although there was no proof that such was the case when Reichenbach published his hypothesis. Still, Reichenbach's explanation of the intimate intermixture of constituents, of the broken crystals, and of the presence of spheroids visible in most ordinary chondritic meteorites was novel and imaginative, based as it was on the supposed processes occurring within a comet. His contention that different types of meteorites came from different "workshops" in space—presumably either different nebulae or widely separated areas within the same nebula—was a clever stroke. And his speculation that, after formation, the silicate pieces encountered an atmosphere of gaseous iron—whereby he sought to explain the fine intermixture of nickel-iron particles and the Widmanstätten structure—was certainly original. However, Reichenbach did not present a convincing picture of the mechanism of the consolidation process that produced meteorites. He barely mentioned how meteoritic irons came into existence and gave no explanation of their markedly different structures. Furthermore, he failed to make clear how and by what means meteorites were dispersed from comets.

Reichenbach's contemporaries, in particular Wilhelm Haidinger, perceived these defects. Reichenbach and Haidinger disagreed quite frequently, and the question of the origin of meteorites was not an exception. Meteorites, Haidinger wrote in 1861, appeared to have come from the same source. He stressed that "a long series of intermediate forms" connected the "almost pulverent" meteoritic stones, such as Cold Bokkeveld, with the "highly crystalline meteorites of Chassigny, Juvinas, [and] Shalka." In like manner, "a long series of structural transitions connect the non-crystalline irons," such as

Cape of Good Hope, with the "beautifully crystalline varieties" of Hraschina, Elbogen, and Lenarto, "ending with the most perfect type, that of Braunau." The crystals of olivine contained in the meteorites of Hainholz, Brahin, Atacama [Imilac], and Krasnojarsk "prove the power of crystallization to have remained active during a long period of time." These types of crystals, Haidinger asserted, "could not possibly have come into existence except under the action of *high temperature* combined with *powerful pressure*." The compressive force of the crust of a planet on the matter in its interior was "quite adequate" to produce the necessary heat and pressure, despite the fact that it was rotating in the intense cold of cosmic space. The globules [chondrules] were merely mobile particles embedded in the interior mass of crystallizing liquid. Thus, they formed "within an *already* pre-existent and voluminous agglomeration," and not, as Reichenbach maintained, "*within the vaporous dust* freely dispersed through cosmical space." The generation of gases within the shell of a planet, Haidinger suspected, and the development of extremely high pressure, could have caused an explosion that threw the planetary fragments into "progressive or retrogressive, elliptical or parabolical comet-orbits—the greater number of them into hyperbolical ones, so that, after their first perihelion, they would disappear for ever from our system." The characteristic form of meteorites, together with their cosmic velocity, he concluded, "leaves no room for any other solution."[15]

Reichenbach and Haidinger's theories represent the prototypes of late nineteenth-century hypotheses of the origin of meteorites. In one view, meteorites were fragments or ejecta from a preexistent parent body—a planet or asteroid—and had experienced a formative process there for a long period of time prior to their entry into space. In the other view, the constituent particles of meteorites had crystallized or precipitated at high temperatures from a gaseous or fluid medium, after which they experienced repeated collisions, consolidations, fractures, high temperatures, and reconsolidations until they finally survived their fiery journey to earth. We now examine the evidence used to support each of these views.

PLANETARY OR ASTEROIDAL ORIGIN

There were several scientific findings in the 1860s that influenced theorists concerned with the origin of meteorites. We have already described the crucial steps by which scientists ascertained that the orbits of some meteor streams and certain short-period comets coincided, and their conclusion that meteorites were larger pieces of a comet or stream which had survived passage through the atmosphere (see chap. 3). One drawback to this idea was that meteorites had never been reported to have fallen during a meteor shower. Alexander Herschel wrote to Maskelyne in 1876:

> It is very curious that two large fireballs clearly tracked on their paths last Sept. & shown to spring (both of them) from a good radiant point for the beginning of that month of *common shooting stars* (September "Pegasids") appear certainly to have *been detonating*

> fireballs! I never met with a case before of a detonating or so-called "aërolitic" meteor having been shown to belong to a *meteor shower* in the very satisfactory way that these two very considerable ones turn out to have belonged; & I suppose we will have to admit that shooting stars are sometimes small aërolites after all, tho' the flickering sparks & bolides of the August & November star showers have (quite oppositely) never shown the slightest symptoms of detonating or depositing a Meteorite even after the most glorious display of light, streak, sparks, etc. in its regular fireball course.[16]

It was difficult for the proponents of the cometary theory to explain just why a meteorite had never apparently fallen on such occasions.

Huggins's discovery that some nebulae were gaseous gave some comfort to adherents of the theory that meteoritic particles originated in nebulae, because a high-temperature atmosphere, partially hydrogen, could provide the reducing atmosphere necessary for the production of partially oxidized or unoxidized nickel-iron.

Thomas Graham's report that he had detected substantial quantities of indigenous hydrogen in meteoritic irons was, for two decades, a third significant influence. Jean-Baptiste Boussingault, a professor of chemistry at Paris, had determined in 1861 that the Lenarto iron contained nitrogen. Five years later, Graham, a chemist at the University of London, made a more complete investigation of the occluded gases in Lenarto and found that, when heated to redness, the gas evolved from the meteorite contained 86 percent hydrogen, 4.5 percent carbon dioxide, and 9.7 percent nitrogen. The Lenarto meteorite, Graham wrote, came from a region where hydrogen prevailed: "This meteorite may be looked upon as having imprisoned within it, and bearing to us hydrogen of the stars. The inference is that the meteorite has been extruded from a dense atmosphere of hydrogen gas, for which we must look beyond the light cometary matter floating about within the limits of the solar system."[17] Coming on the heels of Huggins's detection of hydrogen lines in the spectra of gaseous nebulae, Graham's results appeared to offer reasonable evidence that iron meteorites, at least, originated in gaseous nebulae and were not related to the particulate matter that gave rise to meteors. The required reducing medium, then, was hydrogen. The basis for this theory disappeared in 1886, when Gerrard Ansdell and James Dewar found that hydrogen resulted from the chemical reaction of water and other constituents within the meteorites.[18]

The diligent astronomical study of the asteroids in the 1850s and 1860s yielded another important influence on theory. By 1865 astronomers had found eighty-five asteroidal bodies, and they felt reasonably certain by then that the masses could not be fragments of a single planet. Also, the American astronomer Daniel Kirkwood had in the early 1860s investigated the orbits of known asteroids and found that there were gaps in the distribution of their mean distances from the Sun. Kirkwood determined that if there had been asteroids in these gaps, their periodic times would have a simple fractional relation (i.e., 1/3, 2/5, 1/2) to the time of Jupiter's orbital revolution. Conjunctions with Jupiter, Kirkwood postulated, had acted to remove asteroids into other orbits, and during this process collisions could have occurred:

> If the distribution of matter in the zone was nearly continuous, as in the sense of Saturn's rings, it would probably break up into a number of concentric annuli. On account,

however, of the great perturbations to which they were subject, these narrow rings would frequently come into collision. After their rupture and while fragments were collecting in the form of asteroids, numerous intersections of orbits and new combinations of matter would occur, so as to leave in the present orbits but few traces of the rings from which the existing asteroids were derived.[19]

In 1850 Boisse inferred from the differing densities of meteorites that they were fragments of a planet or comet that had been destroyed. He had based his theory on the probability estimates of two French scientists, Boucheporn and Boubée, that collisions between celestial bodies had occurred on several occasions, and also on the assertion of Arago that the nuclei of several comets were solid and opaque.[20] The discovery of the Kirkwood gaps provided new evidence for the hypothesis that meteorites were remnants of destructive collisions of asteroidal or planetary bodies.

The discoveries of Huggins, Graham, Schiaparelli, and Kirkwood influenced scientists favoring the two leading theories, but there was also much attention to the chemical and mineralogical constitution and to the structure of meteorites. Beginning in the mid-1860s, Gabriel A. Daubrée was tireless in his efforts to produce meteorites artificially, and in the early 1880s, his compatriots Ferdinand Fouqué and Auguste Michel-Lévy aided this research.[21] Daubrée commenced this work by melting separately the insoluble fragments of over thirty meteoritic stones. To prevent oxidation, he fused these at the melting point of platinum (1773° C) in carbon-lined crucibles. Upon cooling, the ordinary chondrites yielded large crystals of olivine and enstatite. He then fused a number of terrestrial magnesium silicates, from which olivine and enstatite also crystallized, and in the presence of charcoal yielded iron with a slight nickel content. This result led Daubrée to the conclusion that in meteorites some of the olivine had been reduced, producing enstatite and unoxidized nickel-iron. He was also able to simulate the dark veins found in many meteoritic stones by rubbing together fragments of the iron-bearing residue of the terrestrial rocks he fused. He attempted to simulate meteoritic iron by melting terrestrial iron together with nickel and iron phosphide, and though the resulting sections displayed distinct figures when etched, these figures were not as regular as the Widmanstätten structure. Alexander Herschel, in a letter to Maskelyne from Paris, made some comments on Daubrée's early work:

Mr. Daubrée showed me friction planes in the interior of a great number of freshly broken aërolites which he compares to friction surfaces, or surfaces of striation, which occur sometimes, so I understood him to say, *without a fault taking place* in the mass of strata, in some terrestrial rocks. He finds that these internal friction surfaces in aërolites are coated with very thin films of metallic iron, spread out by friction and pressure to a kind of polished vein or like varnish upon the rubbing surfaces.

He also showed me his specimens of fusions of aërolites in carbon crucibles. The crystallization of the enstatite in prisms in some of them is very beautiful and perfect. It is clear that this is not the way in which the original meteorites consolidated, particularly as the iron collects itself into buttons instead of being scattered here and there through the mass.[22]

Daubrée's efforts rested on his assumption that meteorites originated by an analogous though somewhat different process than had terrestrial rocks.

Despite his inability to synthesize meteorites completely successfully, he felt justified at this early stage in offering several tentative conclusions. Carbon was not the reducing agent present during the formation of meteorites, because carbon in the form of graphite was rarely found in them. Nor was hydrogen, because its presence ought to produce water and hydrates, and these had been found only in the Orgueil carbonaceous meteorite. Meteorites, Daubrée wrote, had experienced a process of scorification accompanied by incomplete oxidation, in which such active elements as magnesium and silicon were completely oxidized, leaving such less active elements as iron and nickel partly or totally uncombined or alloyed with each other. The Earth's crust had been subjected to a much more complete process of scorification than had meteorites, resulting in the production of phosphates and sulfates instead of phosphides and sulfides, and in the absence of metallic iron.[23]

After Graham's publication concerning the presence of hydrogen in iron meteorites, Daubrée acknowledged that hydrogen was undoubtedly present during the formation of meteorites and had acted as a reducing agent. He also expressed his assumptions and views more explicitly. The study of meteorites permitted comparison of the constitution of celestial bodies with that of Earth, and thereby clarified the Earth's origin. The evolution of planetary bodies from nebulous matter appeared to have been similar everywhere, so that "the history of our globe is like a recapitulation of the general history of the universe." Moreover, meteorites were fragments of a destroyed celestial body. The sequence of the formation of meteoritic stones was (1) the solidification and crystallization of mineral particles, in which small grains of nickel-iron clustered around the silicate globules; (2) the formation of breccias and conglomerates; (3) the production of friction surfaces or veins; and (4) the production of incoherent fragmentation such as would occur after explosion or violent shock. The absence in meteorites of sedimentary rocks, granite, gneiss, or mica could be explained in two ways: they came from the interior of planets constituted like Earth, with iron meteorites having formed the core, or, alternatively, the planets that produced meteorites lacked these types of rocks.[24] Fouqué and Michel-Lévy improved upon Daubrée's techniques to the extent that they were able to replicate eucrites and howardites. However, their attempts to simulate the brecciated structure in meteorites failed. The frequency of brecciation in meteorites, they thought, was due either to the explosive movement that dispersed them into space, to the high temperature that accompanied their formation, or to the enormous pressure to which they were subjected when they entered the atmosphere.[25]

Gustav Tschermak also held the view that meteorites originated on a planetary body or bodies, but his hypothesis was quite different from that of Haidinger or Daubrée. The collision of two even small asteroidal bodies, he wrote, would cause their disintegration, but the resulting pieces would be far larger than any known meteorites. This would also be the case if a planetary body disintegrated following an internal explosion. Instead, Tschermak thought that "active volcanic agencies" could eject meteorites from any cosmic mass small enough that its gravitational force was "insufficient to attract again to its surface all the fragments which have been hurled from it." Tschermak was well aware of the fact that his idea was a variation of the "Laplacean" hypothesis that meteorites were ejecta from lunar volcanoes. However, he insisted that many cosmic bodies might meet the requirements, thus accounting for the fact that meteorites fell to earth from many directions. For Tschermak, though, the chondrules provided the most compelling evidence that a volcanic agency was involved:

> These chondra bear no indications of having obtained their spherular form by crystallization. . . . They resemble rather the spherules which are frequently met with in our volcanic tuffs, . . . more especially the spherules of olivine of Kapfenstein and Feldbach in Styria . . . [which] we know . . . to be the result of volcanic trituration, and to owe their form to a prolonged explosive activity in a volcanic "throat," where the older rocks have been broken up and the tougher particles have been rounded by continued attrition.
> The character of the meteoritic chondra indicate throughout a similar mode of formation.

The significant amount of brecciation also supported the idea of a volcanic origin. Moreover, Tschermak added, some of the tufaceous meteorites gave evidence of later modification by heat, and others indicated that a chemical change had occurred in a reducing atmosphere, most likely of hydrogen. He admitted, however, that the volcanic ejecta from a planet consisted only of disintegrated solid rock, because no meteorite resembled a volcanic slag or lava. Meteorites, then, were catapulted from numerous cosmic bodies as a result of the sudden explosive action of volcanic gases, among which hydrogen must have been prominent.[26]

Tschermak's hypothesis caught the attention in 1879 of Sir Robert S. Ball, Astronomer Royal of Ireland. He calculated that the largest asteroid, Ceres, would have barely enough energy to put a volcanic fragment into an Earth-crossing orbit, and that it would have to expel over fifty thousand fragments for one to reach Earth as a meteorite. Nevertheless, Ball thought Tschermak's idea had merit, and he was particularly impressed by Tschermak's description of the similarity of meteoritic stones and volcanic tufa. The planet that ejected meteorites millions of years ago, Ball wrote, must have been Earth. Volcanic ejecta from Earth with sufficient velocity would orbit the Sun and cross the Earth's orbit at each revolution. The perturbation of their orbits would cause the fragments to travel for thousands of years before returning to Earth.[27] Though Ball held a prominent position, his theory found no support in scientific quarters.

Despite Ball's probability estimates, Tschermak was not discouraged from thinking that meteorites of the same constitution had been ejected from the same parent body and continued to move in the same Earth-crossing orbit until they finally entered the Earth's atmosphere. He attempted to correlate the dates of fall of certain structural types of meteorites, in particular the howardites and eucrites. He turned for help to the Viennese astronomer Edmund Weiss, who, when furnished with the fall dates of the Stannern, Jonzac, Juvinas, and Petersburg eucrites, worked out the orbital details and gave Tschermak a formula for the relationship of the longitude of the orbital node and the date of fall. On the basis of this information, Tschermak predicted that a eucrite would fall toward the end of October 1899. One did fall on 24 October 1899 at Peramiho (Tanzania), East Africa. This event, of course, reassured Tschermak, and resulted in a slight revision of the formula. However, Gustav von Niessl, astronomy professor at Brünn and an authority on meteor orbits, attempted to compute the original orbits of these eucrites, but found that there was no common point of origin. The Peramiho fall, then, was just pure coincidence. Also, Tschermak had to admit that the ordinary chondrites presented difficulties for his theory; those that had fallen in mid-February, for example, displayed considerable structural differences.[28] Tschermak's hypothesis, therefore, bore no immediate fruit.

Another hypothesis, expressed in the 1870s by two prominent scientists, held that meteorites came from comets but also linked comets to a shattered planet or to disintegrated asteroids. Johann C. F. Zöllner, professor at Leipzig and a pioneer astrophysicist, objected to Schiaparelli's theory on the grounds that it did not explain adequately the light phenomena associated with the passage of comets. If a planet exploded, Zöllner postulated, the debris would include atmospheric and internal gases and surface water. These substances would freeze immediately in the intense cold of interplanetary space, and would move in the same orbit as the solid fragments. When close to the Sun at perihelion, the gases would vaporize, burning intensely in the nucleus and streaming out to form the comet's tail. Thus the light phenomena could be readily explained.[29]

Arthur W. Wright, professor at Yale University, arrived at a somewhat similar conclusion through extensive investigations of the gases in meteorites. His chemical analyses and spectroscopic studies showed that hydrogen was much more abundant in iron meteorites than in meteoritic stones, the latter containing much more carbon dioxide and carbon monoxide.[30] Further, he determined that the spectra of all of the occluded gases in meteorites were "clearly identical" with those of several comets. The volume of gases contained in a large meteorite or in a cluster of such bodies, Wright declared, was "sufficient to form the train as ordinarily observed." A comet was merely a meteorite of considerable size or a swarm of many smaller ones, containing large quantities of carbon dioxide and some hydrogen and carbon monoxide, and emitting these gases as a result of solar heat. The gases in the train of a comet were visible due partly to reflected sunlight and partly to light resulting from their combustion.[31] Wright went on to suggest that some of the short-period comets might have derived from asteroids. He cited Kirkwood's discovery of the gaps in the asteroid belt and postulated that the effect of Jupiter's gravity had thrown the missing bodies into highly eccentric orbits around the Sun: "It is not difficult to see that these great changes of temperature in a mass of considerable absorptive and low conducting power must give rise to powerful stresses, and that under the intense action of the sun near the perihelion the action may be sufficiently energetic to cause the splitting up of the bodies themselves." Some of the pieces of an asteroid became short-period comets that continued to disintegrate, resulting in the evolution of large volumes of gas and in the production of "spent fragments," which were meteorites.[32]

In his article, Wright did not refer specifically to the "Roche limit." In 1848 Edouard Roche had formulated a mathematical expression denoting the minimum distance that must separate a satellite from its planet to prevent disintegration. Roche's proof apparently came to the attention of many astronomers in the 1870s, and it was used during the next decade to explain the dispersion of cometary matter at perihelion. Breakup of an asteroidal body by the tidal forces of the Sun or of a major planet, then, became another possibility. It is somewhat surprising that theorists at this time did not consider the probability of a grazing collision of asteroidal bodies. Telescopic observations had revealed the fact that some of the asteroids were not spherical, but scientists used this fact to support the idea of the disintegration of a major planet. Perhaps there were so relatively few asteroids known to exist at the time that scientists felt there was very little probability of such occurrences, given the vast space separating the orbits of Mars and Jupiter. To many scientists, however, nebulae and comets seemed to offer a more reasonable source of meteorites than either the partial or total disintegration of planetary bodies or intense volcanic action within them.

ORIGINS FROM THE SUN, NEBULAE, OR COMETS

The results of spectroscopic studies in the 1870s and 1880s were the principal foundations of hypotheses that placed the source of meteorites in nebulae or in the Sun, and they also gave further support to the theory of meteorites' cometary origin. The spectroscope, originally considered an instrument for terrestrial chemical research, soon became indispensable in astronomical investigations. In 1861 Kirchhoff and Bunsen mapped the solar spectrum, and when they compared it with laboratory flame and spark spectra, they discovered the elements caesium and rubidium. Three years later, Huggins, as we have noted, discovered gaseous nebulae. Then in 1868, Norman Lockyer detected a new yellow line in the Sun's spectrum which he attributed to a hypothetical element helium, found twenty-seven years later by William Ramsay in the Earth's atmosphere. Despite these triumphs, however, scientists in the early decades of spectroscopy often found themselves at odds concerning interpretations of their results. It was an arduous task to examine a faint nebula in any detail before the introduction in the 1880s of dry plate photography. There was continual controversy about the precise wavelengths of lines before Henry Rowland in 1887 replaced the prism of the spectroscope with the concave diffraction grating. Also, scientists found that it was very difficult to measure the wavelengths of the observed lines of such a rapidly moving object as a meteor. With respect to meteorites, hypotheses about their origin depended largely upon a scientist's perception of the identity of the spectra of a number of meteorites and the spectra of certain self-luminous celestial bodies. Given the uncertainties, it is little wonder that different theories about meteorites emerged, and that some scientists viewed the results of spectroscopic studies with a good deal of skepticism. Spectroscopy did not settle matters, but it did add valuable new evidence.

Henry Clifton Sorby's publications on the origin of meteorites illustrate how spectroscopic studies influenced hypotheses. Sorby's initial series of articles appeared in the mid-1860s. He first challenged Reichenbach's idea that pallasites could never have been in a state of fusion. According to Reichenbach, if this were so, the iron would have sunk to the bottom of the melt and the olivine would have risen to the top. The difference in density, Sorby wrote, depended upon the gravitational force: "On the surface of a small planetary body, or towards the interior of a larger, iron and olivine might remain mixed in a state of fusion long enough to allow of gradual crystallization." Pallasites, then, "should be considered evidence of fusion where the force of gravitation was very small."[33] His next communication touched on the results of his microscopical studies. Olivine, he said, contained "glass-cavities," just as lava did; hence it had once been in the state of igneous fusion. It also had "gas-cavities" similar to those in volcanic minerals, which indicated the presence of a gas during solidification. To see either type of cavity, he wrote, one must resort to the preparation of thin sections and study these at a "magnifying power of several hundreds." In many meteorites "the constituent particles have all the characters of broken fragments," which "sometimes give rise to a structure remarkably like that of consolidated volcanic ashes." Sorby continued: "It would therefore appear that, after the material of the meteorites was melted, a considerable portion was broken into small fragments, subsequently collected together, and more or less consolidated by mechanical and chemical actions amongst which must be classed a segregation of the iron, either in the metallic state or in combination with other substances." This process of consolidation, wherein all evidence of a previous history was obliterated, Sorby termed "metamorphism."[34] In some cases, Sorby thought, the fragmentation occurred after crystallization, but in others the detached globules might have been the result of a disintegration while the material was still molten. Certain peculiarities, then, connected meteorites with volcanic rocks, while in other respects they differed markedly, and these facts had to be kept in mind "in forming a conclusion as to the origin of meteorites."[35]

From the foregoing, one might surmise that Sorby was at least leaning toward the hypothesis that meteorites were planetary fragments, but in 1866 he voiced a quite different opinion.[36] From his examination of the microstructure of over one-half of the known meteorites, he concluded that they were "formed by the aggregation of small particles, which at an earlier period, existed as drops of melted glass," or as "ultimate cosmic globules," since that appeared to be the earliest condition of matter. Previously, these particles had been gaseous, and they condensed just as water vapor condensed to form raindrops and snowflakes. Thereafter, the particles aggregated into masses that experienced metamorphic action and were broken "by mechanical violence." The larger planets, Sorby conjectured, might have been formed by the aggregation of such particles, and the small masses "moving between Mars and Jupiter" might be called large meteorites that had not yet united to form a single planet. Thus Sorby's conclusion concerning the origin of meteorites, though not spelled out in detail, was very close to that of Reichenbach. However, it was not Reichenbach's speculations but Huggins's discovery of gaseous nebulae that influenced Sorby. The existence of gaseous nebulae gave new life to Kant's century-old theory. The title of Sorby's article, "On the Physical History of Meteorites in Connection with the Nebular Theory," reveals this fact.

In 1877 Sorby returned to the subject. The cometary hypothesis, he pointed out, could not explain certain facts.[37] To begin with, meteorites must have had an igneous origin, because he had never detected in them any fluid cavities, and further, some constituent particles were originally detached glassy globules, like drops of fiery rain. The brecciated structure that commonly occurred was the result of mechanical force of great intensity. The subsequent consolidation was a "great difficulty in the way of Reichenbach's cometary theory," since there was not sufficient gravitational force in a comet to effect consolidation. After consolidation, meteorites must have been heated to temperatures just below the melting point, causing metamorphism in the meteoritic stones, after which some were again broken and reconsolidated with fresh material. To produce the Widmanstätten structure, the meteoritic irons must have been kept for a very long time just below the melting point. "All these facts," Sorby wrote, "agree so admirably with what we know must now be taking place near the surface of the sun, that I cannot but think that, if we could only obtain specimens of the sun, we should find that their structure agreed very closely with that of meteorites." On the one hand, meteorites could well be fragments of the Sun detached from it in the past by violent disturbances. On the other hand, meteorites might be "residual cosmic matter, not collected into planets, formed when the conditions now met with only near the surface of the sun extended much further out from the center of the solar system." But these conclusions, Sorby emphasized, must be viewed as "only provisional."

One of the most thoughtful contemporary summaries of available data for use in the formulation of a theory of meteorite origin was never published. In 1875, Nevil Story-Maskelyne wrote a series of articles on meteorites, which appeared in *Nature,* and subsequently he decided to use them as the basis of a book on the history of the science of meteorites.[38] In 1879, however, upon the death of his father, he inherited substantial properties in Wiltshire, and in 1880 he resigned his position as keeper of minerals at the British Museum. Thereafter, politics and estate management became his chief occupations, and he never saw the manuscript through to publication. The sixth and last chapter focused on the origin of meteorites, and in contrast to Reichenbach, Tschermak, and Sorby, who were concerned with the structure of meteorites, Maskelyne concentrated on their chemical and mineralogical composition.[39]

Maskelyne thought that three closely allied theories deserved the most consideration: (1) the Sun or the stars had thrown meteoritic materials into space; (2) the slow cooling of nebular vapor had formed a "meteoric star dust," with nickel-iron being the last of the constituent elements to condense; (3) meteoritic dust had revolved, cometlike, in a circumsolar orbit and experienced changes in its constitution by concentrating into aggregations, which had periodic encounters with a glowing atmosphere of hydrogen mingled with gaseous iron and other metals. These theories, he wrote, raised "baffling questions" that no one would have thought of asking seriously twenty years before. Maskelyne was optimistic that the next generation might see their solution, since there had been "substantial steps in obtaining the answers."

Maskelyne then listed more than a dozen inferences that could be made, given the chemical and mineralogical compositions of meteorites. Meteoritic matter had existed at some period at an "enormous temperature," but had cooled under conditions that permitted complete crystallization. The same elements that predominated in terrestrial rocks composed meteorites, although the elements did not have the same relative importance in the Earth's crust and in meteorites. The most remarkable differences were (1) the elements iron and nickel, which were "in excess in meteorites"; and (2) oxygen and calcium, which were present in larger relative amounts on the Earth's surface. In these respects, he wrote, meteorites "hold a middle place between the terrestrial rocks and the fervid elemental gases on the surface of the sun and of the hotter stars." In the latter, however, the absence both of silicon and silicates and of oxygen, phosphorus, sulfur, and chlorine, which were present in small amounts in meteorites, complicated the problem, "while the presence in such large amounts of magnesium in the solar chromosphere, would not accord with an identity in composition of the purely iron meteorites with the gaseous solar envelope." Here Maskelyne was referring to the results of early spectroscopic studies, in which investigators estimated the relative importance of elements in the Sun or stars by reference to the brightness of the lines.

The nature of meteoritic minerals, he continued, indicated that the oxygen originally present was sufficient to oxidize such elements as magnesium, calcium, aluminum, and silicon, as well as all of the small amounts of chromium and usually some portion of the iron. However, phosphorus, sulfur, and most of the carbon must have "remained in dissociation from oxygen under conditions (probably of enormous temperature), where even some of the iron passed into oxidation in the form of ferrous silicate." The iron thus oxidized to form the silicate, he wrote, was "taken up at the expense of a certain amount of nickeliferous iron, which thereby became richer in nickel" and remained mechanically entangled in the silicates. In this statement, Maskelyne came close to enunciating the "rules" voiced in 1916 by one of his successors at the British Museum, George Prior. They were: first, "the less the amount of nickel-iron in chondritic stones, the richer it is in nickel," and second, "the richer in iron are the magnesium silicates."[40]

Maskelyne suspected, as Reichenbach had, that in the meteoritic stones the chondrules and crystalline silicates had formed prior to the time that the metallic nickel-iron particles aggregated in their midst. With respect to iron meteorites, he thought that the troilite and the nickel-iron had consolidated simultaneously, and that the varieties of schreibersite had crystallized at the time of solidification. The types of minerals formed depended, of course, on the presence of various elements: olivine, enstatite, and quartz depended on the presence of silicon and magnesium; augite on the presence of calcium; and anorthite and the other feldspars on the presence of aluminum and the alkalis. But at the same time, these minerals "took up into their composition more or less of corresponding ferrous silicates, according to the excess which the original supply of oxygen afforded for the oxidation of the metallic iron."

Among the many dilemmas that meteorites presented, Maskelyne found three to be more perplexing than the rest. One was the existence of what he termed "chladnitic aërolites," now called enstatite achondrites or aubrites. "The aspect of the minerals," he wrote, "the absence from them of any appreciable ingredient of ferrous silicate, while nevertheless nickel-iron is present in some small amount, leaves the impression that either they have undergone some powerful reducing action subsequent to their original formation, (which the small amount of iron present renders improbable), or that they have not become charged with iron and ferrous silicates as others and especially the chondritic aërolites have been." The second puzzle was that some meteoritic irons contained occluded hydrogen, which, he said, lent plausibility to Daubrée's view that the nickel-iron in meteorites resulted from a reducing process. But other irons did not contain hydrogen, and moreover, some stony irons contained occluded carbon dioxide. Thus the only conclusion that could be made was that "different kinds of meteorites have different histories." Finally, there were just a few meteoritic stones, "generally black from the presence of carbon," having "a crystallisable paraffin-like substance contained in the pores of the stone, but external to its component particles of crystallisable minerals." The quantity of the substance was so minute, Maskelyne stated, that one might suppose it had been absorbed while the stone was in "an atmosphere composed of such a hydrocarbon in the state of vapour." At this rather crucial point, the manuscript ends. Perhaps Maskelyne thought that it would be straining credibility to suggest that while some of the "meteoritic star dust" encountered a "glowing atmosphere of hydrogen," other portions passed through clouds of gaseous hydrocarbons, and still others through regions of carbon dioxide. In any event, it is a pity that he never completed and published the manuscript.

Nevil Story-Maskelyne and Sir Joseph Norman Lockyer offer a contrast in scientific styles. Both were good though not outstanding scientists. But whereas Maskelyne was content to assemble and ponder over data and finally get around to presenting his findings (his work on enstatite was published belatedly), Lockyer put his speculations into print immediately.[41] Sometimes there was a handsome reward, as in the case of the discovery of helium in the Sun. In other instances, however, Lockyer's haste resulted in faulty generalizations. His "Meteoritic Hypothesis" sought to give a single explanation for celestial astronomical phenomena, and it failed. To present all of his work and ideas on the subject would lead us far astray, but we should review those parts of the hypothesis that bear directly on the subject of meteorites.

In the 1860s Lockyer mapped the spectra of numerous meteorites, but only for the purpose of aiding his investigations of the solar spectrum. It seemed to him that an element gave a different spectrum when investigated alone than when it was part of a mixture of other elements. An adherent of Waterston's theory, he thought that the influx of meteoritic matter into the Sun produced sunspots, and he supposed that the mixture of elements in meteorites would show a spectrum similar to the mixture in the solar spectrum. He therefore proceeded to compare the laboratory spectra of meteorites with the solar spectrum.[42] Beginning in 1874, however, the purpose of his spectroscopic investigations of meteorites changed drastically. During a summer trip to northern England and Scotland, he stayed for a few days with Peter Guthrie Tait, professor of natural philosophy at Edinburgh. Lockyer, not only as an astronomer but also as editor of *Nature*, was familiar with Tait's ideas about comets, a subject of timely discussion because of the appearance in that year of Coggia's comet. In 1869 Tait had published a mathematical analysis of the changing contours of comets, on the assumption that a comet was a cloud of small stones and fragments of iron. His study apparently showed that the sudden development of tails, the occurrence of comets with numerous tails, and the lack of relation between the direction of a comet's tail and its solar radius vector were due to the differing motions of the small fragments. Further, Tait postulated, the motions of these fragments caused their mutual impact, which created heat and caused the release of luminous gases, accounting for part of the comet's light.[43] Tait's analysis impressed Lockyer, particularly after he made detailed telescopic observations of Coggia's comet at the home of a friend on his way back to London. There Lockyer learned, probably from Maskelyne, about Reichenbach's cometary theory. His summer activities stimulated him to inaugurate a comprehensive program to study the spectra of meteorites and of all self-luminous celestial phenomena, which culminated in 1890 in the publication of *The Meteoritic Hypothesis*.[44]

Lockyer's general conclusion was that "all self-luminous bodies in the celestial spaces are composed of meteorites, or masses of meteoritic vapour produced by heat brought about by condensation of meteor swarms due to gravity. . . . The existing distinction between stars, comets, and nebulae rests on no physical basis."[45] The heat-producing collisions of meteorites, the average distances separating them, and the numbers of meteorites constituting a swarm were the three most important factors that created the distinction, because these conditions governed the temperature of the space occupied by the swarm and hence the type of spectrum that the luminous object produced. Meteorites that fell to earth were artifacts of the processes occurring in celestial bodies. They were formed by the condensation of gases released by collisions, grew by the fusion brought about by additional collisions, and in some cases attained such a size that the heat of impact was not sufficient to volatilize the entire mass. The heterogeneous structure of a meteorite reflected its history. Further, because the only distinction between a shooting star and a fireball that dropped a meteorite was that of size, all meteorites had once been members of a still-active cometary swarm, or of one that had become dispersed in space.[46]

When Lockyer began his intensive studies of the spectra of meteorites, he was convinced of the validity of the assumption that the luminosity of comets was due in large part to the heat generated by the collisions of meteorites, either from internal motions or from encounters with other meteorites. His comparison of the spectra of meteorites at low temperatures—that is, with the Bunsen burner and oxy-coal gas flame as heat sources—and of the spectra of comets appeared to confirm his point of view. He did detect the change of a comet's spectrum to a higher temperature form as it approached the Sun, and ascribed this change to an increased number of collisions stemming from the Sun's tidal action. What Lockyer did, however, was to combine the spectra of the most important elements in meteorites until they matched the spectrum he had mapped of a given comet.[47] When the spectroscopic observations of others, such as Alexander Herschel in England and Miklós von Konkoly Thege in Hungary, corresponded with his own, Lockyer cited them approvingly. But when other astronomers disagreed with his results, Lockyer attributed the mismatch to the difficulties inherent in making observations or to the sudden changes in the spectra of comets.[48]

Lockyer's hypothesis rested largely on his interpretation of the so-called chief nebular line. Huggins in 1864 had detected this line in gaseous nebulae and thought initially that it was due to nitrogen. After he found that the nitrogen line and the chief nebular line did not have exactly the same wavelength, he suggested, following Lockyer's lead with respect to helium, that it was due to a hypothetical extraterrestrial element called nebulium.[50] Lockyer, however, identified the line with the remnant of the magnesium band of a "gently glowing meteorite," and brought into evidence the results of the spectroscopic observations of other astronomers to buttress his own (fig. 25). Huggins disagreed with Lockyer's interpretation, which the latter duly noted in his publications, while emphasizing that magnesium was a constituent of olivine, one of the most important minerals in meteorites.[51] The detection of the chief nebular line in a nova in the constellation Cygnus in 1877 caused Lockyer to postulate that novae resulted from the collisions of huge meteoritic swarms, the maximum brilliancy occurring when the denser parts of the swarms met.

Lockyer gained support from other quarters. P. G. Tait in 1879 returned to the subject of comets. He assumed that a comet's head was a swarm of meteorites, varying in size from a marble to boulders 30 feet wide, which were uniformly distributed in a sphere 20,000 miles in diameter, and that the total mass was 0.001 that of the Earth. He demonstrated first that if each meteorite described an elliptical orbit, collisions must occur. When the swarm was subjected to the tidal force of the Sun, collisions would increase, and the heat of impact would be sufficient to volatilize a portion and render the outsides of the stones white-hot. The collisions would also cause the evolution of gas and produce extremely fine material, which would cause the phenomena of the tail and reflect sunlight. Further assumptions relative to the size of meteorites and the average distance separating them served to explain the luminosity and the transparency of a comet. Lockyer was pleased with Tait's analysis, although he thought that the mass of a comet probably never exceeded 0.0002 that of

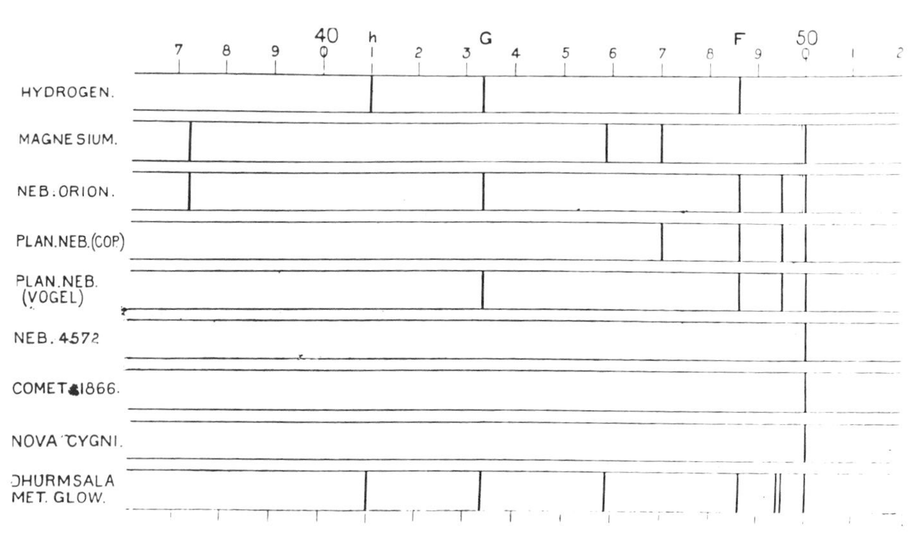

Fig. 25. Spectra of nebulae compared to spectra of hydrogen, cool magnesium, and meteorite glow. From Lockyer (1887). Originally white on black, this figure has been reversed for clarity.

the Earth, and that the average distance between meteorites, assumed by Tait to be 17 feet, was too small.[53]

Lockyer also received help from George H. Darwin, son of Charles Darwin and one of the finest applied mathematicians in England. In 1888 Darwin published an article in which he suggested that the fluid pressure essential to Laplace's nebular hypothesis was a result of countless impacts of meteorites. Glancing collisions could provide the necessary energy with some attendant volatilization of some of the meteorites, and when the gases cooled they condensed into "metallic rain," which fused with the molten surfaces of others. During the process, however, the meteorites fell "towards a centre of aggregation"; the supply of meteorites at the exterior of the swarm became sparse; and the swarm gradually contracted.[54] Lockyer cited Darwin's work in *The Meteoritic Hypothesis,* although he neglected mention of the fact that Darwin had indicated that meteorites reaching earth were "the dust and refuse of the system."[55] Fedor A. Bredikhin's analysis of comets in 1890 was a further aid to Lockyer. Bredikhin proposed that explosions in the head of a comet ejected meteorites from the mass, throwing them into orbits which eventually crossed that of the Earth. "Although the connection between comets and meteorites is not a matter of doubt," the abstractor of the article wrote, "[Bredikhin's] investigation demonstrates it from a new point of view."[56] Finally, Alfred Lord Tennyson, England's poet laureate, testified to the widespread knowledge of Lockyer's theory:

> Must my day be dark by reason, O ye Heavens, of your boundless nights, Rush of Suns, and roll of systems, and your fiery clash of meteorites?[57]

Meunier was one of Lockyer's most outspoken opponents. "Meteorites," he wrote in 1887, "are aggregations, often very complex, of different minerals, in which their mode of existence necessarily supposes the successive play of perfectly determinable actions." Lockyer's idea that meteorites originated from the condensation of vapors produced by collisions and that they grew from the fusion of additional particles did not apply to the history of the meteorites with which meteorite scientists were familiar. The Krasnojarsk meteorite, for example, "necessarily presupposes the prior existence of a huge mass of which it was a part. Apart from its regular structure, it includes minerals that are quite unequally fusible." Meunier was enthusiastic about the great conquests of spectroscopy, but thought that the comparison of the spectra of meteorite gases with solar, stellar, and cometary radiations proved just one fact: the unity of the chemical composition of the universe. It did not prove that meteorites were in space close to the Sun, or within the masses of comets. Conclusions were being drawn with a haste that was not sufficiently scientific; "prudence should restrain us."[58] Yet it was Huggins's persistence in refusing to accept that the chief nebular line was due to magnesium that led to the collapse of Lockyer's hypothesis. Huggins asked James E. Keeler, a young astronomer at the Lick Observatory in California, to measure the chief nebular line. In the early 1890s, Keeler was mapping nebular spectra with unprecedented accuracy by using Henry Rowland's concave diffraction grating. He found a slight but nonetheless distinct difference between the wavelength of the remnant of the magnesium band and that of the chief nebular line.[59] This result removed the principal foundation of Lockyer's hypothesis.

H. A. Newton was not a partisan of Lockyer's theory, but he did maintain that meteorites originated in comets. In his 1886 address, he attempted to establish this viewpoint by arguing that every other theory of meteorite origin faced insurmountable difficulties. The recurrent heating and cooling that a cometary mass experienced in its solar orbit was sufficient, Newton thought, to account for the structures of meteorites, and those which fell to earth were very probably ejected from a comet by the Sun's action near perihelion. Granted meteorites had not fallen during star showers. However, there was a chance of this occurring only about two or three times each century. The failure, then, did not constitute proof that meteorites were not cometary debris. "The comets alone," Newton concluded, "give us specimens to handle and analyze."

At the end of the nineteenth century, there was no sufficiently convincing factual data that favored the theory that meteorites originated in nebulae and/or comets, or that they were pieces of a planetary parent body. Some support for the cometary theory accrued from the fall of the 4-kg Mazapil meteoritic iron during a star shower on 27 November 1885, since the fragments of Biela's comet, previously dispersed, were thought to have caused the shower. In 1901, however, A. G. Högbom, a Swedish scientist, plotted meteorite falls of known dates according to the day of the month and the petrographical nature of each (i.e., eucrites, howardites, etc.). He arrived at only one positive conclusion—namely, that meteorite falls reached a maximum about mid-June and a minimum about the end of October, a finding that agreed in most respects with Greg's 1860 analysis. But Högbom observed no correlation between meteorite swarms and star showers, although he believed that there was considerable probability that the Earth periodically encountered swarms of meteorites of a particular petrographical nature. It was obvious that more data were required before any certain conclusion could be reached.[60]

LIFE IN METEORITES AND ON OTHER PLANETS

Beginning in the 1860s, the existence of carbonaceous meteorites served to combine previous speculations about two subjects: whether life existed on other planets, and how life had originated on Earth. Camille Flammarion, then an apprentice astronomer and calculator at the *Bureau des Longitudes,* focused popular attention on the matter in his book, *La Pluralité des Mondes Habités.* In the second edition of this work, published in 1864, Flammarion devoted considerable space to a description of the various types of meteorites. He noted that the principal constituents of iron meteorites were kamacite, taenite, and plessite, demonstrating that he was familiar with the studies of Reichenbach, published only three years previously. He also stressed that certain meteorites contained carbon, oxygen, and hydrogen, as well as water combined as a hydrate. This chemical composition, he wrote, gives proof of the existence of life on the surface of other worlds; outside of our globe, there are chemical elements of a vegetable kingdom similar to ours. He cautioned, however, that meteorites ought not be confounded with the rains of seeds, grain, and so forth that were occasionally reported. These were terrestrial substances that were probably carried by high winds. Meteorites were cosmic substances, within which it was difficult to identify the vestiges of vegetation or animal life.[61]

Flammarion's ideas did not attract scientific comment, probably because the book was intended for popular consumption. But that changed in the spring of 1871, when Hermann von Helmholtz, one of the most respected scientists in Germany, suggested that comets and meteors might have scattered "germs of life wherever a new world has reached the stage in which it is a suitable dwelling place for organic beings."[62] By implication, Helmholtz was proposing a hypothesis that life had originated on Earth far in the past through the agency of meteorites. A few months later, Sir William Thomson, Helmholtz's counterpart in England, offered a similar conjecture that was much more explicit:

> Because we all confidently believe that there are at present, and have been from time immemorial, many worlds of life besides our own, we must regard it as probable in the highest degree that there are countless, seed-bearing meteoric stones moving about through space. If at the present instant no life existed upon this Earth, one such stone falling upon it might, by what we blindly call *natural* causes, lead to its becoming covered with vegetation. . . . The hypothesis that life originated on this Earth through moss-grown fragments from the ruins of another world may seem wild and visionary; all I maintain is that it is not unscientific.[63]

As with Flammarion, the statements of Helmholtz and Thomson rested on the analyses of carbonaceous meteorites. By 1871 seven had fallen and been recovered. However, most scientists did not believe that the Renazzo or Ornans stones were of the carbonaceous type because of their color and texture, and the Grosnaja fall was not reported in the scientific literature until 1878. Thus the analyses of only four meteorites—Alais, Cold Bokkeveld, Kaba, and Orgueil—prompted the hypothesis. Berzelius, Wöhler, and Marcellin Berthelot, three of the best chemists of the period, analyzed these meteorites.[64] However, the theoretical foundations of organic chemistry and the techniques of analysis were still too rudimentary to permit them to present unequivocal reports of their findings. This uncertainty led to disagreement as to whether the organic material in meteorites had a biological origin, or whether it was simply a mixture of hydrocarbon compounds that were synthesized in an unknown manner. The hypothesis that meteorites brought life to Earth was a leap taken by some who adhered to the view of biological origin.

Berzelius reported his analysis of the Alais meteorite in 1833. Destructive distillation yielded a blackish substance, indigenous water, carbon dioxide gas, a soluble salt containing ammonia, and a blackish-brown sublimate, which Berzelius confessed was unknown to him. His further analysis of the blackish substance detected silica, magnesia, iron oxide, alumina, and an insoluble carbonaceous residue, which amounted to 12 percent of the substance. He had prefaced his report by asking whether the presence of humus or a trace of other organic (*organischen*) compounds might give a hint of the presence of organic productions on other celestial bodies. His conclusion was that "the presence of a carbonaceous substance in the meteoritic earth has an analogy with the humus contained in terrestrial earth, but it is likely that it was added in a different manner, that it has different properties, and that it does not justify the conjecture that it has an analysis analogous to the carbonaceous matter in terrestrial earth." At the end of his lengthy report on several other meteorites, Berzelius returned to the Alais stone: "The carbonaceous substance, which is intermixed with this earth, appears not to justify the conclusion that organic nature is present at the place of this earth's origin."[65] Berzelius's remarks may well have been influenced by his belief that meteorites were ejecta from volcanoes situated in two different areas of the Moon and his knowledge that the Moon lacked sufficient atmosphere to support life. There was, however, no immediate response to his conclusion.

Friedrich Wöhler, in addition to studying under Berzelius in Sweden, had translated many of Berzelius's works into German, and it was to Berzelius that Wöhler first announced in 1828 his preparation of urea from ammonium cyanate. Wöhler, therefore, knew about Berzelius's analysis of the Alais meteorite. In 1858 Wöhler analyzed the Kaba meteorite, and in 1859 he collaborated with Moritz Hörnes, custodian of minerals at the Vienna museum, in an analysis of the organic substances in the Kaba stone. Moreover, in 1859 Wöhler supervised the analysis of the Cold Bokkeveld meteorite by Elijah Harris, which formed a part of the latter's doctoral dissertation. Wöhler announced the results of Harris's analysis in 1859, and in 1860 published further remarks, which included a phrase generally overlooked: "*According to what we know at present,*" Wöhler wrote, "this organic substance can only be produced from other organized bodies" (emphasis added).[66] He described the material as being similar to such hydrocarbons as ozocerite or scheerite.[67] Harris, in his dissertation, wrote: "Prof. Wöhler has found in the stone from Kaba an organic substance which is decomposed by heat with the separation of coal. This same thing I have found in the meteorite from Cape of Good Hope." In the reported analysis, Harris also listed the organic material as "bituminous substance."[68]

The references to coal or coal derivatives are important because they reveal why some scientists believed that the organic substances in meteorites had a biological origin. The belief that coal was a product of the decomposition of organic matter was well entrenched at the time, although the mechanism of its formation was unknown. Haüy, for example, writing in 1801, stated that the majority of naturalists considered that coal originated from the animal and vegetable kingdoms, because of the imprisoned fossils, the imprints of the leaves of different plants—particularly ferns, and the occasional presence of petrified wood.[69] Many nineteenth-century scientists also believed that petroleum oil was a decomposition product of organic matter. Reasoning by analogy, if the same type of substances were found in meteorites, they in all probability had a biological origin.

The analysis in 1864 of the Orgueil meteorite by François Stanislaus Clöez, professor of chemistry at the Muséum d'Histoire Naturelle, also focused on decomposed organic matter. Clöez determined that the organic matter was a completely amorphous, homogeneous substance resembling a humus of certain combustible earths. It contained, he wrote, carbon, hydrogen, and oxygen, and its composition was very similar to peat from the Somme valley or to the lignite of Ringkohl near Kassel. Thus it was different from the material found by Wöhler and Harris in the Kaba and Cold Bokkeveld meteorites, but similar to that found by Berzelius in the Alais stone. Evidently because he had not read Berzelius's report in its entirety, Clöez added that Berzelius had stated that the presence of a humuslike substance suggested the presence of organic substances on celestial bodies.[70] This misstatement gave more force to Clöez's own view of the biological origin of the organic substances in question.

In 1866 Marcellin Berthelot, professor of organic chemistry at the Collège de France, also analyzed the Orgueil meteorite. Eight years before, Berthelot had, by the employment of a two-stage process, successfully synthesized numerous hydrocarbons, and at the time had ridiculed the idea that a vital force distinguished organic from inorganic substances.[71] Berthelot wished to determine from what substances the organic residues in the meteorites originated, but he admitted that this goal was beyond the current capabilities of organic chemistry. Still, in his analysis of the Orgueil meteorite he completed what he considered to be the first step. He employed the technique of hydrogenation, whereby the organic materials were transformed into corresponding hydrocarbons: "I reproduced, in effect, although more laboriously than with coal, a considerable proportion of saturated hydrocarbons, $C^{2n}H^{2n+2}$, comparable to petroleum oils."[72] That there were organic substances in meteorites was beyond question, but in Berthelot's view, this fact did not signify a biological origin. He had announced this conclusion two years before, stating that the hydrocarbons in meteorites were produced in the same manner as analogous terrestrial bodies—namely, by the direct combination of the composing elements without the mediation of organic life.[73] It was a belief that Berthelot held throughout his long and productive scientific career.

These reports contained all of the pertinent published information on carbonaceous meteorites available to Helmholtz and Thomson. It is obvious from the content of their lectures that both were well acquainted with the various theories of the origin of meteorites; for example, both mentioned results of the spectrographic analyses of comets. Yet in advancing the hypothesis that meteorites carried life, each man gave a different message to his audience. Helmholtz sought to mitigate the bleak and dismal conclusion, reached by many physicists, that at some time in the distant future Earth would not be a fit habitation for life. Nevertheless, Helmholtz wrote, life was precious; carried by meteorites, it would continue to exist in the universe, and we should work to "ennoble the lives of our posterity." Thomson, in contrast, wished to eradicate the materialistic implication of Darwin's evolutionary theory—namely, that life on Earth had originated spontaneously from inorganic matter—and at the same time wanted to emphasize that "overpoweringly strong proofs of intelligent and benevolent design lie all round us . . . [and] that all living beings depend on one ever-acting Creator and Ruler."

Thomson advanced the hypothesis in his presidential address at the Edinburgh meeting of the British Association for the Advancement of Science. One of his biographers wrote that it was "received with great applause," but added that "many of the orthodox Scottish clergy . . . were aghast," and that "others seemed to regard the hypothesis of the meteoric introduction of life as a huge scientific joke."[74] Joseph Dalton Hooker, the foremost botanist in England, wrote in a letter to Charles Darwin that "the notion of introducing life by Meteors is astounding and very unphilosophical. . . . Does he [Thomson] suppose that God's breathing upon Meteors or their progenitors is more philosophical than breathing on the face of the earth?" Thomas Henry Huxley, the eminent zoologist and paleontologist, called Thomson's theory "creation by cockshy—God Almighty sitting like an idle boy at the seaside and shying aerolites (with germs) mostly missing, but sometimes hitting a planet!"[75] Johann Zöllner, the critic of Tait's cometary theory, also attacked Thomson. The hypothesis was unscientific, he wrote, first because it merely converted the question of the origin of life on Earth to the question of the origin of life on the meteorites, and second because a meteorite entering the atmosphere became incandescent, a state that no organism within the meteorite could survive.[76]

Helmholtz, in responding to Zöllner's criticism in 1873, also made known the fact that he had proposed the hypothesis a few months before Thomson. He pointed out what several scientists such as Wöhler and Haidinger had reported—that only the outer surface of larger meteoritic stones became heated during their atmospheric passage, while the interior remained very cold. Any germs on the surface could be blown away on the meteorite's initial contact with the upper atmosphere, while those in the interior would remain unaffected by the high temperature of the surface. Further, Helmholtz defended Thomson's view that the germ-bearing meteorites might be fragments of a former planet by stating that some pieces might have escaped the excessive temperature attendant upon an explosion. Helmholtz considered the hypothesis a proper use of the deductive method in science: "if failure attends all of our efforts to obtain a generation of organisms from lifeless matter, it seems to me a thoroughly correct scientific procedure to inquire whether there has ever been an origination of life, or whether it is not as old as matter and whether its germs, borne from one world to another, have not been developed wherever they have found a favorable soil."[77]

Thomson, who became Lord Kelvin in 1894, apparently never relinquished the hypothesis. In 1877, at the Plymouth meeting of the British Association, he requested in a section meeting that three questions be considered: "Was life possible on a meteorite moving in space? Was life possible on a meteorite while falling to the earth's surface? and, Could any germs live after the meteorite had become embedded in the earth?" Thomson maintained that each question could be answered affirmatively.[78] In 1894 Thomson permitted his 1871 presidential address to be reprinted with only minor revisions that did not change the sense of his remarks. In 1904 he insisted that there was no "prospect of a process being found in any laboratory for making a living thing, whether the minutest germ of bacteriology or anything smaller or greater."[79]

Other scientists were also intrigued with the idea of germs in meteorites. Ferdinand Cohn, a respected German botanist and bacteriologist, and Louis Figuier, a French science writer, espoused the hypothesis, and Louis Pasteur was reported to have investigated unsuccessfully the Orgueil meteorite, seeking evidence of organisms.[80] Walter Flight, a mineralogist at the British Museum

(Natural History), though admitting that "there is not a particle of evidence to prove the persistence of living germs on meteorites during their passage through our atmosphere," thought that the parent bodies of meteorites "may very probably have on their surface some forms of organized beings." Flight cited in support of Thomson's hypothesis the experiments of Gottlieb Haberlandt, a Hungarian botanist who found that the seeds of a variety of plants germinated after extended periods at low temperatures and in a vacuum.[81]

There were some scientists who looked for evidence of organic life in meteorite thin sections. In 1878 Carl A. Gümbel, a geologist at Munich, reported that he had used a bleaching compound on thin sections of the Kaba and Cold Bokkeveld carbonaceous meteorites in an effort to discern organic structures, but without success. "Perhaps," he wrote, "by employing this method on more fertile material of these meteorites or on thin sections of other carbonaceous meteorites, one may be successful in demonstrating the presence of organic life from extraterrestrial planets."[82] Almost immediately there was a report that fossil organisms had been detected, - not by inspection of carbonaceous meteorites, but instead in thin sections of ordinary chondritic meteorites, stony irons, and irons. Otto Hahn, a lawyer by profession, who made the claim, had also pursued a doctorate in geology at the University of Tübingen with an investigation of *eozoon canadense*. The controversy over *eozoon* had been raging since 1863, when William Logan, chief of the Canadian geological survey, postulated that *eozoon*, found in Archean limestones in Canada, was one of the earliest recognized fossils. The most vocal supporter of the hypothesis was the Canadian geologist John W. Dawson, who maintained that *eozoon* was the fossil of a giant foraminifer. Among those opposed was the distinguished German zoologist Karl A. Möbius, who considered *eozoon* a product of mineral metamorphism, a conclusion now universally accepted. Hahn, who published *Die Urzelle* . . . in 1879, took the position that similar forms in several types of rocks, and particularly those in meteoritic stones and irons, were in fact plant fossils. The plants which produced the chondrules in the Knyahinya meteorite (consisting principally of olivine and hypersthene), for example, bore a close resemblance to algae and ferns, and Hahn named these *Urania Guilielimi* to honor Kaiser Wilhelm I. The Widmanstätten figures in the Toluca meteoritic iron, Hahn wrote, were nothing other than fossilized plants, and he called them *Astrosideron Quenstedti* to honor his professor at Tübingen, Friedrich Quenstedt, who had helped Hahn acquire the thin sections.[83]

Hahn pursued the subject further in 1880 in another book, *Die Meteorite (Chondrite) und ihre Organismen*. He realized, he wrote, that the evidence presented in his previous work was not convincing because the fossils were not clearly portrayed. The photographs in the present book, he said, were more representative of the fossil sponges, corals, and crinoids he had identified in about four hundred thin sections of twenty meteorites. None of the meteorites, it should be emphasized, were carbonaceous meteorites. Ninety-five percent of the thin sections were from the Knyahinya chondrite, and most of the rest came from the l'Aigle, Pultusk, Siena, Tabor, Borkut, Parnallee, and Monroe stones. Hahn raised the question as to whether the fossils might have a terrestrial origin—that is, whether the meteorites might have been projected into a terrestrial orbit as the result of a collision. He pointed out, however, that most arguments favored the cosmic origin of meteorites as being debris from a planet destroyed by a collision with another cosmic body. Hahn feared that in the future Earth would meet the same fate.[84]

Hahn sent his collection to a nearby German zoologist, David F. Weinland, who verified Hahn's findings and published several tracts in support. Weinland wrote that all of the coral structures were petrified, having become silicates of magnesia, and named one prominent coral, *Hahnia meteorica,* in honor of its discoverer. Weinland considered the sponges to be fully developed and identified three different genera. He was uncertain whether the forms thought by Hahn to be crinoids belonged to that class, because he judged that some were certainly sponges. All of the corals that Hahn had identified, Weinland pointed out, were extraordinarily small in comparison to the corresponding terrestrial forms. Weinland concluded that all of the chondritic meteorites that had been examined seemed to have come from a single planetary body, which had apparently burst apart, and whose atmosphere was quite similar to that of the Earth.[85]

There were immediate responses to the claims of Hahn and Weinland. Carl Vogt, professor of zoology at the University of Geneva, wrote that the sponges alleged by Hahn to be in meteorites had neither the form nor the structure of known sponges, that there was no similarity between the forms that Hahn claimed to be corals and those of true corals, and that none of Hahn's illustrations corresponded to the external forms of crinoids. He continued: "One does not know at what to be most astonished, the author's complete ignorance of the laws of evolution or the audacity with which he presents his views in terms worthy of the oracle of Delphi." Vogt suggested that Hahn was misled by optical illusions, because the transmitted light of the microscope had a peculiar effect on the shapes of transparent crystals.[86] Arnold von Lasaulx, professor of mineralogy at Bonn, termed the fossil organisms found in meteorites "completely hypothetical," and thought that Hahn and Weinland's theory ought to be eradicated root and branch.[87] J. Lawrence Smith also took Hahn to task. He had probably examined more thin sections of meteorites microscopically than any other person, Smith wrote, and he had never discovered organic remains in any of them. Besides, the chemical composition of meteorites was "averse to the existence of any such remains," since no carbonate of lime had ever been found in the interior of a meteorite. The two common minerals, enstatite and bronzite, might resemble vegetable or other organic forms, Smith said, but they would not confuse even a tyro. Smith concluded that Hahn was an observer whose "imagination has run wild with him."[88]

Neither Hahn nor Weinland responded to this caustic scientific criticism. However, their ideas were reviewed favorably in at least one article in the popular press.[89] Thereafter, the question as to whether meteorites contained germs or organisms became moot, although it was revived again, as we shall see, in the 1930s and 1960s (see chap. 9).

Scientists who study meteorites are, in an important sense, historians. Meteorites bear the scars of their fiery passage through the atmosphere, and some reveal by corrosion or by isotopic decay a great terrestrial age. Internally, they also carry vestiges of events and processes that occurred in the far distant past. In a manner similar to that of the historian who accumulates evidence of social or political events from written documents, folk tales, art objects, and the like, the scientist, in investigating the history or the origin of meteorites, must elicit data from astronomy, chemistry, mineralogy, and petrography. In the late nineteenth century, scientific understanding of important physical processes relevant to the study of meteorites—for example, radioactivity—was lacking. In addition, certain techniques such as X-ray diffraction, which would have aided the search for evidence, were not known. Nineteenth-century scientists, then, were unaware of a great deal of evidence contained in meteorites, which gradually became available in the twentieth century. The circumstances are comparable to the discovery by historians of untouched material in a dusty and forgotten archive.

In addition to the availability and reliability of evidence, there is also the question of its interpretation. In this review of late nineteenth-century theories of the origin of meteorites and of their organic constituents, it is apparent that while one scientist might place great weight on certain evidence, another might find the same evidence of little value. Lockyer, for example, thought that the results of his spectroscopic studies were incontrovertible, whereas Meunier considered them inconsequential insofar as the origin of meteorites was concerned. Thus no leading theory of the origin of meteorites emerged at the time, both because there was insufficient evidence of their past history and because interpretations of what evidence was available were not sufficiently convincing to a majority of the scientists who studied meteorites. It is significant in this regard that Emil Cohen, who wrote his comprehensive work on meteorites at the turn of the twentieth century, made no mention of his own view as to their origin.

Meteoritic stones and irons that contain vestiges of the very distant past lie in public museums and private collections throughout the world. Even before most scientists agreed that meteorites had an extraterrestrial origin, mineral collectors and dealers sought to recover these rare objects either for their own cabinets or for sale or trade to museum curators, who prized them for their value to science or for the enhancement of their mineral displays. We turn now to this activity, for it sheds light on how some of the great meteorite collections were amassed, and also depicts the workaday world of many prominent meteorite scientists of the nineteenth century.

6

Curators and Collectors

Chladni, in his *Über Feuer-Meteore* (1819), decried the fact that some meteorites in collections were discarded in the eighteenth century, terming their removal "Enlightenment vandalism." These acts, he wrote, occurred at five places: Dresden, Vienna, Copenhagen, Bern, and Verona; and they involved meteorites from two sixteenth-century and four seventeenth-century falls.[1] Chladni's indictment, which was intended to embarrass those scientists who two decades previously had scoffed at his theory that meteorites actually fell to earth, has been employed over the years by authors desiring either to underscore the obstinacy of eighteenth-century scientists or to disparage as pretentious science's claim to objectivity. One recent critic took the museums to task in writing: "Museums could have played a major role in the meteorite controversy. The sad fact is, however, that they allowed their precious specimens of meteorites to become lost or even threw them away."[2]

Several points should be made in response. To begin with, museums combining public exhibits and the scientific study and arrangement of collections of natural historical objects came into existence only at the turn of the nineteenth century. The Vatican, the Palais Luxembourg, and the British Museum gave public or quasi-public entry to their collections beginning in the mid-eighteenth century, but they were not at the time museums of natural history in the modern sense. In the latter half of the eighteenth century, there was an increasing number of private mineral collections, amassed by members of the nobility or by the wealthy, but the owners of such collections or their curators decided what should be collected and who should be permitted entry. For example, in 1767, the extensive mineral collection of Pedro Francisco Davila, a wealthy Peruvian resident in Paris, contained a piece of clayey earth renowned for the cure of cancers, another stone to which was attributed the efficacy of driving away rats, and a third that was thought to render women fertile.[3] Sir Hans Sloane, whose massive accumulations of plants, stuffed animals, insects, minerals, ethnographic specimens, antiquities, and works of art provided the core of the British Museum's original corpus, also collected similar objects. A number of these eighteenth-century mineral collections possessed meteoritic irons and stony irons—for example, the Campo del Cielo, Steinbach, and Krasnojarsk meteorites—but these pieces of "native" iron baffled scientists and were preserved only because of their rarity.

Moreover, a substantial number of meteoritic stones were preserved and not discarded. The Ensisheim stone, which fell in 1492, is a special case, since it was suspended for centuries in the parish church. Yet of the twenty-seven falls in the eighteenth century that are now considered to have actually occurred, specimens of eighteen (two-thirds) still exist in collections. Our review in chapter 1, however, demonstrates that chemical and mineralogical knowledge in the eighteenth century was on such a rudimentary level that the mere presence of meteorites in collections could hardly have informed the question of whether they had a cosmic or a terrestrial source. Finally, there is evidence, sparse to be sure, that private individuals were probably more responsible than museums or collectors for the loss to science of meteorite specimens. Brezina, the curator of the Vienna collection, acquired two fragments of the Verona [Vago] meteorite (the loss mourned by Chladni) as a gift in 1888 from Count Marco Miniscalchi, whose family had preserved them for over two centuries.[4] A 27-kg piece of the Limerick meteorite, which fell in 1813, was finally turned over to the National Museum in Dublin in 1946. It had passed through the hands of family members for decades, and was recovered through "remembered stories."[5] A portion of the Schellin meteorite, which fell in 1715—and apparently the largest remaining fragment—was in 1897 in the possession of a family in Brandenberg.[6] In earlier centuries at least, pieces of meteorites were particularly prized by some individuals, a subject to which we return in chapter 7.

The development of private natural history and mineral collections in the eighteenth century did significantly aid the recovery of meteorites for scientific study, beginning in the early nineteenth century. Mineral dealers opened businesses in many major European cities and established contacts with knowledgeable agents in countries around the world in order to provide rare mineral specimens for their wealthy clients. Explorers and travelers were always on the alert for rare or unusual objects, and returned to Europe with boxloads of natural historical and mineral specimens, which they sold to dealers or dispensed to museums. In the early 1800s, then, there was the semblance of an organized network of collectors and dealers, who were prepared to cater to the desires of scientists or curators of museums wishing to purchase meteorites reported to have fallen or been found in remote places around the globe.

The desire to possess unusual types of meteorites, a specimen of a fall having a low total weight, or a sizable piece or all of a large meteoritic iron always fueled appetites. After the mid-nineteenth century, nationalism entered the picture. An extensive and superb meteorite collection in a national museum became a symbol of a nation's scientific prestige and development, just as colonial empires, fleets, and standing armies signaled political power. Also, farmers, trappers, and hunters—the people most likely to find meteorites or to trace the paths of their falls—became aware relatively rapidly of the interest in meteorites and parted with them only for substantial prices. Collectors and dealers were always relieved when finders had little knowledge of the market value of a meteorite, or when they yielded readily to the argument that a meteorite should be turned over to a museum or university for the sake of science. The cost of acquisitions was a lasting impediment to museum curators, and stimulated efforts in several directions. Curators or dealers attempted to persuade wealthy patrons to purchase specimens and make them gifts to a museum. Curators sought to exchange meteorite specimens with other museums, private collectors, or dealers in return for others they did not possess. In a few instances, curators pooled their resources to purchase a meteorite, which was then divided among them. These activities, the exchanges in particular, yielded benefits for science. Meteorite specimens were distributed around the world, and the scientific investigation of meteorites commenced in India, Australia, Japan, and in the Americas as well as in Europe.

By about the turn of the twentieth century, travelers and explorers had located and described or brought back to major museums some of the largest meteorites that had fallen to earth. By this time also, the legal rights of the finder of a meteorite, either on another person's private property or on public lands, had been settled in most countries. However, even the most opulent collectors and the most well-endowed curators of the final decades of the nineteenth century balked at what they considered to be outrageous asking prices for meteorites. Just before 1900, Ernst A. Wülfing, later a professor of mineralogy at Heidelberg, made the first attempt to establish reasonable prices for meteorite specimens. Though not accepted universally, Wülfing's formulae did introduce a measure of reason into what had been a chaotic market situation.

Chladni, as might be expected, was the first to collect meteorites systematically, and in the early 1800s, others soon joined this activity. By about 1810, the eminent German chemist Martin Klaproth owned ten meteorite specimens; Johann F. Blumenbach, anthropologist, natural historian, and professor of medicine at Göttingen, had acquired eleven; and the theologian, mystic, and physiognomist Johann Kasper Lavater of Zürich had ten in his collection. About 1820 the Marquis de Drée in Paris had accumulated twenty-six meteorites; Chladni possessed thirty-three; and the collection of the Vienna cabinet numbered twenty-seven stones and nine irons.[7] Vienna clearly held the lead, so we first describe the activities there and also those at the British Museum during the first half of the nineteenth century.

THE VIENNA AND BRITISH MUSEUM COLLECTIONS TO ABOUT 1860

Karl von Schreibers instituted an active policy for the acquisition of meteorites in 1806, when he became curator of the imperial natural history cabinet of Vienna. At the time, the cabinet had five meteorites: the Hraschina iron, sent in 1751 along with the bishop's report of the fall; the Tabor stone, which the imperial treasury turned over to the cabinet in 1778; a specimen of the Krasnojarsk stony iron, which came in 1782 from the Theresian academy upon its dissolution following the death of Maria Theresa in 1780; a piece of the Eichstädt stone, about which Abbé Stütz scoffed in 1789, but which he nevertheless preserved; and a specimen of the Mauerkirchen fall, obtained from Chladni.[8] Schreibers's background gave no hint that he would become an avid collector of meteorites. He was born in Pressburg in 1775 and was educated in Vienna, where he received a medical degree in 1798. His interest in natural history, particularly zoology, blossomed during a two-year grand tour of Europe from 1799 to 1801, but he then settled in Vienna and practiced medicine until his appointment as curator. After the Stannern fall in 1808, which Schreibers investigated as a member of an imperial commission, he availed himself of every opportunity to acquire meteorite specimens, and their number in the natural history collection grew steadily.

Schreibers remained head of the natural history cabinet until his retirement in 1848. However, in 1827, a minerals cabinet that included meteorites was established as a separate entity within the natural history cabinet. Friedrich Mohs, one of the most respected mineralogists in Europe at the time, received a joint appointment as custodian of minerals and to the chair of mineralogy at the university. Almost immediately, Mohs acquired the fine mineral collection of Jacques Friedrich von der Null, a Viennese banker, within which were a number of meteorites. To assist him in arranging the imperial mineral collection, Mohs secured the appointment of Paul Partsch, who was well acquainted with the mineral cabinet. Partsch was born in 1791, the son of a government official, and was educated in philosophy, law, and natural history at the University of Vienna. By 1815 he had amassed one of the best private mineral collections in Vienna, and in that year he also arranged the collection of the imperial mineral cabinet. During the following twelve years, he traveled throughout Europe meeting the foremost European mineralogists, prepared a geognostic map of lower Austria in 1823, and wrote reports on geological matters for several imperial commissions to which he was appointed.[9] When Mohs became imperial counsellor of the exchequer in charge of mining and monetary affairs in 1835, Partsch became custodian of the mineral cabinet. During his twenty-one-year tenure, the number of meteorites in the imperial cabinet increased from 56 to 136.

Some of these acquisitions were gifts. In 1842, for example, Benjamin Silliman, Jr., curator of the Yale College cabinet, wrote to inquire: "Has M. Partsch a piece of the Meteorite which fell in 1837 in the Cold Bokkeveld in South Africa. . . . If he has it not the Yale Cabinet will with pleasure send it to the Cabinet in Vienna."[10] In 1855 W. E. Logan, chief of the geological survey of Canada, wrote (in French) to Partsch advising that he was forwarding a 200-gram piece of the Madoc iron meteorite, found in October 1854, and stating that "in order to make the famous collection of the imperial museum of Vienna more complete, I ask you to please accept the fragment that I am sending."[11]

Partsch appears to have been the first curator to publish a complete list of the meteorites in a museum collection. In 1843, Partsch wrote, the collection had 94 localities with 258 specimens, and the Hraschina and Elbogen irons, placed side by side "are undoubtedly the most valuable and the most noteworthy of the specimens in the imperial meteorite collection."[12] Such publicity, as collectors and curators soon learned, elicited inquiries regarding possible exchanges. C. U. Shepard, then in the early stages of building his extensive collection, wrote in 1847 to Partsch with such a request.

> In place of receiving a cash compensation of $75. for the Little Piney stone, I shall much prefer to receive specimens of meteorites, at present wanting in my cabinet. Although my collection embraces nearly 60 Localities I am still without any other German specimens save that of Stannern in Moravia! The Agram [Hraschina] iron in particular, is a good desideratum with me.[13]

Shepard did receive a 5-gram fragment of the Hraschina meteorite, and he must have been satisfied with the exchange because he visited Vienna frequently in the following decades; in fact, it was Shepard who delivered the Madoc specimen to Vienna in 1855. Partsch and the later custodians of the Vienna collection actively encouraged collectors and dealers to propose exchanges, and this practice was a major reason why the Vienna meteorite collection became the largest and most extensive in the course of the nineteenth century.

In October 1848, during the revolutionary disturbances in Vienna, fire destroyed much of the natural history collection, but Partsch was able to save the most valuable specimens of minerals and meteorites.[14] Schreibers was crushed to see his life's work go up in flames; he died a few years later, in 1852. In the governmental reorganization that followed the revolutions, an imperial geological survey was established with Wilhelm Haidinger as its first director. Haidinger was born in Vienna in 1795 and studied mineralogy under Mohs at Graz and Freiberg. In the 1820s he visited Paris, where he met Biot and Cuvier, and settled for several years in Edinburgh, where he was employed to arrange Thomas Allan's fine mineral collection in accordance with Mohs's system. Simultaneously, he translated Mohs's major work, *Grund-Riss der Mineralogie,* into English as *Treatise on Mineralogy* (1823). Subsequently, he worked at a porcelain factory in Elbogen, Bohemia, for a dozen years, and then returned to Vienna, where he was a governmental mining inspector.[15] Haidinger's intense interest in meteorites commenced in 1847 with the fall of the Braunau iron, a large piece of which came to the Vienna collection and which Haidinger described. In his position as head of the geological survey of the sprawling Austro-Hungarian Empire, Haidinger was able to aid substantially in the acquisition of meteorites and minerals for the Vienna cabinet. After Partsch's death in 1856 and the appointment of Moritz Hörnes, a promising young mineralogist, as custodian, Haidinger's activity with respect to the meteorite collection was such that he was considered to be joint custodian until illness forced his retirement in 1866.[16] Haidinger, for example, immediately turned over to the museum the Kakowa meteorite, which fell in Romania in 1858, and which the provincial governor sent to the geological survey. "The collection of meteorites, both stones and irons, in the imperial mineral cabinet," he wrote in his Kakowa report, "is a veritable jewel, a manifestation of the zeal, knowledge, and perseverance of our Vienna, our Austrian fatherland. This was the view guiding me, when I handed over the Kakowa stone."[17] During Hörnes's tenure as custodian, the cabinet acquired 109 meteorites, so that at his death in 1868, the number of localities totaled 245.

Reichenbach's feud with Hörnes and Haidinger marred this period. Reichenbach was undoubtedly brilliant and he was a good scientist, but he did have peculiarities. He was born in Stuttgart in 1788, and studied science at the University of Tübingen, from which he later received a Ph.D. in 1821. Technically adept, Reichenbach made substantial improvements in the charcoal furnaces of an ironworks at Blansko, Moravia owned by Count Hugo Salm, who made him a partner in the firm. Reichenbach's business activities then extended into iron mining and the production of beet sugar. But he also published reports of his chemical analyses, mainly on the constituents of beechwood tar, and gave the names paraffin and creosote to those substances. In 1840 he fell out with Count Salm's heirs, but by this time he was a wealthy man and purchased the castle Reisenberg outside of Vienna. His later ventures into the manufacture of silk and of iron and steel were failures.[18] A shower of small stones at Blansko on 25 November 1833 stimulated Reichenbach's interest in meteorites; he directed the search, which yielded eight stones weighing a total of 350 grams. He became an enthusiastic collector, and when he died in 1869, his collection numbered 418 large pieces and 94 smaller fragments from 204 localities. In 1844, however, Reichenbach began to study another phenomenon, which provoked scorn and criticism in the scientific world. He became convinced that there existed a universal force, apart from electricity and gravity, termed *Od*. This odylic force, or animal magnetism, displayed itself as an aura of light that could be observed only by "sensitive" persons. When Justus Liebig, the editor of the prestigious *Annalen der Chemie,* permitted the publication as a supplement to Volume 53 (1845) of a lengthy article by Reichenbach on the subject of animal magnetism, Friedrich Wöhler objected strenuously. Liebig replied that he was ashamed, but had to consent to its publication because of enormous pressure.[19] Reichenbach never flagged in his conviction that Od existed, and died at Leipzig, where he journeyed to convince Gustav Fechner, the great experimental psychologist, of its reality.

The Vienna curators were undoubtedly dubious about Reichenbach's commitment to science, given his devotion to the odylic force and the fact that he had published little about meteorites before 1857. There had been friction between Partsch and Reichenbach, but his violent attack on Hörnes in 1859, interjected in the middle of an article on the chemical constitution of meteorites, brought the enmity into the open. Hörnes, Reichenbach wrote, had refused him permission to study the Vienna meteorite collection. Reichenbach was in error, Hörnes purportedly told him, if he thought the collection had a scientific purpose. It was instead a treasury of rarities, and his position would be in danger if he permitted Reichenbach to touch a meteorite. Reichenbach could only look at the meteorites for a few minutes under Hörnes's supervision. He had given the Vienna cabinet two pieces of the Blansko stone, Reichenbach wrote; now he could only look at them through a glass case. Gustav Rose in Berlin packed and shipped meteorite specimens to him upon request. Such was the difference between Vienna and Berlin.[20]

Haidinger sent an immediate reply to J. C. Poggendorf, the editor of the *Annalen der Physik*. When Partsch had been curator, Haidinger wrote, he had permitted Reichenbach to take a number of the most valuable meteorites from the cabinet to study, but obtained their return only after sending Reichenbach several reminders. Hörnes could not be blamed for refusing the requests of someone who had earned bad credit. With all due respect to Reichenbach's good scientific work, Haidinger concluded, his personal reputation in Vienna was not such that one had to suffer his presumptions.[21] Reichenbach did not let the matter rest. In a lengthy footnote in a subsequent article, he retorted that Haidinger had made an insulting attack on him when he was only attempting to expose the deplorable and disgraceful regulations of the mineral cabinet. Haidinger had no official relation to the cabinet, so why should he meddle in matters that were not his concern? Instead of addressing himself to the duties of a curator vis-à-vis the scientific public, Haidinger had resorted to a bitter personal attack.[22]

The feud was important. Had Reichenbach and the Vienna meteorite scientists been on excellent terms, it is conceivable that Reichenbach might have given his collection to the Vienna cabinet. As it was, in 1858 Reichenbach commenced negotiations with the University of Tübingen to accept and preserve his collection after his death. In 1869 it went to Tübingen, forming a substantial basis of one of the great European collections.[23] The hostility between Hörnes and Reichenbach continued. In 1861 Hörnes wrote to Maskelyne: "I hear that Baron Reichenbach is parting [*sic*] for Paris & London, and beg you and your friends *not* to charge this gentleman with commissions for our Museum."[24] And in 1867, when Reichenbach had to sell the Menow meteorite because of business losses, he did not even approach the Vienna museum, but offered it to the curators of the Berlin, Paris, and London collections.[25]

During the first half of the nineteenth century, the meteorite collection of the British Museum lagged well behind that of the Vienna imperial cabinet in terms of both numbers of localities and quality—that is, in the size or weight and in the rarity of the specimens. A major reason for this was the apparent lack of adequate funds for acquisitions. The museum was, first and foremost, the state library, and it housed many other important collections, such as antiquities, rare coins, and drawings. The purchase of meteorites and of other natural historical objects, and the provision of facilities or the trained staff necessary for their arrangement and study, did not receive a high funding priority until the 1850s and 1860s. There was growth, but it was limited and depended largely upon gifts. Nevertheless, there were a few notable purchases.

In 1807, when Charles (Karl) Konig was appointed assistant in charge of minerals in the museum, there were seven meteorites in the collection. These were a specimen of the Krasnojarsk pallasite, presented in 1776 to the British minister to Russia; a piece of the Campo del Cielo iron, given by Don Rubin de Celis to the Royal Society in 1778 and turned over to the museum; two fragments of the Siratik meteoritic iron, which General Charles O'Hara, commandant at Goree, Senegal, from 1766 to 1779, had brought back; samples of the Wold Cottage, Siena, and Benares stones, which were donated in 1802 and 1803 by Sir Joseph Banks; and several pieces of the l'Aigle meteorite, sent to the museum by J. B. Biot in 1804.[26] In 1810 a parliamentary grant secured for the museum the magnificent mineral collection of Charles Greville, which included seven meteorites; however, the Tabor stone was apparently the only new locality among them. Konig, born in 1774 in Brunswick, Germany, and educated at Göttingen, was a knowledgeable mineralogist and had emigrated to England in 1800. In 1813, he was appointed keeper of the Natural History Department which was then a single unit. In 1827 a separate department of botany, and in 1837 one of zoology were established, leaving Konig as keeper of minerals and paleontological specimens. At his death in 1851, the mineral collection included seventy meteorites.

A gift of several Eskimo knives in 1819 was an intriguing early acquisition. Captain John Ross, commander of an expedition searching for a Northwest Passage, encountered a band of Eskimos, whom he termed "Arctic Highlanders," on 11 August 1818 far north in Baffin's Bay close to Cape York. Edward Sabine, later a noted geophysicist, who accompanied the expedition, noticed the knives carried by each Eskimo man and wondered about the source of the iron fragments that were affixed to the blade and formed the cutting edge. After repeated questioning through an interpreter, Sabine learned that the pieces had been broken off two large "stones" located on a hill in the vicinity of the coast. He suspected that they were meteoritic iron, and an analysis by William Hyde Wollaston after the return to England revealed a nickel content of 3 to 4 percent. It was not until 1894, however, that Commander Robert Peary found the "stones."[27]

The 634-kg mass of the Campo del Cielo iron was a major acquisition. George Canning, British Foreign Secretary, had in 1826 dispatched Sir Woodbine Parish to Argentina to acknowledge Britain's recognition of Argentina's independence from Spain. As a gesture of thanks, the Argentine government presented the large piece to Parish.

> Sir W. P. had a big box made to bring it home but no ship would bring it as the sailors said it was a loadstone & would draw all the bolts of their ship & derange the compasses. At length a man of war brought it home & Sir W. sent it to [Sir Humphry] Davy with the request that he would place it where he thought best.
>
> He presented it in *his own*! name to the British Museum. Sir W. P.'s friends however interfered and König [*sic*] put the proper notice on it.[28]

The sailors' fears were a reflection of bygone years when it was believed that the compass pointed to a gigantic lodestone near the North Pole, which, if approached too closely, would pull iron nails and bolts from a ship's timbers. The story about Davy may be apocryphal; many of his scientific contemporaries thought he was vain and brash. In any case, the mass was a prize for the museum.

Letters from John Henry Heuland to Konig in the 1840s illustrate some of the difficulties involved in the acquisition of meteorites, in particular the high prices of American meteoritic irons, which were beginning to be found in increasing numbers. Heuland (1778–1856) was a London mineral dealer who had traveled extensively in Europe, was a fellow of the Geological Society, and had amassed his own sizable mineral collection.[29] In August 1843 Heuland advised Konig that he was awaiting the arrival from America of three meteoritic irons, and wished to offer them to the British Museum before approaching Vienna. Both Shepard and Silliman tried to get them, Heuland wrote, but he was successful because of his friendship with Gerard Troost, their owner. The total price was $8 per pound, or £500, plus the freight cost to England. This figure was, Heuland stated, an "unheard of low price . . . I consider it an extraordinary triumph to have secured them, for no entreaty was spared in the United States to keep them there."[30] Evidently Konig was not similarly impressed, for in October Heuland wrote: "I beg to state that having offered the three meteoric irons *at cost price,* it is utterly impossible for me to make any reduction, deeply regretting the necessity of sending them abroad."[31] A few days later, nevertheless, Heuland wrote Konig that he was willing to sell the three meteorites individually: the 145-pound Alabama [Walker County] mass for £285; the 60-lb Cocke County [Cosby's Creek] iron for £120; and the 55-lb Sevier [Cosby's Creek] iron for £110. Heuland continued: "The Alabama mass I intend having cut for Vienna, Berlin, Paris, Prince Lobkowitz of Prague, and Baron Reichenbach, . . . and if you desire a portion of it, please to bespeak it, assuring you though that the Pound will cost you its worth being positive that I cannot realize less than £500 for the whole mass."[32] Heuland did divide the Walker County meteorite, selling the major portion to Reichenbach and smaller fragments to the other museums. Konig purchased, apparently at cost, a 48.4-lb piece of it and also the two Cosby's Creek specimens.[33] The transaction illustrates another problem of the period: pieces of the same American meteorite were frequently sold to European dealers and museums under different names, and decades passed before the confusion was cleared up.[34]

Gerard Troost (1776–1850), a native of Holland, was educated in natural history, pharmacy, and medicine at Leyden and Amsterdam. In 1807 he was given the task of collecting valuable mineral specimens by Louis Bonaparte, Holland's ruler, and studied mineralogy in Paris under Haüy for two years in preparation. In 1810, when Troost was in Philadelphia on his way to Java, he learned of Louis Bonaparte's abdication, so he settled there and applied for U.S. citizenship. He was an accomplished scientist, and in 1812 was a founding member of the Academy of Natural Sciences of Philadelphia. He left Philadelphia in 1825 as a member of the utopian group that settled in New Harmony, Indiana, but he left the colony in 1827 and settled in Nashville, where he taught geology and mineralogy at the university for the rest of his life. Troost and Heuland were old friends. They had met in Rotterdam in 1806, and Troost realized that Heuland's intimate knowledge of the European market would enable him to dispose of all of the minerals and meteorites that Troost could send.[35]

In February 1846 Troost notified Heuland that he was having a large piece of the Carthage iron sawed, but owing to difficulties it would be another month before he could send a 250-lb end piece, whose surface would be a foot square. He also promised Heuland a piece of the Charlotte meteoritic iron.[36] Heuland immediately offered the former meteorite to Konig, but his first approach was unsuccessful. In early April, his note to Konig stated that he was "aware that the Trustees would not grant £800 for the meteoric mass from Carthage, Tennessee, I sounded M^r [Richard] Simmons if he would not make you a present of it, but I did not succeed."[37] Heuland, however, was not discouraged. In July he wrote Konig that he was sorry that the British Museum had refused his offer for the Carthage iron because Troost had sent it particularly for that collection; however, he did have another iron weighing 5 lb 10 oz from Smithland, Kentucky.[38] The trustees did approve the purchase of the Smithland meteorite at a cost of £10, but declined Heuland's offer to sell the 239-lb piece of the Carthage iron for £600, or for £500 plus a 40-lb slice of the Campo del Cielo meteorite.[39] Finally, in February 1847, Heuland advised Konig that he had cut the Carthage mass, and had two pieces remaining: one weighing 70 lb 15 oz and another weighing 54 lb 3 oz. "The last is by far the most important of the two, and you can select the one you prefer at my ultimatum of £200. You allowed £70 for a 12p$^{\underline{d}}$ specimen of the Lockport [Cambria] iron infinitely inferior to the Carthage iron in every respect, which is the most instructive, the most interesting and at the same time the scarcest of all the meteoric irons known."[40] This time, the trustees approved the purchase of the smaller piece.

In June 1848 Konig purchased a 6-lb, 3-oz piece of the Murfreesboro iron, which Troost had sent to Heuland.[41] However, shortly thereafter, Heuland and Konig had a falling out. In November 1848 Heuland wrote that "although you wished me never to offer you another Meteorite, or anything else, I am compelled to send you the enclosed on account of the Burlington iron, now on the road from Professor Shepard. I always thought to treat you in the most conscientious manner possible, and having taken £10 less for the Rutherford [Murfreesboro] meteoric iron, than I could have had from the Vienna Museum, is a proof of it."[42]

What caused the dispute is unknown, but the reason may be that Heuland had begun to act as Shepard's agent, thus closing off another source of American meteorites. Shepard had begun a correspondence with the British Museum as early as 1824, when he sent a box of minerals and requested an exchange.[43] By the 1840s, however, Shepard was also dealing in meteorites, and it was Shepard who sold the Cambria meteorite to Konig in 1845. Konig, evidently, could not recover from the shock of having had to pay £70 for the 12-lb specimen, for Shepard wrote him:

> Notwithstanding your incredulity, I am satisfied that the Lockport [Cambria] mass would have realized to me £100, though I have no complaints to make at the disposition made of it. The Br. Museum has the best of the bargain this time, & they may well congratulate themselves on the acquisition, for sure I am that they will never make so good one again at as small a sacrifice of money. I confess I thought the fragment too interesting to be slit up & this was one inducement with me to propose it where its integrity would never be impaired.[44]

Heuland's arrangement with Shepard was short-lived, however, because in 1850 ill health compelled Heuland to retire to Sussex, where he died in 1856.

After Konig's death in 1851, George R. Waterhouse became keeper of minerals and fossils. Waterhouse was a paleontologist and gave little attention to the acquisition of meteorites or minerals. In fact, a specimen of the Madoc iron, given to the British Museum by William Logan simultaneously with his gift to Vienna, was the only major acquisition during the six years of Waterhouse's tenure. Thereafter, however, a new era dawned, in which the British Museum challenged the supremacy of Vienna, insofar as meteorites were concerned.

THE MAJOR EUROPEAN COLLECTIONS IN THE LATE NINETEENTH CENTURY

In a letter written to Heuland in 1846, Troost asked: "What did become of the little collection of Chladni? I became intimate with him—he had specimens, some of them not larger than a green pea."[45] The answer was that Chladni had willed it to the University of Berlin, where it went after his death in 1827. The university, founded in 1810, incorporated a mineralogical museum, which received all of the minerals previously located in the royal mineral cabinet, housed in the Berlin Mining Institute. Dietrich Ludwig Karsten (1768–1810), who taught mineralogy at the institute and was curator of the cabinet, died soon after the transfer of the minerals to the museum, and Christian Samuel Weiss, professor of mineralogy at the university, was his successor. Gustav Rose, in 1863, summarized the early history of the collection.[46] He did not know just how many meteorites had been transferred to the museum along with the minerals, but there were at least two: a 890-gram specimen of the Krasnojarsk pallasite, given to King Frederich Wilhelm III in 1803 by Tsar Alexander I; and a 544-gram piece of the Durango [Chupaderos] iron, donated in 1807 by Alexander von Humboldt after his return from his extensive travels in the Americas.[47]

Weiss, who made significant contributions to the science of crystallography in the first two decades of the century, became intrigued with meteorites and negotiated the purchase of Klaproth's mineral collection, which included 17 new meteorite localities, after the latter's death in 1817. Also, Weiss, who had struck up a friendship with Chladni, paved the way for his acceptance by the Berlin scientific community, and persuaded Chladni that his collection would be most useful if it were given as a bequest to the Berlin museum. Chladni's collection numbered 31 stones and 10 irons; of these 18 were localities that were new to the Berlin collection. Weiss also made purchases and exchanges, and there were additional gifts from Humboldt, so that when Weiss died in 1856, there were 90 meteorite localities in the collection. His successor, Gustav Rose, made a most significant addition in 1862, by the purchase and exchange of a large number of meteorites with C. U. Shepard, so that in 1863 Rose was able to list, exclusive of doubtful meteorites, 176 stones and irons in the collection. After Rose's death in 1873, the acquisition of meteorites did not receive such a high priority. At the end of the century, the collections at Vienna, London, and Paris far surpassed that at the Berlin museum.

There exist in the Muséum d'Histoire Naturelle at Paris early catalogs of the meteorite collection. The first lists meteorites chronologically by the date of fall: Ensisheim (1492) heads the list, followed by Mauerkirchen (1768), Sena (1773), and so forth. The collection numbered 28 meteorites in April 1815, when the emperor of Austria presented a specimen of the Stannern stone, but 13 of these were duplicates of the l'Aigle fall, deposited by Biot.[48] This chronological catalog soon became obsolete as finds became interspersed with falls, and about 1841, possibly as the result of a sizable exchange of specimens with Partsch of the Vienna cabinet, a revised list was established. Not all of the meteorites in Paris, though, were in the museum collection. Some came quite naturally into the hands of the professors of mineralogy at the museum, who used them in lectures, and the Geological Service acquired a substantial number. When the museum collection came under Gabriel Daubrée's administration in 1861, the catalog listed about 100 meteorites, but about half of these were duplicates.[49]

Daubrée, as previously noted, viewed meteorites as providing important evidence that could be employed in formulating theories of mineral formation and geological processes on a universal scale, and soon attempted to enhance the museum's collection. He was only partially successful in unifying the separate collections; it was only in 1926, for example, that the museum professors, at the urging of Alfred Lacroix, agreed to transfer the meteorites in the Geological Service collection to mineralogy.[50] But Daubrée actively sought exchanges, and he encouraged officials in France's overseas possessions and colonies to be on the lookout for meteorites. This strategy paid handsome dividends. The two stones comprising the Aumale meteorite, which fell in Algeria in 1865, were sent to Paris by the chief engineer of mines of Algeria, who in the same year presented the only known fragment (72 grams) of the Dellys meteoritic iron. In 1867 Marshall MacMahon, then governor-general of Algeria, sent to the museum the largest stone (ca. 6 kg) of the Tadjera meteorite, which fell that year.[51] In 1866 General Achille Francois Bazaine, then in Mexico during Emperor Maximilian's brief reign, dispatched the 780-kilogram Charcas meteoritic iron from San Luis Potosi to Emperor Napoleon III, who turned it over to the museum. Also, Daubrée, in an exchange with the Vienna cabinet, retrieved a fragment of the Lucé meteorite, which members of the Académie des Sciences had unsuccessfully analyzed nearly a century before.[52]

Stanislaus Meunier succeeded Daubrée as curator of the meteorite collection in 1892 and held the position until his retirement in 1920. His enthusiasm for the acquisition of meteorites equaled that of Daubrée, so that in the late 1890s the Paris collection numbered just over 400 meteorite localities, trailing Vienna and London, but still the third largest in the world. The records of purchases and exchanges are much more complete at the British Museum and at the Vienna Natural History Museum; a more detailed description of their collecting activities illustrates how these extensive European collections came into being.

The appointment in 1856 of Richard Owen as superintendent of the natural history department at the British Museum proved to be crucial in the development of the natural sciences in England in the nineteenth century. Owen (1804–1892) developed an early interest in dissection, trained under the best anatomists in Scotland and England, and became a professor at the Royal College of Surgeons in 1836. His scientific studies were prodigious; his early work centered on invertebrate anatomy; after he came to the British Museum he focused more on paleontology. His lengthy report in 1859 recommended the move of the Natural History Department to a separate building, which, after years of debate, became a reality in the construction of the British

Museum (Natural History) in South Kensington from 1873 to 1880. Given his scientific background, Owen immediately recognized that mineralogy and paleontology should become separate departments, and in 1857 he arranged the appointment of Nevil Story-Maskelyne as keeper of minerals at the museum (fig. 26). Story-Maskelyne's mother was the only child and heiress of Nevil Maskelyne, astronomer royal; his father, Anthony Story, took the surname Story-Maskelyne in 1845. Since he was known informally as Maskelyne, we refer to him by that name herein. Educated at Oxford, he had an early interest in science, and he agreed in 1850 to give lectures in mineralogy at Oxford only if he were provided a laboratory for chemical and mineralogical analysis. He became professor of mineralogy at Oxford in 1856, and retained this position when he was appointed keeper. As at Oxford, he insisted as a condition of his acceptance that a well-equipped laboratory be established at the museum. Within a few years, Maskelyne turned his attention to the acquisition of meteorites.

Fig. 26. M. H. Nevil Story-Maskelyne (1823–1911). Courtesy of the British Museum (Natural History), London.

A complicated arrangement involving Robert P. Greg brought approximately 65 new meteorite localities to the British Museum in 1862 and 1863. Greg (1826–1906) came from a wealthy commercial family in Manchester, and after attending the University of Edinburgh, settled on his country estate in Hertfordshire.[53] His father in 1835 had purchased the fine mineral collection of Thomas Allan of Edinburgh (arranged according to the Mohs system in 1823 by Haidinger), and Greg made such choice additions to it that it became the best private mineral collection in England. In collaboration with W. G. Lettsom, Greg in 1858 published the *Manual of the Mineralogy of Great Britain and Ireland,* at the time and in later editions an authoritative work. But also in the 1850s Greg's interest turned to meteors and meteorites. As noted in chapter 3, he wrote several articles and translated from the German others by Haidinger for publication in English journals. In 1860 Greg sold his mineral collection to the British Museum and sought to augment his meteorite collection. In 1861–62, Greg and Maskelyne negotiated the purchase from August Krantz (1809–1872), a mineral dealer at Bonn, of a sizable collection of meteorites, in which Greg received the duplicate specimens and also contributed funds toward the purchase, in return for which museum duplicates were given to him in 1863 and 1864.[54] An incentive to Maskelyne may have been Krantz's promise: "If you take the collection I will produce [*sic*] to you in a short time many more localities to beat the Vienna Collection."[55] It was a promise that Krantz was never able to fulfill, but the purchase brought many rare meteorites to the museum collection. Greg's interest in collecting meteorites soon diminished; in 1865 he sold his collection for £500 to the Calcutta museum.

In a peculiar way, the Civil War in the United States contributed to the enhancement not only of the British Museum collection but also to those at Berlin and Paris. The careers of both Charles U. Shepard and J. Lawrence Smith were disrupted by the war. Shepard (1804–1886), who was born in Newport, Rhode Island, and educated at Amherst and Harvard, had taught natural history and chemistry at Yale and at Amherst, and in 1852 accepted a position as professor of chemistry at the Medical College of Charleston, South Carolina (fig. 27). When talk of secession began in 1859, Shepard as a Yankee evidently felt uncomfortable in the city and beat a hasty retreat northward. In November 1860 he deposited his meteorite collection—consisting of 88 stones, 63 irons, and several doubtful specimens—in the mineral cabinet of Amherst.[56] However, Shepard apparently retained many duplicates, because while in Europe in 1862 he sold numerous meteorites to Maskelyne and Daubrée, and Gustav Rose reported that he purchased Shepard's entire collection.

In contrast to Shepard, Smith (1818–1883) was a Southerner. He was born near Charleston, South Carolina, studied science and medicine at the University of Virginia, and after receiving his M.D. from the Medical College of Charleston, completed three years of graduate work at Giessen and Paris. From 1844 to 1846 he was state assayer in South Carolina, and then for four years became involved in mining ventures in Turkey. After teaching for brief periods at Louisiana University and the University of Virginia, he married into a prominent Louisville family, and in 1852 was appointed professor of chemistry in the medical department of the University of Louisville. The residents of

Fig. 27. Charles Upham Shepard (1804–1886). From Merrill (1916).

Kentucky, a border state, were bitterly divided by the war, and in 1861 and 1862 eastern Kentucky became a battleground between Union and Confederate forces, the prize being Louisville, an important rail junction and Ohio river port. Smith was despondent when he visited London in the fall of 1862, writing to Maskelyne: "I wish to sell my entire collection of meteorites for situated as I am in the U. States in the territory between the contesting parties in my country, I have little hope in my life time of ever establishing a cabinet of minerals in Louisville."[57] Less than a week later, Smith met "old C.U.S.," as he characterized Shepard to friends, and sold him his collection. Shepard was elated. He wrote immediately to Maskelyne telling him that he had become the owner of Smith's collection and that he wished to sell some of the meteorites.[58] The British Museum and Daubrée in Paris did buy a few from Shepard, but Gustav Rose in Berlin purchased most, including a 13.4-kg New Concord stone.

Maskelyne lost no time in publishing catalogs of the meteorite collection. Konig, in 1847, had compiled a three-page list of the specimens, which numbered at the time 26 irons and 35 stones. Between May 12, 1861 and August 1, 1863, Maskelyne prepared four catalogs, noting in each the startling growth of the collection.[59] All of the eleven catalogs that he eventually compiled through November 1, 1877 had the same format. Entitled *Catalogue of the Collection of Meteorites Exhibited in the Mineral Department of the British Museum,* the list was divided into "aerolites," "siderolites," and "aerosiderites." If Maskelyne's purpose in publishing the catalogs was to attract requests for exchanges, he was completely successful. During the years 1861 through 1864, he made exchanges with museum curators at Copenhagen, Madras, Calcutta, Heidelberg, Vienna, Yale, Paris, Debrecen, Berlin, and Göttingen, and with dealers or private individuals at Turin, Moscow, Santiago, and Boston.[60]

Another strategy that Maskelyne (or Owen) adopted was to ask the museum trustees to prod government officials to instruct administrators in Britain's far-flung possessions to keep an eye out for meteorites. This idea may have occurred to him as a result of the lengthy dispute over the 3,500-kg Cranbourne meteoritic iron, found in Victoria, Australia, which the Colonial Office became involved with before it arrived at the museum in 1865.[61] The strategy was particularly successful in India, for between 1859 and 1870 nine meteorites that had fallen in British India came to the museum through the office of the secretary of state for India.[62] The Foreign Office also aided in the acquisition of meteorites. Earl Granville, the foreign secretary, became involved in 1882 in obtaining as a gift from the shah of Persia a piece of the Veramin meteorite, which fell in 1880. Granville, in 1883, also supervised negotiations that brought from Japan over 4 kg of the Ogi meteorite, which fell in 1741 and had been preserved in private hands.[63]

Haidinger and Hörnes in Vienna were well aware of Maskelyne's intention to make the meteorite collection at the British Museum the finest in the world. At first, Hörnes refused his offers to exchange, and accused him of "piling up Meteorites" that would serve no useful scientific purpose.[64] Hörnes's indictment is odd, given the fact that Reichenbach had leveled the same charge against him just a few years before. But it was also unfounded because Maskelyne himself, his assistants, and his students at Oxford were all engaged in research work on meteorites. In late 1862, however, the atmosphere changed, and Maskelyne negotiated a sizable exchange of specimens with the Vienna cabinet. In 1867, Haidinger acknowledged that, in numbers of meteorite localities, the British Museum collection nearly matched that of the Vienna cabinet. He wrote: "This is an expression of the power and influence of Great Britain in all directions of its colonial possessions, but it is also evidence of the zeal of the director and of the very considerable resources which were placed at his disposal."[65] While he gave a grudging compliment to Maskelyne, Haidinger still believed that money and power had been the most important factors in the British Museum's success.

Hörnes died in 1868, and Gustav Tschermak became custodian of the Vienna mineral collection, simultaneously receiving an appointment as associate professor of petrography at the university. Tschermak (1836–1927) was a native of Litovia, Moravia, and studied chemistry and crystallography at the University of Vienna before receiving his Ph.D. at Tübingen in 1860 (fig. 28). During the 1860s, while teaching chemistry and mineralogy as a privatdozent at the university and serving as Hörnes's assistant, Tschermak carried on important petrographical research and also completed his classic work on the soda feldspars, which provided the basis for the acceptance of isomorphism.

Fig. 28. Gustav Tschermak (1836–1927). Courtesy of the Naturhistorisches Museum Wien.

In chapter 4, we described the significant research he accomplished on meteorite thin sections during the period when he was custodian. However, Tschermak also spent time on meteorite acquisitions; during his eight years as custodian, he added 55 new localities to the Vienna collection, bringing the total to 299 in 1878.

Tschermak's relations with the British Museum were very cordial. He effected an exchange in 1869 and another sizable one in 1870. In 1871 he wrote to Maskelyne:

> I am still continually busy with meteorites, chiefly with an investigation of the Shergotty stone. I have found therein an augite and a colorless silicate that crystallizes in the tesseral [isometric] system. I shall analyze the latter in the next few days. It is probably something new, and I shall give it the name Maskelynite, in your honor since you were the first to have undertaken the analyses of meteoritic minerals in the most salutary way.[66]

Tschermak was referring to the important studies, published in 1870 and 1871, in which Maskelyne used the polarizing microscope to identify a number of the minerals present in meteorites. Tschermak's initial judgment concerning maskelynite was somewhat off the mark; it is a plagioclase, labradorite, converted by a preterrestrial impact shock to an isotropic or imperfectly isotropic glass.

Daubrée and his assistant Meunier also got on famously with Maskelyne. In fact, Daubrée and Maskelyne must have met frequently on social occasions in London or Paris, for Daubrée never failed in his letters to send his and his wife's best wishes to Maskelyne's wife and daughter. In 1867–68 there was an exchange which brought specimens of the Charcas and Sierra de Deesa [Copiapo] irons and of the Molina, St. Mesmin, and Aumale stones to the British Museum, and another major exchange occurred in 1875.[67]

Maskelyne also corresponded with Adolf E. Nordenskiöld, the brilliant Finnish geologist. The Russian administration of Finland forced Nordenskiöld to leave the country in 1858 because of his liberal political views, but he immediately received an appointment as chief of the mineralogy division of Sweden's National Museum in Stockholm.[68] In 1864 Maskelyne requested Nordenskiöld to negotiate an exchange with the University of Helsinki for a specimen of the Luotolax meteorite, a eucrite, since at the time he was interested in studying the achondrites. Nordenskiöld replied that A. E. Arppe in Helsinki "dared not" give away a part of the stone because it belonged to the university.[69] Despite this disappointment, Maskelyne continued to write to Nordenskiöld and was rewarded by receiving a piece of the Hessle stone, which fell at Uppsala, Sweden, on New Year's Day, 1869.[70]

In November 1870, Nordenskiöld, who was actively involved in Arctic exploration for scientific purposes, wrote Maskelyne that he had just returned from Greenland and had discovered a meteorite "the most northerly fallen meteorite hitherto found . . . and as having fallen *during* the *miocene period*."[71] Maskelyne immediately wrote that he wished to purchase a large piece, and Nordenskiöld replied that he was sending an 87-kg mass for a price of £435. As it turned out, the purchase of this Ovifak iron was undoubtedly the most unfortunate that Maskelyne made. In January 1871 Nordenskiöld wrote:

> The beautiful and interesting meteorite I brought with me from Greenland makes me much trouble. It is necessary to cover the stone with *damara varnish* to prevent them from breaking. Please let it be done on my account as soon as possible and keep it, as I do, enclosed hermeticalay [*sic*] in a glass (over quicksilver). . . . I am quite unable to explain the bursting of the stone.[72]

There was further correspondence about the constituents of the iron and additional instructions for its preservation. By January 1876 the interchange indicates that controversy had already begun concerning the meteoritic origin of the Ovifak iron. The relatively high nickel content of these iron masses, whose combined weight totaled 35 tons, persuaded a number of scientists in the late nineteenth and early twentieth centuries that they were meteoritic. However, early on, the weight of opinion favored their terrestrial origin, and recent studies have confirmed this judgment.[73] On Nordenskiöld's invoice for £435 in the files of the British Museum (Natural History), there is noted: "[specimen no.] 44434. This mass has gone to pieces."

Tschermak resigned his custodian's position in 1878 to become head of a newly established institute for mineralogy and petrology at the University of Vienna, and Maskelyne relinquished his keepership in February 1880. Their replacements, Aristides Brezina and Lazarus Fletcher, were both excellent scientists and able administrators. Brezina (1848–1909) was born in Vienna and displayed an early bent for science, studying crystallography and mineralogy at the imperial mineral cabinet in 1862, while auditing the courses of Eduard Suess in geology and Tschermak in petrography at the university (fig. 29). He studied under Gustav Rose in Berlin in 1868, and while pursuing his doctorate at Tübingen, which he received in 1872, worked intermittently as an assistant at the Vienna cabinet. He then taught at Graz and at Vienna before being appointed custodian of the Vienna collection.[74] Lazarus Fletcher (1854–1921) received his early education at the Manchester Grammar School, and in

Fig. 29. Aristides Brezina (1848–1909). Courtesy of the Naturhistorisches Museum Wien.

1872 won a science scholarship at Balliol College, Oxford, where he distinguished himself in mathematics and physics. He was assistant to Professor Robert B. Clifton in the Clarendon Laboratory from 1875 to 1877, and in 1877 he was appointed Millard Lecturer in Physics at Trinity College (fig. 30). Maskelyne, who learned from Clifton of Fletcher's keen interest in crystallography, invited him to visit the British Museum, with the result that Fletcher became Maskelyne's chief assistant from 1878 to 1880.[75] Aided undoubtedly by Maskelyne's recommendation, Fletcher was named keeper of minerals in May 1880, at the age of twenty-six.

Important changes were underway both at Vienna and London. The imposing, domed Renaissance-style Künsthistorisches Museum and Naturhistorisches Museum in Vienna, identical in design and facing each other across a park containing a large statue of Maria Theresa, had been in the process of construction since 1872. After their completion, Brezina had fine new offices, laboratories, library facilities, and display halls for the collections of minerals and meteorites. The vast Romanesque natural history museum in South Kensington, designed by Alfred Waterhouse as a modern cathedral to shelter the "Wonders of Creation," opened its doors in 1880. One of Fletcher's first tasks was to oversee the plans for the space allocated to the mineralogy department and to make arrangements for the move.

Fig. 30. Sir Lazarus Fletcher (1854–1921). Courtesy of the British Museum (Natural History), London.

Fletcher and Brezina had a good working relationship for over a decade, although it was apparently never cordial. The first instance of their cooperation involved the Estherville mesosiderite, which fell in Iowa on 10 May 1879 and had a total weight of about 700 pounds. Charles P. Birge of Keokuk came into possession of the largest piece, which weighed 437 pounds, and wrote in September 1879 to the Vienna, Paris, and London museums to advise that he would entertain offers. In late October Brezina wrote Maskelyne, who was then still keeper, that "We should agree on some common behaviour, be it that we buy the meteorite together and deal it afterwards, or that any other arrangement take place." Maskelyne agreed, and Brezina replied immediately, advising that Daubrée also consented to any common action. He added, however, that he could only afford 1,000 florins (equivalent to £86), that Daubrée could only spare 4,000 francs (£160), so that Maskelyne would have to be the chief bidder. Brezina also stated that he had written to Birge concerning their joint action, and proposed that the meteorite be sent to London, where models could be prepared before it was cut. On 30 November 1879 Maskelyne offered Birge £450 for his piece, which he accepted, and turned the division of the meteorite over to Fletcher. Cutting the meteorite proved to be extremely difficult, and several years passed before Brezina and Daubrée received their portions. Initially, Fletcher employed a stone sawyer, who worked for 380 hours without completing the job, using up a large number of copper saws in the process. The job was then turned over to an engineer, who completed the first cut in 59 hours, using a hardened steel circular saw, powered by a steam engine. He learned that the cut could be made more easily if the saw was kept cool and lubricated by a mixture of oil, soft soap, and water, so that the second section required only 53 hours of labor. Fletcher expended over £35 in sectioning the meteorite, but the experience stimulated curators and dealers to search for less expensive methods of cutting iron and stony iron meteorites.[76]

Brezina took the initiative in proposing exchanges with the British Museum, but in these Fletcher had to be on his guard, for Brezina was a clever trader. At the very beginning of his tenure, for example, Brezina received an offer from Henry A. Ward of Rochester, New York, to sell a 4,840-gram piece of the Augusta County meteoritic iron, and almost simultaneously one from B. Stürtz, a Bonn mineral dealer, of a 1,150-gram piece of the same meteorite. In comparing the asking prices, Brezina found that Stürtz's price per gram was over three times that of Ward, with the result that he accepted Ward's offer.[77] Fletcher must have sensed Brezina's acuity, for he was very careful when Brezina offered in mid-1881 to trade specimens of the Chulafinnee iron and the Tieschitz chondrite in exchange for samples of the Rowton iron and the Dyalpur ureilite. Brezina had just purchased the 14.7-kg main mass of the Chulafinnee iron from William E. Hidden, collector and mineral consultant to Thomas A. Edison, and Vienna also possessed the main mass of the 28-kg Tieschitz stone, which fell in Moravia 15 July 1878. In contrast, the original weight of Rowton, the only iron meteorite to fall in England in modern times (20 April 1876), was only 3.109 kilograms, and the total weight of the Dyalpur stone, a relatively rare ureilite that fell in British India in May 1872, was only 284 grams. When Fletcher agreed to the exchange, he pointed out to Brezina that there was a disparity in the original weights of Chulafinnee and Rowton.

He asked, therefore, that Brezina give two falls in exchange for the Rowton specimen, and added the proviso that Brezina not publish any report on Rowton until the paper of Dr. Walter Flight, a staff mineralogist, on the Rowton iron had appeared. Brezina gave ground but not without a protest. "We shall give a higher weight of Chulafinnee against Rowton," he replied, "but you surely know that it is not a custom to exchange *proportional* to the fallen weight; so I may propose 39.5 [gram] Rowton [and] 14 [gram] Dyalpur against 60 [gram] Chulafinnee [and] 17.3 [gram] Tieschitz." Fletcher accepted this compromise and wrote to Brezina: "Of course in pointing out the original weights I merely wished to indicate that the exchange value is connected to some extent with the amount of the original fall & therefore with the amount which is to be shared amongst the terrestrial inhabitants, though that value is far from being in a direct ratio. I shall be glad to recommend the Trustees . . . to agree to the exchange you propose."[78] Fletcher, despite his youth and inexperience, was also a keen trader. He realized the importance of the original weights, and these figured into the formulas for meteorite exchange values, which were proposed two decades later.

In acknowledging receipt of the Rowton and Dyalpur specimens, Brezina advised Fletcher that he was forwarding two reprints of papers he had published on the orientation of the Reichenbach lamellae in the Widmanstätten figures, and that they included "the representation of Irons through Phototypie which is beginning to give satisfactory results."[79] These were among the first microphotographs of meteoritic structures to be published in professional journals. In 1886 Brezina sent Fletcher "a paper of Dr. Victor Goldschmidt, a friend and former auditor of mine. I hope you will find it to be a useful work and I should be indebted to you if you could conveniently give a short notice about it in Nature of any other serial destined for English readers."[80] Stimulated by Brezina, Goldschmidt went on to become one of the founders of modern crystallography.

Brezina and Fletcher negotiated further exchanges in 1882, 1884, and 1891, the last involving four years of bargaining and a substantial number of specimens. At the conclusion of this exchange Fletcher wrote: "I may say distinctly that in giving up our greatest rarities & as good spec^ms^ as are ever likely to leave the Museum, I feel that I am not acting wisely from the purely commercial exchange point of view, & that my only justification is the desire not to put any obstacles that can be avoided in the way of a scientific exchange."[81] In this note, Fletcher was alluding to a practice that Brezina had instituted and that Fletcher deplored—namely, acting as a dealer by selling to other dealers or collectors slices or pieces of meteorites that he had acquired. Fletcher did not wish to hear of specimens of the British Museum's precious meteorites being dealt with in this manner.

Brezina undoubtedly adopted the practice because of financial exigencies of the Vienna museum. It is clear that he wished to maintain the status of the Vienna meteorite collection as the best and the largest in the world. When he assumed the position as custodian, he was appalled by the loose accounting practices of his predecessors. In March 1886, for example, he requested Fletcher to check and advise him of the weights of all meteorite specimens that Vienna had sent to the British Museum as exchanges in the years 1863,

1869, and 1870, because "the former registrations [of meteorites] under Hörnes and Tschermak have not been made with sufficient exactness," and he warned Fletcher that the weights of the meteorites sent from Vienna might not be correct.[82] As a meticulous bookkeeper, Brezina set a value on every meteorite specimen that left or was acquired by the museum in exchange, and it was a rare occurrence when the value of the specimens received was lower than the value of those surrendered.

Brezina's ambitions, however, were constrained by a very slim budget. The total annual allocation for the acquisition of minerals and meteorites in the period 1848 to 1866 was approximately 2,000 florins.[83] In 1867 there was no allocation, probably as a result of the disastrous war with Prussia. In 1870 the budget returned to 2,000 florins annually, but it was again sharply reduced in the period 1878–1884, possibly owing to the financial and monetary problems of the Austro-Hungarian Empire. At the time, 2,000 Austrian florins amounted to only about $780 U.S., and Brezina evidently conceived of his strategy because of the meager budget. He persuaded some wealthy Austrians to establish a loan fund under the administration of the museum director, upon which Brezina could draw for the purchase of minerals and meteorites, and to which he would turn over funds received from their sale. As a result, the total monies expended for the acquisition of meteorites skyrocketed: approximately 24,000 florins in 1887; 7,100 in 1888; 6,200 in 1891; and 17,000 in 1896. Brezina did not register the meteorites at the actual date of acquisition, but only after he had made the decision whether the whole piece or part should be sectioned for eventual sale and this process was completed. Thereafter, he registered the meteorite and assigned values to each individual section. Among the more costly meteorites that were registered in 1887 were: a 5.11-kg piece of the Misteca iron; the entire 20.84-kg Kendall County iron; five pieces, including the 32.65-kg main mass, of the Eagle Station pallasite; the 41.608-kg Charcas (Descubridora) iron, which Brezina labeled Catorze; a 32.487-kg piece of the Nelson County iron; the 59.90-kg Hex River Mountains iron; the 42.625-kg Kokstad iron; and a 2.478-kg piece of the Cleveland iron, which Brezina labeled Dalton, Whitfield County. Of these, the Charcas iron was the only one not cut up for sale.[84]

Clarence S. Bement, a Philadelphia collector who made exchanges of meteorites with the British Museum, wrote to Fletcher in 1893:

> At present I am exchanging with Brezina, and have secured from him a goodly number of desiderata. I might add, parenthetically, that Dr. B. is somewhat responsible for the present high prices for meteorites which now prevail in this country. As an enthusiast he has no superior, or perhaps no equal, in this line of collecting, and many times I have been told that specimens which I had hoped to procure would be sent to him because of the high valuation he placed upon them.

Fletcher noted in the margin of this letter what his reply would be. "Making not slightest attempt to compete with Vienna. Could not think of adopting the Vienna process—purchasing & selling to recoup the expenditure. This is an interference with the market which is unsatisfactory in itself & its effects."[85]

It was, however, Brezina's purchase of the 384-kg Mount Joy meteorite from Bement for over 17,000 florins in 1892, and more particularly the fact that Brezina had established a private meteorite collection, that caught the attention of the Austrian government authorities. There was an investigation of Brezina's transactions in 1896, at the conclusion of which both Franz von Hauer, director of the museum, and Brezina were forced to retire. One Vienna newspaper speculated that Brezina's removal was politically inspired because he was an active member of a social-political union.[86]

It is doubtful that Brezina was guilty of anything more than poor judgment in dealing simultaneously for the museum and for his own collection. He made donations to the museum collection; in 1887, for example, he gave the museum four pieces of the Fort Duncan [Coahuila] iron, including one 12.140-kg specimen with a total value of 1,105 florins. His resignation did not injure his good name. Fletcher and such dealers as Henry A. Ward continued to make meteorite exchanges with him as a private collector. The government appointed him director of the Jubilee in 1898, on the occasion of the fiftieth anniversary of the founding of the Austro-Hungarian Empire and of Franz Joseph's becoming emperor. He continued research and publication on his favorite subject, the structure of meteoritic irons, and when he died of a heart attack in 1909, he left his substantial collection of tektites to the Vienna museum.[87]

When Brezina left his post, the Vienna collection numbered 482 meteorite localities and that of the British Museum 476. Fletcher was just as anxious as Brezina to augment his meteorite collection, but his approach was different. His policy was not to make any offer for any meteorite, but instead to request the owner to set an asking price, which Fletcher would either accept or refuse. Also, although Fletcher wished to have large meteoritic irons, he does not appear to have been as intent on acquiring them as Brezina.[88] He maintained a substantial correspondence with dealers, and British officials and resident scientists in British India, Australia, Canada, and Central and South Africa were always on the lookout for meteorite falls and finds. One method Fletcher used to encourage public interest was to produce an entirely new type of catalog. At intervals over a period of 33 years—from 1881 to 1914, after he retired as keeper—he published twelve catalogs, all entitled *An Introduction to the Study of Meteorites Represented in the Collection,* London, British Museum (Natural History). The catalog listed, of course, the siderites, siderolites, and aerolites in the collection, and in later editions Fletcher included a history of the collection together with the gifts made to it, and appendices listing telluric irons, pseudometeorites, and the casts of meteorites located in the museum. What made the catalog unique was Fletcher's lucid survey of the development of the science of meteorites, his account of contemporary research, and his exposition of theories of the origin of meteorites. The number of pages that he devoted to this section increased substantially as the years passed, and complimentary references to the catalog in his incoming correspondence give evidence that his efforts were appreciated.

Fletcher's reports on the Tucson meteorite, on the meteorites found in the Atacama desert, and on Mexican meteorites reflect not only his interest in history but also his belief that historical data might be useful in the solution of scientific problems. Mariano Barcena, a Mexican scientist, had made a statement in an 1876 report that provoked Fletcher's interest in the latter subject. Barcena wrote:

> That peculiar property, difficult to explain, which the Mexican soil has in attracting the meteoric irons, is even noticed at present; numerous are the shooting stars which cross the atmosphere of that republic, and more especially in the months of August and November. This phenomenon, which is also observed in other parts of the world, I have seen on various occasions in my country. Lately, one of those shooting stars came against a summer-house in the State of Puebla, causing much damage to the occupants.[89]

Fletcher agreed that many masses of meteoritic iron had been found in Mexico, but he did not believe that the Mexican soil possessed any peculiar property to attract meteoritic irons or that iron meteorites fell during Perseid or Leonid meteor showers. He disagreed with J. Lawrence Smith, who, after studying the meteorites of northern Mexico, concluded that masses dispersed from a single fireball had fallen over hundreds of miles in northern Mexico. He also doubted the hypothesis of Josiah D. Whitney, which Oliver W. Huntington of Harvard supported, that falls from a single fireball had spread over a distance of more than a thousand miles in Mexico and the United States. Fletcher granted that one large fireball might scatter many iron masses, but he thought that their fall locations would not be too far apart. The masses, he believed, could have been moved for either religious or utilitarian reasons, and he pointed out that the Toltecs were quite capable of transporting heavy weights over long distances. Fletcher was only partially successful in this endeavor. He did identify several iron masses that had been given different names as portions of the Coahuila meteorite, but in other instances he labeled as identical several irons since proven to be from different showers. The methods for determining the identity of meteoritic irons were still too rudimentary in 1890 to ensure complete success.[90]

Fletcher kept excellent records of his correspondence during his long tenure as keeper. One vignette is well worth recording—his acquisition of the Crumlin meteorite—for it turned out to be a happy affair. The 4.2-kg stone fell on 13 September 1902 in County Antrim, a dozen miles from Belfast, where the annual meeting of the British Association for the Advancement of Science was being held. Fletcher hurried to Crumlin, but was initially unable to persuade Andrew Walker, the owner of Crosshill Farm, which was the site of the fall, to part with the meteorite. Somehow, Fletcher sensed or learned that Miss A. Black, Walker's niece, had considerable influence on her uncle, and that she dearly wished to have an organ. Fletcher sent his own check for £10 for the organ, and the meteorite came to the British Museum at a cost of £105. Later, in 1908, Miss Black wrote Fletcher that she was now married and wished to apply for the post office position in Crumlin, which was vacant. She requested the use of Fletcher's name as a reference, since it was a government position. There is no record of Fletcher's reply, but it seems likely that he gave his consent (fig. 31).[91]

MAJOR U.S. COLLECTIONS IN THE LATE NINETEENTH AND EARLY TWENTIETH CENTURIES

Until the twentieth century, the meteorite collections in Europe far surpassed those in the United States both in numbers of localities and in quality. The

Fig. 31. The "recovery" of the Crumlin meteorite. From the *Irish Weekly Independence and Nation,* 16 November 1902. Courtesy of the British Museum (Natural History), London.

farmers and prospectors who found many iron meteorites during the westward population expansion in the United States were well rewarded from their sale to dealers, who dispersed them to what appeared to be an eager and unlimited European market. Most of the main masses of the numerous falls that occurred in the United States in the nineteenth century met the fate exemplified by the Estherville meteorite, purchased jointly by Vienna, Paris, and London. During most of the nineteenth century, there were only a half-dozen individuals or institutions in the United States that were interested in establishing substantial meteorite collections. In contrast to the national museums in London, Vienna, and Paris, the Smithsonian Institution in Washington displayed little interest in meteorites until the 1880s. This relative lack of activity in the United States reflected to a large extent the greater interest among European scientists in meteorite research. This situation began to change in the last decades of the nineteenth century as more U.S. scientists turned their attention to meteorites. Between then and World War I a significant trend became evident in the United States that was experienced later in Europe—that is, the acquisition by public or quasi-public institutions of substantial collections, either through purchase, gift, or legacy. Thereafter, the main centers of meteorite research were the museums and universities, which housed these large collections.

Barbara Narendra has recently described the history of the Yale University collection, which was the first in the United States.[92] By 1854 when he retired as professor of chemistry, Benjamin Silliman had acquired about 30 meteorites, the first being the Weston meteoritic stone. From about 1842 his son, Benjamin Silliman, Jr., negotiated for the acquisition and exchanges of meteorites with other collectors, such as C. U. Shepard and J. Lawrence Smith, and with the Vienna, Berlin, and London museums, even though James Dwight Dana, Silliman's son-in-law, was nominally curator of the collection. In 1869, just after George J. Brush, Jr., became curator, there were 103 meteorites listed in Yale's earliest catalog. By the end of the century Edward S. Dana, who became curator in 1874, had increased the number of localities to 200; a major acquisition was the gift by his heirs of the 65 meteorite localities of Hubert A. Newton, who had been so convinced that they were cometary debris. This is one instance of a moderately sized meteorite collection coming into the possession of a university by way of gift.

Another example involved the collection of J. Lawrence Smith. We recall that in 1862, despondent about the course of the American Civil War, Smith sold his collection to C. U. Shepard. However, Smith was soon collecting meteorites again with more than ample funds. He had, in 1866, resigned from his professorship at the University of Louisville to become president of the Louisville Gas Works. In 1870, on the occasion of an exchange of meteorite specimens with the British Museum, he wrote to Maskelyne: "Here I am president of a large gas company, executor of a large estate, administrator of sundry charitable institutions, chemist, mineralogist, etc."[93] The cache of meteorites that Smith desired most was buried in the substantial mineral collection that Gerard Troost had left at his death in 1850. In 1863, Smith had reported to Maskelyne that Troost's collection was stored in a stone building at the University of Nashville, but that it was encumbered by a mortgage, adding that therefore "old C.U.S." did not have Troost's specimens.[94] Years later, in 1875, Smith wrote to Maskelyne that he had succeeded in persuading a Louisville institution to purchase Troost's entire cabinet for £4000, which he said was a low figure. "They won't separate the meteorites," Smith stated, "but I have succeeded in getting the management of it to make exchanges."[95] Thus the meteoritic irons that Troost had collected came gradually into Smith's possession. By 1875, Smith was also wealthy enough that he could send a 539-lb mass of Bolson de Malpimi [Coahuila, Butcher No. 4] to the British Museum, and was "willing to accept your promise as to the payment of the £80 next year at this time."[96] Later, he wrote to Maskelyne: "The fact is I do not believe that there is any one who has collected as many meteorites as myself & have parted with them as freely and for so little consideration."[97] At Smith's death in October 1883, his collection of 250 meteorite localities was purchased by Harvard University through the agency of a subscription fund. Among the subscribers were J. Lawrence Smith, several Harvard professors, and a number of wealthy Bostonians. Smith's widow gave the proceeds to the U.S. National Academy of Sciences to establish an endowment that would supply funds for research on meteorites.[98] Monies from the Smith Fund are still used for this purpose. At the time, Harvard possessed about 50 meteorite specimens, collected largely through the efforts of Josiah P. Cooke, Jr., professor of chemistry. The addition of Smith's collection brought the total number of localities to 270.

In his boyhood Charles Upham Shepard began to assemble what was to become perhaps the finest mineral collection in the United States in the nineteenth century. His meteorite-collecting activity commenced about 1840; as already noted, he was in the mid-1840s selling and exchanging meteorite specimens with the Vienna cabinet and the British Museum. The deposit of about 150 meteorite localities at Amherst College in 1859 was in no way an indication that he had lost interest. In 1880, he wrote to Fletcher at the British Museum: "Since making over my meteorites to Amherst College I have been endeavoring to aid my son (Prof. C. U. Shepard, jr. of Charleston, S. Car.) in the formation of another cabinet, in which enterprise you will see that he has met with much success by the enclosed catalogue."[99] The younger Shepard, who was a professor of chemistry at the Charleston Medical College and also owner of a chemical laboratory and a phosphate fertilizer business, had indeed done well under his father's tutelage and with the aid of his considerable negotiating skills. At that time the collection numbered 68 stones and 57 irons.[100] Advancing age was no impediment to the elder Shepard's activities. He arranged exchanges with Fletcher in 1881, 1882, and 1883, and with other collectors in America and Europe, and he succeeded in obtaining a specimen of one of the initial Glorieta Mountain pallasite masses in 1885 from H. G. Torrey of New York.[101] When the elder Shepard died the next year, the meteorite collection numbered 217 specimens. For the previous six years, the younger Shepard had been corresponding with Frank W. Clarke, honorary curator at the Smithsonian Institution, relative to the disposition of the collection, and after his father's death Shepard deposited the collection there. In 1915 it was bequeathed to the Smithsonian.[102]

In the early years of its existence, the Smithsonian Institution was a focus of debate among several groups who were concerned about its ultimate purpose. Secretary Joseph Henry saw its role as that of a research institution, sponsoring programs in meteorology, geological and geographical exploration, and North American ethnology, and encouraging statistical studies, chemical analyses, and the determination of physical constants.[103] Others wished to have the Smithsonian incorporate a national library, and there was considerable support for the idea of a national museum. Income from the Smithson endowment was modest, so that Henry, apart from his view of the mission of the institution, opposed the proposed library and museum for financial reasons.

In accordance with this policy, Henry did not attempt to emulate the meteorite-collecting activities of the European museums. He did, however, support meteorite research. James Smithson, along with the monetary endowment used to establish the institution, bequeathed his mineral collection, which included 14 meteorites. Henry's 1854 report mentioned that J. Lawrence Smith analyzed these meteorites in the laboratory of the institution. Henry did not identify the meteorites, and they were apparently lost in the fire that destroyed the entire Smithson mineral cabinet in 1865.[104]

Budgetary stringencies, the outbreak of the Civil War, and Henry's policy toward collecting activities probably combined to account for his attitude toward the purported Port Orford meteorite. John Evans, a geologist employed by the U.S. government to make a survey of the Washington and Oregon territories in the mid-1850s, described in his journal in late July 1856 the route taken and the details of the terrain he observed in the mountainous area near Port Orford, Oregon. The journal made no mention of finding a meteorite.[105] However, in 1859 Charles T. Jackson of Boston found a meteoritic fragment among minerals that Evans had collected during his survey. Letters from Jackson brought replies from Evans that he had come upon the meteorite in the Rogue River mountains about 40 miles from Port Orford, that it resembled the Pallas meteorite, that its dimensions were about 5 by 3 feet, with about 5 feet projecting above the ground. He added that he hoped that the Boston Society of Natural History would make efforts to have the government recover the meteorite and place it in the Smithsonian Institution. Jackson and his colleagues responded immediately by urging Joseph Henry to recover the mass and by making the meteorite's existence known in several publications. The news caught the interest of Maskelyne in London, who queried Henry as to what steps were being taken to recover the meteorite. Henry replied that the meteorite was located on the interior side of the Coast range, that it weighed several tons, and that it would cost $20,000 to transport it. He had been advised, he added, to disinter the meteorite, photograph it, and measure it, so that it could be reproduced in plaster, and to procure a specimen for the British Museum. Henry concluded: "I think it probable that these suggestions will in time be acted upon."[106] Henry, of course, could not possibly have persuaded the governmental authorities to dispatch a ship to recover the meteorite in early 1861, when the secession of the southern states was already in process and war loomed large on the horizon. Evans, the person who could have presumably pinpointed the site of the find more closely, died in May 1861. Countless search expeditions have failed to locate the Port Orford pallasite.

Despite Henry's opposition to a national museum and to its incorporation into the Smithsonian, Congress in 1858 passed enabling legislation. Spencer Baird became head of the museum and assistant secretary of the Smithsonian, though Henry retained control over the museum's disbursements until 1872.[107] This circumstance may explain Baird's action with respect to the Butcher irons of the Coahuila shower. Dr. H. B. Butcher in 1868, depending upon reports of previous explorers and upon local guides, recovered eight iron masses in the state of Coahuila, Mexico, just south of the great bend of the Rio Grande. Butcher immediately put them up for sale. Baird apparently made no attempt to obtain them for the National Museum, but instead sent Maskelyne, at Butcher's request, copies of a circular offering the "Mexican irons" for sale. Baird wrote that he hoped the British Museum would secure part or all of the group, and conveyed the information that the Vienna museum "has expressed a desire to obtain all of it, but have made no definite proposition."[108] Maskelyne and Tschermak corresponded about a joint purchase, but nothing came of it.[109] Through J. Lawrence Smith, each later obtained one of the large masses.

The disappearance from view for several decades of the 635-kg Tucson meteoritic iron ring, now a showpiece of the Smithsonian meteorite exhibit, and of the 1,545-kg Casas Grandes iron meteorite is more difficult to understand and explain. Settlers in the Tucson area knew about the meteorite for years. After the territory came under U.S. control following the Gadsden Purchase in 1850, Dr. B. J. D. Irwin, an army surgeon, arranged to have it shipped to Washington, and it arrived at the Smithsonian in 1863.[110] Stored probably in some basement, it must have been forgotten, since George P. Merrill did not mention it in his manuscript catalog of 1880.[111] However, when interest in meteorites began during the next decade, it was retrieved for exhibit. A similar situation occurred with the Casas Grandes iron. Natives who were excavating the ruins of an ancient temple in the north of the state of Chihuahua a few years before 1867 came upon the meteorite. A Texan, August Santleben, made arrangements with the Mexican government to secure the meteorite, and had it transported to the U.S. Centennial Exhibition in Philadelphia in 1876. At the close of the exhibition, the Mexican government donated it to the British Museum, but instead it went to the Smithsonian, although Henry had requested only a sample that could be analyzed for its gaseous content.[112] Again, Merrill did not mention this meteorite in his 1880 catalog. The Smithsonian staff evidently found this meteorite in the 1890s. Merrill sent the meteorite to Henry A. Ward in 1901 to have it sliced, and Ward made use of this opportunity not only to cut a slice for himself, which Merrill had promised him, but also slices for sale to the Vienna cabinet and to the British Museum. Ward's action came to Merrill's attention, and he wrote an acerbic letter to Fletcher, complaining that "Mr. Ward had no moral right to let you have any of the iron." Fletcher noted on the file: "We keep the Casas Grandes. U.S. ought to have given us half the mass for I told them the history of it. . . . L.F."[113]

Following Baird's appointment as secretary of the Smithsonian Institution in 1878, the prospects of establishing a major meteorite collection brightened. George B. Goode, a biologist and ichthyologist, became head of the National Museum in 1880, and he appointed George P. Merrill as head of the petrology and geology section in 1882, and made arrangements in 1883 to have Frank W. Clarke, the chief chemist of the U.S. Geological Survey, serve at the National Museum as honorary curator of minerals. Clarke, as noted earlier, saw to it that the Shepard collection of meteorites and minerals came to the museum, and by the end of the century, he and Merrill had acquired over 171 meteorite specimens, a very substantial increase over the 6 meteorites cataloged by Merrill in 1880. The combined collection of Shepard and the National Museum numbered 348 meteorites in 1902.[114] Merrill's biggest disappointment came a few years later, when he was unsuccessful in securing the Ward-Coonley collection, an account of which will be given shortly.

Ernst Wülfing's 1894 report on meteorite collections listed the American Museum of Natural History in New York City as having only 19 meteorites. A dozen years later, however, the museum could boast of having one of the largest meteorite collections in the world, and displayed two treasures: the 31-ton Ahnighito iron brought by Captain Robert Peary from Cape York in the Arctic Circle, and the 15.75-ton Willamette iron, the largest mass ever found in the United States. This radical change came about as a result of the purchase of individual meteorites or collections by wealthy patrons, who presented them to the museum.

The first major addition was the magnificent mineral collection of Clarence S. Bement of Philadelphia, which was purchased and donated to the museum in 1900 by J. Pierpont Morgan, and which included about 400 meteorites. Bement was the son of an industrialist, and the merger in 1899 of William B. Bement & Son with two other companies created the firm of Niles, Bement & Pond, one of the foremost machine tool manufacturers in the United States. Clarence Bement began to collect meteorites about 1882, and he was soon on cordial terms with dealers, collectors, and curators on both sides of the Atlantic. He increased his collection substantially in the early 1890s by the purchase of Henry A. Ward's first sizable meteorite collection. Fletcher at the British Museum corresponded frequently with Bement, undoubtedly because Bement appeared to be frank and honest in his dealings, and also because he provided a fund of information about meteorite falls and finds and prices. In addition, Bement succeeded in obtaining the entire masses of two American falls: the Ottawa and Pricetown stones. As a result, exchanges with Fletcher brought such prizes to Bement's collection as specimens of Newstead, Nejed, Gibeon, Cowra, Youndegin, Greenbrier County, San Francisco del Mesquital, Mantos Blancos, Barranca Blanca, Ogi, Udipi, Khairpur, Mornans, Pirgunje, Yatoor, and Durala.[115] In March 1898 Bement wrote Fletcher that he was intending to sell his collection, because it had grown too large to be stored in his home. A year later Bement was still vacillating about the sale, but Morgan's offer proved decisive.[116]

The other major meteorite collection acquired by the museum was that of Stratford C. H. Bailey. The museum had purchased a large mineral collection from Bailey in 1875, and Bailey had also donated about 150 meteorites to Adelbert College in Cleveland, Ohio. As with Shepard, however, as an elderly man Bailey could not resist the urge to collect minerals and meteorites, and in the mid-1890s he listed over 260 meteorite specimens. Ward ranked Bailey's collection as one of the best in the United States, and after Bailey's death, his niece sold the entire collection to the American Museum of Natural History in New York. The minerals in it were disposed of to dealers, but Bailey's meteorites were retained to augment Bement's collection.[117]

The Cape York meteorites were the source of the Eskimo iron knife blades, concerning which Captain John Ross had reported in 1818. Another Arctic explorer, Isaac Hayes, brought back additional fragments, and A. E. Nordenskiöld thought that he had discovered the source when he located the Ovifak iron in 1870.[118] The fact that by the turn of the century most authorities agreed that this iron was pseudometeoritic cast doubt on the authenticity of the later Cape York finds. Captain Peary, accompanied by Hugh J. Lee on the North Greenland expedition in 1894, persuaded two Eskimos to guide them to the source of the iron, and located on the northern shore of Melville Bay three masses: the 3-ton "Woman," the 400-kg "Dog," and the 31-ton "Tent" or Ahnighito, the first two showing marks of repeated chipping and hammering. The legend surrounding these meteorites is described in chapter 7. Peary returned in 1895 and loaded the two smaller pieces aboard his vessel, the *Kite,* though not without difficulty. His crew, aided by Eskimos, moved "Woman" onto a timber sledge, which was transported on iron rollers over a plank tramway laid on a rush filled roadbed. At the shore, they shifted the sledge to a large, thick cake of ice, which was floated across open water to the firm ice of Melville Bay, against which the vessel was moored. Again, the crew moved the sledge on rollers across the ice, but this time aided by the ship's winch. As the winch began to lift "Woman" aboard, the ice beneath the sledge broke, partially submerging the piece. Fortunately, the piece had been firmly secured, so it was lifted aboard safely. At this time, Peary partially excavated "Tent," estimated its weight to be 100 tons, and made plans for its recovery.

Though "Tent" proved to weigh only 31 metric tons, its transport and loading were much more formidable tasks than Peary had anticipated. In August 1896 he arrived at Cape York in the 370-ton vessel *Hope,* and hired every able-bodied Eskimo in the village to help. Parenthetically, Peary was not depriving the natives of their source of iron; for decades whalers and expedition ships had furnished all of their needs. Severe storms forced Peary's early departure, but by that time the meteorite had been lifted from its hole by four huge jacks, moved on rails and timbers to the brow of a hill overlooking the bay, and rolled down to the water's edge. In August 1897 Peary returned and constructed a massive wood bridge from ship to shore. The ship was stabilized to withstand the heavy weight, and the meteorite, loaded on a timber car, rode on heavy steel rails placed on the bridge onto *Hope* (fig. 32). A 100-ton floating crane at the Brooklyn Navy Yard lifted it from the hold of the vessel, and there it rested until 1904. In the interim, R. P. Whitfield, chemist and curator at the American Museum of Natural History, made analyses of each of Peary's three masses to determine the nickel content. Morris K. Jesup, president of the museum, solicited the opinions of Fletcher, Brezina, and E. A. Weinschenk of Munich as to the authenticity of the pieces, before Mrs. Jesup purchased them from Peary for $40,000 and gave them to the museum.[119]

The Willamette iron came to the American Museum in April 1905 as a gift from Mrs. William E. Dodge, II, of New York, who purchased the meteorite for a reported price of $26,000, and who stipulated that it should remain a single mass. Ellis Hughes, a farmer living on the outskirts of Willamette, Oregon, found the meteorite in 1902 on land owned by the Oregon Iron and Steel Company. Over a period of several months in 1903, Hughes succeeded in moving the 15.5-ton mass over densely wooded and rough terrain three-quarters of a mile to his home. He fashioned a timber wagon with wooden wheels, onto which he jacked the meteorite, and moved the wagon gradually by attaching it by a wire rope to a capstan, which was anchored to trees in its path, and turned by horsepower (fig. 33). Hughes built a shed around the meteorite next to his house, and charged 25¢ admission to the numerous people who came from the surrounding area to see the curiosity. The Oregon Iron & Steel Company won judgment in the ensuing lawsuit over ownership of the meteorite, although the public sided with Hughes because of his ingenious feat of moving the huge mass, helped only by a horse and his fifteen-year-old son. The company exhibited the meteorite in 1905 at the Lewis and Clark World's

Fig. 32. The largest Cape York meteorite, Ahnighito, being pulled across the massive wooden bridge from the shore to the hold of the *Hope*. From Peary (1914), 581.

Fair in Portland, after which Mrs. Dodge purchased it for the American Museum of Natural History.[120]

Henry A. Ward traveled from Chicago to see the Willamette iron while it was still in Hughes's possession, and persuaded Hughes to sell him several fragments for scientific purposes.[121] Ward was undoubtedly the most enthusiastic and the shrewdest of the nineteenth-century meteorite collectors. He is best known by his activities in helping to establish and develop more than a hundred natural history museums in colleges, universities, and cities across the United States. Ward advised his clients on fund raising, on the building and layout of museums, on the proper methods of preparing, mounting, and displaying zoological and geological specimens, and on the preservation and care of the exhibited material. In return, Ward's Natural Science Establishment of Rochester, New York, received orders for specimens, museum furnishings, and supplies.[122] To provide display material for his clientele, Ward traveled all over the world, acquiring such specimens as the *Megatherium cuvieri,* a massive South American land sloth.

Ward evidently possessed an unusually engaging and magnetic personality. A letter from C. Hicks, a British entrepreneur residing in Coronal, Chile, to Lazarus Fletcher in 1889 demonstrates how Ward impressed people:

> I received a visit here from Mr. Ward, an American Naturalist, who appears to be making almost a speciality of the study of Meteorites, he was then on his way to England after

Fig. 33. The Willamette iron. Ellis Hughes, who built the crude wagon on which it rests, and his fifteen-year-old son, who helped him move the meteorite, are shown in the photograph. From Miller (1923).

having visited all the district of Atacama and even penetrated into the interior of Bolivia *Hunting aerolites*. . . . I am now trying to obtain a specimen [of the Campo del Cielo meteorite] for him. He seemed a very intelligent and nice man, and I took the liberty to introduce him to you with my card."[123]

Similarly, in 1898 Ward was able to persuade the shah of Persia to part with a specimen of the Veramin meteorite. In 1904 the officials of the town of Santa Rosa de Viterbo, Colombia, gave Ward the 612-kg main mass of the Santa Rosa iron in return for his promise to have a statue of a local hero erected in the marketplace. The Colombian government, however, stepped in to prevent its export, although Ward did leave the country with a 150-kg end piece. And despite Merrill's displeasure in 1901 over Ward's selling slices of the Smithsonian's Casas Grandes meteorite, he was soon on friendly terms with Ward again. Ward's success in the early 1900s in assembling the most extensive meteorite collection in the world was in large measure the result of his winning personality, although he was also knowledgeable, an experienced trader, and had sufficient monetary backing.

Ward's intense interest in collecting meteorites began in the 1880s. After 1893, when financial difficulties caused the stockholders of Ward's Natural Science Establishment to reorganize the business and take over control from Ward, all of Ward's energies were devoted to the acquisition of meteorites. After he sold Clarence Bement his first collection, Ward amassed another, consisting of 170 meteorites, which formed a part of the natural history exhibit that his company displayed at the World's Columbian Exposition at Chicago in 1893. It was the sale of this entire exhibit for $100,000 to Marshall Field that prevented the company's complete collapse, and it was this exhibit that formed the principal basis of the Field Museum of Natural History.[124]

Ward was then a widower, and he met and married Lydia Avery Coonley, a wealthy Chicago widow. In 1897 Lydia Coonley Ward and Ward's good friend, Clarence Bement, purchased a choice collection of meteorites that Ward for several years had been setting aside at the Natural Science Establishment. The following year, Avery Coonley, Ward's stepson, purchased Bement's half interest for $1,261, and he and Ward's wife promised to advance Ward sufficient funds to enable him to transform it into a major collection.[125] This was the genesis of the Ward-Coonley collection; all monies advanced to Ward became part of its cost, and the cost increased each succeeding year by the addition of 6 percent annual interest. With this free rein, Ward was indefatigable. Between September 1898 and April 1899, he traveled through England, France, Germany, Austria, Hungary, Serbia, Norway, Russia, and Persia (where he acquired the Veramin specimen), and augmented the collection by 152 new specimens. The total cost of this venture was $1,800, since he negotiated the sale of portions of some meteorite specimens in the collection. In 1901 he made an agreement with the widow of James Gregory, a major London dealer in minerals and meteorites, to sell, for a commission of 10 percent, the 335 meteorites that Gregory had amassed over many years. Ward, of course, kept all or portions of the rarest of the Gregory meteorites. In this way, the Ward-Coonley collection acquired the 141-kg Youndegin iron, the 61.7-kg Wabar [Nejed] mass, and a 4-kg piece that was one-third of the original mass of the Pipe Creek chondrite. In 1904 Ward bought the entire collection of Count Julien Siemaschko of St. Petersburg. Siemaschko had acquired about 350 meteorites, among them the major portions of a number of Russian and Siberian meteorites. Thus the Indarch, Azerbaijan chondrite; the Mighei, Ukraine carbonaceous chondrite; the Pavlodar, Siberia pallasite; and the Ochansk, Russia chondrite came into Ward's possession.

Ward had printed and mailed flyers to mineralogists, geologists, and dealers throughout North and South America announcing his desire to purchase meteorites, and also subscribed to a clipping service so that he could be informed immediately about any meteorite that fell or was found. It was in this way that he learned about the finding of the Willamette iron, and that he was able to acquire the 244.6-kg St. Genevieve County iron, when its owner was willing to part with it in 1900, and also the 80.3-kg Bath Furnace stone soon after its fall in November 1902. It was Ward's normal custom to slice specimens of meteorites that he purchased for sale or exchange to other collectors or museum curators, and during the period 1898 through 1904, he recovered almost $15,000 from such sales.[126]

In 1900 the Ward-Coonley collection numbered 424 meteorites, and Ward began to consider its sale. In April 1901 he signed an agreement with the American Museum of Natural History for the deposit of the collection, although Ward paid transportation costs and furnished the cabinets for its display. The agreement specified that the museum was to have the first refusal of the collection, the value of which was placed at $30,000 the first year, and $31,600 plus the cost of any specimens added during the second year. The available records show that at that time the net cost of the collection plus interest was approximately $22,000. The museum did not exercise its option during the period of the agreement, which was a relief to Lydia Coonley Ward, who thought that the value Ward had placed on the collection was far too low. It remained on deposit at the museum, however, for over a decade.

Ward, aged 72, was struck by an automobile in Buffalo on 4 July 1906, and died immediately from injuries. The Ward-Coonley collection at that time numbered 620 specimens, the most extensive in the world, and its net cost was about $30,000. Mrs. Lydia Coonley Ward engaged Chester G. Gilbert, a consulting geologist attached to Lehigh University, to appraise and supervise the collection. He estimated its value to be $149,500 as of 9 September 1907. The reason behind the disparity between cost and valuation of the collection becomes evident when individual meteorites are considered. For example, Ward purchased the entire 244.6-kg St. Genevieve County iron for $500, and sold or exchanged over half of it. Gilbert appraised the 106-kg remaining in the collection at $12,727. Similarly, Ward's cost of the 150-kg end piece of the Santa Rosa iron was the expense of his trip to Colombia, and he sliced and sold or exchanged one-third of it. The value placed on the remaining 98.2 kg was $9,820. Gilbert also supervised the selection and withdrawal from the Ward-Coonley collection of about 250 duplicate specimens, which formed the basis of yet another Ward collection, this one called the Rochester collection.[127]

Mrs. Ward acted immediately to dispose of the Ward-Coonley collection. She corresponded with George P. Merrill at the Smithsonian Institution, who expressed great interest in its acquisition, and she traveled to Washington, D.C. to urge members of the Illinois congressional delegation to introduce a bill that would provide funds to the National Museum for its purchase. The Panic of 1907, however, caused the failure of these efforts, and in the same year, the American Museum of Natural History declined to act on Mrs. Ward's offer to sell the collection for $105,000.[128] In March 1912 the Field Museum of Natural History in Chicago acquired the collection at a cost of $80,000.[129] After Mrs. Ward's death in 1924, Ward's Natural Science Establishment purchased the Rochester collection and broke it up for sale, Harvard University having the first pick, and the University of Rochester the second.[130] The vast majority of the meteorites that Henry A. Ward collected so assiduously now rest in public museums or institutions of higher learning.

PRICES AND EXCHANGE VALUES

Curators of museum collections complained continuously from the 1840s forward that the asking prices of meteorites were too high. There was always, however, a modicum of rationality in what seemed to be a chaotic market. The prices of such huge iron masses as the Willamette meteorite were always negotiated, since the number of potential buyers was very limited. In most instances, however, the total weight of the find or fall influenced the price. Brezina, as already noted, protested to Fletcher that the total weight of the Rowton fall ought not enter into exchange negotiations, but he certainly knew better. The very rare Chassigny achondrite always commanded a high price, whereas a specimen of the Pultusk shower, which rained an estimated hundred thousand stones near Warsaw, sold at a minimal price. Similarly, the type of meteorite was a factor. A eucrite or howardite cost more than an ordinary chondrite. The number of owners of a given meteorite also was an influence. If one person owned an entire meteorite, the asking price for the sale or exchange of a specimen was always high. For example, S. H. C. Bailey, who was the sole owner of the 1.5-kg Tomhannock Creek chondrite, reported to James Gregory: "The first piece that I disposed of (13.5 grams) brought me four rare stones. . . . The 2nd slice of 14 grams brought me 15 falls . . . a little larger slice will fetch me a new iron of which no cutting has heretofore been disposed of."[131]

The prices per gram asked by dealers for any given meteorite, however, varied considerably; the reason was that the total weight available for sale or exchange was very often unknown. Christian L. O. Buchner made the first attempt to remedy this situation. In his *Die Meteoriten in Sammlungen* (Leipzig, 1863), Buchner listed not only the numbers of meteorite localities held in various public and private collections but also the total weights of 263 meteorites that were represented. In 1893 Ernst A. Wülfing, then a privatdozent of mineralogy and petrography at Tübingen, decided to bring Buchner's tabulation up to date, feeling that the high prices and limited availability of meteorite material was an impediment to research. He sent inquiries to 350 public institutions and private collectors, requesting information as to the name and weight of every meteorite in their collections. Wülfing soon received over 100 replies, but he delayed compilation and publication of the information in the hope that he could obtain more complete data. He also conceived the idea that the results would permit the determination of the exchange value of a meteorite.[132] Wülfing's belief was that if an exchange value could be established on a rational basis, then an exchange between collections could be more easily made and a more extensive distribution of the various types of meteorites would be possible. As a preliminary step, he divided meteorites into groups paralleling Brezina's classification, and began to ascertain the total weight of each group.[133]

Wülfing published the results of his survey and his meteorite exchange values in 1897.[134] He identified 536 localities, and voiced his belief that he had included over 80 percent of the meteorites that were in collections—that is, the total reported weight in collections was 32,412 kg, whereas the total reported weight of all meteorite falls and finds, exclusive of those still in place, was 39,912 kg. He listed all known meteorites in alphabetical order, noted those concerning which details were either lacking or unsatisfactory, and gave the names and locations of all collections. The latter section showed not only the number of localities but also whether the collection possessed a substantial fraction of any particular meteorite. For example, at the time Brezina had 90 percent of La Primitiva and over 96 percent of Sao Juliao de Moreira.

Wülfing thought that eight factors had to be considered in determining a meteorite's exchange value: the amount of preserved material; the total weight of the petrographic group in which the meteorite was classified; the number of owners of a meteorite locality; the addition of a new fall or find of substantial weight to a group of limited weight; the cost of acquisition; the state of preservation; the historical interest; and whether the meteorite was an observed fall or a find. He used the first three factors in a formula to determine the exchange value, W:

$$W = \frac{1}{\sqrt[3]{GNB}}$$

where G = group weight; N = amount of preserved material, and B = number of owners. He compiled four tables for determining the exchange value: for 1–3 owners; for 4–8 owners; for 9–20 owners; and for over 20 owners. In each table, he listed group weights vertically and amounts of preserved material horizontally. A figure listed at the intersection of the vertical and horizontal columns in each table showed the exchange value per gram. The value of the Canyon Diablo iron, of which over 4,100 kg were at the time preserved in collections, was taken as unity. To estimate the exchange value of a new meteorite, one found the group to which it belonged, added the authenticated weight to the group weight, and determined the number of owners. Finally, Wülfing appended a systematic review of all meteorites and listed their exchange values.

Many curators, collectors, and dealers welcomed Wülfing's massive effort (his book, including appendices, numbered 500 pages), but not without reservations. Emil Cohen at Greifswald compared Wülfing's theoretical exchange values with the current asking prices of eight dealers. He found that substantial differences resulted because some relatively large amounts of a meteorite were held by collectors who preferred not to sell or exchange samples, and also because specimens of some rare meteorites only infrequently appeared on the market. Cohen believed that the area of a meteorite slice was an important factor, and insisted that the number of owners who were willing to part with a specimen was much more important than just the total number of owners.[135] In reply to Cohen, Wülfing emphasized that his tables were intended to furnish approximate exchange values, not the market values of individual meteorites.[136] There were, nevertheless, some attempts to correlate Wülfing's tabulations with market prices. In 1904 Ward tried to determine a relation between Wülfing's values and market prices for over 360 meteorites, pointing out in his article that if only a small meteorite mass fell, "it is like a piece of the true Cross."[137] Warren M. Foote, a Philadelphia mineral dealer, made the most ambitious attempt in 1913, when he listed the market values of 465 meteorites and found that Wülfing's exchange values, if multiplied by a factor of 3.2, approximated the asking prices. Foote concluded: "It is certain that Wülfing's work has lessened the absurd variations in value, which abound in exchanging, and that it has also aided in regulating trade prices."[138] Foote expressed the hope that Wülfing would publish a revised edition of his book, but Wülfing, by then a professor at Heidelberg, was not inclined to do so.

There were others, however, such as Merrill at the Smithsonian, who rejected Wülfing's estimates out of hand. In 1913, Merrill wrote Oliver C. Farrington, curator of the Field Museum in Chicago, requesting a specimen of the Pavlovka meteorite, on which he wished to make a chemical analysis. Farrington replied that the Field Museum had 161 grams, which had "by Wülfing's formula the value of $5.79/gram," and that he could not part with a specimen for less than $5.00 per gram. Merrill wrote back an angry letter, asking why Farrington quoted "Wülfing's utterly absurd calculation. In what lies the value of a meteorite excepting for purposes of investigation?" Somewhat shaken, Farrington pointed out that Pavlovka was a howardite, that only 1,123 grams of it were extant, and that he would reduce the price to $3.50 per gram or negotiate an exchange. Merrill, however, still balked at the price, stating that he might consider the matter later.[139]

World War I caused a prolonged collapse in the market for meteorites. At its end, many of the institutions and museums in central Europe, which had previously been eager to acquire meteorite specimens, lacked the funds to purchase them, and even those in western Europe were financially strapped. Meteorites from new falls and finds, of course, came on the market regularly, which increased the supply of almost all meteorite types. Further, the introduction of new techniques permitted analysis of much smaller, even minute, samples, which caused a reduction in demand (see chap. 9). Thus, over many decades, the chaotic market in meteorites, which prevailed in the late nineteenth century, gradually became stabilized.

We mentioned above the Clackamas Indian legend about the Willamette iron, the myth of the Eskimos concerning the three masses of the Cape York meteorite, and also the uses that the Eskimos made of the iron fragments that they chipped and chiseled so laboriously from them. The myths, legends, and folklore surrounding meteors and meteorites are significant. They shed light on the cosmographical ideas of some peoples, indicating that the phenomena of meteors and meteorites were among the first of nature's mysteries that humans attempted to solve. They show that religious beliefs influenced explanations of the appearance of meteors and of the fall or finding of a meteorite. Meteorites engendered terror among some peoples and awe among others, and these emotional reactions may, to a certain extent, explain why none of the meteorites mentioned in ancient texts survive. Peoples through the centuries also used meteorites in various ways. This fact raises questions about the probable equitable distribution of falls of iron meteorites over the Earth's surface, about the malleability of the various types of meteoritic iron, about the beginnings of iron and steel metallurgy, and about the influence, in modern times, of meteorites on the development of nickel alloy steels. We now address this subject matter.

7

Folklore, Myth, and Utility

It was not until the late nineteenth century that scholars realized that myths and folk beliefs about meteors and meteorites existed not only among ancient peoples but among contemporary primitive and civilized peoples as well. Centuries before, certain myths that had prevailed in classical antiquity had been made available through translations. As Aeschylus documented in his *Prometheus Unbound,* when Zeus became angry, he threw a shower of stones down to earth (fig. 34). According to Vergil in the *Aeneid,* a fiery bolide was a sign from the gods of the imminent fall of Troy, and it sufficed to persuade the aged Anchises to leave. St. Paul encountered trouble at Ephesus, when some of the populace thought that the new Christian religion was responsible for the neglect of their Temple of Artemis, in which reposed her image that had fallen from the sky. Lenain de Tillemont in 1693 wrote in astonishment about the black conical stone that Elagabalus had brought from Emesa to Rome in A.D. 218, and which he placed in a magnificent temple to be worshiped by the citizens.[1]

Evidence concerning folk beliefs about meteors and meteorites accumulated from two sources: (1) reports of popular reactions to meteorite falls and finds under circumstances indicating that they had been objects of worship, which increased in numbers through the century; and (2) disclosures about folk beliefs relating to shooting stars, thunderstones, and meteorites, which the new discipline of folklore gradually made available. Both demonstrated that myths and folk beliefs were by no means confined to the area of the Mediterranean when the Greek and Roman civilizations flourished there.

Folklore became a recognized academic discipline by the mid-nineteenth century, under the guidance of the brothers, Jacob and Wilhelm Grimm. Their *Deutsche Mythologie,* published in 1835, was an exemplar, and the spirit of German nationalism it displayed stimulated scores of German scholars to investigate the folk beliefs of the various territories or *Länder* whose peoples spoke a common tongue. Understandably, then, a great deal of the folklore studies of the nineteenth century concerns Germany. Within decades, however, the boundaries of the discipline expanded. Rising nationalism in other Euro-

Fig. 34. Zeus hurling a thunderstone. Statue in the National Archaeological Museum, Athens, dated ca. 450 B.C. Photograph courtesy of Vagn F. Buchwald, Technical University of Denmark.

pean countries aided the process of collection. In England, Sir Edward Tylor's works founded the science of social or cultural anthropology, and Sir James Frazer's *The Golden Bough,* first published in 1890, gave it widespread prominence. Today the study of folk beliefs is a worldwide endeavor, and this chapter draws on material from both nineteenth- and twentieth-century sources.

There is no consistency in beliefs about meteors and meteorites. A culture may have several divergent ideas about, say, the significance of the appearance of a shooting star. Two different cultures may have opposite views about the fall of a meteorite, one considering it a good omen, the other a bad sign. Also, peoples in two widely separated cultures may have identical beliefs, either

Photomicrograph of a thin section (3.6 mm wide) of the Pulsora olivine-bronzite chondrite exemplifies why early investigators had difficulty identifying the mineral constituents. At the upper left, olivine fragments are mixed with silica-rich glass. At the lower right, the normal matrix, containing numerous chondrules and fragments, is mostly fine-grained olivine-pyroxene. The olivine and pyroxene display pastel tones; the glass is pink; and the nickel-iron and troilite are dark brown. The spherule at the top left is metal/troilite. From Fredriksson et al. (1975). Courtesy of Kurt Fredriksson, Smithsonian Institution.

Olivine within pallasites showing true color and internal fractures and reflections. At left, Krasnojarsk, U.S.N.M. no. 1057, width of section, 1 cm. At right, Imilac, U.S.N.M. no. 6272, width of section 1.5 cm. Courtesy of Roy S. Clarke, Jr., Smithsonian Institution.

Plate 1.

A 75-g slice of the Bustee aubrite, U.S.N.M. no. 5622, 9 cm wide. The large, light-colored inclusions are enstatite, and much of the rest of the specimen is fragmented enstatite. It was the study of this meteorite, among others, that led Nevil Story-Maskelyne (1870) to suggest that enstatite was a member of a solid solution series. Courtesy of the Division of Meteorites, NMNH, Smithsonian Institution.

Coarsely crystalline plagioclase in a polished thin section (2.2 mm wide) of the Serra de Magé eucrite, U.S.N.M. no. 839. Transmitted light photograph with crossed Nicols. Courtesy of Roy S. Clarke, Jr., Smithsonian Institution.

Two views of a piece of the Bishopville aubrite, which is almost completely enstatite. This specimen, U.S.N.M. no. 971, is 10 cm wide and weighs 573 g. It is a portion of a 1090-gram specimen that came to the Smithsonian in the Shepard collection. Shepard (1846, 1848*c*) named the mineral chladnite but was unsuccessful in his analysis of it. Courtesy of the Division of Meteorites, NMNH, Smithsonian Institution.

Plate 2.

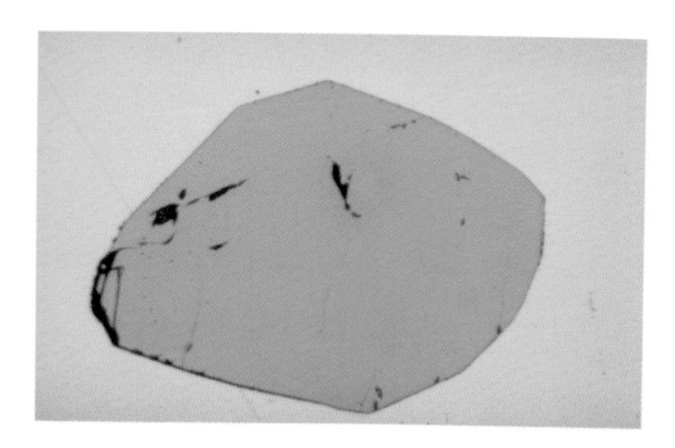

Troilite in a matrix of kamacite in a metal-rich part of the Reckling Peak A79015, Antarctica, mesosiderite. The section (560 μm. wide) is lightly etched. Courtesy of Roy S. Clarke, Jr., Smithsonian Institution.

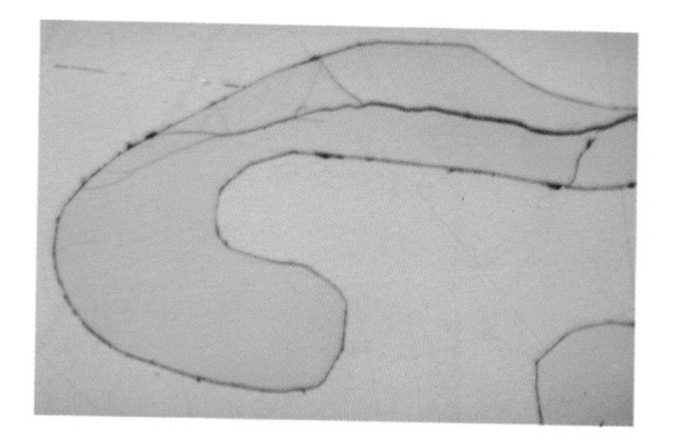

Schreibersite in a matrix of kamacite in a metal-rich part of the Reckling Peak A79015, Antarctica, mesosiderite. The section (560 μm. wide) is lightly etched. Courtesy of Roy S. Clarke, Jr., Smithsonian Institution.

Plate 3.

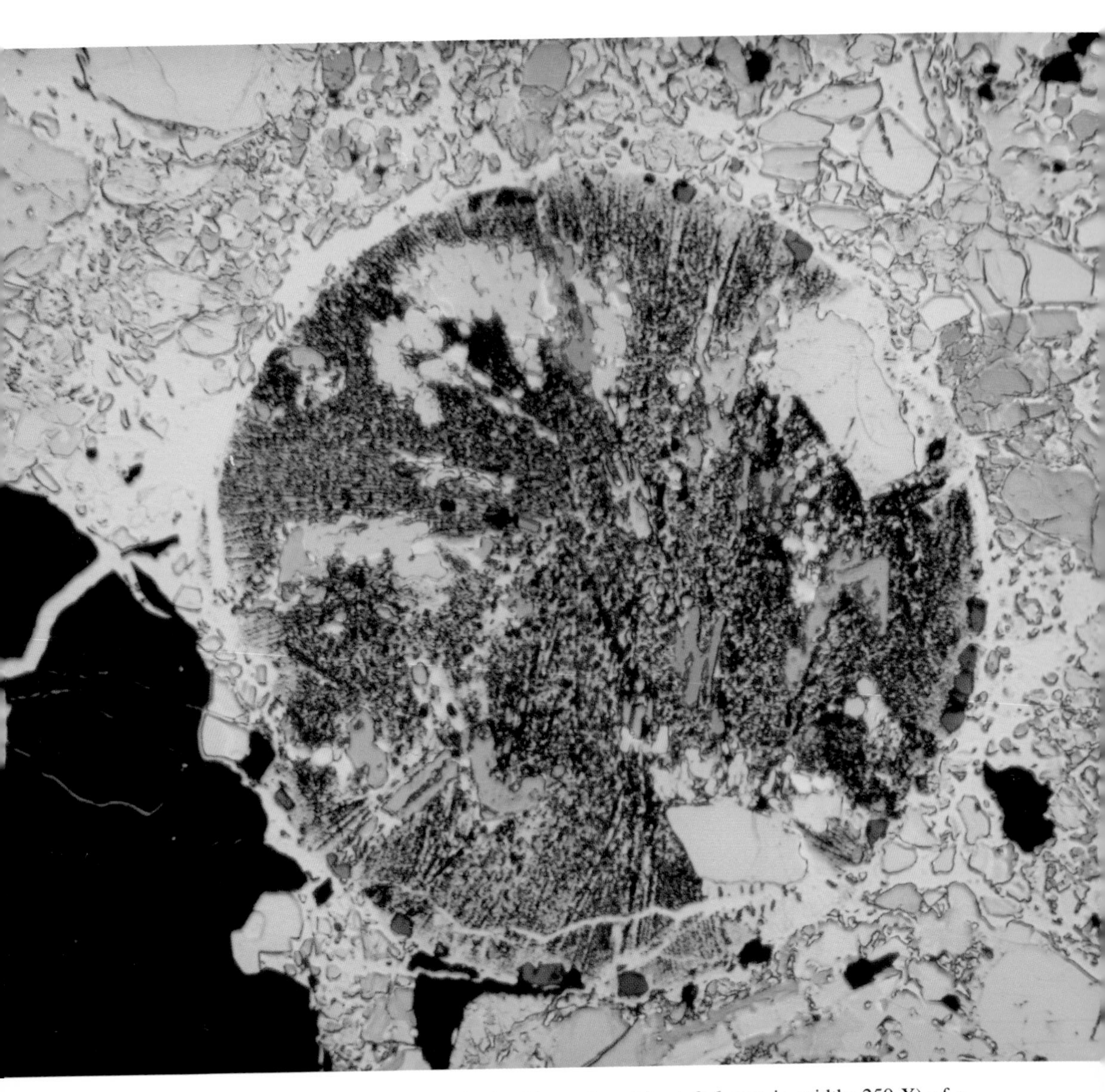

Photomicrograph of a polished ultrathin section (about 0.6 mm in width; 250 X) of a chondrule in the Allegan olivine-bronzite chondrite. The olivines are purple, blue, and yellow, and chromites are dark brown. The radiating pyroxene is very fine-grained. Detailed studies of the chemical composition and structure of chondrules have given rise to numerous theories as to how and when they formed. Photo by R. H. Beauchamp. Courtesy of Kurt Fredriksson, Smithsonian Institution.

Plate 4.

because of the diffusion of ideas or because the phenomena of meteors and meteorites may give rise to the same concept in different localities. Folklorists think that some of the beliefs originated in the distant past, since they gained credence in Europe before Christianity arrived. Others appear to have a distinctly Christian flavor, although it is quite possible that an original pagan belief acquired a Christian interpretation over time.

The folklore and myths about meteors and meteorites have an intrinsic interest in that they demonstrate the extraordinary range of the human imagination. A number of these beliefs suggest a reason why relatively few meteorite falls and finds were reported in Europe before the end of the eighteenth century. Peasants in remote areas regarded meteorites as possessing singular attributes or powers, or as containing within them gold or silver. A meteorite was a prize to be treasured, and its fall or disposition were not matters to be divulged to authorities who would seize it. There is sufficient evidence to make this idea plausible.

During the course of the nineteenth century, data accumulated that confirmed the prehistoric use of meteoritic iron. Farmers plowing fields, engineers digging excavations for public works, and explorers in remote areas found iron meteorites that had obviously been chiseled or hammered. Archaeologists turned up artifacts that had been fashioned from meteoritic iron. Chronicles written in past centuries also came to light, which recorded the fabrication of meteoritic iron into knives, swords, and scimitars; twentieth-century scientists have made thorough studies of a few of these that were deposited in Chinese tombs well over two thousand years ago. About 1900 the question as to whether the use of meteoritic iron gave impetus to the genesis of iron metallurgy began to be debated. Most authorities now agree that the existence of iron meteorites had no effect on the beginnings of iron metallurgy. But there does appear to be at least a tenuous connection between the increasing knowledge of the chemistry of iron meteorites and the development of nickel alloy steels in the late nineteenth century.

BELIEFS ABOUT METEORS

A folk belief common in Europe, Eurasia, and North America is that if one makes a wish when one sees a falling star, it will come true or be fulfilled.[2] In some areas there is the caveat that one must make the wish before the starlight is extinguished. Some scholars speculate that this belief originated at a time when people thought that the gods occasionally opened the dome of heaven to see what was occurring on Earth, thereby releasing a star. If one made a wish while there was still light and before the doors slammed shut, the gods would hear the wish and see to it that it was fulfilled.[3]

In some areas, shooting stars are favorable omens signifying that the viewer will enjoy good luck. In Switzerland, a meteor was considered to possess the power of God, so that it could ward off pestilence and hemorrhage.[4] Swabians believed that a shooting star presaged a year of good fortune.[5] In a few localities, however, a person must perform some action to ensure the subsequent good luck. In Chile one must pick up a stone simultaneously, and in the Philippines, one must tie a knot in a handkerchief before the light is extinguished.[6] The direction of the falling star also assumes importance. This was the case in the ancient Near East; and among the modern Hawaiian Japanese, if a meteor comes in your direction, you must open the collars of your kimono to admit the good luck.[7]

Much more frequently, however, ancient and modern peoples considered meteors to be signs of misfortune or death. Shakespeare, in several passages, echoed beliefs that appear frequently in folklore literature. In the opening scene of *King Henry IV, Part I,* the king voices relief at the end of the internecine strife that has gripped England, and

> Which, like the meteors of a troubled heaven,
> All of one nature, of one substance bred.

Later in the same play, when Henry's supporters, the Earl of Worcester and his son, Hotspur, rebel, the king pleads with Worcester:

> . . . be no more an exhaled meteor.
> A prodigy of fear, and a portent
> Of breached mischief to the unborn times?

In *Julius Caesar,* Calpurnia, Caesar's wife, warns him that "Fierce fiery warriors fight upon the clouds," and that "The heavens themselves blaze forth the death of princes."[8]

The earliest literature extant gives evidence that meteors have long been considered omens of disaster and death. In 1973, Bjorkman made a comprehensive summary of the content of Mesopotamian writings on this subject, which were collected and systematized in the second millenium B.C. With a few exceptions, meteors were evil omens, signifying the occurrence of serious injury, death, rebellion, or the fall of kingdoms.[9] It is a common belief throughout Europe that a falling star signifies that someone has died, as modern folklore studies amply demonstrate.[10] There are, however, variations on this theme. In Swabia, in the late nineteenth century, the belief was that only a person seeing three meteors in one night was doomed to die.[11] In Lithuania, people thought that Fate, a spinner, began to spin the thread of each person's life at birth and attached the thread to a star; at the moment of death, the thread broke and the star fell.[12] Similarly, in other eastern Baltic countries and in central Europe, people believed that everyone had a personal star, which fell at his or her death.[13]

The idea that shooting stars were omens of misfortune or bad luck was prevalent worldwide in the nineteenth and early twentieth centuries: in Europe, the British Isles, Iceland, Africa, the Orient, and in North America among both white and black communities and several Indian tribes.[14] However, in North Carolina, Kansas, and black communities in the southern United States, bad luck occurred only if one talked about a falling star.[15] In many areas of the United States, it was believed that one experienced misfortune only if one pointed to a meteor.[16] The Seneca Indians thought that pointing would cause the star to reveal one's hiding place.[17] Other peoples believed that with certain precautions, bad luck could be avoided. In Spain and among Hispanic Americans in New Mexico, uttering the words, "God guide it," would avert misfortune, whereas in Russia, one said "Amen."[18] The Baronga peoples in Lourenco Marques spat on the ground when a meteor appeared and cried out, "Go away, go away, all by yourself."[19] One folklorist hypothesized that saying such words is a manifestation of primitive defense, as when people danced or sang to exorcise spirits.[20]

In a number of Roman Catholic communities, shooting stars are identified with human souls, either expiating their sins in purgatory or wandering about the world. In Polish Silesia, Baden, several areas of France, and in parts of Kentucky, Louisiana, and Texas, people believe that meteors are souls leaving purgatory to go to heaven.[21] A more general opinion in France and Catholic Germany is that a shooting star is a suffering soul entreating prayers from those who see it. If one recites "Rest in peace" three times before the meteor is extinguished, the soul will be delivered from purgatory.[22] At Santiago, Chile, falling stars are thought to be wandering souls, and country folk, upon seeing one, call out, "May God guide you to a good path."[23] Shooting stars in the Philippines are considered to be the souls of drunkards, which return to earth at night and sing, "Do not drink, do not drink." Each day they attempt to climb to heaven, but each night they fall down again.[24]

Peoples have often identified shooting stars with unusual actions by the inhabitants of heaven. In the ancient Near East, meteors were messages from the gods to mankind; in Mecklenburg, they were angel messengers sent by God; and the Skidi Pawnee thought they were messages from friends.[25] In Ireland, people spoke of the Perseid meteor showers as the burning tears of St. Lawrence, who was burned alive on a gridiron in 258 during the reign of Valerian, and whose feast day is August 10. In Thessaly, folk believed that heaven opened on the night of August 6, the feast of the Transfiguration, and that meteors were torches that appeared in the opening.[26] Swabians and Bohemians thought that shooting stars appeared when angels pinched out the ends of the wicks of heaven's lights.[27] Aelfric, a Saxon who lived ca. A.D. 1000, had a more naturalistic explanation: meteors were fire struck off the heavenly bodies like the sparks shooting from a fire.[28] Moslems in the Middle East believed that shooting stars were fires that the angels rained on devils.[29] Russians thought they were demons who transformed themselves into stars and were driven out of heaven by angels.[30]

In certain areas peoples compared shooting stars and bolides to animals. In the ancient Near East they were called lions, dogs, and jackals; in Greece they were goats; and in the Germanic territories they became snakes and particularly dragons.[31] In both Mesopotamia and Germany, people thought they were demons or devils with an animallike form. In Mecklenburg, Austria, and Iceland, special precautions were thought necessary when a dragon with a bare head and fiery tail passed over the house. Children were brought inside, covered, and sprinkled with holy water.[32] If one was traveling in a wagon with one of the wheels reversed when a dragon appeared, it would dive into a house and set it on fire.[33]

These fiery dragons, people believed, carried all sorts of things with them. It was thought in Hesse that if one insulted a dragon, it would throw foul cheese and stinking rubbish, and in Mecklenburg it would drop excrement.[34] Brunswick residents believed that the dragons could be influenced. If one stood under the eave of a house and shouted, "Fire dragon, share with me," the dragon would drop a ham or a side of bacon. The dragon also carried money, which it occasionally let fall, making many people rich.[35] In Mecklenburg the dragon brought money only if it was illuminated; if it was dark, it carried vermin. A fireball was a dragon with a great deal of booty stolen from elsewhere.[36] Stones from fallen meteors, the Swabians believed, contained cores of genuine gold and silver.[37] In Polish Silesia, it was thought that in the place where a meteor fell one would find either a treasure, a gelatinous mass, or cow dung.[38]

Folk beliefs, then, taught that shooting stars were transient links between heaven and earth. They were the vestiges of departed souls; they signaled bad luck, but with proper precautions, good fortune might follow and one's wish would be fulfilled. They were to be feared, but if something that was not negative fell from them, it was likely to be a treasure, perhaps gold or silver.

MYTHS AND FOLK BELIEFS ABOUT METEORITES AND THUNDERSTONES

Studies indicate that in the past within a few societies little or no significance was attached to falls or finds of meteorites. Bjorkman reported that no evidence yet exists to show that the Mesopotamians venerated meteorites or considered them to be remarkable in any way. As was the case with eighteenth-century Europeans, the Mesopotamians did not understand the connection between fireballs and meteorites. They knew that stones might fall from the heavens, but they did not link these objects to the phenomena of meteors and bolides, which they believed to be messages from the gods.[39] The Eskimos living in northwestern Greenland within the Arctic Circle, who found several masses of the Cape York meteorite, did not regard them with fear or awe. They told Peary that the three pieces—a woman, her dog, and her tent—had been hurled from the sky by the evil spirit, Tornarsuk, but they did not hesitate to hammer and chip the masses and carry away the head of "Woman" to a more convenient location.[40] Henry A. Ward was critical of the Kansas farmers, who, he reported, knew full well that the pieces of the Brenham pallasite recovered in 1882 were meteorite fragments, yet were using them for such purposes as to hold down a rain barrel cover, to weight the top of a haystack, and to plug a hole in a pigpen.[41] Taking unusual natural phenomena in stride is apparently a timeless human attitude among certain types.

In other times and places, however, meteorites were thought to be omens of evil or misfortune. An early sixteenth-century letter from the Grand Master of the Knights of Rhodes to a correspondent in Venice described a shower of stones near Cairo and expressed his fears that the meteorite shower was an omen of the end of all things.[42] People in Switzerland took a less dire view of such an event, believing that an abundant fall of stones meant that war was coming.[43] To some Mexicans, it is a more personal matter; they offer a prayer to St. Bartolomé to protect them from falling stones.[44] When the Jhung chondrite fell in 1873 in West Punjab, Pakistan, some natives believed that it was a warning from heaven because of the evil deeds of mankind.[45] Tanzanians had known about the Mbosi meteoritic iron for generations, but did not report its existence until 1930 because it was taboo.[46] Australian aborigines in 1900 were "deadly afraid" of the Warbreccan masses of the Tenham meteorite, a correspondent reported: "They cover them in the bush with kangaroo grass, twisted gidga bark and mud, and then by boughs over the top. Their idea is if the sun sees them more stones will be shaken down" to kill them.[47] When the Durala stone fell in 1815 in the Punjab, India, the people were terrified, and the Rajah, viewing the event as an evil omen, gave an Englishman permission to take it away, although some Brahmins wished to build a temple for its veneration.[48] Here, the opposite attitude of the Brahmins and the other natives demonstrate that there is often a fine line between intense fear and reverential awe.

The worship of meteorites appears to have been a much more common reaction. It was prevalent in the Greco-Roman world, as is evidenced by the extant writings of historians and chroniclers, and by the existence of a plethora of betyl coins or medals that date from approximately 300 B.C. to A.D. 300. Classicists have differed on the precise meaning and derivation of the Greek word, *baitylos,* but they agree that it was either a sacred stone or represented a sacred stone or meteorite that had fallen from heaven.[49] The obverse of a betyl coin characteristically features an omphalos, that is, a cylindrical pedestal whose upper surface is a hemisphere or a cone. Gerald A. Wainwright, an Egyptologist and archaeologist, wrote a series of articles about 1930 in which he noted the presence of omphaloi at temples in ancient Egypt dedicated to the worship of Ammon. He postulated that an omphalos at a subsidiary temple represented the sacred meteorite at the chief temple; that the identification of Ammon with Zeus and with the Semitic deity, Yahweh, occurred about the tenth century B.C., thus spreading the worship of meteorites to Greece and to Asia Minor; and that "meteorite, omphalos, and thunderbolt were all one and the same thing in religion."[50]

Critics have since pointed out that Wainwright's identification of thunderbolt, meteorite, and omphalos is not universally the case. Forbes concluded that "Wainwright's reasoning stands only as an attractive hypothesis."[51] Yet H. A. Newton of Yale, who in the 1890s documented Greco-Roman myths about meteorites, was convinced that the creases and depressions that appeared on the surfaces of almost all omphaloi were artists' attempts to simulate the regmaglypts on the surfaces of meteorites, which are frequently the result of their atmospheric passage. O. C. Farrington, then curator of the Field Museum's meteorite collection, supported Newton's opinion.[52] A few examples of the legends and beliefs and illustrations of the betyl coins representing them help indicate the complexity of the scholarly problem.

The temple at Delphi was one of the most important shrines in the Greco-Roman world. Pausanias (ca. A.D. 180) described the omphalos there as a moderate-sized stone that was anointed every day and covered during festivals with newly shorn wool. The myth surrounding its location at Delphi was that Saturn (Cronos) had devoured four sons that Cybele (Rhea) had borne him, but that when Zeus was born she gave him a stone in lieu of Zeus, which he proceeded to swallow. After Saturn's dethronement, he disgorged the stone, and either he or Zeus threw it from heaven to a place that was considered the center of the earth. Later, the site became Apollo's temple, but the stone or omphalos remained (fig. 35a).

We made brief reference earlier to the image of Artemis (Diana) at Ephesus. In the second century B.C. the magnificent temple of Artemis at Ephesus was named one of the seven wonders of the world. Worshipers paid homage to the image of Artemis therein, which was said to have fallen from the sky. According to Wainwright, the omphaloi in the subsidiary temples of Artemis at Perga, Pogla, and Andeda replaced and represented the meteorite

a. Silver coin of Mithradates I, king of Parthia ca. 171–138 B.C., representing the omphalos of Delphi.

b. Silver coin of Vespasian, Roman emperor A.D. 70–79, representing Venus (Aphrodite) of Paphos, Cyprus.

c. Silver coin of Elagabalus, Roman emperor A.D. 218–222, representing the stone of Emesa.

d. Bronze coin of Elagabalus, Roman emperor A.D.218–222, representing the stone of Astarte.

Fig. 35. Ancient coins depicting worshiped stones or omphaloi. Courtesy of the Naturhistorisches Museum Wien.

at Ephesus.[53] Silversmiths at Ephesus made copies of the stone from precious metals and minerals, and fearing that their craft was threatened by Christianity, they created trouble during St. Paul's mission.

There were three temples of Aphrodite (Venus) in Asia Minor that were connected to a meteorite: the first was at Aphaca, a sacred place not far from Byblos, where the second was located, and the third was at Paphos on the nearby island of Cyprus. Aphaca was sacred because Aphrodite Urania fell at the place as "fire from a star." Aphrodite, both at Byblos and Paphos, was represented by an omphalos—that is, by an elongated cone within a temple of two columns (fig. 35b).[54]

We also mentioned (both earlier in this chapter and in chap. 1) Elagabalus and the stone of Emesa. At age twelve, Elagabalus became high priest at Emesa of the temple of the sun god, Heliogabalus, who, in the form of a black stone, resided therein. Because of the lavish ceremonies over which he presided, his manly beauty, and a rumor that he was the illegitimate son of the just-murdered Caracalla, to whom in fact he was related, the Roman troops at Emesa named the youth emperor under the name of Marcus Aurelius Antoninus. When an attempt to put down the rebellion failed, the Senate recognized him as emperor in A.D. 218. He took the name Elagabalus and made a triumphal entry into Rome, bringing with him the stone, which he insisted become an object of public worship. According to the contemporary historian Herodian, the idol was a large black conical stone (fig. 35c). Elagabalus soon lost his popularity because of his profligacy, and was slain by the praetorian guard in 222. It is not known what happened to the stone.

Eusebius (ca. A.D. 260–340), citing a very old Phoenician chronicle, wrote that Astarte, the Semitic nature goddess, donned the head of a bull and went to look for a "star fallen from the sky." When she found it, she brought it to the island of Tyre to be worshiped. The inhabitants of neighboring Sidon worshiped a second stone of Astarte, and some coins of Tyre exhibit both stones (fig. 35d).[55]

There are at least half a dozen other legends of the Greco-Roman period that speak of gods falling down from heaven and that were commemorated by temples and by betyl coins. The worship of meteorites died out about the time of Constantine, who, legend has it, was converted to Christianity in 312 by the vision of a flaming cross in the sky at midday. Four centuries elapsed before large numbers of people venerated (rather than worshiped) another stone, the legendary black stone built into the wall of the Ka'ba, the holy shrine of Islam in Mecca.

The black stone in the wall of the Ka'ba is a holy relic. Muslim religious leaders know its origin and history through oral tradition and written records, and they have cooperated with inquisitive Westerners to the extent of providing this information and giving a cursory description of the stone.[56] Thus, although many have speculated since the early nineteenth century that the black stone was a meteorite, there is no proof that such is the case. Recent studies, in fact, discount its meteoritic origin.

Paul Partsch, curator of the Vienna cabinet, published the first comprehensive history of the black stone in 1857.[57] He relied on the travel accounts of Carsten Niebuhr (1772), J. L. Burckhardt (1814), and Ali Bey (1807), and also corresponded with Ritter von Laurin, the Austrian general consul in Egypt. In his official capacity, von Laurin knew Mohammed Ali, viceroy of Egypt, who had in 1817 defeated the fanatic, heretical sect of Wahhabis and retrieved the holy stone, a fragment of which he kept. Von Laurin saw this fragment and described it in his letter to Partsch, adding that an English resident who had also viewed the stone considered it to be a meteorite. Although Partsch was cautious, he favored the stone's meteoritic origin, and authorities accepted this opinion.

The legend is that the stone came from paradise. In one version, it was initially Adam's guardian angel, who was transformed into a stone as punishment for Adam's fall. The angel Gabriel gave the stone to the patriarch, Abraham, to build into his house or into the first Ka'ba. There was indeed a temple on the site, dating from about A.D. 200 and housing idols, which the Arabs worshiped before the time of the prophet Mohammed (570–632), but it was destroyed, possibly by fire, during his lifetime. Mohammed placed the stone in the wall at the northeast corner of the shrine when it was rebuilt. It was subsequently broken on at least four occasions: once by a fire; twice by fanatic sects who took possession of the stone for a time; and once by invading Egyptian troops, whose leader shattered it with a maul. Reports state that on each occasion the recovered pieces were cemented together with mortar, and the whole bound with a silver hoop.

It is impossible to estimate the original size of the stone or even its present dimensions. One observer in the early tenth century wrote that it had a length of 1 cubit (slightly over 2 feet). Another, who saw it during the remodeling of the wall in the early seventeenth century, stated that it measured 1.5 by 1.33 yards. Ali Bey stated that it was 42 inches high, and Mohammed Ali reported that it was 2.5 feet long and 1.5 feet high. At present, the exposed face, which is surrounded by a wide oval frame of silver, measures 20 by 16 cm—approximately the same dimensions of the face recorded by Ali Bey. Burckhardt wrote that the face was composed of a dozen smaller stones of various sizes and shapes; at present eight small pieces comprise the face, the largest about the size of a date.

The criteria for judging what mineral species the stone contains have been the color, texture, and estimated specific gravity. According to one legend, the angel gave Abraham a transparent hyacinth; according to another, it was originally pure white and became black either because it was kissed by a sinner or because of the sins of mankind. The exterior face of the stone is black and highly polished, due to its having been rubbed by millions of pilgrims. Modern observers report that there are a few white or yellow dots on the face, and an official record states that it is white with the exception of the face. Von Laurin described the fragment he saw, which was purportedly carried away by Mohammed Ali, as having a pitch-black exterior and a silver-gray, fine-grained interior, in which tiny bottle-green cubes were embedded. Burckhardt wrote that it was difficult to judge the quality of the stone, but that it appeared to be lava. The English resident Lyons, who, according to von Laurin, thought the stone was a meteorite, remarked that it was heavy. Another report, however, stated that it floated on water; this quality permitted the identification of the pieces recovered from its initial theft.

Partsch evidently favored a meteoritic origin of the stone, both because of von Laurin's description of the black exterior of the fragment he viewed, its interior texture, and its purported heaviness, and because Muslims said that it came from heaven and venerated it as the Greeks and Romans venerated similar stones not too far distant in time and place. In 1974 Dietz and McHone emphasized that the Muslims do not claim that the stone is a meteorite. They postulated that the stone is an agate, because of the high polish it displays among other physical attributes and because an Arab geologist, who studied the stone carefully during a pilgrimage to Mecca, reported that "diffusion banding is clearly discernible within the stone."[58] In 1980, however, Thomsen presented a different hypothesis. She suggested that the stone may be a chunk of impactite glass, mined from one of the meteorite craters at Wabar in the so-called Empty Quarter of central Saudi Arabia, about 1,100 km from Mecca. She pointed out that the "whiteness may derive from an exposure of the interior white core of a bomb or . . . from a large fragment of white glass or sandstone," and that the whiteness remains only where it is protected by cement. Further, she wrote: "The yellow and white spots may be remnants of glass and/or sandstone. The porosity which allows it to float is due to vesicles in the glass, and the resistance of the material to abrasion due to the hardness of the glass. The blackness results from the nickeliferous iron spherules captured from an explosion cloud of Ni and Fe." Thomsen also thinks that ancient Arabs may have observed the meteorite fall, estimated to have occurred about six thousand years ago, and that natives later carried the impactite glass to Mecca along a caravan route.[59]

Thus, there is now considerable doubt that the black stone of the Ka'ba is a meteorite. Whether or not the stones worshiped by the Greeks and Romans were meteorites is also a question, since none of these objects have survived the centuries. There are only the considerable number of legends. In contrast, archaeologists, explorers, and farmers have found about a dozen meteorites or fragments of meteorites in North America in circumstances which indicate that they were worshiped, venerated, or at least prized, but for most, no legend exists. Two of these are pallasites; the rest are irons. What is intriguing is that for most, the degree of corrosion gives evidence of a very long terrestrial age, meaning that the falls occurred thousands of years before Indian natives populated North America. The Morito iron, which the Indians of Chihuahua, Mexico, venerated in pre-Columbian times, probably fell within the last two thousand years, according to Buchwald, and may, therefore, have been witnessed.[60] There are only a few clues as to why the meteorites were venerated, and one can only surmise why the natives believed that the objects in question had fallen from the sky.

The Hopewell Indian culture flourished from about 500 B.C. to A.D. 500. The Hopewells settled in villages and did primitive farming, and they expended substantial labor in constructing large burial mounds and related earthworks. Within several of these mounds in southern Ohio and central Illinois, archaeologists have found ornaments either made from or overlaid with meteoritic iron. A separate fragment weighing 767 grams, which was found in Mound No. 4 of the Turner group in the Little Miami Valley, Ohio, was a pallasite, and scientists have since identified it as a transported piece of the Brenham pallasite.[61] The site of this find is a thousand miles distant in Kiowa County, Kansas, which adds to the previous evidence that the Hopewell Indians carried on trade over long distances. In addition, twenty-two beads that were found in Mound No. 9 of the Havana, Illinois, group were fashioned from a fine octahedrite, and the location of the main mass from which this material was obtained is not known.[62] The presence of meteoritic iron objects in these burial mounds indicates that they were probably treasured possessions, but evidence that the objects were venerated is lacking. The occurrence along with them of artifacts made of copper and of shells, which were much more common material, supports the idea that they were personal belongings.

Other Indians besides the Hopewells buried meteorites in graves. The Oktibbeha County meteorite was in an Indian tumulus near Columbus, Mississippi. An Indian grave at Livingston, Montana, consisting of a pile of rocks over the remains, yielded a 16-kg iron meteorite in addition to stone tools, arrowheads, and pieces of pottery.[63] The Camp Verde 61.5-kg iron is a transported piece of the Canyon Diablo meteorite. In 1915 G. A. Dawson opened a stone crypt in an ancient Indian building and found the meteorite inside wrapped in a feather cloth. Archaeologists have estimated the age of associated pottery at about 800 years.[64] Similarly, maguey cloths enveloped the 1,545-kg Casas Grandes meteoritic iron, which was found before 1867 in a multichambered tomb in northern Chihuahua, Mexico. Other chambers contained human remains, which were wrapped in the same way.[65]

The Indians who buried these meteorites must have regarded them not only with reverence but also as possessing supernatural powers. What legends do exist support this view. The finder of the first mass of the Navajo, Arizona, meteorite, which was buried under rocks, reported in 1921 that the Navajo Indians had known about the piece for three centuries, but because it was sacred they had covered it with rocks to conceal it from white men and other tribes. Its weight of 1,500 kg probably prevented its transport. The second 683-kg mass, found five years later just 48 meters from the first, was also hidden under rocks, above which was a marker stone.[66] The Skidi Pawnee Indians, whose ancestral home was in east-central Nebraska along the Platte River, called meteorites the children of Tirawahat, their chief god. They wrapped objects believed to be meteorites in bundles that they considered sacred and that belonged either to individuals or to the tribe. A wonderful being named Pahokatawa came from the sky as a turtle-shaped stone, as legend predicted, and the tribe carried it with them in a bundle. When the warriors offered prayers and smoke to it before battle, they were successful, and there was no disease when the stone was in the camp. When they moved to Indian Territory (Oklahoma), they left the stone on a high hill in western Nebraska.[67]

Other Indian tribes had very similar beliefs about the Willamette, Wichita County, and Iron Creek meteoritic irons. We noted in chapter 6 that the Clackamas Indians of northwestern Oregon venerated the Willamette meteorite as Tomanowos, or "Visitor from the Moon," and that before battles the warriors washed their faces and dipped their arrows in the water that collected in the hollows of the iron. The Kiowas, Comanches, and Apaches in Indian Territory venerated the Wichita County meteorite, whose original site was just across the Red River in north Texas. They believed that it came from the Great

Spirit, and well-worn trails indicated frequent visits to it by these tribes.[68] The Cree and Blackfeet Indians of Alberta, Saskatchewan, and Montana thought that the Iron Creek meteorite had fallen from heaven, and venerated the iron as a "medicine-stone." When white men removed it about 1870 to a mission house 60 miles distant, an old medicine man predicted that war, disease, and a dearth of buffalo would result. In only a few months famine, plague, and war did come to the Indians.[69]

Various North American Indian tribes, inhabiting an area well over a million square miles, appear to have had strikingly similar beliefs about meteorites and to have carried out similar rites or practices. Despite antagonism or enmity between tribes, it seems that they communicated tales and legends about these strange objects, so that a common folklore developed around meteorites. Given this circumstance, knowledge of any observed fall and a description of the physical condition of the resulting object would have been rapidly broadcast. Peary was convinced that the Eskimos of Melville Bay observed the fall of the Cape York meteorite, "else how could these rude natives have obtained any idea of their heavenly origin, and why should not the brown masses have been to them simply *weeaksue* (rocks) like all the others in their country?"[70] But such is not necessarily the case. The Plains Indians of the North American West and Southwest knew their terrain, and heavy metal masses or even blackened stones are not common features of the countryside. Indians might well have attributed a heavenly origin to such rare objects when they came upon them.

Peoples on other continents also venerated meteorites. A fanciful and pleasant myth explained the fall in the 1740s of a stone at Ogi, Japan, which was placed in a family temple and protected by priests for more than two centuries. An annual festival celebrated the rendezvous of the goddess Shokuja and her consort, who are identified with the constellations Lyra and Aquila and separated by the river of heaven, the Milky Way. No bridge spans the river, but on the festival night a huge jay spreads his wings across it and permits the two constellations to meet. Stones, once used to steady the loom of Lyra, the weaver, fell from the shores of the Milky Way to earth.[71]

An equally imaginative Chinese legend concerns the "jewel of Ch'en-pao." An ancestor of the Emperor Shih Huang Ti, who unified China in 222 B.C., found a remarkable stone on a mountain, which he recognized as a god and to which he made offerings. Thereafter, the stone flew here and there, at times over the city, appearing as a meteor. Sometimes, it took the form of a male pheasant and made noises during the night, which all of the female pheasants in the city would answer. The god had no special name; he was called merely the "jewel of Ch'en-pao."[72]

When the Sabetmahet chondrite fell in 1855 in the Gonda district, United Provinces, India, the natives decked the stone, estimated to weigh over 2 pounds, with flowers and burned sandalwood to it. They left only 3 grams for scientific investigation.[73] When the submagistrate of the district of Andhra Pradesh, India, arrived at the village where the Nedagolla iron fell in 1870, he found the natives worshiping the meteorite, believing it to be God's blessing on them.[74]

The Tamentit, Morocco, meteoritic iron, recognized in 1863, fell, according to Lacroix, about the end of the fourteenth century. The Arab legend was that a block of gold fell between Noum en Nas and Tittaf. The inhabitants came immediately to the site and quarreled about its possession, whereupon God changed it to silver. When there was still disagreement, God changed it iron, after which the people lost interest in it. Sheikh 'Amr had it conveyed to the oasis of Tamentit, where it was placed in front of a mosque.[75] The Durumna meteorite, which fell in 1853 in Mombasa, Kenya, was initially worshiped and then rejected. German missionaries attempted to purchase the stone from the shepherds who had recovered it, but the nearby Wanikas would not permit this, considering the meteorite to be a god because it had fallen from heaven. They built a temple for the 577-gram stone, anointed it, clothed it, and placed pearls on it, and they would neither sell it to the missionaries nor let them see it. Three years later, a Masai raid resulted in the destruction of villages and loss of life, and when famine followed, the Wanikas lost all respect for the stone and sold it to the missionaries.[76] When natives in the British Central African Protectorate [Nyasaland] brought the Kota-Kota meteoritic stone in 1905 to a magistrate, they declared that evil would follow them for taking away God's stone. It was, according to them, only part of a large stone that had been held sacred for years.[77]

Another attitude distinct from worship is regarding a meteorite as a favorable omen or as possessing preternatural powers. The Uwet meteoritic iron is one example. The natives in southern Nigeria stated that the mass fell about 1825, and they considered its presence beneficial for the general welfare of the community. When British authorities removed the meteorite in 1903, an outbreak of smallpox occurred that decimated the population, which attributed the plague to the absence of the iron. When the meteorite was returned, the community prospered. An unnamed American attempted to make off with the 55-kg meteorite in 1904, but the government prevented its departure when the natives complained. In 1907 the people permitted the cutting of a piece for scientific study.[78] In 1925, the Queen's Mercy chondrite fell at night, accompanied by detonations and a bright light, about 15 miles from Matatiele, Cape Province, South Africa. The frightened natives ran into the bush, and when they returned in the morning, they found the largest stone "smoking hot." A medicine man declared that unless someone touched the stone there would be a cattle plague, but whoever did so would drop dead. An old woman, who said she had not long to live anyway, volunteered and only burned her hand. The medicine man then said that if the stone was used as a talisman, it would bring good luck to the wearer, so that the stone was broken up on the spot and distributed to all of the villagers. Two other, smaller stones from the shower were, fortunately, recovered.[79] The Ensisheim stone, which fell in 1492 in Alsace, also belongs in this category. Holy Roman Emperor Maximilian assembled his council to help determine the meaning of the fall of the meteorite; the verdict was that it was a favorable omen for the success of his continuing wars with France and his efforts to repel the Turkish invasions.[80] He instructed the populace to hang the meteorite with an appropriate inscription in the local church.

A somewhat odd coincidence occurred recently. In China, after the fall of a great dynasty, annalists habitually recorded the astronomical phenomena that took place prior to the event, which were interpreted as portents or omens of the demise of the regime. As Joseph Needham wrote, "There was a stylized pattern of what was supposed to take place at the catastrophic close of a dynasty."[81] No one, it appears, has recorded the following sequence of events. On 8 March 1976, the largest stone meteorite ever recorded as having come to Earth fell at Jilin (Kirin), Manchuria. The total weight was approximately 2 metric tons, and the mass of the largest piece was 1.77 metric tons. On 9 September 1976, Mao Tse-tung died, and within a few weeks Chiang Ch'ing and the "Gang of Four" were arrested. The Great Cultural Revolution had ended.[82]

Thunderstones were, of course, not meteorites, but people did think that they fell from the skies, so that it is appropriate here to describe the folklore surrounding them. As noted in chapter 1, thunderstones are neolithic or paleolithic stone tools for the most part, but others are petrified echinites or unusual marcasite or pyrite crystals. It is significant that nineteenth- and twentieth-century anthropologists did not learn of any thunderbolt myths in areas where the Stone Age still existed—that is, among the Tasmanians, Australian aborigines, Melanesians, and Polynesians.[83] Peoples of the ancient civilizations, according to Wainwright, confused fossils and Stone Age tools with meteorites. There is a fossil known as *Lithodomus,* shaped like a belemnite, and these abound at Akhmim, where the Egyptians erected a temple to Min, the god of thunder. Another god identified with thunder, Horus, had a temple at Letapolis, where a fossil named *Nerinea requieniana,* also shaped like a belemnite, is found. Wainwright also pointed out that stone axes were deposited as votive offerings at the sanctuary at Delphi and at the temples of Artemis at Ephesus, Perga, Pogla, and Andeda.[84] There was just as much confusion between meteorites and Stone Age tools in Asia, Africa, and South America as there was in Europe. Needham reported that about A.D. 660 a meteorite was presented to the Chinese emperor and was called the stone axe of the thunder god.[85] Peoples just could not conceive of the original purposes of the pieces of fashioned stone, so they believed that they were extraterrestrial. The actual fall of meteorites, of course, strengthened this idea.

Such stones, according to ancient belief, fell during thunderstorms, and lightning drove them into the earth, from which they gradually rose to the surface.[86] In Germany, they penetrated seven fathoms and came to the surface in seven years, whereas in Hungary the depth was 7 feet and the time required to reach the surface was seven days. In Burma, people placed an earthenware pot over a hole where lightning had supposedly struck, so that the stone would rise into the pot after one year. The Burmese tested the authenticity of a thunderstone by placing it, surrounded by rice, on a plate; if the stone was authentic, birds would not venture to eat the rice.

Everywhere, peoples thought that thunderstones possessed extraordinary or magical powers. Hence the stones were passed on from generation to generation and concealed under the roofs of houses, on the tops of beds, and in cupboards. A common belief was that a thunderstone would protect a house from being struck by lightning. The stones were also believed to have remarkable medicinal properties. In Germany, when rubbed on the affected site, they cured erysipelas, inflammations, and cramps. In Ireland, it was thought that drinking water in which a thunderstone was placed would cure the sick. The Japanese ground the stone to powder, which was used in a compress to open virulent sores. In Nigeria, the stones were employed to protect against rheumatism, and in Tanganika the stones were hung on the gates of villages to ward off disease. The peoples of South Africa, India, Cambodia, the Malay Peninsula, and Java used the stones as amulets. There were also more mundane uses. In Hungary, housewives placed a thunderstone in the back of a baking pan to make the bread taste better, and farmers thought that the presence of a stone would keep milk from becoming sour.

Possession of a thunderstone, then, was highly desirable. The Slavs thought they brought good fortune and prosperity in all affairs. In parts of Germany, it was believed that the stones had the power to cause invisibility. On the Guinea coast, a stone supposedly made one invulnerable, and one sought treasures with it. Brazilians thought that a thunderstone would indicate where there was a deposit of gold. In Moravia, people believed that on the day of the Passion, a stone had the power to reveal the place of a buried treasure.

To summarize, peoples universally regarded thunderstones as treasures to be preserved. In many places the veneration of meteorites was a common practice, and in others they were considered to be favorable omens. There were, to be sure, localities where the inhabitants viewed meteorites as evil omens, or where there was no expressed belief one way or the other. It also appears certain from the literature that people everywhere confused meteorites and thunderstones.

Given the general attitude of country folk toward thunderstones and meteorites, it is possible that peasants, woodsmen, and miners in remote areas did not report genuine meteorite falls to the authorities because they wished to keep the stones they recovered. We know that this happened in Nigeria in the case of the Uwet iron, and natives did not reveal the finds of the Kota-Kota or Thunda meteorites for years. But no evidence exists that a fall went unreported in eighteenth-century Europe. If there were such instances, the statistical dearth of falls, noted in chapter 1, would be explained. The fact that a substantial percentage of the original material of some of the older meteorites cannot be accounted for lends credibility to the idea that people wished to possess a meteorite. Conrad Gesner, as noted in chapter 1, mentioned in 1565 that visitors had chipped away portions of the Ensisheim stone. Today, less than 69 kg of the original 127-kg mass is preserved. The Siena shower in 1794, according to contemporary reports, consisted of about 200 stones, the largest weighing about 3 kg. Only about 2.25 kg of this meteorite are in collections. Narendra reported that she could account for less than 50 pounds of the Weston meteorite, which weighed about 350 pounds when it fell.[87] One might argue that it was merely the desire to own curiosities that prompted people to take fragments of meteorites. This is indeed the case with the hundreds of pieces of the Canyon Diablo meteorite that have disappeared since about 1900. Nevertheless, it may not be legitimate to extrapolate our attitude toward meteorites back to a time when people believed that they contained precious metals or possessed magical powers.

USES OF METEORITES

Archaeologists have found objects determined or thought to have been fashioned from meteoritic iron in less than a dozen Near Eastern localities that date from before the first millenium B.C. (table 4). This is not surprising because iron implements, if unprotected, would in the interim have corroded into a pile of rust. It is difficult, in fact, for archaeologists to identify with certainty the intended use or function of many of the objects listed in table 4, because of the degree of corrosion they have undergone. It is unfortunate, however, that experienced meteorite scientists have carefully examined and analyzed relatively few of these objects. For example, there has been no published analysis of any of the eighteen items found in the tomb of Tutankhamen which are presumed to be meteoritic iron, although these appear to be in good condition. Nor does anyone seem to have examined critically the alleged 20-lb meteorite fragment found in Crete at Hagia Triada. Those archaeologists who are interested in meteoritic iron objects agree that some analyses have been less than satisfactory and that a number of other pieces have not been analyzed at all.[88]

In 1884 Andrée cited an Egyptologist, Franz von Lauth, as proposing that the first iron that was worked into tools by the Egyptians was of meteoritic origin. By association to the Coptic word *banipe* (*ferrum*), in which the first component is the old Egyptian *ba,* von Lauth in 1868 tried to demonstrate that *ba* meant "iron." Further, Andrée wrote, von Lauth found the word *ba* with the supplement *ne-pe,* meaning "of heaven," and from this combination, *ba ne-pe,* concluded that it referred to "metal of heaven," or meteoritic iron. Andrée did not quarrel with von Lauth's etymological analysis, but he did object to what he considered was an unwarranted extension from the fact that primitive societies had used meteoritic iron to the assumption that its use resulted in the development of iron metallurgy. Other peoples, such as the Greenland Eskimos visited by Ross and Sabine, Andrée insisted, had fashioned meteoritic iron into tools without thereby having been led to the smelting of iron.[89] Von Lauth, of course, had not found or seen any meteoritic iron artifacts; his hypothesis was based on his studies of the meaning and origin of certain words. Nevertheless, the idea attracted some metallurgists, as a means of explaining the transition from copper and bronze to iron metallurgy in the Near East, a phenomenon that archaeologists were just beginning to explore in the late nineteenth and early twentieth centuries.[90]

In 1911 Wainwright discovered nine iron beads in two predynastic Egyptian graves at Gerzeh, 50 miles south of Cairo. It was not until 1928 that an analysis was made, but then C. H. Desch determined that one contained 7.50 percent nickel. Immediately thereafter, Wainwright began to publish his series of articles on thunderbolts, meteorites, and omphaloi, among which was one entitled "Iron in Egypt."[91] In it, Wainwright revived von Lauth's hypothesis (without citing this source), but used slightly different transliterations—namely, *bỉȝ* instead of *ba,* and *bỉȝ-ŋỉ-pt* instead of *ba ne-pe*. In addition to emphasizing the nexus between thunderbolts, meteorites, and omphaloi men-

TABLE 4
IRON OBJECTS PRESUMED TO BE OF METEORITIC IRON, FOUND IN NEAR EASTERN LOCALITIES DATED PRIOR TO 1000 B.C.

No.	Country	Approx. date B.C.	Location	Object(s)	Analysis
1	Iran	4600–4100	Tepe Sialk; habitation	3 small, hard, heavy balls	Ni unstated; showed Widmanstättan structure (a)
2	Egypt	3500–3300	Gerzeh; two graves	9 tubular beads	3 contained 7.5 percent Ni (a)
3	Iraq	3100–2800	Uruk; Warka Anu Ziggurat	fragment	Ni unstated (a)
4	Iraq	ca. 2500	Ur; royal cemetery	broken disc (?); 3 fragments	disc, 10.9 percent Ni; structure of two fragments meteoritic; third probable (b)
5	Anatolia	2400–2200	Troy; treasure L	macehead or finial	(1) 2.44 percent NiO; (2) 3.91 percent NiO; (c)
6	Iraq	2900–2400	Al 'Ubaid; temple	fragment	Ni unstated; forged from meteorite at low temperature (c)
7	Anatolia	2400–2200	Alaça Hüyük; grave	crescent-shaped plaque; two pins with gold heads	plaque, 3.06 percent NiO; pins 3.44 percent NiO (c)
8	Egypt	2050–2025	Deir-el-Bahari; tomb	thin blade from amulet	Ni/Fe ratio, 1/10 (c)
9	Egypt	ca. 1340	Thebes; tomb of Tutankhamun	dagger blade; 16 miniature blades; model headrest	Ni analysis by Dr. Iskandar unstated; all deemed meteoritic (c)
10	Syria-Palestine	1450–1350	Ugarit; Ras Shamra	ax head	3.25 percent Ni (c)
11	Crete	1600–1400	Hagia Triada; Minoan palace	20-lb piece of meteorite	no analysis, but deemed meteoritic (a)

References: (a) Waldbaum (1980), 69–79; (b) Hutchison et al. (1977), 244; (c) Bjorkman (1973), 124–127.

tioned earlier, Wainwright produced new evidence. In the important preburial ceremony known as the "Opening of the Mouth," priests in predynastic times used a forked flint knife, which was replaced by a stone knife in the early dynastic period. Later, they used a chisel named "the *mdtft* of *biȝ*," and Wainwright pointed out that the miniature ceremonial tools found in the tomb of Tutankhamen were chisel blades—replicas of the actual iron chisels employed in the rite. Wainwright did not state that the Tutankhamen tomb blades were made from meteoritic iron, but he implied that they were, and he also strongly suggested that the widespread use of meteoritic iron preceded the coming of iron metallurgy. He concluded:

> Thus, then, iron in Egypt as in many other countries was obtained from meteorites long before the Iron Age set in. Moreover, the word *biȝ* proves to have stood primarily for

> iron, or rather meteoric material in general, whether iron, or stone, or a conglomerate of the two. From this proceed the uses to which *bỉꜣ* was put by the priests, and the secondary meanings which the word took on. From it are derived the New Kingdom expression *bỉꜣ-ƞỉ-pt* and finally the Coptic *ⲃⲉⲛⲓⲡⲉ*, which merely mean "iron" without any thought of the other meanings of *bỉꜣ* .[92]

R. J. Forbes was a leader in refuting Wainwright's hypothesis. "Wainwright," he wrote,

> never brought forward one text in which *bỉꜣ* must have meant iron, the old and recognized translation copper fits in just as well. . . . There is no reason not to suppose that the Egyptians thought their meteoric iron just a form of black unrefined copper and that they recognized its celestial origin much later. . . . Only when they had attained the full Iron Age and iron was smelted and worked in Egypt itself copper and iron were distinguished as separate metals.[93]

Later Forbes said that "Meteoric iron could never be a great factor in the rise of iron metallurgy, for, in the absence of chemical knowledge, the connexion between it and iron ores must have remained unknown."[94] More recent scholars have agreed with Forbes.[95]

Scientists have examined several Chinese iron artifacts much more thoroughly. In 1971, Gettens, Clarke, and Chase published a comprehensive study of two early Chinese weapons.[96] One was a broad axe with a bronze tang and rusted remains of an iron blade. The other was a dagger axe with a bronze blade and the remains of an iron point. Simultaneous microprobe analyses of the nickel and iron content of the oxidized blade of the broad axe demonstrated that taenite and kamacite lamellae had been present, so that it was possible to conclude that the blade had been fabricated from a meteoritic iron octahedrite. Analysts also found remnant metal in the point of the dagger axe to have the composition of kamacite, which suggested the strong probability that the point was also made from meteoritic iron. They could not, however, determine the structure of this iron, nor could they conclude that both blades had been manufactured from the same meteorite. Both weapons date from the early Chou dynasty, about 1000 B.C. They probably came from the tomb of Marquis K'ang, Prince of Wei, the area in Honan province in which the Shang capital was located. There is textual evidence of the presence of Marquis K'ang at the funeral of King Ch'êng in 1004 B.C., which tends to corroborate the dating of the weapons. The fabrication of the blades occurred about four centuries prior to the widespread appearance of cast iron metallurgy in China.

In 1979 Chung made a similar study of a Shang dynasty bronze yüeh-axe with an iron blade.[97] Archaeologists unearthed this weapon in Hopei Province and dated it about the fourteenth century B.C. Again, the iron blade was totally oxidized. The analysis indicated a nickel content of 3.4 percent, but Chung postulated that the metal had originally contained approximately 6 percent nickel, the balance having been lost by corrosion. Another test gave evidence of an original lamellar structure, and Chung concluded that the blade had been made of meteoritic iron.

The Hopewell Indians of the Ohio Valley fabricated knives, chisels, ear ornaments, and buttons by hammering or cold-working meteoritic material. Crushed fragments of olivine or interstices in the metal from which the olivine had been lost revealed that at least some of the artifacts had been fashioned from a pallasite. George Kunz in 1890 remarked that the meteoritic nuggets found there greatly resembled the Brenham pallasite, and although Brezina agreed with this opinion, other scientists did not. Recently, Wasson and Sedwick concluded from their analysis of the nickel and trace element composition that the Ohio material was virtually identical to the Brenham pallasite.[98] The Indians at Havana, Illinois, fabricated the beads found there, which varied in diameter from three-sixteenths to five-eighths of an inch, from sheets or strips of meteoritic material that were fashioned into cylinders with a lapped seam on one side.[99] However, Buchwald determined that the Indians must have intermittently annealed the strips during the cold-working process. The microstructure he observed indicated that the annealing temperature was about 650° C, and the slightly distorted appearance of the kamacite grains showed that cold-working followed the last annealing process.[100]

We have already described the theory advanced by C. U. Shepard in the mid-nineteenth century that there was a broad zone stretching across the United States and northern Europe in which the majority of meteorites fell or were found (see chap. 3), and also noted that Mariano Barcena in 1876 suggested that the soil of Mexico had the peculiar property of attracting meteorites (see chap. 6). Maskelyne, too, pointed out that there were more falls in that region of India to the north and west of Calcutta than elsewhere, and he threw out the suggestion that the attractions of huge mountains (the Himalayas) aided in "drawing to this neighborhood meteoric matter." The same idea also appealed to Oliver C. Farrington. In 1915, he emphasized that there had been numerous falls of irons in the southern Appalachians and that the "largest iron meteorites of North America are found in the Cordilleras."[101] The concern here was the apparently uneven distribution of iron meteorite finds, more having turned up in North and South America, Africa, and Australia. However, the finds of meteoritic iron artifacts in Europe, the Middle East, and China demonstrate that such iron was often put to use by our ancestors. Also, Bjorkman postulated that about 1900 B.C. an Assyrian trading outpost in Anatolia may have been marketing meteoritic iron, and that the colony disappeared when the supply was depleted.[102] If future research supports her hypothesis, further use of meteoritic iron would be demonstrated. Moreover, the recent finds of iron meteorites in Antarctica offer evidence of the random distribution of iron meteorite falls.

Another question that has puzzled scientists is the forgeability of meteoritic irons. According to a manuscript written in 1723, the Sultan of Ghazni [Afghanistan] ordered that a portion of an iron meteorite that fell in the area about A.D. 1000 be made into a sword, but that it was impossible to carry out his command.[103] The same manuscript mentions another iron meteorite, weighing about 2 kg, which fell near Jalandhar, Punjab, India, in 1621. The smith whom the king ordered to make a sword and dagger from it stated that it could be forged only if one part of terrestrial iron were mixed with three parts of the "lightning iron." The smith made two swords, a dagger, and a knife, and the knife is now in the Freer Gallery of Art, Washington, D.C. (fig. 36). In contrast, James Sowerby reported that in 1814 he had no difficulty in forging a sword from the Cape of Good Hope meteoritic iron, which he presented to

Fig. 36. Knife allegedly forged from the Jalandhar meteorite that fell 10 April 1621 in Punjab, India. Length 26.1 cm. Note the damascene pattern. Courtesy of the Freer Gallery of Art, Smithsonian Institution, Washington, D.C. (acc. no. 55.27).

Czar Alexander of Russia in gratitude for the latter's stand against Napoleon. The blade was hammered at a red heat, he reported, out of a piece 1 inch thick, and then ground and polished and hammered cold. The iron worked easily and the entire process took ten hours. The length of the sword was 2 feet, its width 1 ⅜ inches; it was slightly curved, pointed, and sharpened on both edges. The surface, Sowerby admitted, was slightly blemished owing to the spread of inclusions, and it had a whitish color with a silvery luster.[104] Also, the sultans of Solo at Soerakarta, central Java, preserved the Prambanan meteoritic iron, known before 1797, in the courtyard of their palace for the specific use of making *kris,* or patterned weapons. Berwerth in 1907 described five kris that had been presented to the Emperor of Austria. The curved kamacite and taenite figures, he explained, furnished the background for the damascened pattern, adding that there were no previous reports of such workmanlike and artistic treatment of meteoritic iron.[105]

In his lectures on meteorites, Henry A. Ward was fond of saying that "the farmer's plow is the greatest friend and the blacksmith's forge the greatest enemy of science." Ward's statement is not quite accurate, because many of the farmers who unearthed meteorites took them immediately to the blacksmith. Because of the weight and texture of the masses, the finders in many instances thought that they were made of silver. This was the case with the Bacubirito, Campo del Cielo, Smithland, Zacatecas, Walker County, and Carthage meteoritic irons. After the farmer's dreams had been dispelled at the forge, the blacksmith usually turned out horseshoes, nails, knives, and agricultural implements from the meteorite iron. Large pieces such as the Tucson ring were used as anvils.

The forging properties of meteorites also interested scientists. Chladni had a knife forged from the Elbogen meteoritic iron; Partsch made a sword from Lenarto; and Neumann a penknife from Elbogen. Ludwig Beck experimented with the Toluca iron and found that although it welded satisfactorily, it did not harden and could be filed only with difficulty. In the forged state, he reported, it was harder than forged iron but less malleable, and he concluded that it was not suitable for forged tools. J. W. Mallet in 1872 made extensive tests of the Staunton iron. He forged one piece into a very good knife blade, but he had difficulty in working a second piece that had been heated to a red heat in a partial vacuum. A third specimen, heated in vacuum to a white heat, shattered when forged. Mallet ascribed the differences to the escape of absorbed gases, to the decomposition of the sulfur and phosphorus, and to the melting of these decomposed products. A. W. Wright at Yale found that Magura was as hard as steel when drilled, whereas Cross Timbers [Red River] was softer, and Knoxville [Tazewell] softer still. Huntington at Harvard determined that the Jamestown iron was almost as malleable and forgeable as lead, such that pieces of it could be rolled cold in the form of ribbons.[106]

Controversy resulted because of these reports. Whereas in 1868 J. D. Dana stated that all meteoritic irons without exception could be forged, St. John V. Day in 1877 refuted this claim. In 1893 Emil Cohen came to the conclusion that many meteoritic irons were not too malleable and could be easily forged only under certain conditions. Thus a Mexican blacksmith, according to Beck, reported that the Toluca iron had to be forged while white hot, not red hot, and had to be heated often during the forging process; further, charcoal and not pit coal must be used.[107] In an attempt to clarify matters, G. P. Merrill at the Smithsonian had pieces of eight different meteorites forged in the 1920s. The Nejed [Wabar] and Coahuila irons forged well, he reported, but neither could be tempered "to give more than ordinary cutting power" (fig. 37). Yet the Canyon Diablo, Mount Joy, Casas Grandes, and New Baltimore irons were malleable even when cold. When they were heated to a red heat, all could be hammered to a knife edge without any indication of cracking. But the Toluca and Williamstown [Kenton County] irons, he stated, proved to be more refractory.[108]

The phosphorus and sulfur content, the size, number, and distribution of inclusions, the amount of corrosion, the structure, the temperature to which the iron is heated, and the length of time it remains at temperature—all influence the forgeability of a meteoritic iron. However, blacksmiths and scientists have forged or attempted to forge many; Buchwald lists almost a

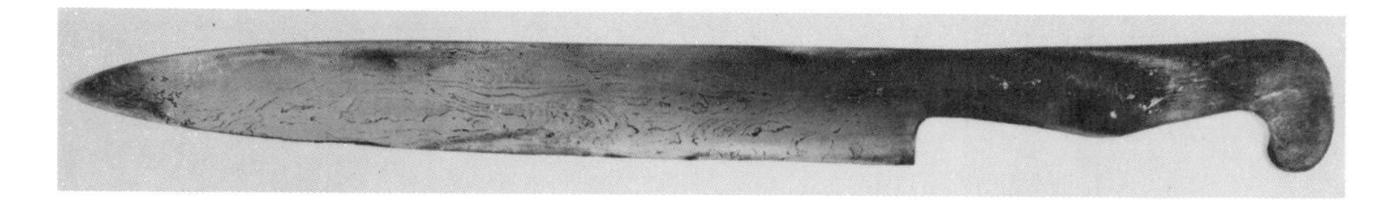

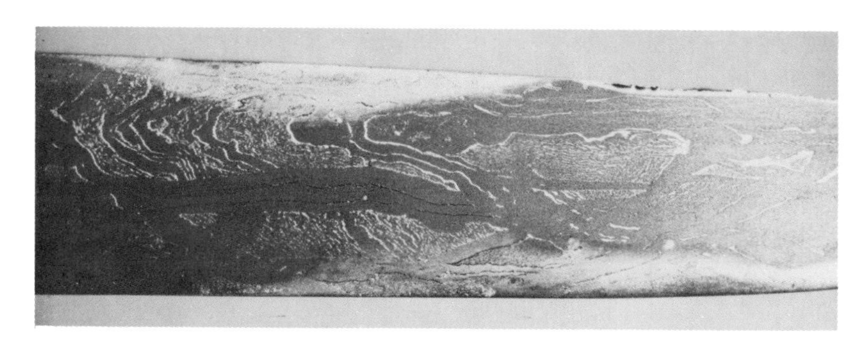

Fig. 37. Knife forged at the Smithsonian Institution from the Wabar (Nejed) meteorite. Length 35.5 cm. Note the damascene pattern. Courtesy of the Division of Meteorites, NMNH, Smithsonian Institution.

hundred that show unmistakable evidence of artificial reheating.[109] There has been one beneficial consequence of these labors of blacksmiths and scientists. For decades, scientists were puzzled when they saw a granular microscopic structure both in some hexahedrites and in others that, because of their nickel content, ought to have displayed the Widmanstätten pattern. They classified such irons as "nickel-poor ataxites." Buchwald, in particular, has studied these irons extensively. While in some meteorites the structure resulted from cosmic experiences, in many others "artificial reheating effectively destroyed the original cosmic structures." Thus Buchwald maintains that the class "nickel-poor ataxite" is superfluous.[110]

F. B. Howard-White, author of a recent history of nickel, goes far in attributing the development of nickel alloy steels to investigations of the properties of meteorites by Faraday and Stodart and later experimentalists, writing that Faraday's "imaginative mind . . . led him to wonder why meteoritic iron had unusual qualities. Little did he realize that his experiments would inspire investigations which, in the subsequent century and a half, would have a prodigious effect upon the demand for nickel."[111] Howard-White's theory is attractive, but it is not entirely convincing.

Michael Faraday and James Stodart, a master cutler, in 1820 conducted several experiments on alloying nickel with iron as a result of their investigations of a type of steel made in India, known as *wootz,* which in the manufacture of cutlery was far superior to British steel. A high carbon content and slow cooling gave wootz its exceptional properties, but Faraday and Stodart did not discover this. Faraday, in a chemical analysis of wootz, found traces of silica and alumina that he thought provided the key, but efforts to duplicate wootz by adding silica and alumina to iron failed. However, the idea that something had been added to produce wootz persisted, and they proceeded to experiment with adding platinum, iridium, rhodium, osmium, and palladium to various batches. Faraday or certainly Stodart would have known about the sword that Sowerby forged from the Cape of Good Hope meteorite, so in the course of their tests they produced two samples of nickel-iron, one 3 percent nickel and the other 10 percent. The former was very malleable when forged, and they speculated that it had better corrosion resistance than pure iron. The latter was "less malleable, being disposed to crack under the hammer," but they determined that it oxidized less than pure iron. They also noted that "It is a curious fact, that the same quantity of the nickel alloyed with steel, instead of preventing its rusting, appeared to accelerate it very rapidly."[112] This was the extent of Michael Faraday's experiments with nickel-iron alloys, although he maintained his interest in them.

Howard-White pointed out that in 1824 a Swiss metallurgist, J. W. Fischer, produced meteoritic steel by the addition of nickel, after being impressed by the structure of the polished surfaces of meteorites and of the damascened blades he saw in the Vienna museum. He received an Austrian patent on this superior steel, and in 1825 visited London, where the hardness and elasticity of the samples he carried with him impressed Faraday. Two Englishmen, J. Martineau and H. W. Smith, capitalized on Fischer's discovery by securing a British patent on "meteor steel," which was manufactured by adding to blister steel a "meteor powder" composed of twenty-four parts zinc, four parts nickel, and one part silver. Several other individuals in England and the United States in the 1840s and 1850s also experimented with the addition of nickel and/or zinc to cast iron, which in some cases resulted in patents.[113]

It was not until the 1880s, however, that superior ferro-nickel alloys and true nickel steels were developed. Henri Marbeau in France and James Riley in Scotland were the leaders in this development, and there is no evidence that either the presence of meteoritic irons or the earlier experiments influenced their work. Further, it appears that scientists made the first mechanical tests on meteoritic iron only after nickel steel was produced in quantity. Buchwald reported that about 1900 the Krupp laboratories made tests on samples of the Gibeon meteorite. It was very ductile, sustained a 180° bending test without fracture, and its tensile strength was 41.4 kg/mm.[114] Thus it is difficult to support the idea that knowledge of the properties of meteoritic irons contributed to the development of nickel steels. Studies of the Widmanstätten structure in the twentieth century did, indeed, contribute greatly to the understanding of alloy transformations, of precipitation hardening, and of the hardening of steel. We refer briefly to these developments in chapter 8.

8

New Directions 1900–1950

In the first few years of the twentieth century, several scientific discoveries and developments occurred that had far-reaching implications for meteorite research. Some had an almost immediate effect in clarifying ideas concerning the chemistry, mineralogy, and structure of meteorites. Others, though they initially appeared to offer easy solutions to important problems, demonstrated to scientists that unraveling nature's secrets was much more difficult than the state of knowledge and the methods and instrumentation of the early twentieth century would permit. By mid-century, largely owing to the scientific research and technological developments necessitated by World War II, scientists were prepared to tackle these problems with more confidence and more assurance of success.

The most significant discoveries and developments occurred in the areas of atomic physics, X-ray crystallography, and chemical thermodynamics. With respect to atomic physics, there was the phenomenon of radioactivity. Within a few years of its discovery, scientists began investigations to determine whether meteorites contained radioactive elements, and later to try to ascertain their ages by the calculation of radioactive half-lives. The discovery of isotopes led to attempts to determine whether the isotopic composition of elements in terrestrial rocks and meteorites was the same or different. And when theories of the structure of the nucleus of the atom appeared to require clarification by determining the evolution of the elements and their relative abundances, scientists turned to investigations of the chemical composition of meteorites for possible answers.

X-ray crystallography had its beginnings in 1912, when Max von Laue, professor of physics at Munich, discovered that the atoms or ions within crystals diffracted X-rays and thereby revealed their lattice arrangements. Within a decade, scientists were able to determine the space lattices of the phases of meteoritic iron, and by 1930 Sir William Lawrence Bragg and his colleagues elucidated the structure of many common silicate minerals. X-ray crystallography permitted the measurement of the sizes of atoms and ions, and thereby gave more exact knowledge of the phenomenon of isomorphism. Also, advances in the understanding of X-ray phenomena led in the 1920s to the development of quantitative X-ray spectroscopy, which resulted in the discovery of trace elements present in meteorites and what was at the outset a crude measure of their concentrations. These values did provide more extensive knowledge of the relative abundances of the elements present in meteorites, however.

The phase rule, enunciated in 1876 by J. Willard Gibbs, provided the foundation for research into chemical thermodynamics. About 1900 scientists began to apply the phase rule to the alloys of metals and to construct equilibrium or constitutional diagrams that showed under what temperature conditions and what composition ranges certain phases would be present or be transformed. An iron-nickel equilibrium diagram soon appeared, promising a clarification of the Widmanstätten structure. This diagram—and several others that replaced it—proved to be erroneous, and decades passed before the phase changes in meteoritic irons were fairly well understood. But Gibbs's phase rule was also applied to the study of the chemical transformation of rocks and the genesis of minerals, and this significantly advanced geochemistry. Scientists now proposed much more sophisticated models of the Earth, and also turned their attention to the size and composition of the body or bodies from which they surmised that meteorites had come.

Thus scientists began to employ meteorites to gain new insights into or to substantiate physical, chemical, or geological hypotheses, and this activity enlarged the horizons of the science of meteorites considerably. In addition, scientific attention to two other sets of phenomena—craters and tektites—extended its boundaries. The study of craters had for decades been propaedeutic fieldwork for budding geologists, and theories about the origins of the craters of the Moon had interested armchair geologists since the end of the eighteenth century. Then about 1900, the crater surrounded by Coon Butte, Arizona, provoked such a controversy about its origin that by mid-century an entire subdiscipline had come into existence—namely, meteorite impact theory. About 1900 controversy also began concerning the origin of the peculiarly shaped glass objects that F. E. Suess called tektites. Tektites had been found in only a few limited areas on the Earth's surface, but where they had been found, they occurred in very substantial numbers. One hypothesis after another failed to survive scientific inspection, and in 1950 the origin of tektites was still a puzzle.

Scientific interest in pursuing the study of meteorites in these new directions did not imply abandonment of traditional methods—that is, estimating the velocities of meteors and meteorites, recording the direction and time of falls, reporting the location and condition of finds, chemical analysis, and the microscopic study of thin sections and etched surfaces. Even in these areas early twentieth-century scientists made advances, to which we now turn our attention.

NEW INSIGHTS INTO OLD PROBLEMS

The Genetic Relationship of Meteorites and a New Classification

George T. Prior had a distinguished undergraduate career in the natural sciences at Oxford, and in 1887, after a brief period in Germany where he studied analytical chemistry, he joined the staff of the British Museum (Natural History) as a chemist in the Mineralogy Department. His daily activity diaries indicate that his intense and continuous involvement with meteorites did not commence until about 1900. The final entry on 26 July 1909 reads: "Vacation (for King's birthday) (Appointed Keeper)."[1] Prior succeeded Lazarus Fletcher, but his new responsibilities did not diminish his scientific productivity. In 1905 Oxford had awarded him the D.Sc. in recognition of his record of significant publications, but his most important research papers on meteorites appeared in the decade following his elevation to the keepership.

An important influence on Prior's research activity during this period was an article published in 1878 by Adolf E. Nordenskiöld. At that time, Nordenskiöld's results did not attract attention, and they apparently escaped the notice of Emil Cohen when he compiled his comprehensive history at the end of the nineteenth century. Whether Prior learned of Nordenskiöld's article independently or because of an article published in 1910 by W. A. Wahl that referred to Nordenskiöld, is not clear. It was, nevertheless, a starting point.

Nordenskiöld singled out for study the chemical analyses of nine chondritic meteorites, and decided to compare the percentage amounts of their constituents after excluding oxygen and sulfur from their reported compositions. He calculated the following range of percentages for those elements reported in all of the analyses:

Element	Percent range
Si	23.66–28.02
Mg	20.18–23.61
Fe	41.37–47.82
Ni	1.58– 4.71
Al	0.23– 2.12
Na	0.38– 1.78

These results, Nordenskiöld claimed, did not indicate an accidental similarity but an actual identity. All of these chondrites, which fell between 1812 and 1876, formed a natural group with a common origin. Many other stones probably belonged to this group, he surmised, and if trustworthy analyses were available, he would be able to arrange several other groups. Nordenskiöld believed that the slight differences in the chemical composition of the group were due to their having been heated after their formation under either oxidizing or reducing conditions. Metallic iron was the most recent constituent, having been the result of the reduction of iron silicate.[2]

In 1910, Walter Wahl at the University of Helsingfors investigated a large number of chondrites, and published evidence tending to confirm Nordenskiöld's hypothesis; there was a uniformity in the chemical composition of chondrites when their analyses were calculated on an oxygen-free basis.[3] Prior was obviously intrigued by the concept of the comparison of analyses, but decided to make the comparisons in a different way. He surveyed published chemical analyses and in 1913 presented the comparative results of forty-one chondrite analyses. His table listed the percentages of nickel-iron, troilite, olivine, and insoluble silicates in the stones, and also the ratios of iron to nickel in the nickel-iron, of magnesium to iron in the olivine, and of magnesium to iron in the bronzite (orthopyroxene). Leaving aside the nickel-iron, he wrote, the table exhibited "the remarkable similarity" of the stones, leading to the conclusion that almost all chondrites were "practically identical in chemical and mineral composition, the identity extending in the main even to the chemical composition of the olivine and pyroxene."[4] From the table, one could elicit the approximate percentage mineral composition of this common type:

Nickeliferous Iron (in which Fe:Ni is about 10)	9
Troilite	6
Olivine (in which Mg:Fe = 3)	44
Bronzite (in which Mg:Fe = 4)	30
Feldspar (Oligoclase)	10
Chromite, &c	1
	100

Almost immediately, however, Prior reported that the ratio of iron to nickel of this ideal composition should be lowered from 10 to 6. In many chondrite analyses, the amount of nickel had been underestimated owing to the incomplete separation of the iron and nickel.[5]

In 1916 Prior returned to the subject of chondrite comparisons, but with a different conclusion. He had not, he wrote, appreciated the significance of the variation in the amount of nickel-iron in chondrites. The fact was that "the less the amount of nickel-iron in chondritic stones, the richer it is in nickel and the richer in iron are the magnesium silicates." These were Prior's "rules," and by way of illustration, Prior proposed that chondrites could be divided into four groups, each denoted by a specific meteorite, and which contained, in order, decreasing amounts of nickel-iron. The four specific meteorites were Daniel's Kuil, Cronstad, Baroti, and Soko-Banja, and their principal mineral characteristics were as follows:

Type	Percentage of nickel-iron	Ratio of Fe to Ni in nickel-iron	Ratio of MgO to FeO in olivine	Ratio of MgO to FeO in pyroxene	Ratio of total Fe atoms to Mg atoms
Daniel's Kuil	25½	13	—	∞	1·03
Cronstad	18½	11	4	6	1·22
Baroti	9	6½	3	4	1·46
Soko-Banja	4	3	2½	3	1·80

Prior emphasized that the four divisions were arbitrary, and that the chondrites were "probably a perfectly continuous" series in which his rule was exhibited. There were 4 or possibly 6 meteorites of the first type, 34 of the second, 25 of the third, and 26 of the fourth. An additional 25 chondrites did not correspond to any of the types, and Prior suspected that the analyses of most of these were erroneous in one respect or another. The achondritic stones and stony irons also fitted into Prior's scheme because some, such as the pallasites and mesosiderites, contained large percentages of nickel-iron, while others, howardites and eucrites, for example, had little or none.

The genetic relationship between meteorites, Prior wrote, led to the conclusion that "all meteorites have had a common origin from a single magma." The composition of the Daniel's Kuil meteorite most nearly represented the composition of the original magma: a high percentage of nickel iron, a high ratio of iron to nickel in the nickel-iron, the magnesium silicate nearly pure enstatite, little or no lime, feldspar close to oligoclase, and calcium sulfide (oldhamite) a characteristic accessory constituent. Without specifying the oxidizing agent, Prior proposed that progressive oxidation of the nickel-iron produced increasing amounts of olivine and bronzite. He did not mention the origin of iron meteorites, but he did include them in a classification based on chemistry and mineralogy, with the nickel content as the principal criterion.

Prior realized that the validity of his concept of a genetic relationship among meteorites depended upon the accuracy of chemical analyses. Although he was a skilled analytical chemist, it was impossible for Prior to make new analyses of all meteorites, so he culled all previous analyses to find those he considered trustworthy. In his articles he referred occasionally to some analyses that were stated ambiguously, to others that appeared to be erroneous, and to the necessity of having to use analyses that were eighty years old. He expressed his concerns more frankly in an exchange of correspondence during 1917 and 1918 with George P. Merrill of the Smithsonian Institution. Merrill had just completed a report on the chemical and mineralogical composition of meteorites, designed primarily to verify or discount reports of the presence of minor constituents, and had paid J. E. Whitfield, a Philadelphia chemist, to make many of the analyses.[6] Prior wrote: "My theory, I feel, must rest on my own [analyses] & those of a few others such as Fletcher, Penfield, Hillebrand, Börgstrom, & Mingaye and only on the accumulative effect of the rest. Even your list of 59 analyses I should considerably curtail." He noted what he considered were defects in a dozen of the analyses listed in Merrill's report (most of them Whitfield's), and severely criticized the results of J. Lawrence Smith, whose "method of analysis was certainly imperfect."[7]

Merrill was sympathetic. He replied:

> I will agree to almost anything you may say regarding the unsatisfactory nature of most analyses made by others than members of the little group you mention. . . . We must not be unreasonably harsh, however. Smith looked only for what he considered essential, and methods were not perfected nor chemicals pure. . . . Whitfield received his early training in the laboratories of the U.S. Geological Survey and undoubtedly is the best qualified man available.

Merrill placed most of the blame, however, on the insufficient amount of material available for analysis. "Another point which I have frequently harped upon," he wrote, "is the almost fanatic greed of collectors who regard their material as too precious for use. Think of Gümbel basing his analysis on 1.5 grm. of the Homestead material! . . . The great trouble . . . is the insane notion that a meteorite is something too precious for investigation on a true petrographic basis."[8] As it turned out, Prior's judgment was almost completely vindicated in 1953, when Urey and Craig selected the "most trustworthy" analyses of meteoritic stones. All seventeen of Whitfield's analyses and all thirteen of J. L. Smith's analyses were discarded, as were two of

Prior's, one of Fletcher's, and one of Mingaye's. Urey and Craig, however, accepted eleven of Prior's analyses among the ninety-four they used for their classification.[9]

Merrill was not impressed by Prior's theory, however. Earlier, in 1915, he had written to Stanislaus Meunier in Paris: "I have to confess that I have not paid as much attention to meteorite classification as perhaps I ought. . . . It has seemed to me as though the time was hardly ripe as yet for the evolution of a really satisfactory classification but perhaps I am wrong in this."[10] Now, Merrill wrote to Prior: "The difficulty I shall have in accepting the relative amount of Fe Ni as a basis for classification lies not merely in the inaccuracies of determination, but the fact that it (the Ni) is but a minor constituent at best."[11]

Merrill's assessment upset Prior, who wrote:

> Your classing Ni as a minor constituent of meteorites is a sad blow & cuts me to the heart, after I have been trying to show that it is all important and a determinant of all the other characters in a meteorite. In the case of irons, the Ni content practically determines the present classification. . . . My classification simply extends this distribution according to the Ni-content of the nickel-iron to the chondrites and achondrites, and this process is prevented from being an arbitrary one by the relationship subsisting between the Fe:Ni ratio and the composition of the magnesium silicates. As to the inaccuracy of the Ni determination, this should be removed in future by the dimethylglyoxime method of separation from the iron.[12]

Despite Merrill's discouraging stance, Prior persisted in his attempt to develop a rational classification, and in 1920 published one that gradually gained a measure of acceptance in the following decades.[13] In a lengthy introduction, Prior dwelled on the defects of the Rose-Tschermak-Brezina system and was particularly critical of the changes instituted in it by Brezina. Those of Shepard, Daubrée, and Meunier, he pointed out, had not been widely accepted. None of the classifiers, with the exception of Farrington in 1911, had paid sufficient attention to chemistry, but Farrington's system, Prior stated, was vitiated by his use of unreliable analyses. Prior's main classes were irons, stony irons, and stones, with the latter being subdivided into chondrites and achondrites. He distinguished four groups within these classes, based on the ratio of iron to nickel, the type of magnesium silicate, and the type of feldspar. There were two noteworthy innovations. First, Prior revived the idea, proposed in 1864 by Alfred Des Cloizeaux, that the pyroxene be named enstatite, bronzite, or hypersthene on the basis of the ratio of MgO to FeO that it contained. Second, he proposed that the terms clinoenstatite, clinobronzite, and clinohypersthene be extended to cover all of the monoclinic pyroxenes, including those having varying amounts of calcium, for which Wahl in 1907 had proposed the name "enstatite-augite."[14]

There were defects in the classification, which Prior freely admitted. The scheme did not accommodate the calcium-rich achondrites very well, and the mesosiderites could be placed in one or the other of two groups, depending upon one's preference. Nevertheless, Prior's classification did draw attention to the puzzling chemical relationships that appeared to exist among meteorites, and also to the crucial importance of accurate chemical analysis. Even Alfred Lacroix, in 1927, though adhering to Daubrée's nomenclature in his classification, employed Prior's iron:nickel ratios and declared that his own experiments confirmed Prior's theory that all chondrites were derived from the same original magma.[15]

Petrology and Structures

Merrill and Prior made different approaches to the investigation of meteorites, which resulted in different emphases. Prior saw chemistry as the key to their origin and to their differentiation into masses having diverse mineral constituents. "The Rose-Tschermak-Brezina scheme of classification," he wrote, "now generally adopted, in the case of chondrites is somewhat complicated and divides them into groups which do not appear to have much relation to their chemical and mineral composition."[16] Moreover, "the introduction of subdivisions based upon brecciation and the presence of veins is a quite unnecessary complication and overelaboration."[17] Merrill, in contrast, aware of the lack of accuracy of chemical analyses and of the fact that samples of the same meteorite yielded different results, stressed the petrological approach. He wrote to Prior: "I never allow myself to enter into elaborate calculations of mineral composition without first studying thin sections."[18] Training and experience, of course, influenced the preference of each man.

Merrill devoted much study to the structural constituents of meteorites, in particular chondrules, and gave a great deal of thought to the origin of the chondrules and to what Sorby had termed the "metamorphism" of meteoritic stones. In 1920, for example, Merrill described the Cumberland Falls meteorite, which fell on 9 April 1919 in Whitley County, Kentucky, and which Merrill, perhaps smugly, stated had no proper place in the prevailing scheme of classification.[19] The composition of the Cumberland Falls stone is that of an enstatite achondrite, but it contains inclusions of a black chondrite as well. It is therefore termed a polymict breccia, although this designation sheds no light on its origin, which remains a problem. Merrill called the meteorite a breccia that was composed of fragments of two quite dissimilar stones. There were definite indications of compression, he wrote, "manifested by small slickensided surfaces and by the crushed and optically distorted conditions of the pyroxenes as shown both in hand specimens and in thin sections." Whereas the original fragmentation might have been the result of an impact or of an explosion, the meteorite was subsequently subjected to a substantial pressure that further crushed and distorted the particles and then compressed them "into a firm rocklike mass." The texture, Merrill said, resembled a terrestrial rock that had been "subjected to dynamic metamorphism." He thought that the fragments constituted direct evidence of the destruction of a preexisting planet.

In 1921 Merrill returned to the subject of metamorphism.[20] All chondritic meteorites, he wrote, had a volcanic or tuffaceous origin, but their different textures were due to metamorphism in which both heat and pressure had played a role. He thought that the most perfect chondrules were present in those meteorites which displayed the greatest fragmentation and in which the crystallization was confused. He considered interstitial glass to be the product of metamorphism—that is, it was the last mineral to congeal in a stone that had experienced a series of changes. He judged that the metal in chondrites was also a later addition because of its distribution and the manner in which it surrounded or enveloped the chondritic material; thus it had a secondary origin.

Just before his death in 1929, Merrill wrote a summary of the theories of the origin of chondrules that had been advanced by Reichenbach, Sorby, Tschermak, Daubrée, and others (see chap. 5), and then presented his own views.[21] "True" chondrules were consistent with Sorby's idea of an origin as drops of fiery rain. These were of glass or were "cryptocrystalline or radiating enstatites," and they displayed a round or oval form having a smooth rindlike crust and surface, often with one or more saucerlike depressions or protuberances. Other chondrules, Merrill continued, were of a compound, holocrystalline nature, or were porphyritic because of the development of olivine or pyroxene polycrysts in a more or less glassy base. These did not have smooth exteriors, and they were irregularly round, giving "unmistakable evidence of an origin of form through mechanical attrition." These, he stated, should be called "chondroidal forms" instead of chondrules. Meteoriticists today can appreciate Merrill's careful observations of chondrules. They have, of course, proceeded further than he did in discriminating the various types of chondrules. The differences Merrill noted contribute to the alternative theories of their origin, which are reviewed in chapter 9.

As for the origins of other structures in meteoritic stones, Merrill continued, some scientists, Brezina and Wadsworth included, had thought that rapid crystallization followed by a crushing action had produced "obscure and confused structures." Following Sorby, Tschermak, Berwerth, and Wahl, Merrill believed that these resulted from their tuffaceous origin, accompanied in many instances by metamorphism. The origin of the structures of some meteorites, such as Nakhla and Shergotty, was perplexing; possibly they were a result of direct cooling from a molten magma. Impact or compression undoubtedly produced clastic structures. Metamorphism, Merrill insisted, gave rise to crystalline types. Fragmented structures contained the most perfect chondrules, and as the structure became less fragmented and more crystalline, the chondrules became "more highly altered, often merging imperceptibly into the groundmass." The St. Michel meteorite (classed later by Van Schmus and Wood as a recrystallized L6 chondrite) was a good example of a structure showing evidence of metamorphism in which heat was the primary factor, Merrill wrote. The transformation of a normal plagioclase feldspar into the mineral maskelynite was also clear evidence of the action of heat. Here Merrill was off the mark, since shock has been shown to be the cause of this transformation.

The state of science in Merrill's day was such that he was unable even to surmise within what ranges of temperature or pressure metamorphism might bring about noteworthy changes in the structures of meteoritic stones. Yet he did attempt, using what data were available, to indicate how nickel-iron might have invaded and surrounded granules of pyroxene and olivine. The pyroxenes and olivines fused at about 1400° C. The melting points of iron and nickel were higher: pure iron at 1530° C, and pure nickel at 1430° C. However, the old Catalan furnaces produced spongy iron at a temperature of only about 700°–800° C, and therefore it was possible that, given moderate pressure and a sufficient length of time, the visible structures had resulted.[22]

Wahl, in 1910, had written that most theories about the production of the structures of meteoritic stones were more correct than the proponents of other theories would admit. The polemic literature, he pointed out, was due in large part to a lack of knowledge, since the structures were not the result of a single event but of several partially obliterated events.[23] It was to Merrill's credit that he persisted in attempting to trace the sequence of these events. His emphasis on thermal and dynamic metamorphism was an important contribution, for the study of these phenomena as they related to the petrology of meteoritic stones became a pressing research objective after 1950.

In the 1920s, while Merrill was preoccupied with the effects of metamorphism on the microscopic structure of stony meteorites, William Lawrence Bragg and his students at Victoria University in Manchester were investigating the atomic structure of silicate minerals by X-ray diffraction methods. In the late nineteenth and early twentieth centuries, many chemists and mineralogists sought to find the key(s) to the perplexing and apparently intractable problem of the constitution of silicate minerals (see chap. 4). One of the most dedicated was Frank W. Clarke, chief chemist of the U.S. Geological Survey, who began to investigate the silicates in 1886. Clarke was a partisan of the multiple acid theory, so that he designated the acids required to incorporate all of the silicates and listed the acid, basic, and normal salts that formed the silicate compounds. He worked out structural formulas for many silicate minerals, a number of which, particularly with the alumino-silicates, were necessarily complicated. The real problem encountered by Clarke and other chemists was that they could neither synthesize the hypothetical acids nor isolate the acids in solutions of the minerals of which they supposedly formed a part. Another approach, employed by Vladimir I. Vernadsky and W. and D. Asch, was to make an analogy between the silicon in the silicates and carbon in the organic compounds, and to postulate the existence of hexagonal ring structures. In the most basic structure, the silicon atoms were placed at the apexes of the hexagon, each attached to an adjacent oxygen atom and also linked by another oxygen atom to another silicon atom at the nearest apex. These hypothetical structures failed to explain the composition of the more complex silicate minerals, however. Attempts to determine chemistry from the study of optical properties were similarly unavailing.[24]

"The problem of the constitution of the silicates," Berman wrote in 1937, "is a problem in crystal chemistry, because the silicates are with but few exceptions solid crystalline compounds as we know them in nature. To bring a silicate into solution necessarily destroys the arrangement we are trying to study."[25] X-ray diffraction methods investigated the substance as the crystal that it is in nature, and less than two decades after the discovery of X-ray crystallography the central problems of the constitution of the silicates were solved. William Henry Bragg and his son, William Lawrence (both subsequently knighted) were among the first to employ the new methods to study the atomic structure of crystals, and by early 1913 they had succeeded in relating the wavelengths of the diffracted radiation to the distances between the atomic planes in crystals. Following the end of World War I, the elder Bragg studied organic crystals, while his son commenced the investigation of silicates. By 1930 the younger Bragg had worked out the crystal structures of over forty silicates. In that year, he published his classic paper on the subject with illustrations of the linkages of the silicon-oxygen tetrahedra (fig. 38).

Bragg explained: "(SiO_4) groups can be linked together to form silicon-oxygen complexes with indefinite extension in space. It is this feature which gives rise to the variety of silicate structures. . . . It explains the difficulty often encountered in the past in writing formulae for the silicates. Such structural formulae must take into account the spatial extension of the silicon-oxygen groups."[26] Bragg's results were subsequently modified, as was the nomenclature of the silicate groups. Nevertheless, his work was a significant contribution to the sciences of mineralogy and geochemistry and, by extension, to the scientific study of meteorites.

X-ray diffraction technique and the application of the principles of physical chemistry aided clarification of the origin of the Widmanstätten structure, but progress was by no means straightforward. Among the many scientists who studied the phases and phase transformation in alloys at about the turn of the century, Floris Osmond in Paris and Henrik Roozeboom in Amsterdam were probably the most influential. Osmond, who was trained as an iron and steel metallurgist, employed the Le Chatelier pyrometer soon after its invention in 1886 to determine the critical points in steel, and over the course of time identified the alpha, delta, and gamma phases. Roozeboom's background was more theoretical. In his doctoral dissertation at Leiden in 1884, Roozeboom investigated the pressures and temperatures at which certain hydrates change from solid to liquid to gas. Shortly thereafter, Johannes van der Waals at Amsterdam pointed out to Roozeboom the significance for such investigations of the phase rule of J. W. Gibbs, published in 1876. Roozeboom then commenced a lifetime study of heterogeneous equilibriums. He worked out the phase transformation of the binary cadmium-tin system and the ternary cadmium-tin-bismuth system, and in 1900 applied the phase rule to iron and steel. At this point, the different approaches of Osmond and Roozeboom converged.

In 1904 Osmond and his colleague, G. Cartaud, in consultation with Roozeboom, published the first phase-equilibrium diagram of the iron-nickel system. Osmond and Roozeboom collaborated in 1911 to produce a modified diagram; figure 39 is a composite of these two early diagrams. It is instructive to compare several of its details with the recent diagram published by A. D. Romig and J. I. Goldstein, illustrated in figure 17 (p. 122).[27]

We first note that the line AC in the Osmond diagram (fig. 39), marking the beginning of the transformation from the gamma phase to the gamma-plus-alpha phase, is lower than the line HL in the Romig-Goldstein diagram (fig. 17). Also, the line AB, indicating the transformation from the gamma-plus-alpha to the alpha phase, is in the higher temperature range lower than the line HK, and at B corresponds with a nickel content of about 9 percent. These discrepancies are a result of the relatively crude instruments available to the early investigators. Gibbs's phase rule does not predict the rates of transformation. In the iron-carbon system, the transformations are relatively rapid, whereas there are considerable lags in the iron-nickel system. These lags caused continuing difficulties for investigators who attempted to determine the

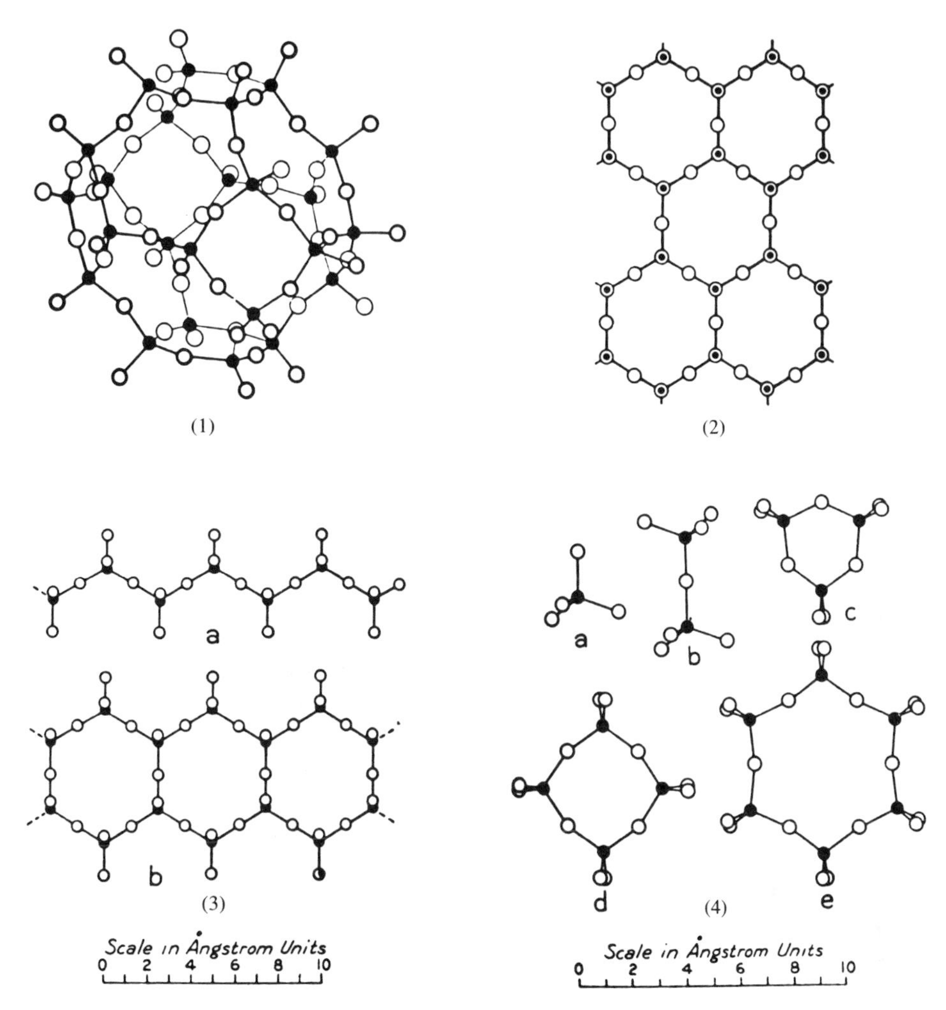

Fig. 38. Bragg's illustration of the linkages of silicon-oxygen tetrahedra. Black: Silicon, with or without aluminum; White: oxygen. (1) Silica type, SiO_2; three-dimensional linkage. (2) Disilicate type, Si_2O_5; two-dimensional linkage. (3) Metasilicate type: (a) SiO_3, single chain linkage; (b) Si_4O_{11}, double-chain linkage. (4) (a) Orthosilicate type, SiO_4; independent tetrahedra; (b) Pyrosilicate type, Si_2O_7; paired tetrahedra; (c) Ring-linkage, Si_3O_9; (d) Ring-linkage, Si_4O_{12}; (e) Ring-linkage, Si_6O_{18}. From Bragg (1930).

exact temperatures at which transformations occurred at any given percentage of nickel. The line CE in the Osmond diagram has no counterpart in the modern diagram and was the result of a misinterpretation. Osmond mistook the beginning of ferro-magnetism on cooling as a phase change, an error that was eventually corrected. More serious was Osmond's assumption that the line BCD at about 360° in the iron-nickel system was a eutectoid, corresponding to the eutectoid line at 720° C in the iron-carbon system. Over the following three decades, most investigators supported Osmond's ideas with some modifications.

There were two major reasons for the support of the concept of a eutec-

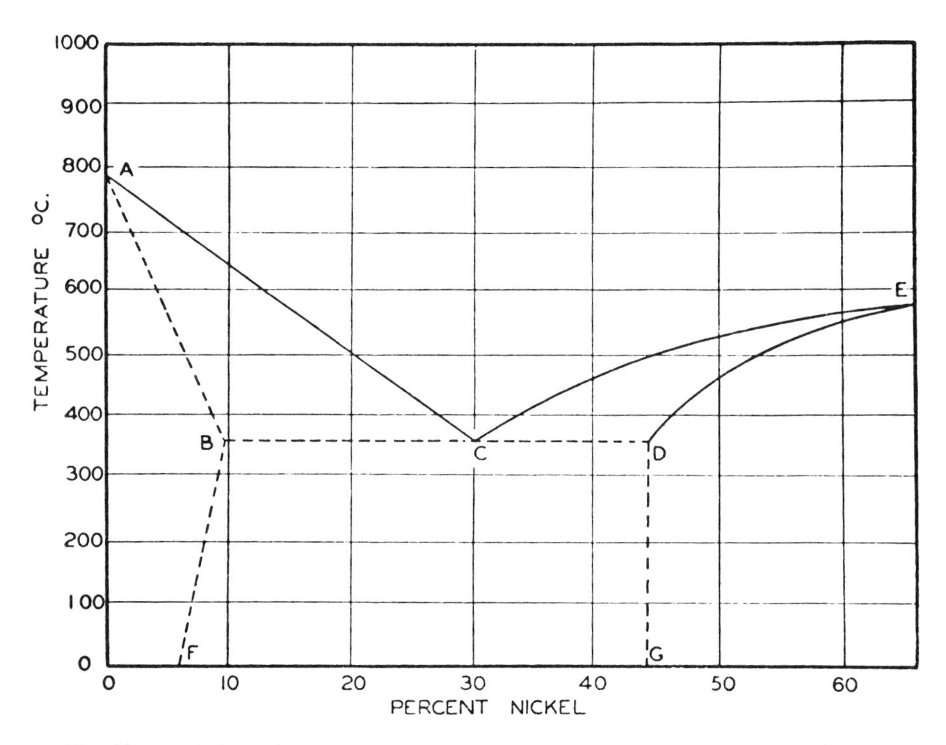

Fig. 39. Early iron-nickel phase diagram. Perry drew this composite from the diagrams published by Osmond and Cartaud in 1904 and by Osmond and Roozeboom in 1911. From Perry (1944).

toid. First, as mentioned in chapter 4, many chemists, beginning in the 1860s with Reichenbach, believed that kamacite and especially taenite were compounds having definite proportions of iron and nickel. Second, there seemed to be distinct analogies between the iron-carbon and iron-nickel systems, not only in the features of their phase-equilibrium diagrams but also in the appearance under the microscope of the transformation products. In the iron-carbon system, ferrite and the compound cementite, Fe_3C, formed the eutectoid pearlite. By analogy, in the iron-nickel system, kamacite and the compound taenite formed the eutectoid plessite. There had been controversy about the formula of taenite, and this continued. However, in 1917 E. Pfann reported his conclusions after examining the San Cristobal meteoritic iron. This meteorite has a very unusual structure, consisting of an aggregate of taenite and troilite crystals; according to Buchwald it is unique.[28] Pfann, on the bases of analyses of the taenite that showed a nickel content of about 26 percent, considered the taenite to be almost pure, and placed the saturation point of kamacite in taenite (Osmond's line DG) at 27 percent. He moved the eutectoid point, C, to about 18 percent, and set the maximum nickel percent of the kamacite phase at 7 percent.[29] To support his conclusions, Pfann published several photomicrographs of pearlite, which he had found in the San Cristobal iron and which he considered to be a decomposition product of the eutectoid. The idea of a eutectoid was attractive, because it apparently explained the actual changes detected in meteoritic iron as the nickel percentage increased. Below 7 percent nickel, the structure would be that of a hexahedrite composed entirely of kamacite. Between 7 and 13 percent nickel, the iron would be an octahedrite with an increasing amount of plessite. Between about 13 and 18 percent, the structure would consist of plessite with a minor amount of kamacite, and at 18 percent nickel only plessite could be observed. Above 18 percent nickel, the taenite increased and plessite decreased, until at 27 percent nickel the structure would be pure taenite.[30]

In 1877, Sorby postulated that the Widmanstätten structure was due to the "separating out of different compounds of iron and nickel and their phosphides," this process occurring over a long period of time at a temperature just below that of fusion (see chap. 4). Osmond and Cartaud demonstrated by their phase diagram that the structure could only form 700° C or more below the melting point, so that the process must be one involving solid-state diffusion. Also, they surmised correctly that the Widmanstätten structure resulted from the precipitation of kamacite preferentially along the octahedral planes of the original solid solution. In 1926, J. Young in England made an X-ray analysis of the Carlton meteoritic iron. He confirmed that kamacite had a body-centered cubic lattice, that taenite had a face-centered cubic lattice, and that the kamacite plates formed along the octahedral planes of the original face-centered cubic solid solution, although he hesitated in calling the latter taenite. Young went on to state that the process in which the plates thickened was necessarily slow because it involved the diffusion of atoms, that the "composition of the kamacite layers must change continuously as they increase in width," and that "the nickel content of the narrowing layer, which lies between them, must steadily increase."[31]

There was confusion about the Widmanstätten structure both before and after Young's study. A structure resembling the Widmanstätten pattern in meteorites can be produced by the furnace cooling of iron-nickel alloys or, in some instances, by the quenching of low-carbon steels. Sorby found such a structure in 1877 in a sample of Swedish iron that had been slowly cooled, and in 1910 Carl Benedicks reported seeing the structure in an alloy containing 11.7 percent nickel. The work of N. T. Belaiew best demonstrates how scientists were misled. In 1924 Belaiew published what he considered to be a definitive exposition of the Widmanstätten structure in meteorites.[32] The separation of a phase from a solid solution, he explained, could result in three different kinds of structures: a network structure, in which the phase will precipitate out along grain boundaries; the Widmanstätten structure, in which the phase crystallizes in the middle of the grains; and the production of a large crystalline structure. What occurred depended on the rate of transformation. Belaiew included a photomicrograph of a steel containing 0.55 percent carbon, which had been cooled in air, and claimed that it showed a Widmanstätten structure, although it would later be known as a transition structure. From this evidence, he concluded that "a rate of cooling which would cause the largest possible granulation and, afterwards, the quickest separation of the constituents are the conditions most favorable for the production of the Widmanstätten structure. Such or like conditions are, apparently, realized in meteoric irons." In the 1930s Robert F. Mehl and his associates produced Widmanstätten patterns in iron-nickel alloys that had been held at temperatures of about

1100° C for a week and then permitted to cool slowly.[33] During the 1920s and 1930s, investigators also determined that the Widmanstätten structure could be produced in nonferrous alloys by an age-hardening process, followed by an appropriate heat treatment. The nomenclature or, more precisely, the exact definition of the term "Widmanstätten structure" was the source of the difficulty to a great extent. To Belaiew and many others, any structure obtained artificially that resembled the structure of meteoritic octahedrites could be termed a Widmanstätten structure, whereas some scientists insisted that the term be restricted to structures resulting from diffusion.[34]

Berzelius, as noted in chapter 4, postulated that the presence of schreibersite was instrumental in producing the Widmanstätten structure. On the basis of his authority, a number of the nineteenth-century scientists who investigated meteoritic irons supported this view. In the late 1920s, Rudolf Vogel, using the iron-nickel equilibrium diagrams as a guide, began to speculate on the range of temperatures at which schreibersite precipitated out of cooling meteoritic iron, and attempted to construct a ternary iron-nickel-phosphorus diagram, at first theoretically and later experimentally.[35] Although Vogel's conclusions and diagram proved to be incorrect, he made a distinct contribution in calling attention to the role of phosphorus.

The culmination of these early twentieth-century studies of the structure of meteoritic irons, and in particular of the Widmanstätten pattern, came in 1944 with the work of Stuart H. Perry, a newspaper publisher, meteorite collector, and metallurgist whose studies were encouraged by Edward P. Henderson of the Smithsonian Institution.[36] Perry gave cogent reasons why the supposed eutectoid did not exist, and he presented a new equilibrium diagram that was fairly accurate and that consisted of a composite of the recent experimental work of several investigators. Perry stressed that the low-temperature transformation values were conjectural and required revision, that increasing phosphorus content depressed the gamma-alpha transformation, and that "meteoritic structures generally were produced by extremely slow cooling." Perry included a large number of good photomicrographs of the structures and constituents of meteoritic irons, and although some of his interpretations were incorrect, they represented a systematic study that was of considerable value to later scientists. After 1950, the major areas of investigation were those stressed by Perry: the nickel content at low-temperature transformation; the role of phosphorus; and the rate of cooling necessary to produce the true Widmanstätten structure of meteoritic irons. From the cooling rate, in particular, scientists could infer the size of the parent body from which the meteorite had come.

Origins and Orbits of Meteorites

In the first half of the twentieth century, the question as to whether meteorites originated within or outside of the Solar System took on a new urgency for scientists specializing in some areas apart from the study of meteorites per se. The spectroscopic methods of astrophysicists began to yield data concerning the relative abundances of elements in the Sun. If meteorites were smaller masses belonging to the Solar System, their compositions might provide a check on the accuracy of spectroscopic results. But if meteorites entered the Solar System as strangers from distant galaxies, the value of a comparison was doubtful. Cosmologists were also concerned. Thomas C. Chamberlin and Forest R. Moulton published their "planetisimal" hypothesis in 1905, which was designed to replace Laplace's theory of the origin of the Solar System. They argued that the original nebula had cooled rapidly and solidified into masses (planetisimals) that formed the planets by a process of accretion. If meteorites were masses that had not accreted or were fragments of destroyed planetisimals, their compositions might shed light on the validity of the hypothesis. Physicists, now aware that there were isotopes of elements, might determine whether the isotopic composition of an element was identical or dissimilar in meteorites versus terrestrial rocks. The question was also vital for geochemists who postulated models of the Earth based on meteorite compositions. If meteorites originated in outer space, the conditions of formation might be very different from those prevailing within the Solar System.

In the early 1900s, despite the arguments of Tschermak, Daubrée, Meunier, and others, the proponents of the theory that meteorites were fragments of a destroyed planet were on the defensive. In 1901 Chamberlin came to their support, postulating that a small planetary body might have been disrupted by coming within the Roche limit of another planet.[37] The disintegration would occur, Chamberlin wrote, without the creation of a large amount of heat, thereby retaining the various structures present. The stresses and fissuring existing in the small planet would aid the disruption, and the release and eruption of the highly compressed gases present in the core could result in "extremely minute division." If the fragments remained clustered, he added, they might constitute comets; if they became dispersed, they would become meteorites.

Many scientists, however, favored the hypothesis that meteorites were fragments of comets, which had entered the Solar System from cosmic space and had gradually disintegrated, forming meteor streams that periodically encountered the Earth. For example, in a letter to Merrill, Prior wrote that his theory of meteorites originating in a common magma was "more in favour of the broken-up planet idea of the origin of meteorites than the more generally accepted view which connects them with meteors and comets."[38] H. A. Newton, one of the most articulate defenders of the cometary theory, had died in 1896. Since 1890 Schiaparelli had been completely absorbed in the research and writing of his monumental history of ancient astronomy, although in 1910, the year of his death, he emphatically repeated his contention that meteorites came from outer space as parts of comets and hence traveled with hyperbolic velocity.[39] There were replacements for these elder scientists who were just as dedicated to the cometary theory, and just as convincing.

Apart from the visible evidence that the Earth periodically encounters the fragments of short-period comets producing the Leonid and other meteor showers, proponents of the cometary theory depended upon the results and data from two other sources: the spectrographic analyses of meteors and, more important, the computed heliocentric velocity of meteors, fireballs, and some meteorites. In two articles published in 1918 and 1922, C. C. Trowbridge, a Columbia University physicist, seriously undermined the former. Beginning in the 1860s, Alexander Herschel in England, Angelo Secchi in Italy, and Miklos von Konkoly Thege in Hungary had analyzed the spectra of bright meteors and found lines indicating the presence of magnesium, iron, sodium, and other metals found in meteorites. They assumed that these meteoroids did not have sufficient mass to withstand atmospheric passage. Trowbridge carefully reviewed their analyses and considered that they were subject to considerable error, concluding that:

> The motion of the meteor through the atmosphere produces an exceedingly high temperature and may bring about chemical and physical changes in the composition of the atmosphere in the track of the meteor; which on reverting to its original state gives out a phosphorescent glow, or the surrounding air may be highly ionized by the vaporizing meteor so that electrical discharges take place great enough to produce an afterglow.

The spectra of the meteor trains observed by Herschel and the others, and attributed to the presence of metallic elements, were undoubtedly afterglow spectra.[40]

The problem of orbits and velocities was much more formidable. The foremost authority at the turn of the century was Gustav von Niessl of Brünn, whose lifework was the computation of the orbits of meteors, fireballs, and a few meteorites whose trajectories had been reported by several reliable observers. Niessl did not mince words about his results. In 1907, for example, he announced that his calculations by the "careful use of observations" of a considerable number of bolides and of fireballs that had dropped meteorites demonstrated that their velocities exceeded that of a mass in an elliptical orbit around the Sun. Therefore, they could not "possibly have originated in the solar system," and must have "entered the solar system from the outside cosmic space."[41] In 1925, Cuno Hoffmeister edited and published all of Niessl's work and added some of his own calculations.[42] He listed data on the orbits of more than six hundred large meteors, of which 79 percent were hyperbolic. Among these, five had dropped meteorites: Treysa, Homestead, Orgueil, Pultusk, and Krähenberg. Confirmation came from another source. Ernest Öpik, in 1932, lent strong support to Hoffmeister's results, for he concluded that at least 30 percent of large meteors were in hyperbolic orbits.[43]

Two aspects of the cometary theory were puzzling. Apart from the Mazapil iron, which fell in 1885, there were no reports of meteorites having fallen during meteor showers. Warren M. Foote, for example, who reported the Holbrook meteorite shower of 19 July 1912, in which some fourteen thousand stones fell, thought that it might possibly be connected to the passage of a meteoroid stream.[44] However, he received a completely negative report from Eric Doolittle, director of the Flower Observatory at the University of Pennsylvania; the meteorite shower did not correlate with the radiants of any streams. The other question was why so few meteorites in hyperbolic orbits fell. Niessl and others ignored this problem, since they assumed on the basis of Schiaparelli's analysis that most meteorites traveling at such high velocities would be burned up during their passage through the atmosphere. In 1922, however, Frederick A. Lindemann and G. M. B. Dobson cast doubt on this assumption.[45] Only meteoroids with small masses are completely consumed in the atmosphere. A relatively high velocity of entry would increase the brilliance of such meteors, but this would be of shorter duration.

The times of falls of meteorites provided the principal foundation for the view that they were members of the Solar System. Most fell in the afternoon or early evening hours, which scientists attributed to their overtaking the Earth in its orbit. Fewer fell from midnight to noon, these being overtaken by the Earth. Other evidence gradually accumulated. Fred L. Whipple analyzed photographs of nine sporadic meteors and determined that at least eight of them moved in elliptical orbits.[46] C. C. Wylie, after reviewing the reported observations and calculations of the orbit of the Pultusk meteorite, concluded that the orbit was elliptical (see chap. 3). In 1934 S. H. Manian, H. C. Urey, and W. Bleakney measured the relative abundance of oxygen isotopes $^{16}O{:}^{18}O$ in several meteorites, including the Homestead stone, which was supposed to have a hyperbolic orbit. They found no difference in the isotopic composition of these meteorites, which was the same as a terrestrial granite from Stonington, Maine, and as laboratory potassium chlorate.[47] In 1939 Fletcher Watson summed up the situation:

> Since the study of photographed meteors, which are comparable to fireballs in brightness, has revealed no hyperbolic velocities, and since the directions of motion of fireballs and meteor streams are practically indistinguishable, it seems probable that the velocities previously assigned to fireballs were too great and that the bodies are permanent members of the solar system.[48]

These findings did not end the controversy. In 1940, for instance, Friedrich Paneth was quite provoked at Hoffmeister, who in a book on meteors, published in 1937, had flatly stated that the interstellar origin of the majority of meteorites was "a well-established fact." The calculations of Niessl, Hoffmeister, and others, Paneth thought, might very well be due to a systematic error.[49] We return to this subject in chapter 9; suffice it to say at this point that most if not all meteorites are members of the Solar System.

THE SCIENTIFIC USE OF METEORITES

Elemental Abundances

In 1901 O. C. Farrington wrote a series of articles on meteorites in the *Journal of Geology*. In his concluding remarks, he pointed out that the chief mineral constituents of meteorites occur in but insignificant amounts in terrestrial rocks. However, he continued, it is uncertain whether this quantitative distinction would appear in the same degree if scientists were in a position to compare the composition of meteorites with that of the entire Earth.[50] We recall that in the 1840s, the different average specific gravities of meteoritic irons, stony irons, and stones had led Boisse to postulate that each of the types might have formed a layer of the parent body from which it came. Tschermak, Daubrée, and others had the same idea when they hypothesized that meteorites were the fragments of a destroyed planet. What was new about Farrington's remark was that he was speculating that the relative abundances of the chemical elements in meteorites and on Earth were the same. His remark did not go unnoticed. Indeed, Emil Cohen thought that it was of such significance that he cited it in the conclusion of the second volume of his definitive work on meteorites.[51]

The first use of Farrington's hypothesis occurred in basic physics. William D. Harkins and his colleagues at the University of Chicago had been developing theories of atomic structures, and by 1916 had presented evidence that the variation in the chemical and in some physical properties of the elements depended upon the arrangement in space and the number of electrons surrounding the nucleus. Harkins assumed, as did other physicists, that the number of electrons depended upon the nuclear charge, and that the nuclear charge was equal to the atomic number. He had also postulated that the nuclei of those atoms having even atomic numbers were built up from helium atoms, with the general formula $n\mathrm{He}'$, where the prime indicated that these elements were "intra-atomic, not chemical compounds," and that the nuclei of those atoms having odd atomic numbers, beginning with lithium, had the formula, $n\mathrm{He}' + \mathrm{H}_3'$. Harkins believed that there were important distinctions between the even- and odd-numbered elements, the most striking one being that the structure of the nuclei of the odd-numbered atoms was less stable. This greater instability, he surmised, should be reflected in the relative abundances of the elements, the even-numbered elements being far more abundant than the odd-numbered ones. It was this hypothesis that Harkins proved in his 1917 paper.[52]

Harkins's source for the determination of the relative abundances of the elements was meteorites. Quantitative analyses of the composition of the Earth's surface were of no value, he stated, because these were representative of material of the "mere skin," which had been altered by geologic processes. Further, the spectroscope could not yield accurate measures of the quantitative composition of the elements at the Sun's surface. The only material available from which the relative abundances of the elements might be calculated was meteorites. "No matter what theory of their origin is adopted," he wrote, "it is evident that this material comes from much more varied sources than the rocks on the surface of the earth."

Harkins took the average of the analyses of 125 meteoritic stones and 318 meteoritic irons that had been compiled by Farrington, and to arrive at a weighted average, he accepted Farrington's assumption that the falls of meteorites were in the ratio of 350 stones to 10 irons. From these data, he calculated the atomic percentage of each of the elements. Despite the criticisms of Prior and Merrill that almost all of the analyses listed by Farrington were inaccurate, they were more than sufficient for Harkins's purpose. His results were as follows:

Atomic number	Element	Atomic percentage (350 stone to 10 iron meteorites) Even	Odd
6	carbon	0.12	
8	oxygen	53.16	
11	sodium		0.62
12	magnesium	13.15	
13	aluminum		1.21
14	silicon	15.35	
15	phosphorus		0.06
16	sulfur	1.46	
19	potassium		0.11
20	calcium	0.97	
22	titanium	0.005	
24	chromium	0.13	
25	manganese		0.06
26	iron	12.79	
27	cobalt		0.04
28	nickel	0.76	
29	copper		0.005
		97.985	2.105

One can sense the elation that Harkins felt when he wrote his summary. The even-numbered elements formed almost 98 percent of meteoritic matter and, by extension, of terrestrial matter as well. In addition, the elements having a low atomic number were much more abundant than those with a higher atomic number, the first 29 elements making up about 99.9 percent of the material. Harkins concluded: "The variation in the abundance of elements as found would seem to be the result of an atomic evolution, which is entirely independent of the Mendeléef periodic system. The formation of the elements seems to be, however, related to the atomic number." And, he stressed, the hydrogen-helium theory of the structure of the nucleus now rested on a firm foundation.[53]

Several aspects of Harkins's paper are worthy of note. First, his values of the atomic abundances for the elements he listed are, with the exceptions of carbon, aluminum, and titanium, surprisingly close to modern values. Second, he eliminated almost a dozen elements that had been reported in some analyses—namely, arsenic, antimony, lead, zinc, tin, gold, tungsten, uranium, barium, strontium, and zirconium—because Merrill had stated that the presence of these elements was either questionable or very doubtful. Third, he did not include platinum, palladium, or iridium, because chemists had found only traces of these elements in just a few meteorites. Finally, there were a sizable number of elements that chemists had failed to detect in meteorites, and, in addition, there were in 1917 five unknown elements, all of which, Harkins hastened to point out, had odd atomic numbers.

It was the search for the missing elements—number 43 and number 75—that led Ida Tacke Noddack and Walter Noddack to confront the problem of elemental abundances. Noddack, who received his Ph.D. in chemistry from the University of Berlin in 1920, was director of the chemical laboratory of the Physikalisch-Technische Reichsanstalt, where Ida Tacke was also employed. They were married in 1926 and collaborated in their research throughout their careers. They began their search for missing elements in 1922, and in 1925 announced that they had isolated rhenium (no. 75) and masurium (no. 43). Controversy surrounded the latter element until 1937, when Carlo Perrier and Emilio Segrè found an isotope of technetium (no. 43), an element that can be produced in measurable quantities only artificially. The Noddacks, however, had learned to detect trace elements by employing X-ray spectroscopy, and thus were able to report minute concentrations of very rare elements. Their 1930 article presented, on the basis of their analyses of meteorites, elemental abundances of 71 elements.[54]

For their analyses, the Noddacks used 42 meteoritic stones and 16 irons. They removed the crusts, separated the nickel-iron in the stones magnetically, extracted all of the inclusions from the irons, and set apart the troilite inclusions from 5 irons for analysis. They then determined the elemental abundances of each of the three components: stone, iron, and troilite. At this point, they were confronted by a problem that has perplexed all investigators who have attempted to determine elemental abundances from the composition of meteorites—namely, finding the most appropriate ratio of stone to iron to be used in arriving at an average composition. The chemical composition of the stones varies considerably, but there are, of course, very substantial differences between the compositions of the stones and the irons. Harkins, as we noted, simply accepted Farrington's estimate that 350 stones fell to every 10 irons, and proceeded on that assumption. The Noddacks took another path. Meteorites, they thought, were independent planetisimals whose orbits were between the Sun and the four inner planets, and whose composition therefore should be similar to the average composition of these planets. They calculated the mean density of Mercury, Venus, Earth and its moon, and Mars, and took this figure, which was 5.1, as the mean density of meteorites. They assigned the figure 3.3 as the mean density of meteoritic stones, and 7.8 as the mean density of meteoritic irons. They admitted frustration about the troilite component; they found that it composed 5.5 percent by weight of the stones, but that it varied considerably in the irons. Consequently, they just assumed that it made up 5.5 percent of the whole. When they computed an average composition of meteorites from these figures, they arrived at the values of 56.3 percent stone, 38.2 percent iron, and 5.5 percent troilite—or, in ratio form, stone:iron:troilite = 1:0.68:0.098. They then arrived at the total abundance of each element by multiplying its abundance in each component by the appropriate factor and adding the three resulting figures.

There are numerous discrepancies between the Noddacks's results and modern values of elemental abundances, but these are understandable, given the relatively crude analytical instrumentation they employed. The abundances of iron, nickel, and cobalt are all substantially higher, indicating that the Noddacks's assumption concerning the average composition of meteorites was too heavily weighted toward the irons. However, all of the elements whose existence in meteorites had been considered very doubtful only a dozen years before appeared in the Noddacks's tabulation. In this respect, their work was a pioneering effort of great future significance.

The Noddacks also drew one conclusion that stimulated future research. Their results showed, they wrote, that iron meteorites contained a large number of lithophile elements, such as beryllium, magnesium, calcium, strontium, barium, and titanium, some of which were in concentrations similar to those found in meteoritic stones. This indicated that meteorites were not fragments of a large planet, because in such a case these elements would have gone into the silicate layer, as on Earth. "There is much more the impression," they concluded, "that meteorites condensed separately at the formation of the solar system."

Just before the appearance of the Noddacks's table of abundances, Henry Norris Russell published his results of the first comprehensive quantitative spectroscopic analysis of the composition of the Sun's atmosphere.[55] Hydrogen was, of course, the most abundant element. Russell stated that oxygen was about as abundant by weight as all the metals together, a finding that correlated with Harkins's results. "The well-known difference between elements of *even* and *odd* number is conspicuous," he wrote, "the former averaging *ten times as abundant* as the latter." His comparison of the abundances of seventeen elements on the Sun's surface and of meteorites showed good agreement. Thus Russell and other astronomers provided a new means of determining elemental abundances, and showed that the composition of meteorites resembled the composition of the Sun much more than it resembled the material of the Earth's crust.

V. M. Goldschmidt was another one of the first scientists who attempted to determine elemental abundances from meteorite analyses. His research, however, encompassed much more of significance than just this problem, and thus is described separately in the next section.

V. M. Goldschmidt, Geochemistry, and Earth Models

Nineteenth-century geologists were necessarily interested in the chemical compositions of the rocks and minerals of the Earth's crust. Geochemistry, however, became a recognized discipline only in the last decades of the century, when a number of geologists began to employ the principles of physical chemistry in their attempts to determine under what conditions rocks and minerals formed and underwent alteration. One of the leaders in this area was Johann H. L. Vogt. Vogt was a Norwegian who pursued graduate studies at Dresden, the University of Christiania, and Stockholm, where he studied geology under the renowned Waldemar C. Brøgger. He became professor of metallurgy at the University of Christiania in 1886, and moved in 1912 to the Technical University of Norway at Trondheim. Vogt began his work in the 1880s with studies of the minerals formed in smelting slags, and made an analogy between these processes and the crystallization of igneous rocks. By the early 1900s, Vogt launched into full-scale investigations of natural silicates, studying the crystallization of the ternary feldspar system (orthoclase-albite-anorthite) and the granite system (quartz-orthoclase-albite).[56] His work was influential not only in Europe but also in the United States.

Frank W. Clarke, during his long tenure as chief chemist of the U.S. Geological Survey, supervised the analysis of thousands of specimens of crustal rocks, providing indispensable data for geochemists. Geochemistry flourished principally at the Geophysical Laboratory of the Carnegie Institution, which was established in 1907 under the directorship of Arthur L. Day. Day has perhaps not received the recognition in the annals of science that he deserves. Under his leadership, the laboratory became one of the foremost scientific institutions in the world. Foreign scientists were welcome to make extended visits to learn the techniques of determining such properties as the viscosity, surface tension, conductivity, and specific heat of silicates and other high melting-point materials in temperature ranges from –50° to 1600° C. Among the respected scientists who were employed at the laboratory in the early decades of the twentieth century were Henry S. Washington, Leason H. Adams, and in particular Norman L. Bowen.

Bowen was born in Canada and did graduate work at Queen's University and the Massachusetts Institute of Technology. Early in his career, he became acquainted with the investigations of Vogt, and his lifework focused on the study of the physical-chemical processes accompanying the differentiation of igneous rocks. Bowen and his colleagues produced some two dozen phase-equilibrium diagrams, among them the plagioclase system and the ternary system, diopside-forsterite-quartz. It was Bowen, probably more than anyone else, who influenced the next generation of geochemists in the United States. The impact of his work did not stop at the Atlantic, for in Norway, Goldschmidt freely acknowledged the debt he owed to Bowen.

Victor M. Goldschmidt was the son of a professor of physical chemistry who taught at Amsterdam, Heidelberg, and the University of Christiania, and this circumstance probably accounts for Goldschmidt's choice of a career. He completed his graduate work in chemistry and geology at the University of Christiania, studying the latter subject, as Vogt had three decades previously, under W. C. Brøgger. But he also went abroad, studying crystallography at Munich under Paul von Groth, and mineralogy at Vienna under Friedrich Becke, who himself had been a student and assistant of Gustav Tschermak. After receiving his Ph.D. in 1911, Goldschmidt spent two years as an instructor at the university, and in 1914 was appointed professor and director of the mineralogical institute. He accepted a chair at Göttingen in 1929, but in 1935 he abandoned it because of the rising anti-Semitism in Germany and returned to Norway, where he immediately received a university post. When the Germans occupied Norway during World War II, Goldschmidt fled to Sweden and then to England. After the war he returned to Norway, where he died in 1947.[57]

Crystallographic studies provided the basis for Goldschmidt's investigation of the geochemical distribution of chemical species and of elemental abundances. Using the X-ray diffraction methods developed by the Braggs, Goldschmidt and his colleagues determined the crystal structures of about 200 compounds of 75 elements, and published the first tabulation of the atomic and ionic radii of many of the elements. As a result of this work, Goldschmidt focused on the phenomenon of isomorphism, which he began to understand was one not just involving physics but related to geochemical principles.

During this period, also, he became interested in determining the differences in the elemental abundances of meteorites and of terrestrial rocks.

In 1922 Goldschmidt published a brief outline of what was later to become a full-fledged theory of the Earth's structure.[58] The different compositions of meteorites—that is, the division between irons, stony irons, and stones—obviously influenced him. Two layers separated the Earth's crust, which was about 120 km thick, and the core, which had a radius of about 3500 km. The zone beneath the crust Goldschmidt termed the "eclogite-shell," following the ideas of Finnish geologist Pentti E. Eskola, who postulated that it was composed of the same materials as in the crust, but very highly compressed. This zone continued to a depth of 1200 km, and beneath it and extending 1700 km to the top of the core was a shell composed of sulfides and oxides—iron sulfide, magnetite, chromite, rutile, and so forth (fig. 40). There were no mineral combinations in meteorites, Goldschmidt wrote, that were equivalent to those of the eclogite layer, because in the small parent bodies from which meteorites came there was insufficient compressive stress to give rise to their existence. The stony irons worried Goldschmidt because they did not appear to fit into the model. He wondered whether it might be possible to produce the pallasites, for example, artificially.

During the 1920s there was interest in the United States also in developing models of the Earth, and H. S. Washington at the Carnegie Institution was at the forefront. In 1924 he collaborated with L. H. Adams in presenting a model that postulated an iron core, a "pallasite" zone, and a silicate shell.[59] The

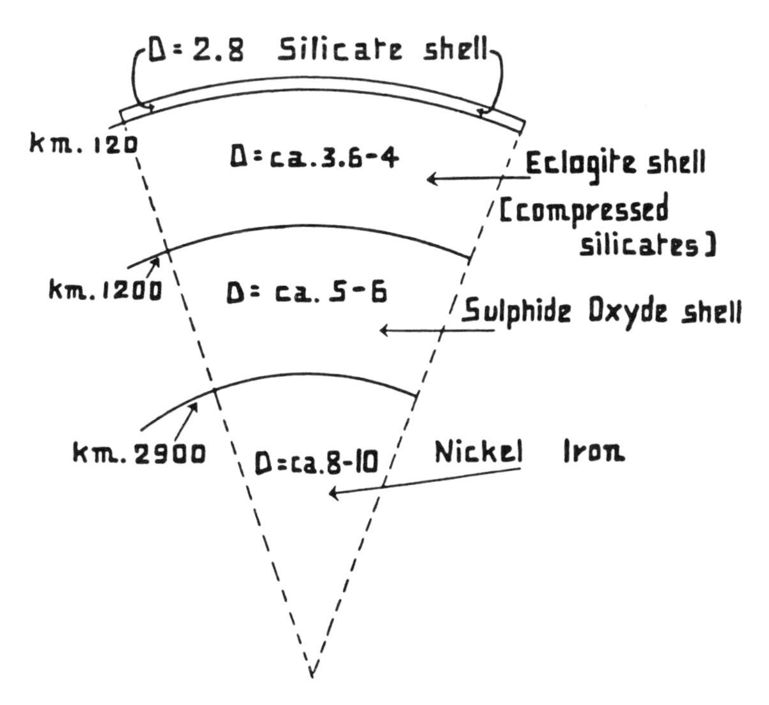

Fig. 40. Section through the Earth in Goldschmidt's model. From Goldschmidt (1922), (1929).

conclusion that a considerable amount of material that was denser than silicate rock existed at the Earth's core, they wrote, was reached by studies of the compressibility of rock and of the velocity of earthquake waves. The reason for assuming that this denser material was iron was the analogy with meteorites, since meteorites were in all probability the fragments of a small planet and many were composed almost entirely of nickel-iron. The existence of a pallasite zone above the core was indicated by earthquake velocities as a function of depth—that is, the velocity increased steadily and rectilinearly down to a depth of 1600 km, and then became almost constant through the next 1400 km. Velocity increases with pressure and hence with depth, they explained, but it is much less in metallic iron than in basic silicate material. The very small rate of change of the velocity below 1600 km depth must be due to an increasing amount of iron mixed with peridotite. The "pallasite continuum" from the core upward consisted, then, of a *lithosporic* region, in which silicate was sporadic in the metal, and then of a *ferrosporic* region, in which the metal was sporadic in the silicate. The Mount Vernon pallasite exemplified lithospor almost perfectly, containing about 32 percent iron by volume and 52 percent by weight (see fig. 7). The Alfianello meteoritic stone, they thought, was a good example of ferrospor, having by weight a content of 8 percent nickel-iron. They concluded: "The Earth, then, is to be conceived of, not as a huge meteorite, but as a body similar to those of which meteorites are but fragments representing different parts of the whole mass. On disruption, therefore, the Earth would yield all of the known kinds of meteorites."

In 1926 Washington returned to the subject, this time from a more quantitative point of view. The radius of the Earth's metallic core was 3400 km; the lithosporic shell, consisting of olivine, pyroxene, and iron, was 700 km thick, as was the ferrosporic shell above it. Then there was a shell of peridotite, extending upward 1540 km, and finally the Earth's crust, which had a thickness of 60 km, the lower portion being basalt and the upper granite. Washington then calculated the chemical composition of the Earth from the amounts and chemical compositions of these constituents. His results were as follows:

Element	Percentage by weight
Iron (core)	31.87
Iron (silicate)	7.54
Oxygen	27.71
Silicon	14.53
Magnesium	8.67
Nickel	3.16
Calcium	2.52
Aluminum	1.79
Sulfur	0.64
Sodium	0.39
Cobalt	0.23
Chromium	0.20
Potassium	0.14

Element	Percentage by weight
Phosphorus	0.11
Manganese	0.07
Carbon	0.04
Titanium	0.02
	100.00

The values for sulfur, phosphorus, and carbon, Washington thought, were probably somewhat low. The total percentage of the approximately 80 other elements was almost negligible; together they constituted about 0.1 percent by weight. According to the calculations of astrophysicists, he said, the 11 most abundant elements in the Earth, as listed in his tabulation, corresponded, with the exception of hydrogen, to the 11 most abundant in the Sun's atmosphere. There were some slight discrepancies, he concluded, but these were explicable by the differences in the physical conditions.[60]

There was a striking contrast between Washington and Goldschmidt in their approaches to geochemistry. Washington arrived at his values by assuming that the lower shells were similar to, if not identical with, meteoritic irons, stony irons, and stones, and based his calculations on their chemical compositions. Goldschmidt, in contrast, had developed a theory as to why the shells that he postulated had formed in the early stages of the Earth's history, and sought to verify his conclusions by reference to the chemical composition of meteorites, in particular the concentrations of the rarer elements they contained.

In 1929 Goldschmidt published a comprehensive account of his views.[61] He first addressed the assumption of the existence of a "pallasite" or a lithosporic-ferrosporic shell, consisting of a mixture of silicate and nickel-iron. Its existence seemed improbable to him, because seismic waves appeared to show marked discontinuities at the boundaries of the iron core and the silicate mantle. Further, it was very improbable that two substances having such different densities as iron and silicate would not, in the molten state, have separated due to the Earth's gravity, even making allowance for the increase of viscosity through pressure. It was very likely, Goldschmidt continued, that this intermediate shell consisted of the sulfides and oxides of heavy metals, especially iron and chromium. Troilite existed as a distinct phase in a very large number of meteorites, and the fact that most meteoritic stones contained 5 to 7 percent troilite indicated a saturation limit of this monosulfide. Molten sulfides segregated as separate sulfide magmas from many of the igneous rocks on the Earth's surface, and such magmas, if they did not solidify or were prevented by a rock barrier, would tend to migrate toward the layer separating the iron core and the silicate mantle. There, most sulfide and heavy oxide compounds would accumulate.

Scientists agreed, Goldschmidt wrote, that the Earth must once have been molten to account for the density distribution that now prevailed. In the early stages, the planet, surrounded by a gaseous envelope, separated into several liquid phases owing to gravity. The principles of physical chemistry then governed the distribution of the chemical elements between the gaseous and the three liquid phases. The volatile elements such as nitrogen and argon, and compounds such as water and carbon dioxide, which had little affinity to the substances in the three liquid phases, entered the primordial atmosphere. Goldschmidt termed these substances *atmophile*. Silicon, the light metals, ionic compounds, and in particular oxygen were the *lithophile* elements that entered the silicate magma. The *siderophile* elements became concentrated in the iron core; these were elements such as nickel, platinum, carbon, and phosphorus and their compounds. Those elements that entered into the molten iron sulfide Goldschmidt called the *chalcophile* elements, such as copper, silver, lead, bismuth, and selenium. Iron was the measure of geochemical affinity, because the partition of all other metals depended upon their affinities to oxygen and to sulfur, compared to their affinities to iron: "If an element is more siderophile than iron, it will enter into the nickel iron core, if it is more chalcophile than iron it will concentrate in the sulphide phase, if its inclination to enter into oxygen compounds is greater than that of iron it will accumulate in the silicate slag."

Nickel, being a siderophile element, Goldschmidt wrote, should be more highly concentrated in the Earth's core than in the crust. The analogy of the core with iron meteorites, which contain about 8 percent nickel, demonstrated that this was indeed most probable. Similarly, metals of the platinum group are siderophile, and whereas surface rocks contain platinum to the amount of about 1 gram in 1,000 tons, analyses of meteoritic irons disclosed its presence in varying concentrations, up to 100 grams per ton. The platinum metals, then, were not very rare elements; their relative rarity at the Earth's surface was a result of their partition between the iron core and the silicate layer. Goldschmidt also surmised that gold and silver might not be too rare, because these elements would have entered the sulfide phase.

The rest of Goldschmidt's 1929 article discussed the results of geochemical studies. He referred to the correlation between the geochemical character of the elements and atomic volume, the atmophile elements having the largest, the siderophile elements the smallest, and lithophile and chalcophile elements having intermediate sizes. He praised N. L. Bowen for having determined through experimentation the process of fractional crystallization in which igneous magmas evolved in a predictable sequence. He described how certain elements and ions having approximately the same atomic radii entered a crystalline phase together, thus producing the phenomenon of isomorphism. His account turned out to be an introduction to the principles of modern geochemistry.

During his few years at Göttingen, Goldschmidt and his associate, Claus Peters, investigated the geochemical properties of several elements, among them the metals of the platinum group and beryllium.[62] In 1937 he published a table of the abundances of about seventy elements, using his own and the Noddacks's results. Commenting on this effort in 1949, Harrison Brown wrote that it was "reasonably accurate," given the analytical methods used by Goldschmidt and the Noddacks. But Brown also pointed out that in 1935 György Hevesy and Hilde Levi had first employed the technique of neutron activation to analyze rare earth elements, and had thereby detected the presence of dysprosium in a sample of yttrium, and also europium in gadolinium, which altered the values of these elements as listed by Goldschmidt.[63] Hevesy had also experimented with the technique of isotope dilution in 1932. After World War II, the gradual perfection of these methods led to increasing accuracy in the determination of elemental abundances in meteorites.

One other Earth model requires description—that proposed in 1943 by Reginald A. Daly, professor of geology at Harvard University. The article in which Daly presented the model was far-ranging and of considerable importance, because in it Daly estimated the mass, radius, and density of the parent planet from which he supposed meteorites came, and also postulated the probable cause of its disruption.[64] Using the most recent data from seismic studies, Daly substantially revised the model envisioned in 1926 by Washington, although he retained its meteoritic character. Instead of the three velocity discontinuities noted by Goldschmidt and the four by Washington, Daly detected a total of eight. Beneath the crust, the Earth's chemical composition changed gradually, but in perceptible stages, from that of meteoritic achondrites to that of chondrites, to that of iron-poor stony iron, to that of iron-rich stony iron, and finally to a fluid nickel-iron core. Daly dismissed Goldschmidt's idea of a sulfide-oxide shell (also proposed in the mid-1920s by Gustav Tamman), stating that there was "no evidence that the sulphide occurred in the original home of the meteorites on a scale comparable to that indicated by the irons or stones of our museums."[65]

To test his hypothesis that the Earth was essentially composed of matter similar to meteorites, Daly sought to determine the physical and chemical nature of the parent body from which actual meteorites had originated. For this purpose he presented a comprehensive review of the most recent scientific data on the chemical and mineralogical composition of meteorites and on the relative abundances and densities of their various types. His comparative data for adjacent planets and the parent planet are shown in table 5 on page 264.

Daly suggested that the parent planet had been originally in a gaseous state at a temperature of about 3000° C. Much of the gas was lost to space because of its volatility at high temperature. As cooling occurred, there was differentiation in the mass owing to density, which resulted in an iron core, an overlying liquid mantle, and a gaseous atmosphere. Crystallized blocks in the liquid mantle sank successively to the depth of the core, again owing to density, creating convection currents in the liquid and a sublayer of silicate, in which were embedded chondrules "inherited from the time of the fiery rain out of the initial body of planetary gas." The process of crystallization and submergence continued until the entire planet was either nearly or completely solidified.

Shortly after solidification, the planet whose orbit was located between Mars and Jupiter made such a close approach to Jupiter that it was greatly distorted by the latter's tidal influence. This event, which possibly occurred more than once, caused crushing brecciation of its constituent crystals and their displacement. Finally, the parent planet approached Jupiter so closely that it "was torn to pieces—the result being the fragmentary asteroids and meteorites of the present day."

This hypothesis, Daly thought, appeared to explain many of the observed characteristics of meteorites, such as the fact that iron meteorites must have

TABLE 5
COMPARATIVE DATA FOR EARTH, VENUS, MARS, AND A "PARENT PLANET" OF METEORITES

	Earth	Venus	Mars	"Parent"
Mean radius (km)	6371	6200	3385	3000
Volume (earth at unity)	1	0.92	0.15	0.105
Mean density	5.52	4.86	3.96	3.77
Radius of iron core (km)	3500	2900	1300	1000
Mean density of mantle	4.5	4.4	3.7	3.6
Mean density of iron core	10.7	10.5	8.4	8.1
Total mass (earth at unity)	1	0.81	0.108	0.07
Mean gravity at surface (earth at unity)	1	0.85	0.38	0.32
Pressure at base of mantle, in thousands of bars	1350	1250	300	225
Pressure at center, in thousands of bars	3500	2700	500	350

Note: Reprinted from Reginald A. Daly, *Bulletin of the Geological Society of America* (1943), 54:417.

formed in a reducing atmosphere, that meteoritic stones had relatively fine grains and confused textures, and that some stones contained chondrules. The crystallization of the iron core, he wrote, "was not sudden," so that the coarse grains of many iron meteorites had the opportunity to form, just as "during a comparatively short time large crystals form in cooling iron ingots from a blast furnace." But melting of the crystallized iron "torn out of the planetary core" might have occurred, resulting, after the radiation of heat to space, in rapid chilling and crystallization, which would produce a "fine grain" in some fragments. Here, Daly confused the dendritic structure formed in pig iron and steel ingots with the coarse-grained structure of meteoritic iron, and also the transition structures of low-carbon steels with the fine-grained structures of iron meteorites. However, Daly was intrigued by the idea of the consequences of rapid cooling. Some minerals, he conjectured, might have been speedily chilled "by the cold of interplanetary space" and converted into glass, as evidenced by the presence of maskelynite and of vitrified silicate chondrules.

Daly did not expect that his model of the meteorite parent body or that his meteoritic Earth model, for which he offered closely reasoned arguments, would be lasting hypotheses. What he hoped was that his work would be carefully studied and subjected to cogent and sympathetic criticism. He concluded by asking astronomers, geophysicists, and geochemists: "How can the model be made more plausible by modification, by taking better account of experimental and other observed data already in hand?" In the following years, certainly, Daly's efforts stimulated the reaction that he desired.

NEW PROBLEMS

How Old Are Meteorites?

Nineteenth-century scientists addressed the question of how old meteorites were only infrequently and then either vaguely or in a comparative manner.

Shepard thought that no meteorites had fallen before the Diluvial period, but he gave no estimate as to when that epoch had begun. Sir Robert Ball surmised that meteorites were ejecta from massive terrestrial volcanoes that had been active early in the Earth's history, but neglected to add how long ago these events might have occurred. Reichenbach, who postulated that meteorites originated from the condensation of a nebula, failed to state how far in the distant past this had happened. Similarly, Tschermak and Daubrée did not mention when the disruption of the planet that produced meteorites had taken place. Sorby suggested that the Widmanstätten structure formed by slow cooling over a very long period of time, without speculating whether this might be a thousand or a million years. As for terrestrial ages, the advanced state of corrosion of a number of iron meteorites led scientists to state in their reports that they must have been on Earth for many centuries.

The ages of meteorites; the time of their formation; the duration of the period that they were in orbit; and how long ago in the distant past some, in particular the irons, had fallen to earth—these subjects constituted a new set of problems for twentieth-century scientists. In the decades before 1950, there were attempts to answer at least the question of the time of origin, but the results served to complicate rather than to clarify matters. It was for this reason that in 1943 Reginald Daly carefully sidestepped the issue as to when the meteorite parent body that he postulated had been disrupted. If the fragments of the destroyed planet, he wrote,

> had the mass of Eros or even one-tenth of that of Ceres, the effect on the earth's crust would have been at least regionally catastrophic. Bombardment on that scale seems never to have affected the crust since the Basement Complex became a constituent of the earth-shell. It would, therefore, be necessary to assume that the larger fragments of the disrupted parental planet were swept up before late-Archean time. Whether space could, thus early, have been cleared of these aberrant members of the asteroid swarm . . . is a problem for an expert in celestial mechanics.[66]

At the beginning of the twentieth century, there was only one scientifically authoritative estimate of the age of the Earth. In 1864 William Thomson (later Lord Kelvin) had calculated, on the assumption that the Earth had cooled uniformly from its initial molten condition, that it solidified not less than 20 million and not more than 400 million years ago.[67] Subsequently, Thomson revised his estimate, and in 1899 computed the age of the Earth's crust as being between 20 and 40 million years.[68] Thomson's figure dismayed uniformitarian geologists and Darwinian evolutionists, because the theories of both schools required more terrestrial time than Thomson would permit. Some geologists, however, supported Thomson. In 1868 Archibald Geikie arrived at an age estimate of 100 million years based on denudation studies, and in 1899 John Joly, after calculating the annual increase of sodium in seawater, concluded that the Earth's age was between 80 and 90 million years.[69] And in 1893 astronomer Simon Newcomb placed the Sun's age at 100 million years, from calculations of the time required to reach its present size, assuming that it had condensed from a nebula and that its radiant energy resulted from gravitational attraction.[70]

Evidence accumulated and interpreted by Ernest Rutherford at McGill

University and Bertram Boltwood at Yale challenged Thomson's estimate less than a decade after Henri Becquerel's 1897 discovery of the phenomenon of radioactivity. After Sir William Ramsay and Frederick Soddy had found that helium was a product of the disintegration of radium and after Boltwood had determined that uranium and radium were genetically related, Rutherford suggested in 1904 that measurement of the amount of produced helium and its rate of formation would give the age of the mineral containing it.[71] Only a few months later, Boltwood published an article in which he called attention to the "notable quantities of uranium and radium in monozite," and stated that this could explain the occurrence of helium in this mineral. The age of monozite was extreme, he continued, since it was present as a primary constituent of some of the oldest igneous rocks. If it were assumed that the mineral contained considerably more uranium than at present, he concluded, then the accumulated helium and the lead could represent the disintegration products of radium "for countless ages."[72]

In the Silliman Lectures that Rutherford delivered at Yale in 1905, he explained how the age of a mineral could be determined by employing the uranium:radium ratio and the analyzed content of helium. As examples, he calculated that the minimum ages of analyzed samples of fergusonite and of a uranium mineral from Glastonbury, Connecticut, were 500 million years. Some of the helium may have been lost, he stated, which was why these ages were the minimum. Rutherford also noted that Boltwood had gathered a large amount of data concerning the possible significance of the presence of lead in radioactive minerals. "If the presence of lead from radium is well established," he wrote, "the percentage of lead in radioactive minerals should be a far more accurate method of deducing the age of the mineral than the calculation based on the volume of helium, for the lead formed in a compact mineral has no possibility of escape." To substantiate the case for lead as an end product, Rutherford stated that if the alpha particle was a helium atom (which was not proven until 1909), then the loss of three alpha particles in the transformation of uranium (atomic weight 238.5) to radium, and the additional loss of five alpha particles in the transformation of radium to the end product, would yield for the end product an atomic weight of 206.5, "very close to the atomic weight of lead, 206.9."[73] In 1907 Boltwood presented a great amount of evidence demonstrating that lead was indeed the final disintegration product.

There were numerous errors in the early age determinations by the helium method and by the lead method, apart from the recognized possible escape of helium. Rutherford's calculations of the uranium:radium ratio and of the half-life of radium were incorrect, and the value of the latter did not become established for a decade. Boltwood, not yet aware of the steps in the thorium transformation series, did not believe that lead was a final disintegration product. Thus the age figures for all minerals containing uranium and thorium were incorrect. Then in 1913 came the discoveries of the isotopes of lead, which meant that only the lead from uranium should be taken into consideration in an age determination using the uranium-lead transformation. Also, there was some delay in the recognition that only samples of unaltered, stable, primary rocks should be used.[74]

Between 1908 and 1910, Robert J. Strutt published the results of several age determinations he had made using the helium method, and estimated the ages of a number of thorianites from Ceylon to be several hundreds of millions of years. Beginning in 1911, Arthur Holmes, who had studied physics under Strutt at Imperial College and had assisted him in experiments, made a determined effort to apply radioactive methods to geology by use of the lead method. Before the introduction of the mass spectrometer in the early 1920s, however, there were still substantial disparities in results, as is evidenced in a 1921 paper by Henry N. Russell.[75] On the assumption that all of the lead was from the disintegration of uranium and thorium, Russell wrote, the calculated age of the Earth's crust was 3.5×10^{10} years. However, Joseph Barrell at Yale had computed the uranium:lead ratios for several minerals of the Middle Precambrian age from widely separated localities, which gave ages up to 1.1×10^9 years. The mean of these figures, Russell wrote, was 4×10^9 years, which, according to the planetisimal theory, was the time of the ejection from the Sun of the material that formed the planets. Russell's estimate, of course, is reasonably close to the modern value, even though his method of determining it was somewhat unorthodox.

Until about 1930 there were a few experiments to determine whether or not meteorites were radioactive but no attempts to determine ages. In 1906 Strutt examined three meteoritic irons and one stone for possible radioactivity and found that only the Dhurmsala chondrite showed radioactivity. In 1917 T. T. Quirke and L. Finkelstein at the University of Minnesota tested sixteen stones and seven irons. They determined that all of the stones displayed radioactivity, but that only the Toluca and Coahuila irons did. They also stated that the meteoritic stones were less radioactive than the average igneous rock.[76] These investigators used the "radium emanation" method for detecting radioactivity; therefore, their results reflected the smaller concentrations of uranium in meteoritic irons than in stones.

Friedrich A. Paneth and his coworkers, beginning in 1928, applied the helium method for determining the ages of iron meteorites. Paneth was a native Austrian, and was educated at Hamburg, Glasgow, and Vienna, where he received the Ph.D. in 1910. From 1912 to 1918 he was an assistant to Stefan Meyer at the Vienna Institute for Radium Research, where he recognized that Radium D and Thorium B were isotopes of lead and that Thorium C and Radium E were isotopes of bismuth.[77] While at the University of Hamburg from 1919 to 1922 and Berlin from 1922 to 1929, Paneth sought to perfect methods of detecting trace amounts of helium, first with the X-ray spectroscope and later with the mass spectrometer. Methods of detection improved to such an extent that, whereas in 1908 Soddy was able to detect 1×10^{-6} cc of helium, Paneth could detect 1×10^{-10} cc, and in 1931 Paneth and William D. Urry were able to measure 1×10^{-8} cc.[78]

In 1933 Urry, who collaborated with Paneth for several years, explained that the helium method had been chosen because it appeared to be simpler. There had to be reasonable certainty that a mineral's composition did not change by admixture from the outside or by losses from the interior and that only such changes occurred as could be ascribed to radioactive disintegration. It was then necessary to determine the masses of uranium, thorium, and helium, to know the disintegration constants of the radioactive elements, and to consider the possibility of the presence of helium at the time of the mineral's formation. As for the constants, that of uranium was well known, and although there was disagreement about the value of the thorium constant, the worst error this uncertainty would produce amounted to about 5 percent, which was less than the experimental error. There was little probability that any helium was present at zero time—that is, at the age of solidification. Further, there was little likelihood of admixture of helium from the outside, so that the main concern was the loss of helium.[79] Although Urry did not mention it, the investigators believed that there was little probability of the loss of helium from compact meteoritic irons.

Urry in 1937 presented the experimental results for 27 iron meteorites.[80] The youngest was 1×10^8 and the oldest was 2.9×10^9 years, with the others showing ages between these figures. Meteorites, then, were not older than the Earth, which calculations had estimated to be between 2×10^9 and 3×10^9 years old. "Using various astronomical arguments," he concluded, "together with these ages, it appears that nearly all iron meteorites are the debris of two planets of our own solar system, which came to grief within the past 100 million years."

In 1939 and again in 1940, Paneth published revised values of his age determinations of irons, which ranged from 3×10^7 to almost 3×10^9 years.[81] He attempted to account for the possible sources of the wide variation in the ages. It was a difficult task, he said, to determine the thorium content. Also, standards for calibration were lacking, so that the absolute accuracy of the values was doubtful, although the relative values were useful as a guide. It was possible, he conjectured, that the parent planet or asteroid from which the young meteorites came had had a molten iron core until its recent disruption. What was more likely, however, was that the disruption occurred in the early history of the Solar System, and that some of the fragments approached the Sun closely enough to become molten, thereby yielding a new age since solidification.

William E. Foshag at the Smithsonian favored Paneth's latter idea. The results achieved by the helium method were erratic thus far, he wrote in 1941, but the differences in ages were probably due to the loss of helium by expulsion during close encounters with the Sun. Robley D. Evans at the Field Museum of Natural History interpreted the results differently. They suggested, he wrote, that the formation of the celestial bodies was a continuing process, one that was possibly still proceeding at a relatively uniform rate.[82]

Paneth and his coworkers at the University of Durham published new and surprising results in 1942.[83] The earlier uranium determinations, they reported, lacked the requisite accuracy. They had now determined the ages of about three dozen meteoritic irons, and these varied from about 2.7 to 6.8 billion years. They believed that the highest ages were correct to within about 1 billion years, and they concluded that the age of the Solar System was about 7 billion years.

Carl A. Bauer at the Harvard College Observatory suggested in 1947 the cause of the substantial variation in ages. Large helium contents had been found only in small iron masses, whereas all large meteorites showed relatively small helium contents. This relation, Bauer wrote, indicated that the helium content was related to the preatmospheric masses of the irons, and he thought it possible that cosmic radiation produced an extra amount of helium in the smaller masses.[84] Additional evidence pointing to the same cause was forthcoming almost immediately. H. E. Huntley reported that a packet of nuclear research plates had been exposed for six weeks at the top of the 11,000-ft high Mont aux Sources in Basutoland, and displayed a large number of cosmic disintegrations. A great majority of the tracks, Huntley stated, appeared to be due to alpha particles. If cosmic rays produced helium, as this experiment showed, he concluded, then the age determinations of meteorites by the helium method might be unreliable.[85]

In 1948 Bauer published calculations on the effects of the mass and the shape of a meteorite on the amount of cosmic radiation it absorbed, which substantiated his earlier observation.[86] The discovery that cosmic radiation produced helium in meteorites in their preatmospheric journeys effectively ended the calculation of their ages by the helium method. However, it also drew the attention of scientists to the fact that the radiation produced nuclides. Later studies of these phenomena resulted in the accumulation of data indicating the duration of the period from the time the meteorite fragment left the parent body to the time it reached the Earth.

Scientists also considered other methods of age determination while the helium method was undergoing trial. Norman R. Campbell and Alexander Wood established the radioactivity of potassium in 1907, and as early as 1912 Otto Hahn investigated the possibility of determining meteorite ages by the use of either radioactive rubidium or potassium. In the 1930s Albert O. Nier and A. K. Brewer independently determined half-life values for ^{40}K, and W. R. Smythe and A. Hemmendinger found its transformation products.[87] The application of the potassium-argon method to estimating the age of meteorites did not occur until after World War II. The scientific and technical knowledge of nuclear physics and nuclear chemistry gained during and immediately after the war led in time to the development of other methods of age determination, also.

Craters

The crater at Coon Butte, in Coconino County, Arizona, later officially named Meteor Crater and now Barringer Meteorite Crater, first came to scientific attention in 1891 (fig. 41). The president of the forerunner of the present Santa Fe Railroad commissioned Albert E. Foote, a Philadelphia mineral and meteorite dealer, to investigate rumors that a substantial vein of iron ore existed near the crater. During his survey, Foote gathered fragments of the Canyon Diablo iron, and after determining that they were meteoritic, he reported his find in the *Proceedings of the American Association for the Advancement of Science*.[88] He included an observation that diamonds were present in the meteorite, but did not explain the presence of the crater. Foote attempted to gain a monopoly on pieces of the meteorite, which were gathered by making arrangements with the owner of the nearby trading post, and began to distribute specimens worldwide, an 835-lb mass being sold to H. A. Newton at Yale for $1,250.[89] Very soon, reports of other scientists confirmed the presence of

Fig. 41. Barringer Meteorite Crater facing west. Canyon Diablo is in the background. Courtesy of the Center for Meteorite Studies, Arizona State University.

diamonds. Estimates of the total weight of those metallic iron pieces of the Canyon Diablo meteorite that have been recovered range as high as 30 tons.[90] Figure 42 shows the distribution of the meteorite fragments that had been found as of 1909.

The early descriptions of Meteor Crater indicated that it was almost circular, but a recent vertical aerial photograph shows that it is square with rounded corners. It is approximately 1220 m in diameter and about 180 m deep. In 1891, Grove Karl Gilbert, chief of the U.S. Geological Survey, asked survey geologist Willard D. Johnson to provide a report on the crater. Johnson noted that meteorite fragments were as far as 8 miles distant from the crater, that fragmented rocks were heaped on the outer slopes, and that the strata of the crater walls were bent back sharply. Because of the meteoritic character of the iron, he wrote, "the general opinion" was that "a big fellow had made the hole": "Appearances in short have at once suggested your theory to the local prospectors." He added, however, that the compass showed no disturbance whatsoever.

It appears, then, that Gilbert initially entertained the idea that the crater was formed by the impact of a meteorite. In 1892 he spent several weeks at the crater, where his survey party made numerous tests in and around the crater in an attempt to detect magnetic variation, and also prepared a detailed topographic map (fig. 43). When he returned to Washington, Gilbert requested chemical analyses of the sand, rock, and meteorite fragments, percolation tests of the sandstone, a computation of the energy required to rupture quartz, and

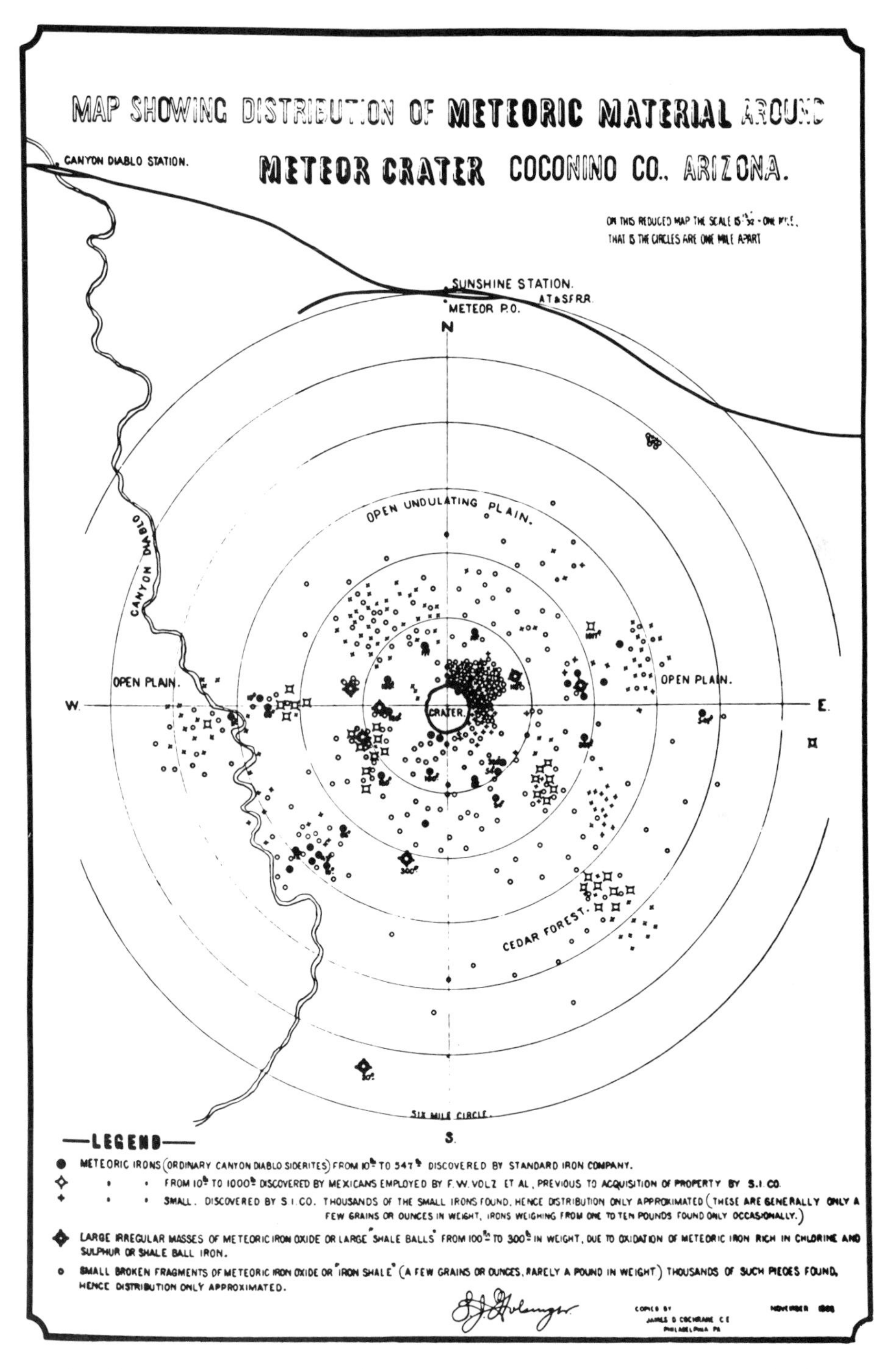

Fig. 42. S. J. Holzinger's map showing recovered meteorite fragments. From Barringer (1909).

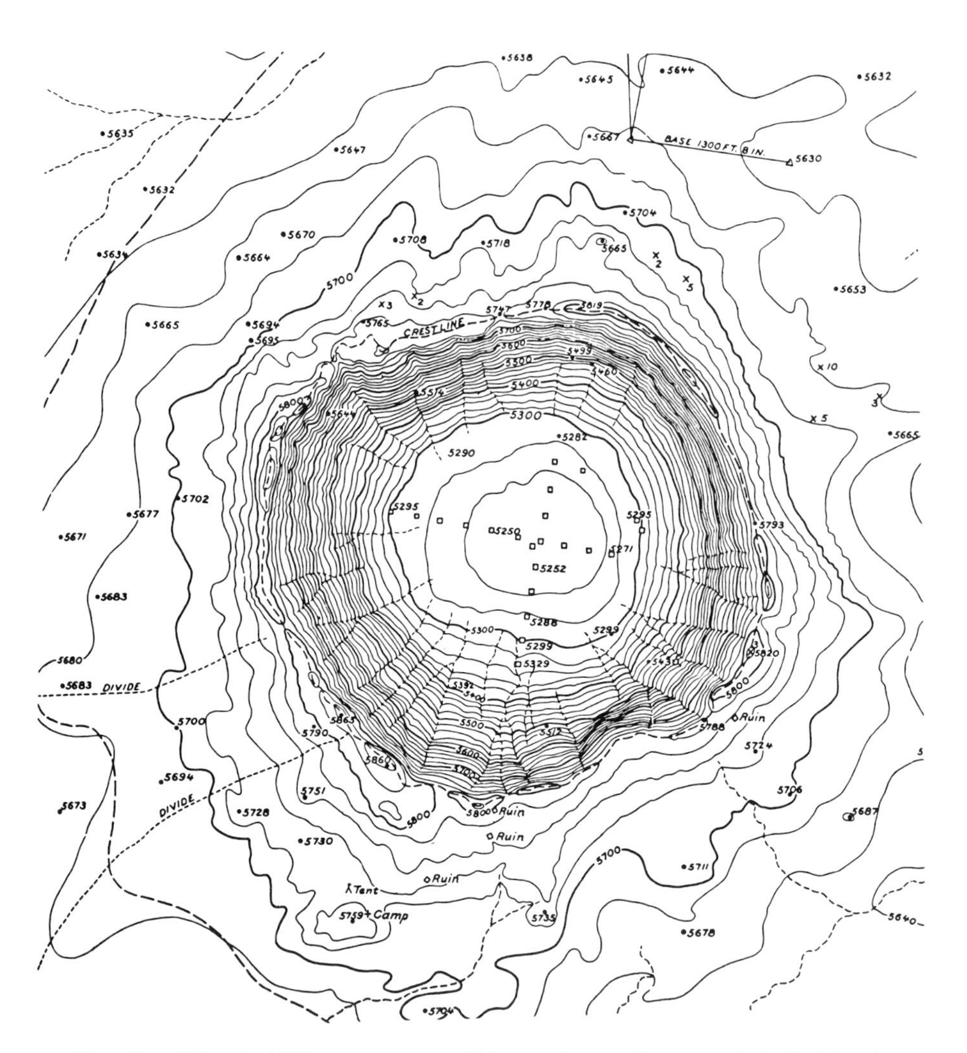

Fig. 43. Gilbert's 1892 contour map of Meteor Crater. Courtesy of the Smithsonian Institution Archives.

a survey of the scientific literature on explosion craters. In consultation with a ballistics expert, he attempted to calculate the energy necessary to produce craters, and he made a large number of experiments to ascertain at what distance a heap of iron chain would disturb a compass needle. Gilbert's conclusion, published in 1896, was that the crater was the result of a subterranean steam explosion and was not created by the impact of a "stellar body." The magnetic attraction, he wrote, was too feeble to show any indication of the presence of an iron mass on the instruments. The amount of debris deposited on the rim of the crater totaled 82 million cubic yards, which very nearly equaled the amount of material that would be required to fill the hole, so that the excess of material "required for the theory of a buried star" was lacking.[91]

In 1902 Daniel M. Barringer, a Philadelphia-born geologist who was part-owner of a lucrative silver mine at Pearce, Arizona, learned of the crater, the large numbers of meteoritic iron fragments surrounding it, and the local belief that a huge mass of iron was buried beneath the crater floor.[92] Soon thereafter, Barringer established the Standard Iron Company, the assignee of the U.S. government mining patent, to explore the property and recover the metals in the hidden mass. The company drilled five test holes at the center of the crater in 1904, recovering some metallic chips but encountering no sizable mass.

Because of the publicity engendered by this activity, Barringer and one of his partners, Benjamin C. Tilghman of Philadelphia, decided to write separate but complementary accounts of the crater and why they thought a huge meteorite was buried there. Barringer described the amount and composition of the meteoritic material, noting that recent analyses had determined that it contained platinum and iridium in addition to diamonds. He also reported the presence both of a large amount of magnetic iron shale containing nickel oxide and of enormous amounts of extremely fine silica powder. The atmospheric heat of passage, he stated, produced the shale, and the tremendous concussion formed the powdered silica. A violent volcanic eruption could not have produced the crater, because lava and sulfur were absent, there had been no volcanic activity in the immediate area, and only the upper layers of strata had been excavated. Also, there was no record of any steam explosion of a magnitude large enough to form such a huge crater and then relapse into quiescence. There were no radial fissures in the vicinity of the crater, and a steam explosion could not have pulverized the sand. Only one other conceivable cause remained, Barringer concluded, and that was the impact of a very large meteorite. It appears, also, that Barringer almost extended himself to antagonize Gilbert and members of the U.S. Geological Survey. Gilbert, he wrote, adopted the steam explosion hypothesis on "what seems to be very insufficient evidence . . . it does not seem possible that any experienced geologist could have arrived at such a conclusion."[93] Tilghman was more restrained. He pointed out that Gilbert's failure to record any magnetic declination was undoubtedly due to the fact that the attraction of shattered magnetic iron was far less than that of solid material, and thought that Gilbert's estimate of the mass of the meteorite was too large, given its probable velocity at impact.[94]

These articles marked the beginning of the first of several scientific controversies that were stimulated by the discovery of Meteor Crater—namely, controversies about whether the impact of a huge meteorite had formed it. Publications supporting and rejecting the hypothesis appeared frequently in the literature for the next three decades. George P. Merrill of the Smithsonian Institution was an early and articulate adherent. In 1908, after a meticulous survey of the crater, Merrill published his reasons for accepting the idea of a meteorite impact. The walls of the crater consisted of strata that had been "crushed and shattered to an extraordinary degree," and at the rim they dipped away outward on all sides at angles from 10° to 80°. The sandstone beds underlying the crater floor were almost unchanged, but where they were disturbed, they apparently dipped downward. What seemed to impress Merrill the most, however, was the extremely fine condition of the silica powder, which he called "rock-flour." It was, he wrote, "unquestionably derived from the sandstone . . . not by simple disintegration, but through some dynamic agency acting like a sharp and tremendously powerful blow." Some of the granules "closely resemble fulgurite glass, formed by lightning striking in siliceous sand." In his concluding remarks, Merrill raised a question that became the focal point of a second controversy provoked by Meteor Crater: The mass, whose size in any event had been exaggerated, might be buried, but it was also possible that upon impact, fragments had broken off and been dispersed and the remainder completely volatilized.[95]

Merrill's article helped gain scientific acceptance of the meteorite impact hypothesis, despite the fact that twenty-eight test holes at the floor of the crater failed to locate the mass. Other respected scientists, Elihu Thomson of the General Electric Company and William F. Magie, professor of physics at Princeton, also gave Barringer advice and continuing support in journal articles. But Nelson H. Darton, a Survey geologist, reaffirmed the steam explosion theory. In 1910 he wrote that the meteorite impact hypothesis was not valid because "no meteor is present," and he found it difficult to accept Merrill's idea that it might have volatilized.[96]

The idea of volatilization was, of course, also anathema to Barringer, and he dismissed it because there were no volatilization products around the crater. By 1914 he came to the conclusion that the impacting mass had not been a single object, but a dense cluster of iron chunks, "possibly the head of a small comet," and he estimated its total weight at about 10 million tons. Also, having inspected the shape of the crater more attentively, he now believed that the bulk of the swarm lay under the south rim of the crater.[97]

By this time, the unsuccessful operations of Standard Iron Company had exhausted the initial investment of the stockholders, and Barringer sought a new source of funds. In 1918 he entered into a lease agreement with the U.S. Smelting and Refining Company. Over several years, this company spent over $200,000 before abandoning the project, when a drill bit in a hole at the south rim became irretrievably wedged at a depth of 1,370 feet. Before this misfortune, however, the drill passed through a zone some 30 feet in extent, from which nickel-bearing fragments were recovered. This layer, Barringer surmised, was the outer shell of the mass.[98]

In 1924 Barringer and his associates planned what turned out to be the final effort to locate the meteorite by drilling. They established the Meteor Crater Exploration and Mining Company, which leased the crater from Standard Iron and offered stock to the public. The estimated value of the iron, nickel, platinum metals, and diamonds that would be recovered, reached a billion dollars. Drilling of a new shaft commenced in 1927, when sufficient funds became available, but within a few months heavy waterflow at a depth of 640 feet prevented further progress. In 1929 Quincy Shaw, the president of the company, in order to reassure stockholders, requested Forest R. Moulton, the respected applied mathematician, to estimate the size of the mass. Moulton's judgment was that the meteorite had either vaporized or was too small to be of value. Shaw immediately ordered a cessation of operations, although a few years later, following a geophysical survey, two unsuccessful test wells were drilled with what money remained. Barringer had personally invested

$120,000 of the $600,000 spent on the exploration project from its inception. He died of a stroke a few months after Moulton gave his opinion.[99]

It is still unknown whether a sizable portion of the mass, recently estimated to have been about 63,000 tons, is buried beneath the crater, although most authorities agree that it volatilized.[100] By the 1930s, however, only a handful of scientists doubted that Meteor Crater was caused by a meteorite impact, and at that time a number of scientists offered evidence that other crateriform features of the Earth's surface were similarly produced. D. M. Barringer, Jr., was among the forerunners in 1928, when he described a crater located 10 miles from Odessa, Texas. This crater was saucer-shaped, having a diameter of 530 feet and measuring 18 feet from the floor to the top of the rim. Barringer found iron shale around the rim, as at Meteor Crater, and located four small pieces of magnetic material.[101]

Leonard J. Spencer, Prior's successor as keeper of minerals at the British Museum, gave widespread prominence to the meteorite impact theory in 1933. In a comprehensive article, Spencer presented evidence that in addition to the Arizona and Texas craters, the group of craters located near Henbury in central Australia and the Campo del Cielo craters in Argentina were formed by the explosive impact of large meteorites.[102] He showed that the cratering effect of a meteorite impact was distinctly different from that produced by the largest mines of World War I. He also cited authorities who believed that the Estonia craters on the island of Oesel and the Ashanti crater, filled by Lake Bosumtwi (Ghana), had the same origin. Spencer described in detail the Wabar craters, discovered in 1932 in the Empty Quarter of the Arabian Peninsula by the explorer H. St. John Philby. "Cindery masses" and "bombs" of silica glass littered the outer slopes of the two distinct craters. The unique development of the silica glass, Spencer explained, was due to the fact that the Wabar meteorites fell on clean desert sand, which was liquefied and partially vaporized, together with a large portion of the irons. The bombs, the smaller of which the natives called "black pearls," were extremely light because of their cellular structure, and the thin black skin that some displayed resulted from the presence of small amounts of nickel and iron in the silica. Philby also found fragments of iron oxide at the site, and several pieces of meteoritic iron, one weighing 25 pounds.[103]

In the late 1930s and 1940s, publications began to focus not on whether meteorite impacts produced craters, but on which terrestrial craters could be attributed to meteorites, and also on the mechanism of crater formation as it influenced the size and internal features. For example, one large group of craters that stimulated continuing discussion among geologists was the Carolina Bays. These consist of hundreds of small, shallow, oval depressions, some of which are filled with water, and which are located along the Coastal Plain in South Carolina. In 1938, F. A. Melton rejected the hypothesis that these were solution sinks and suggested that a widespread shower of large meteorites produced them. "If the supposed impact occurred near the end of Cretaceous time, as the distribution of bays seems to indicate," he wrote, "it will be necessary critically to re-examine the fundamental causes of diastrophism and possibly to rewrite the geologic history of post Cretaceous time."[104]

Discussions concerning the mechanism of crater formation sparked yet another controversy resulting from the discovery of Meteor Crater—namely, the origins of the craters of the Moon. From the time of Robert Hooke's speculations, scientists had believed that volcanic action produced the lunar craters, and this opinion was reinforced by supposed sightings of active volcanoes, which supported the theory that meteorites were ejecta from them (see chap. 2). Two observable facts weighed against the idea of their origin from impact. Meteorites, striking at all angles of incidence, scientists thought, could not produce the circular crater shape, nor could they create the massive craters so easily visible. Grove Karl Gilbert, though rejecting the impact origin of Meteor Crater, was among the first to postulate that lunar craters resulted from impact.[105] Gilbert, however, thought that the missiles were fragments that were torn from the Earth at the time of the Moon's separation. These masses encircled the Earth in a ring and their number gradually diminished as they plunged into the Moon. Gilbert's theory carried with it the assumption that most of the fragments would strike the Moon almost perpendicularly, thereby creating a nearly circular crater. Barringer, in 1924, relying on his knowledge that bullets made circular holes in targets regardless of the angles of incidence, insisted that the mass that formed the nearly circular Meteor Crater approached at an angle of 45°, and went on to suggest that all lunar craters originated from meteorite impact.[106] Spencer also surmised that meteorites produced lunar craters and that their large size was related to the smaller gravitational force of the Moon.[107]

By the mid-1940s, a number of scientists, among them Robert S. Dietz, R. B. Baldwin, and Fred L. Whipple, began to assert that all lunar craters resulted from meteorite impact. Dietz presented a comprehensive review of the features of lunar craters and argued that they were clearly similar to terrestrial explosion craters.[108] The explosion at impact caused their circularity, and their greater dimensions were due to the absence of an atmosphere on the Moon and to its lower gravity. The small mass of the Moon and its low internal pressure, Dietz thought, made it improbable that volcanic action had occurred there. R. A. Daly, however, was more conservative. Following Gilbert, Daly believed that Earth fragments were mainly responsible for the Moon's craters. He agreed that some craters were produced by meteorite infall, but stated that the "Meteor Crater analogy fails for the many lunar 'craters' that have broad flat floors—a form that almost inevitably means former liquidity of the floor material."[109] Lincoln LaPaz also thought that the pendulum had swung too far. LaPaz recalled that Barringer had fought an "uphill battle" to prove that one crater resulted from the impact of a meteorite. Now, Baldwin and Whipple had taken the position that all of the craters of the Moon were produced by meteorites. The lunar craters, LaPaz continued, were not randomly distributed—as they should be if all resulted from the impact of meteorites—and besides, some were polygonal or hexagonal, which indicated that they were a result of the movement of the Moon's crust.[110] It was time, in LaPaz's estimation, to replace speculation with scientific evidence.

Barringer Meteorite Crater, then, was not merely a gaping terrestrial hole, formed by a random giant meteorite. Rather, it was an exemplar that stimulated scientific controversy, because it appeared to offer evidence that meteorites had bombarded the Earth and the Moon and, undoubtedly, other planets of the Solar System for countless centuries.

Tektites

F. E. Suess in 1900 gave the generic name "tektites" to the small, oddly shaped pieces of silica glass that had been found, usually in sizable quantities, in several widely separated localities around the world. The fact that the number of areas where tektites were discovered had begun to increase attracted his interest. The first scientific mention of the dark green moldavites from the valley of the Moldau River in southern Bohemia was in 1787, and another strewn field in Moravia came to light a century later. Charles Darwin, while observing the coral reefs of Australia, first noted the occurrence of the dark brown australites (fig. 44). Nineteenth-century Dutch scientists reported the finds of the teardrop-shaped billitonites on Billiton Island in the Java Sea, and also the javanites of central Java. Suess initially investigated only the moldavites, australites, and billitonites, but in the early decades of the twentieth century, scientists reported the discovery of the philippinites, the indochinites (fig. 45), and the bediasites of Grimes County, Texas.[111]

Fig. 44. Australite. Above: Posterior surface. Below: Side view. The color is medium brown in strong transmitted light, and black in reflected light. From Elbert A. King (1976). Reprinted with permission of John Wiley & Sons, Inc. © 1976 and the author.

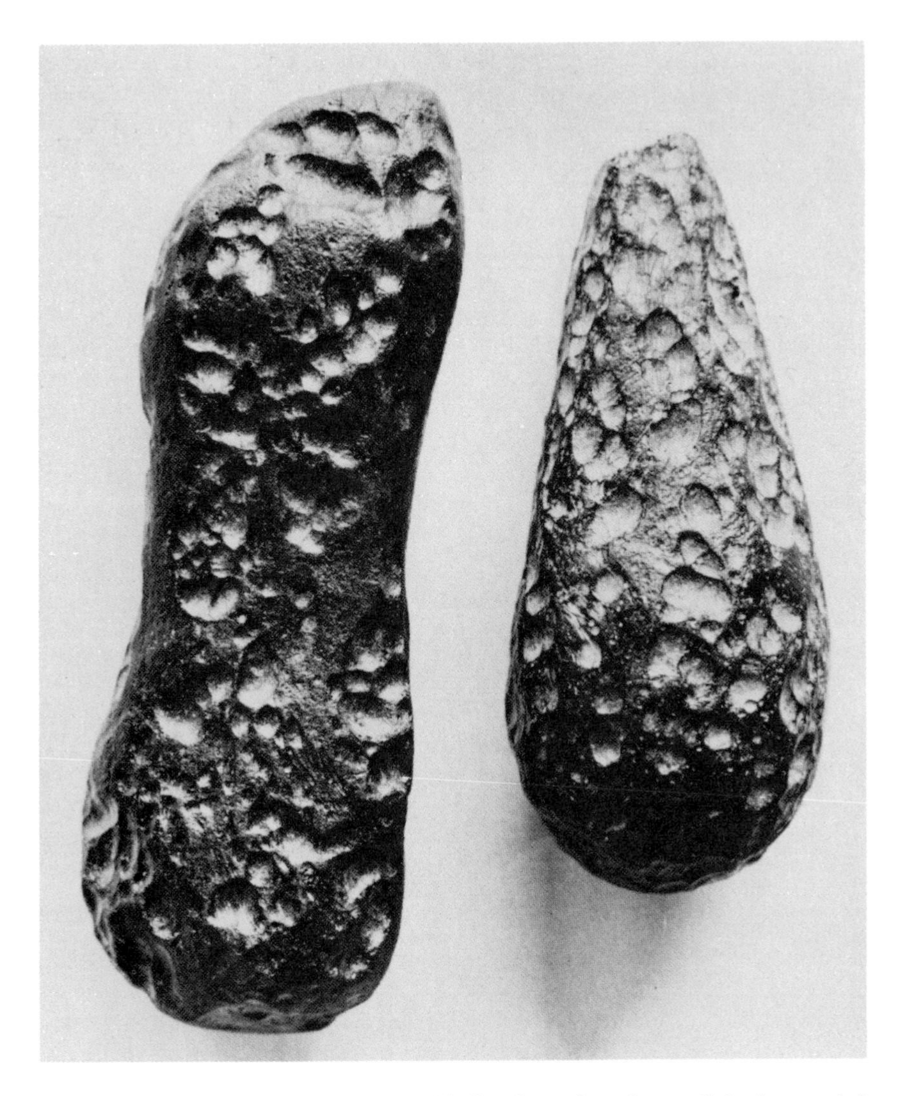

Fig. 45. Indochinites, showing the dumbbell and tear-drop shapes. Color is very dark brown, almost opaque. From Elbert A. King (1976). Reprinted with permission of John Wiley & Sons, Inc. © 1976 and the author.

Suess assumed in his first publication that tektites were a peculiar type of meteorite, but other scientists soon challenged this conclusion.[112] Friedrich Berwerth, probably mindful of the eighteenth-century confusion of true meteorites and Stone Age axes, suggested that tektites might be artificial productions of prehistoric peoples. Berwerth noted that moldavites had been found in paleolithic and neolithic settlements, and that sharp grooves in the edges appeared to have been made with a blunt instrument in soft glass, whereas atmospheric flight should have produced rounded grooves. Berwerth also observed that the chemical composition of each type of tektite was very similar to the composition of the terrain where each had been found. W. Wahl avoided the question whether tektites were of human manufacture, but on the basis of their chemical composition maintained that they must be terrestrial products. Tektites have a chemical composition that is high in silica and alumina and low in soda, magnesia, and iron oxide, with some lime and potash. A fragment having this composition, Wahl stated, could never reach the Earth, because upon entry into the atmosphere it would be ignited and entirely consumed on oxidation.[113]

Wahl's contention gave Suess pause, but with the help of J. Bayer, an archaeologist, he was able successfully to counter the argument that tektites were man-made. Bayer estimated that the moldavites must have fallen thousands of years prior to the age of the Pleistocene peoples. Certainly, the Aurignacians gathered them, just as they did mineral crystals, colored pebbles, and Tertiary period sharks' teeth, and they carried them to their dwellings where they worked them. However, pottery was unknown for many millennia after the time of the Aurignacians, and the first glass objects—beads—were produced in western Europe only in the first half of the first millennium B.C., during the Hallstat period. Tektites, Bayer declared, were not a historical problem. Suess considered it impossible that the difficult technique of producing glass of the highest melting point would have originated independently in three so completely heterogeneous ethnographic areas as Bohemia, the Malay Archipelago, and southern Australia in the far distant past, while peoples in other parts of the world were unaware of it.[114]

Another hypothesis of origin, advanced by I. N. W. Easton in 1921, was that billitonites were not glass but silica-rich colloids that were formed by electrolytic action in granitoid terrain. In certain protected areas, Easton wrote, these gels grew annually in layers, and he believed that he had detected this stratification in many tektites. Suess replied that some water would remain in such silica colloids even though they experienced extreme dessication. Tektites, in contrast, had such a remarkably low water content that they could not have been produced by this process.[115]

Hermann Michel in Vienna attempted to overcome Wahl's argument that tektites could not have a cosmic origin. It was conceivable, Michel thought, that when large masses of iron-poor material that contained unoxidized silicon and other light metals entered the atmosphere, they would not only ignite but disintegrate. Atmospheric oxygen would then combine with the elements, forming silica-rich droplets, which during their further flight would cohere and solidify in the intense cold into glass fragments.[116]

Though Michel's hypothesis incorporated the occurrence of a remarkable sequence of events in a short time period, Suess accepted it, as did Alfred Lacroix in 1932, when he reported the discovery of indochinites. L. J. Spencer, however, influenced by the presence of silica glass fragments at the Wabar and Henbury craters, proposed in 1933 that tektites were the products of explosive meteorite impacts. Completely overlooked in all of the theories of the origin of tektites, Spencer declared, was "the elementary fact that the material of which they are composed is really an impure silica glass." The bombs from Wabar and Henbury were similar in many respects to tektites. The terrestrial rock that a large meteorite struck had to be of the right kind to produce tektites—namely, a sandstone or quartzite—and the composition of tektites indicated that a certain amount of feldspar, mica, or clay was present.

The odd shapes of tektites were due to their spinning through the air after having been ejected by the gaseous explosion. An aerial survey of the districts where australites were found, Spencer concluded, would probably reveal the location of the meteorite craters that produced them.[117]

At the time, scientists did not pay much attention to Spencer's hypothesis. Paneth, in his 1940 review of the theories of the origin of tektites, did not even mention it, although he was critical of Michel's idea. The volume of oxygen required in the rarified upper atmosphere, he wrote, was such a large multiple of the final volume of the tektite that the "mechanism of such a chemical reaction during the rapid flight through a resisting medium becomes entirely incomprehensible." Paneth saw no theoretical reason why a body having the composition of terrestrial granite could not have disintegrated during a close encounter with the Sun. Its fragments would melt, but then, after perihelion, would rapidly solidify into glass globules, which later entered the Earth's atmosphere and spread over large areas of the Earth's surface.[118]

No one, as Paneth emphasized, had ever observed the fall of tektites. This fact undoubtedly gave comfort to those who maintained that they had a terrestrial origin. Yet the extensive strewn fields indicated that they had fallen from great heights, and their chemical composition and physical properties seemed to give evidence of an extraterrestrial origin. It is no wonder, then, that scientists were sharply divided on this issue, and that the controversy continued, giving birth to new theories as the decades passed.

The ideas advanced in the early twentieth century—for example, Prior's belief that more accurate chemical analyses of meteorites would reveal genetic relationships, or Merrill's insistence that meteorites had experienced metamorphism—did not bear immediate fruit. Early twentieth-century scientists also left unsolved problems: the mechanism of the formation of the Widmanstätten structure in meteorites; the appropriate techniques for determining meteorite ages; the origin of lunar and some terrestrial craters; and the origin of tektites. Nevertheless, their successors in the following decades profited by following the avenues that they opened up, one might say almost relentlessly. In the 1950s it seemed apparent to astronomers, geologists, mineralogists, chemists, and physicists alike that meteorite research would lead not just to the solution of the problem of their origin but eventually to the knowledge of when our solar system formed, and also to an understanding of the conditions that caused its birth.

9

Contemporary Meteorite Research and Theories

A goal that has motivated and unified the research of many scientists since the end of World War II is that of understanding the origin and history of the Solar System. Meteorites provide the most reliable evidence in this endeavor. For example, from the chemical compositions of meteorites scientists try to determine the changing conditions of temperature and pressure prevailing in the primordial solar nebula when matter began to condense; to ascertain the mechanism of the subsequent accretion of the primitive matter into planetisimals at various distances from the central mass that later became the Sun; and to discover the alterations that occurred in the proto-planets during the initial tens of millions of years. They attempt to establish the sizes and the probable locations of the parent bodies of the various classes of meteorites, to estimate how and when the breakup of the parent bodies took place, and to determine how the fragments were displaced or perturbed into orbits that intersected that of Earth. They try to measure the temporal flux of meteoroids entering the atmosphere, to estimate their terrestrial ages, and to assess the physical and biological effects that the impacts in past ages of asteroid-sized bodies may have had on our planet. Meteoritics now embraces the investigation of these and numerous other problems. Its practitioners, though still primarily geologists, physicists, and chemists, include, if only for temporary research into a specific topic, scientists from a wide range of disciplines. In fact, a characteristic feature of meteorite research in the past few decades has been the formation of ad hoc interdisciplinary groups to study a problem from different points of view and then report jointly on the results.

Atomic energy programs in many countries, widespread public and private funding for basic scientific research, and in particular the space programs of the U.S. and the USSR have combined to give great impetus to meteorite research. These sources have provided money for the education of graduate students; for the establishment of research centers employing post-doctoral fellows, research assistants, and technicians; and, very importantly, for the expensive experimental equipment necessary to detect, say, the minute concentrations of trace elements present in meteorites. Experimentalists have substantially improved analytical techniques such as isotope dilution and neutron activation, which were known before 1940 but were then of limited value. There have been significant advances in mass spectrometer design, which have yielded increasingly higher sensitivity and resolution. New generations of computers have provided greater flexibility and capacity and more rapid calculating speeds. The electron microprobe was an entirely new instrument, which was developed by Raymond Castaing in France in 1951, and which attained great precision in the early 1960s. Klaus Keil and Kurt Fredriksson considered that the electron microprobe "has revolutionized science," and believed that its "impact on the earth sciences can only be compared to the introduction of the polarizing microscope" in the mid-nineteenth century.[1] With it, an investigator can perform a nondestructive qualitative and quantitative chemical analysis of just a few cubic microns of a polished sample or of a thin section of a meteorite. The equipment, however, is not inexpensive. Together with the associated complex vacuum system, electron optics, counting electronics, and computer array, an electron microprobe costs about half a million dollars.

Fully equipped modern laboratories in which meteorites are studied, then, are expensive and fascinating places that attract and excite research scientists. Meteoriticists, however, can rarely resist the temptation to engage in fieldwork when the opportunity presents itself. Soviet scientists, for example, had in 1947 just commenced a thorough study of the 1908 event that devastated a wide area near the Tunguska River in central Siberia, when the disintegration of a huge iron meteorite, estimated to weigh 70 tons, at Sikhote-Alin northeast of Vladivostok, demanded an immediate field investigation. Similarly, when scientists and collectors in Mexico, Canada, and the United States heard about the meteorite shower on 8 February 1969 at Allende, Chihuahua Province, Mexico, scores of them, including a team from the Smithsonian Institution, hurried there. Some with the blessings of the Mexican government, and others apparently without its permission, gathered over a period of five months about 2 tons of this relatively rare carbonaceous meteoritic material from a strewn field whose area was about 300 square kilometers.[2] Another exciting program, instituted in the 1970s, involves the coordinated search for meteorites in Antarctica by teams of Japanese scientists at the Yamato Mountains and of U.S. scientists at Allan Hills near McMurdo Station (fig. 46).[3] Hunting for meteorites in the Antarctic highlands during the summer season, when temperatures are about minus 15° F, is, Ursula Marvin reported, "an exhilarating adventure"—as it must be.[4] Surely, then, there are numerous scientists who would cheerfully suffer the rigors of extended space travel to get a close look at or, better still, to land on certain asteroids now thought to be the parent bodies of particular meteorite classes. Enthusiasm and wonder about meteorites are emotions that nineteenth-century scientists passed on to their twentieth-century counterparts.

A description in a single chapter of meteoriticists' activities and accomplishments during the past three decades must necessarily be abbreviated. One cannot do justice to the important work of many scientists whose contributions stimulated experimental work and led to theoretical advances. Here, we first describe the introduction of a new system of classification and the methods

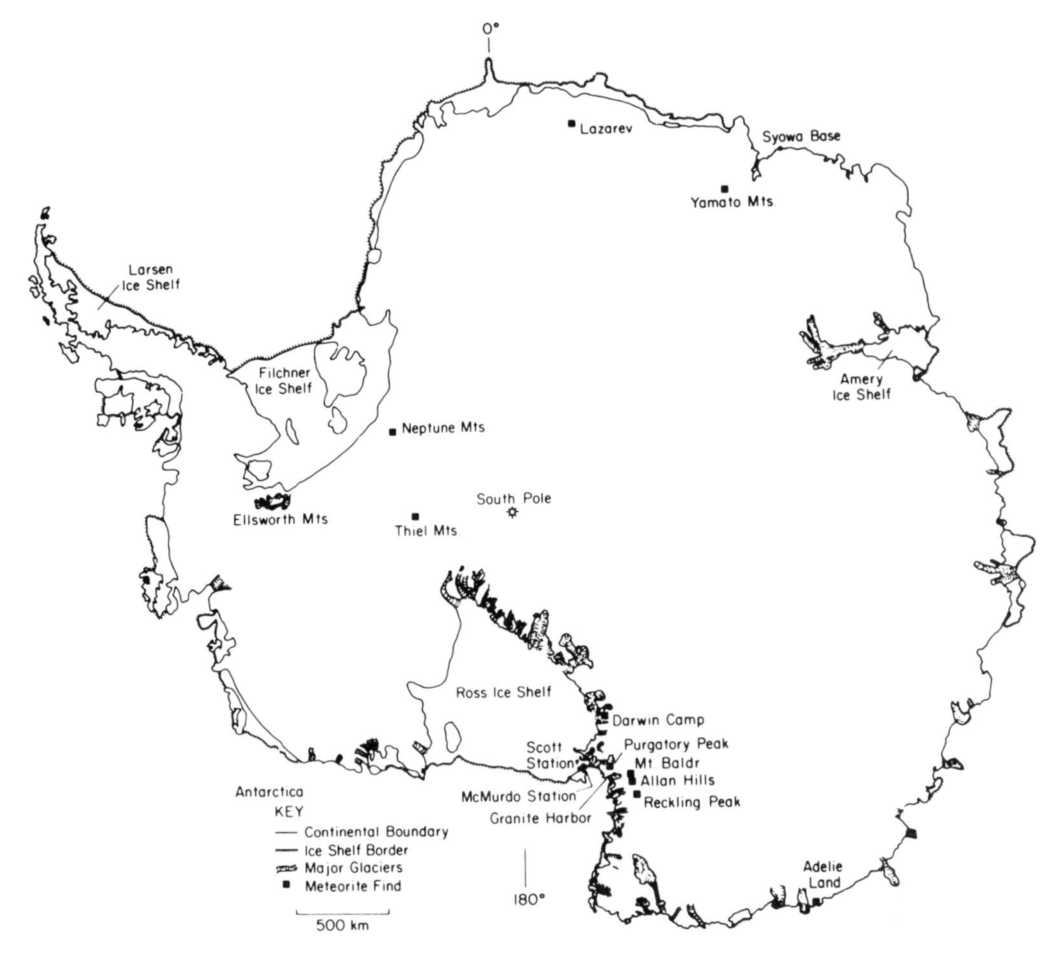

Fig. 46. Outline map of the continent of Antarctica, showing the location of meteorite finds. From *Smithsonian Contributions to the Earth Sciences,* No. 24.

of determining the various ages of meteorites, both of which are crucial underpinnings for theory. Then, we give an account of contemporary theories of the origin of meteorites. Finally, we report on investigations of the possible effects, both physical and biological, that meteorites may have caused on Earth in past ages.

CLASSIFICATION AND AGES

In the late 1940s there were at the University of Chicago's Institute of Nuclear Studies (later the Fermi Institute) several scientists who had been involved in the Manhattan Project during the war, and whose interests thereafter turned to the studies of the planets and meteorites. Two of these men were Harrison S. Brown and Claire C. Patterson, both of whom made significant contributions to meteorite research and both of whom later joined the faculty of the California Institute of Technology. Another was Harold Urey, Nobel laureate, who was probably the most influential scientist involved in meteoritics in the 1950s and 1960s. At Chicago and, after 1958, at the University of California, San Diego, Urey trained numerous graduate students who achieved later prominence in the field. Urey also acted as a catalyst, publishing an apparently unending stream of hypotheses that provoked others to make theoretical calculations or to undertake experimental investigations to test their validity.

In 1947 and 1948 Brown and Patterson published a series of three papers on the composition of meteoritic matter, the last of which won a $1,000 prize for being judged the most significant paper presented that year at the annual meeting of the American Association for the Advancement of Science.[5] To obtain an average chemical composition for the silicate phase of meteorites, they selected over a hundred chemical analyses, most of them from Merrill's 1916 publication. They relied upon Farrington's 1907 compilation for analyses of meteoritic irons, and from these data they confirmed the validity of Prior's rules. If it were assumed, they surmised, that the observed phase distributions formed under equilibrium conditions, then equilibrium conditions must have been established at temperatures up to 3000° C and at pressures up to 10^5 to 10^6 atmospheres. Equilibrium varied from one meteorite to another, such that the greater the content of the metal phase, the greater the temperature and/or pressure; and the greater the distance from the core of the parent body, the greater the content of oxidized iron. The type of meteorite that formed, they concluded, depended upon the depth to which it was buried in the parent body. Brown and Patterson did not estimate the size of the parent body, but it appears that it was about the size postulated by Daly—namely, about 6,000 km in diameter.

Urey found Brown and Patterson's hypothesis unpalatable. He thought that the disintegration of a large planet from an internal cause or from a close encounter with Jupiter was entirely improbable, and that the collision of two large planets would produce only melted fragments and not the complex structures visible in meteorites. Noting defects in the chemical analyses upon which Brown and Patterson had based their calculations, Urey and Harmon Craig, a coworker at Chicago, made a comprehensive review of all published analyses of meteoritic stones.[6]

Judging 286 chondrite analyses on the basis of explicitly defined criteria, Urey and Craig selected 94 as "trustworthy."[7] When they made a plot of the percent of the oxidized iron (almost entirely as silicate) in these superior analyses against the percent of unoxidized iron (as metal or sulfide), they found that the distribution was not continuous, as it should have been according to Prior's theory, but that there were two distinctly separate clusters of points (fig. 47). The two groups contained different amounts of total iron as well as showing differences in the relative proportions of oxidized and unoxidized iron. The H (high iron) group, comprising 41 chondrites, had an average iron content of about 28 percent; the 53 members of the L (low iron) group had an average iron content of about 22 percent. The H group corresponded to Prior's olivine-bronzite chondrites, and the L group to his olivine-hypersthene chondrites. There was no simple mixing process, they stated, whereby one group

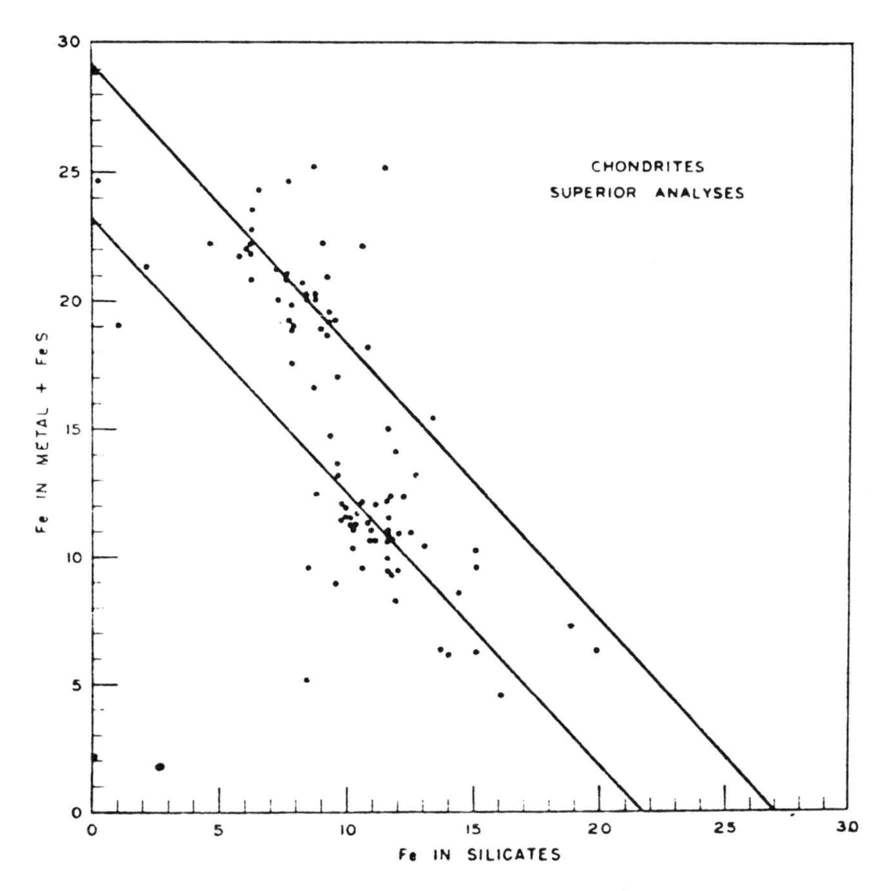

Fig. 47. Silicate phase and metal phase iron in 94 superior analyses of chondrites. Reprinted with permission from Urey and Craig, *GCA* 4 (1953), Pergamon Press, Ltd.

could be derived from the other. Prior, therefore, was wrong. To accomplish this transition, there would have to be a diminution of total iron among the H group chondrites and, simultaneously, either a reduction or an oxidation of the iron.

Considering the other types of chondrites, Urey and Craig noted that the enstatite chondrites appeared to be special because the iron they contained was almost completely unoxidized. The carbonaceous chondrites were also unusual not only because of the presence of hydrocarbons but also because the iron was highly oxidized. Also, the amphoterites contained slightly less iron in a somewhat more oxidized condition than their L group. They pointed out, then, that there might be other chondrite groups, but at this stage they wished merely to establish the existence of the H and L groups.

Urey and Craig gave a brief overview of the achondrites, noting that they could be separated into calcium-poor or chondritic and calcium-rich or basaltic types. The former were similar to the chondrites in general composition, except that they had experienced reheating, perhaps due to a close approach to the Sun or to higher-formation temperatures, and in the process the chondrules had been destroyed and the metal and sulfide phases separated. The basaltic-type achondrites, in contrast, appeared to have experienced complete fusion on the basis of their structure and composition.

To account for the various kinds of meteorites, Urey and Craig proposed that after the low-temperature formation of the proto-planets there had been a high-temperature stage, which they attributed to the adiabatic compression of the surrounding gas. The heat caused local surface melting with some volatilization, producing some of the achondrites. Beneath the surface, a layer of silicate formed, which corresponded to certain achondrites and stony irons. Nickel-iron also separated out into a layer, but there remained at the core primordial material, which retained its volatiles due to a relatively low temperature and which would become the carbonaceous chondrites. During the billion years that it took to form the Widmanstätten structure in the iron layer, repeated collisions with small asteroids produced surface rubble and compacted the varied mixtures of materials. Impacts caused localized melting, and chondrules formed when the tiny molten droplets cooled rapidly. Then, there was a collision between two large asteroids, one having the composition of H chondrites, the other of L chondrites. These bodies were of lunar size or somewhat larger, and their breakup produced the various types of meteorites in our collections.[8]

The significance of Urey and Craig's article was that it established the existence of two distinct classes of chondrites, marking the beginnings of the modern system of classification and offering evidence that meteorites came from more than one parent body. Their ideas as to the origin of meteorites, which they suggested constituted only a "general outline," also stimulated critical thought and research. There were many questions about their model—the source of heat, the mechanism of the differentiation of the original material, the origin of chondrules, the location of the carbonaceous chondrite material, and the size of the parent bodies, to name just a few. Taken together, these problems comprised a research program for meteoriticists for the next decades.

H. Birger Wiik, a Finnish chemist, made the first critical commentary on Urey and Craig's work, and in the process extended the classification of meteoritic stones.[9] He had analyzed 21 stones, among them 11 carbonaceous chondrites, and added to these 9 other superior analyses. He confirmed the existence of the H and L groups, although he pointed out that the total iron in the H group was only about 26 percent. Wiik's analyses of the carbonaceous chondrites were of great significance. All contained about 26 percent iron, demonstrating that they were members of the H group. Wiik also found that they fell into three well-defined subgroups, which he labeled types I, II, and III. These types showed distinct differences in their densities and mineral constituents. Further, Wiik emphasized that meteorite analyses were much more revealing when they were expressed as atomic percentages on an oxygen-, water-, carbon-, and sulfur-free basis, as Wahl had suggested in 1910. When Wiik listed the composition of the carbonaceous chondrites in this manner, he was able to show that they all contained about the same percentages of the major elements. However, this was not the case with minor and trace elements, and Wiik employed neutron activation to detect minute elemental concentrations.

There were disagreements with Urey and Craig's interpretations of their data. In 1961 A. E. Ringwood, a geologist at the Australian National University, Canberra, presented a comprehensive review of the chemical and genetic relationships among meteorites.[10] Ringwood made optical and X-ray studies of the olivine, pyroxene, and metal phases of the chondrites that Urey and Craig had selected or Wiik had analyzed, as a further check to ensure that the analyses were superior, and he added nine more for which optical data were available. Contrary to Urey and Craig, Ringwood considered that his plot of iron as metal or sulfide against oxidized iron proved the validity of Prior's rule. He suggested that the L group was a part of the H group that had lost metallic nickel-iron under slightly oxidizing conditions, possibly due to sinking and segregation in the parent body. There was, then, a definite genetic relationship among meteoritic stones; the chondrites were the "most fundamental and primitive group," and other groups derived from them. If the planets formed from low-temperature material, as astronomers had recently assumed, Ringwood continued, it followed that the primary chondritic material would be highly oxidized and at a temperature of less than 0° C. Carbonaceous material, as found in the Orgueil meteorite, appeared to be representative of the primitive meteoritic matter. Subsequent processes of reduction at temperatures in the range of 700–1000° C, with carbon as the chief reducing agent along with water vapor, resulted in the varying amounts of metal found in the chondrites. Short-lived radionuclides or chemical energy probably provided the source of heat. The mineralogy of the chondrites appeared to be consistent with the chemical differences expected from the operation of reduction processes, and there also seemed to be a distinct correlation between the degree of reduction and the intensity of metamorphism.

Ringwood based his exposition of the high-temperature reduction processes on published phase-equilibrium data, and he went on to explain in detail the manner in which the differentiated meteorites—aubrites, pallasites, mesosiderites, irons, and so forth—probably formed. It seems clear that Ringwood favored a single planet of lunar size as the source of meteorites, although he did concede that there might be other parent bodies. He suggested that disruption could have occurred from an intense steam explosion at the core, or from a substantial reduction in the planet's moment of inertia, resulting in an increase in rotational velocity beyond the limit of stability. The collision of two parent bodies was also a possibility.

Brian Mason, then curator of mineralogy at the American Museum of Natural History, also disagreed with the interpretations of Urey and Craig. In 1960 he stated that the known facts about chondritic meteorites were inconsistent with their origin as massive fragments of a disrupted planet.[11] The minerals in the carbonaceous chondrites gave evidence that they had never been at a temperature above 600° C, and the texture and structure of other chondrites did not display any gravitational segregation of metal and silicates as constituents of large fragments should. Further, the uniformity of the composition of the olivine and pyroxene in individual meteorites demonstrated that they were not random aggregations, as Urey and Craig had suggested, but chemical systems in phase equilibrium. Mason proposed that carbonaceous matter like that in the Orgueil meteorite had been the primary material, and that all other chondritic groups derived from it by dehydration and progressive reduction.

Further, he suggested that the chondrites need never have been parts of a planetary parent body. The achondrites, pallasites, and irons could be explained as fragments of a differentiated planetoid formed by the aggregation of chondritic material.

In 1962 Mason reported on the results of further studies.[12] He selected about 60 chondrite analyses, limiting these to observed falls in order to eliminate the possibility that there had been an extended period during which the metal had oxidized. Also, he included only those in which, as Ringwood had advocated, the ratio of iron oxide to iron oxide plus magnesium oxide in the olivine and pyroxene, as determined by optical methods, was consistent with the ratio in the chemical analyses. His plot of the amount of the iron as metal or sulfide against the amount of oxidized iron now showed five well-delineated groups (fig. 48). Though the enstatite, olivine-bronzite, olivine-pigeonite, and carbonaceous chondrites were all of the H type, they differed markedly from one another in chemical composition. Mason had begun his investigations with the thought that Urey and Craig were in error. His studies, instead, extended the classification.

One of the first highly accurate electron-probe microanalyses of meteorites appeared in 1964. It was the work of Klaus Keil, and Kurt Fredriksson, who carried out their research at the University of California, San Diego.[13] They analyzed simultaneously the iron, magnesium, and calcium content of 20 to 30 olivine and orthopyroxene grains in each of 95 chondrites, taking at least four analyses 3 to 5 microns apart on each grain. Such a technique is routine now, but it was startling at the time. Bulk analyses, they pointed out, could not take into account the variability of composition from grain to grain, and the error in optical and X-ray diffraction methods (used by Ringwood and Mason) amounted to about 5 to 10 percent of the fayalite content of the olivine. They found that each of 86 ordinary chondrites had constant iron:magnesium ratios in the olivine and pyroxene, but they determined that between meteorites the ratios varied among three narrow ranges, corresponding to the H group, the L group, and another, which they named the LL (very low iron) group. The ferrous iron, they learned, was not exactly related to the amount of metallic nickel-iron, so that Prior's rule was not valid, as he had stated it.

Despite Keil and Fredriksson's criticism of the technique of using X-ray diffraction to determine the composition of olivine, Mason continued to use this method as a means of classification because it was rapid and relatively accurate. Olivine, he stated, was present in almost all of the chondrites; it was lacking only in the enstatite chondrites and in a few carbonaceous chondrites. Furthermore, in most chondrites olivine was the dominant mineral and showed a substantial difference in composition, which was directly related to the ratio of iron oxide to magnesium oxide in individual meteorites. The difference in olivine composition was, therefore, an excellent means of classification.[14] By 1967 Mason had, by his X-ray diffractometer technique, analyzed and classified a total of 975 chondrites: 16 enstatite; 430 H-type or olivine-bronzite; 496 L-type or olivine-hypersthene; and 33 carbonaceous. He then proposed that the olivine-pigeonite group, which he had previously posited, be dropped, and that this group be added to the carbonaceous chondrites.[15]

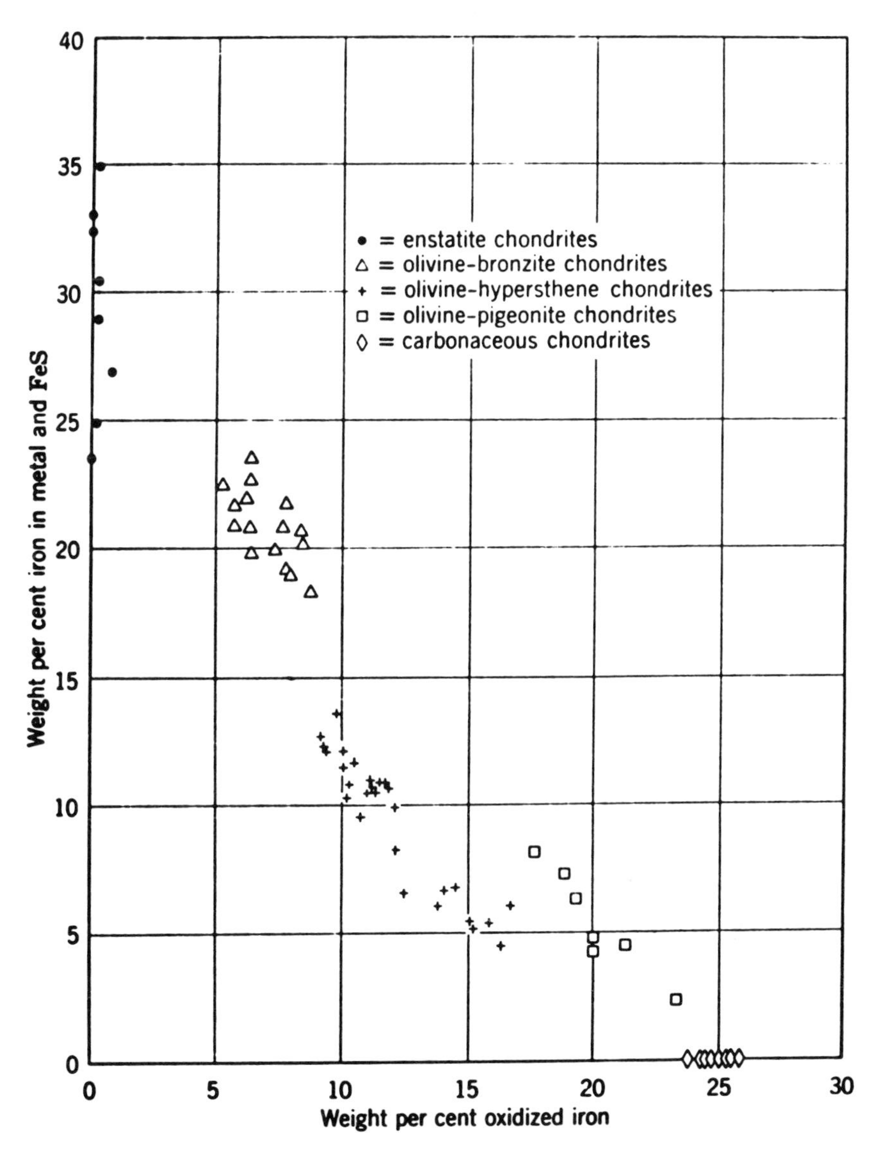

Fig. 48. Relationship between oxidized iron and iron as metal and sulfide in analyses of observed falls, indicating the separation into distinct subgroups and the variation within the subgroups. From Mason (1962*a*). Reprinted with permission of the author.

There were, then, four distinct chemical groups among the chondrites, and probably another—the LL group of Keil and Fredriksson. Mason agreed that this minor group existed, and he had also identified its members, but he thought that the small differences between the L and LL groups did not warrant the establishment of a separate group. However, among all of the chondrites there were very marked petrologic differences. Some displayed heterogeneous olivine and pyroxene crystals indicating substantial disequilibrium, whereas others showed decreasing heterogeneity and hence increasing equilibration. In some the chondrules were sharply outlined, while in others they were so merged with the matrix as to be almost indistinguishable. The ratio of low-calcium monoclinic pyroxene to orthorhombic pyroxene appeared to correlate with the increasing recrystallization of a chondrite, and only completely recrystallized chondrites contained feldspar grains that were easily visible in thin sections. There were also distinct differences in the metal phases, the sulfide phases, and the carbon and water content.

To take into consideration both the chemical groups and these petrologic differences, W. Randall Van Schmus, then at the Air Force Lunar Planetary Research Branch, and John A. Wood at the Smithsonian Astrophysical Observatory in 1967 proposed a comprehensive chondrite classification, which, with some modifications, is now universally accepted.[16] In their scheme, the chondrites were divided into five chemical groups and six petrologic types (table 6). Recent revisions have raised the number of groups to ten and the number of types to seven, although there is some dispute about the value of these changes.[17] This contemporary classification is a reflection of the theory that chondrites come from a number of different parent bodies, and it also demonstrates that they have experienced different thermal histories.

Meanwhile, there had been substantial progress in developing a new classification of the meteoritic irons. The Rose-Tschermak-Brezina system classified irons on the basis of structure—that is, as hexahedrites, ataxites, and octahedrites, with the latter subdivided on the basis of the width of the kamacite bands (see chap 4). Edward Goldberg, A. Uchiyama, and Harrison Brown made the first advance.[18] They used neutron activation to determine the content of nickel and of the trace element, gallium, in over 40 meteoritic irons, and found three distinct ranges of gallium content, which correlated to a large extent with the nickel content and to a lesser degree with the structure. In 1957 another Brown team, led by John F. Lovering, made further progress by finding correlations between the concentrations of gallium and germanium, thereby isolating a fourth group of irons.[19] They found that the structure correlated with the gallium and germanium contents in only one instance—namely, all hexahedrites belonged to Group II. Of the 88 irons they studied, 11 did not fit into any group, and they termed these irons "anomalous." The existence of the four groups, however, indicated that iron meteorites originated in at least four different parent bodies.

In the mid-1960s John T. Wasson, a UCLA geochemist, studied Lovering's classification of irons, and thought that there had been insufficient exploration of those irons that had low concentrations of gallium and germanium. Using improved techniques, Wasson and his coworkers found in Lovering's Group IV two narrowly defined groups, in which the germanium, gallium, and also the nickel concentrations were tightly clustered, and which were also structurally similar.[20] Wasson labeled these groups IVA and IVB, and suggested criteria for establishing a group of genetically related irons based on the extreme range of concentrations of an element, coherent variations of elements, and the existence of similar textures.[21] Almost immediately, Wasson determined that Lovering's Group III actually consisted of two co-genetic groups, IIIA and IIIB, which he suspected originated in contiguous regions of the same parent body.[22] In this study, he began to include analyses of iridium, and over the next few years, he and his coworkers found that over 400 iron meteorites could be assigned to 13 well-defined groups on the basis of their concentrations of gallium, germanium, iridium, and nickel, and their structural similarities. Over 60 irons, however, defied classification by Wasson's criteria, and are therefore labeled anomalous. The different groups testified to the existence of distinct parent bodies, and anomalous irons (more recently designated ungrouped) offer evidence for perhaps 50 more parent bodies. However, the possibility exists that some of the irons could have formed in different localities in the same parent body.[23]

Meteoriticists have studiously investigated the achondrites and the stony irons, but their classification remains about the same as in the Rose-Tschermak-Brezina system. Those classes with only one or just a few members—chassignites, angrites, nakhlites, shergottites—are somewhat puzzling, particularly because all except the one angrite indicate young formation ages. As for the stony irons, Wasson included the pallasites in his investigations of the irons, and proposed that they could be separated into one main group and at least two additional compositional clusters.

Table 7 shows an abbreviated modern classification of meteorites, which includes only well-classified specimens and some Antarctic meteorites. The listed frequency reflects only observed falls, since weathered stones are more easily overlooked than weathered irons. In contrast to the nineteenth-century systems, this classification leads to a number of formational models. The distinct differences in chemical composition indicate that meteorites come from a number of parent bodies; Dodd, for example, considered that an estimate of 70 to 80 parent bodies may prove to be too low rather than too high.[24] The frequency also gives hints as to the compositional types of parent bodies that suffered breakup. The petrologic divisions among the chondrites; the fact that the achondrites, stony irons, and irons underwent differentiation; the brecciation and shock features—all offer evidence of the thermal and impact histories of parent bodies and meteorites. Age determinations provide further clues, although, as with many other meteoritic studies, they do not always provide straightforward answers.

For more than a century and a half, scientists had proposed numerous theories as to the origin of meteorites, and for most of that time they had not considered it worthwhile to speculate as to how old meteorites might be. It is ironic, then, that a relatively precise value of the time when meteorites formed came to hand in the mid-1950s, while scientists still wrestled with the problem of origin. Paneth first attempted in 1929 to determine the ages of irons by the helium method, but his efforts were frustrated when Bauer showed that cosmic radiation produced helium (see chap. 8). The ability to determine the various ages of meteorites—formation age, formation interval, cosmic-ray exposure age, and terrestrial age—reflects to a certain extent the great strides in nuclear physics occasioned by wartime research. However, investigators encountered many pitfalls in attempting to determine ages by various methods, and it was

TABLE 6

Classifications of Petrologic Types and Chondrites

a. Criteria of petrologic types.

	Petrologic types					
	1	2	3	4	5	6
(i) Homogeneity of olivine and pyroxene compositions	—	Greater than 5% mean deviations		Less than 5% mean deviations to uniform	Uniform	
(ii) Structural state of low-Ca pyroxene	—	Predominately monoclinic		Abundant monoclinic crystals	Orthorhombic	
(iii) Degree of development of secondary feldspar	—	Absent		Predominately as micro-crystalline aggregates		Clear, intersti-tial grains
(iv) Igneous glass	—	Clear and isotropic primary glass; variable abundance		Turbid if present	Absent	
(v) Metallic minerals (maximum Ni content)	—	(<20%) Taenite absent or very minor	kamacite and taenite present (>20%)			
(vi) Sulfide minerals (average Ni content)	—	>0·5%	<0·5%			
(vii) Overall texture	No chondrules	Very sharply defined chondrules		Well-defined chondrules	Chondrules readily delineated	Poorly defined chondrules
(viii) Texture of matrix	All fine-grained, opaque	Much opaque matrix	Opaque matrix	Transpar-ent micro-crystalline matrix	Recrystallized matrix	
(ix) Bulk carbon content	~ 2·8%	0·6–2·8%	0·2–1·0%	<0·2%		
(x) Bulk water content	~ 20%	4–18%	<2%			

b. Classification of chondrites. The number of examples of each meteorite type known in 1967 is given in its box.

Chemical group		Petrologic type					
		1	2	3	4	5	6
	E	E1 —	E2 —	E3 1*	E4 4	E5 2	E6 6
	C	C1 4	C2 16	C3 8	C4 2	C5 —	C6 —
	H	H1 —	H2 —	H3 7	H4 35	H5 74	H6 44
	L	L1 —	L2 —	L3 9	L4 18	L5 43	L6 152
	LL	LL1 —	LL2 —	LL3 4	LL4 3	LL5 7	LL6 21

Note: From Van Schmus and Wood (1967). Reprinted with permission of the authors and Pergamon Press, Ltd.

TABLE 7

CLASSIFICATION OF AUTHENTICATED METEORITES

Class or group	Description	Total	Falls	Finds	Fall frequency (percent)
Chondrites					
E	Enstatite	24	13	11	1.4
H	Olivine bronzite	681	276	405	30.5
L	Olivine hypersthene	669	319	350	35.2
LL	Amphoterite	96	66	30	7.3
C (4 types)	Carbonaceous	67	35	32	3.9
–	Anomalous	7	3	4	0.3
–	Unclassified	137	72	65	8.0
Total chondrites		1681	784	897	86.6
Achondrites					
Aubrites	Enstatite	11	9	2	1.1
Diogenites	Hypersthene	15	9	6	1.1
Ureilites	Olivine pigeonite	17	4	13	0.4
Eucrites	Pyroxene plagioclase	55	25	30	2.8
Howardites	Pyroxene plagioclase	24	18	6	2.0
Anomalous	—	10	4	6	0.4
Total achondrites		132	69	63	7.7
Stony irons					
Pallasites	Olivine	39	3	36	0.3
Lodranites	Bronzite olivine	2	1	1	0.1
Mesosiderites	Pyroxene plagioclase	32	6	26	0.7
Total stony irons		73	10	63	1.1
Irons					
IAB		107	6	101	0.7
IC		11	0	11	—
IIAB		68	5	63	0.6
IIC		7	0	7	—
IID		15	3	12	0.3
IIE		14	1	13	0.1
IIF		5	1	4	0.1
IIIAB		197	8	189	0.9
IIICD		21	2	19	0.2
IIIE		13	0	13	—
IIIF		6	0	6	—
IVA		56	3	53	0.3
IVB		12	0	12	—
Anomalous		78	5	73	0.6
Unclassified		115	8	107	0.9
Total irons		725	42	683	4.6
Total authenticated meteorites		2611	905	1706	100.0

Note: Compiled from Graham et al. (1985).

only in the late 1960s that they became confident that they were on sure ground and that they had attained sufficient accuracy to give definitive interpretations of their results.[25]

In the early 1950s three methods appeared promising—the potassium-argon, the rubidium-strontium, and the lead-lead—all involving the measurements of the concentrations of the radioactive nuclide, and of its isotopic decay product, coupled with knowledge of the value of the half-life of the radioactive nuclide. Two Soviet scientists, E. K. Gerling and T. G. Pavlova, employed the potassium-argon method in 1951 to date terrestrial rocks and a few meteorites, and in 1955 G. J. Wasserburg and R. J. Hayden at Chicago's Nuclear Institute used it to determine the ages of the Forest City and Beardsley chondrites, which they reported to be 4.15 and 4.3 billion years old, respectively.[26] In 1956 Leonard F. Herzog and William H. Pinson at MIT, using the rubidium-strontium method, dated the Forest City meteorite at 4.7 billion years.[27] The same year, Claire Patterson announced that the age of several meteorites he had dated by the lead-lead method was 4.55 $\pm$ 0.07 billion years. The lead-lead method was the most accurate, Patterson wrote, and the age of the Earth and of meteorites was one and the same.[28] Patterson's figure is still quite accurate by modern standards; however, recent studies indicate that different classes of meteorites have somewhat different ages of formation.

Each method has its advantages and disadvantages. In the lead-lead method, the decay constants for the radioactive uranium isotopes are known precisely. However, contamination of the lead from terrestrial sources presents a major problem in handling and analysis. Also, late cosmic events may disturb the stability of the uranium-thorium-lead system. For these reasons, it was less popular, but a number of analysts began to use it again in the mid-1970s.[29]

Potassium-40 decays to argon-40, so that this method permits determination of the retention of the gas, which dates either the time that the meteorite parent body cooled or the time that the meteorite was last reheated to a sufficient degree that some or all of the initially entrapped argon escaped. Some meteorites that yield low ages by this method show evidence of shock metamorphism, indicating that gas was lost by reheating, and the friability of others gives a clue that some gas was probably lost in space. One complication of this method is that potassium-40 decays not only to argon-40 but also to calcium-40, so that an investigator must measure the concentrations of potassium and argon and also know the fraction of calcium produced, which is called the branching ratio. In the 1950s and early 1960s, scientists did not know the precise value of the branching ratio, which led to erroneous results.[30] In addition, contamination of the argon and potassium and incomplete extraction of the argon caused difficulties. Although experimentalists gradually overcame these problems, beginning in the 1960s they preferred to use a variation termed the argon-39/argon-40 method. This technique involves neutron activation, followed by selective volatilization of the gas contained in the mineral fractions in a series of temperature steps. Plots of the resulting data, showing the apparent age for each fraction of argon-39 evolved during each step, furnish a profile of the meteorite's experience—that is, whether it retained the gas from the time of its formation, or whether and at what time it was subjected to metamorphic reheating.

The rubidium-strontium system has become more important in age determinations as scientists have developed more precise analytical methods and more sophisticated instrumentation for the measurement of minute isotope concentrations. Radioactive rubidium-87, its decay product strontium-87, and the stable isotope strontium 86 were presumably uniformly distributed in the gaseous phase of the solar nebula when it began to condense to form the meteorite parent bodies. A meteorite's age in this method is the time at which the diffusion of the primordial rubidium and strontium isotopes across the grain boundaries of the various mineral fractions presumably ceased—that is, the time at which the specimen became a closed system subsequent to either melting or thermal metamorphism. The initial ratio of strontium-87 to strontium-86 ($^{87}Sr/^{86}Sr$) existing at that time gradually increased until the present owing to Rb decay. Comparison of the analytical results of ^{87}Rb, ^{87}Sr, and ^{86}Sr of two or more meteorites presumed to have the same age or of mineral fractions of the same meteorite, or the employment of a carefully selected initial $^{87}Sr/^{86}Sr$ ratio, yielded the age.

The method has its problems. One is that ^{86}Sr may contain a nonradiogenic ^{87}Sr isotope, which must be removed to attain accuracy. In the 1950s the half-life of ^{87}Rb was not well known; Herzog and Pinson, for example, used a value of 61.3 billion years, whereas the modern value is 48.8 billion years.[31] Gradual recognition that terrestrial groundwater leached Rb and Sr in different proportions prompted investigators to include only data from fresh falls. Further, in the early 1960s scientists were not quite certain, particularly with chondritic materials, of the nature of the Rb-Sr fractionation process that was being dated.[32]

Graphs called isochrons, which began to be used in the 1950s, show the results of age determinations. Investigators plot the $^{87}Sr/^{86}Sr$ ratio against the $^{87}Rb/^{86}Rb$ ratio. The slope of the plotted line yields the age of the specimens, while the point at which the line intersects the ordinate marks the initial $^{87}Sr/^{86}Sr$ ratio. Some investigators prefer to analyze bulk powder samples from several different meteorites that are chemically and petrographically similar, and plot the results as a "whole-rock" isochron. Early investigators plotted isochrons from only a few samples, and because of the uncertainties in the method, they used analyses of eucrites or howardites to establish an initial $^{87}Sr/^{86}Sr$ ratio as a check. These achondrites have very low rubidium concentrations, so that the values obtained from analysis plot very close to the ordinate of the isochron graph, and only a minor extrapolation is required to arrive at an initial $^{87}Sr/^{86}Sr$ ratio. However, in the late 1960s George Wetherill and his coworkers at UCLA began to publish whole-rock isochrons for a sizable number of meteorites belonging to the same group, and thereby did not have to employ achondrite data to establish the initial $^{87}Sr/^{86}Sr$ ratios. A hypersthene (L) chondrite isochron was the first to appear, and subsequently they published isochrons for bronzite (H) chondrites (fig. 49), LL, C, E, and unshocked hypersthene chondrites (fig. 50). There were minor differences in the ages of the various groups, which were, however, on the same order as the precision of the whole-rock method.[33]

Alternatively, researchers determine isotope concentrations in several different mineral fractions of the same meteorite and plot an "internal" iso-

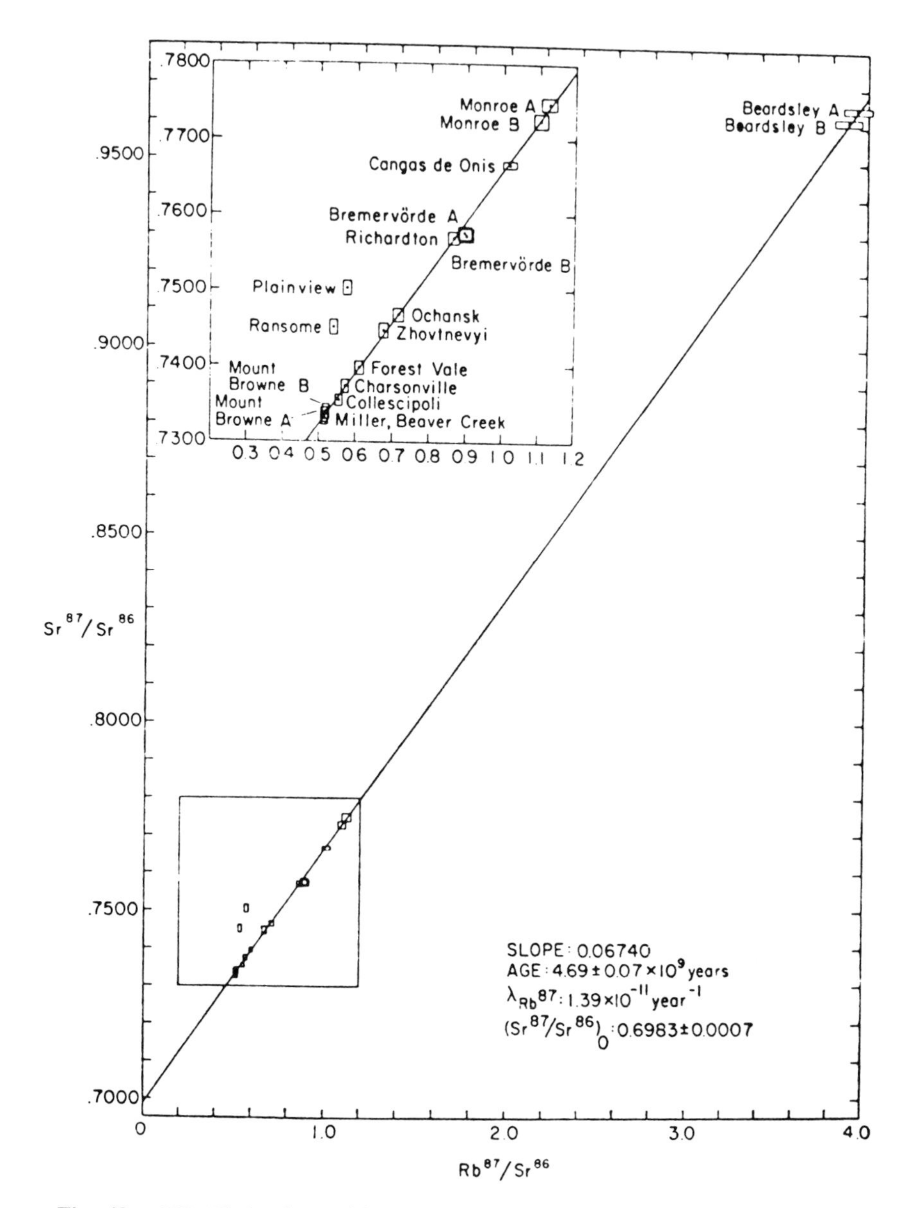

Fig. 49. ^{87}Rb-^{87}Sr isochron of bronzite (H group) chondrites corresponds to an age of 4.69 × 10^9 years and a ^{87}Sr/^{86}Sr initial ratio of 0.6983. From Kaushal and Wetherill (1969), copyright by the American Geophysical Union.

chron. This procedure gives better precision than the whole-rock method unless the meteorite has experienced metamorphism subsequent to its formation. For example, the ^{87}Sr/^{86}Sr ratio is quite different in pyroxene than it is in plagioclase, which yields a considerable spread between the plotted points. However, the task of separating the mineral fractions without contamination is difficult, so that such investigators as Wetherill or Gerald Wasserburg at California

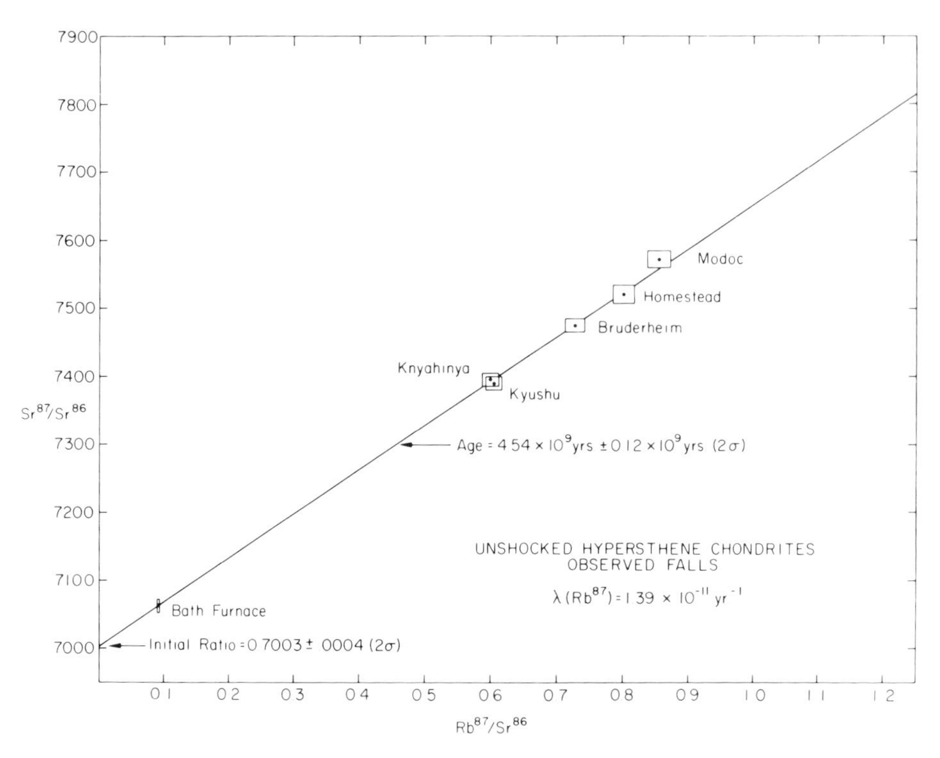

Fig. 50. ^{87}Rb-^{87}Sr isochron of unshocked hypersthene (L group) chondrites corresponds to an age of 4.54 × 10^9 years and a ^{87}Sr/^{86}Sr initial ratio of 0.7003. From Gopalan and Wetherill (1971), copyright by the American Geophysical Union.

Institute of Technology did not begin in earnest to determine internal isochrons of meteorites until the late 1960s.

In 1969 Wasserburg and his group sought to clarify the status of the initial ^{87}Sr/^{86}Sr ratio by its exact determination for a number of meteorites. They first determined that the whole-rock isochrons of six eucrites were almost identical and yielded a precise initial ^{87}Sr/^{86}Sr ratio, which they suggested could serve as a standard for establishing a refined solar system chronology.[34] Later, they determined internal isochrons for several equilibrated chondrites, and they found that these had slightly higher initial ratios.[35] They concluded that in the case of the Guareña meteorite, for example, it had experienced metamorphism 74 million years after its formation, because the initial ratio reflected a higher ^{87}Sr abundance in the solar nebula owing to rubidium decay in the interval. Other workers, however, have interpreted their results differently; for example, the variation in the initial ratio might be due to the inhomogeneity of the solar nebula in different regions where meteorites formed simultaneously.[36]

Another dating method indicates that the formation interval of various classes of meteorites was about 16 million years. Iodine-129, which has a half-life of 16 million years, was decaying to xenon-129 at the time meteorite parent bodies formed. The amount of retained ^{129}Xe provides a measure of the difference in formation time, and investigators find the relative amounts by the same neutron activation and stepwise heating procedures used for the argon-40/argon-39 method. Figure 51 shows results that have been obtained, arbitrarily assigning zero to the position of the Bjurböle chondrite, which indicate a formation interval of about 15 million years. Wasserburg has pointed out that the formation intervals measured by the rubidium-strontium and the iodine-xenon systems are not necessarily the same.[37] The rubidium-strontium method determines the time of formation of matter which is not necessarily cold, while the iodine-xenon method determines cooling time differences among objects with respect to their retention of gas. Further, the data obtained from the latter method are relative, not absolute, ages.[38] Thus, though the age of meteorites is about 4.55 billion years, there is still much to be learned about the duration and about the events that occurred during the formation period.

Soon after the discovery that cosmic radiation produced helium in meteorites, which dashed Paneth's hopes of determining the formation age by the helium method, Paneth and his coworkers realized that the time of breakup of the meteorite parent body could be estimated by determining the helium content. They assumed, however, that part of the helium was the product of radiogenic decay and part the product of cosmogenic decay, which led to erroneous results.[39] During the 1950s, researchers developed other methods of calculating the cosmic-ray exposure age by the measurement and comparison of the amounts of different radioactive and stable nuclides produced by high-energy cosmic rays. Peculiarly, the first reliable age determination has proven to be the most puzzling. Friedrich Begemann and his coworkers, by measurement of the cosmogenically produced helium-3 and hydrogen-3, estimated the exposure age of the friable Norton County aubrite to be 230 million years, which is a far greater age than that determined for a number of hard and compact stones.[40] It is, therefore, exceptional.

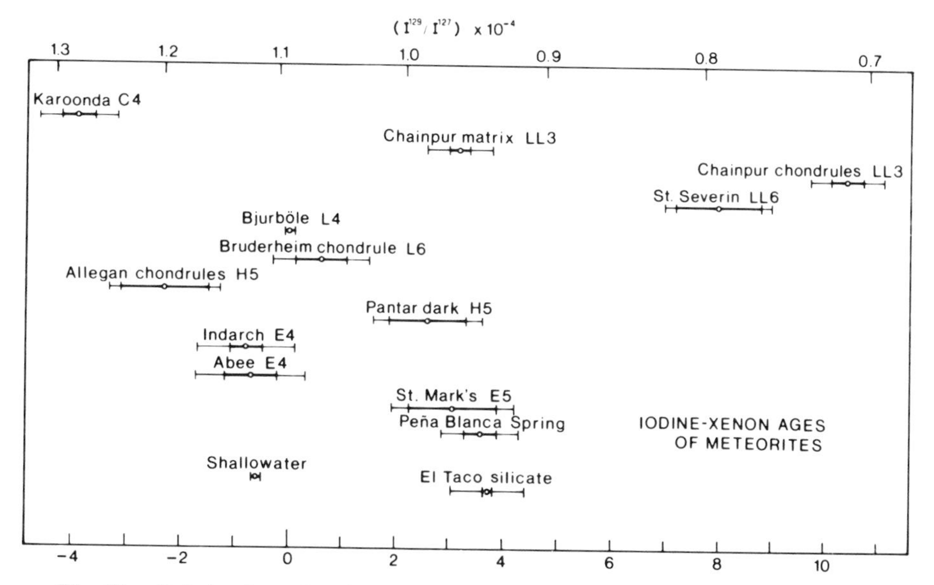

Fig. 51. Relative formation times of meteorites by the iodine-xenon method. Reprinted with permission from Podosek, *GCA* 34 (1970), Pergamon Press, Ltd.

Investigators of exposure ages face a number of complex problems. They must assume that the cosmic ray flux has not changed appreciably in the last 4 billion years, and also that the intensity along the orbits is the same for all meteorites.[41] They must determine, by either empirical or theoretical means, the rates at which the cosmogenic and stable nuclides are produced. Inasmuch as the cosmic radiation has its greatest effect on or close to the surface of the orbiting meteorite, they must make a correction for shielding—that is, for the distance that the specimen was below the surface before its fall. They also have to take into account the possibility of the loss in space of cosmogenically produced gases.

Cosmic-ray exposure ages of several hundred chondrites extend to a maximum of about 60 million years. Among the various chondrite classes, only the H group shows a significant peak at about 4 million years, which indicates the breakup of a parent body at that time.[42] There are several clusters over the 30-million-year range of the L group, and the LL group displays a slight peak at about 8 million years. The ages of meteoritic irons range from 200 million years to 1 billion years. Group IIIA,IIIB irons, with one exception, cluster at about 650 million years, and Group IV irons, with two exceptions, show ages of about 400 million years. Breakup of the parent bodies of these two groups undoubtedly occurred at these times.[43] There is still room for conjecture with respect to the difference in the cosmic-ray exposure ages of stones and irons. Obviously, the stones are more liable to erosional disintegration and collisional destruction. An early suggestion of George Wetherill was that the difference reflected the dynamic lifetime of certain Earth-crossing orbits, and could be used to infer these orbits and the location of the chondrite parent bodies.[44] However, it is now generally accepted that the dominant factor determining the ages is the meteoroid's resistance to destruction in the space environment.

The atmosphere shields a freshly fallen meteorite from high-energy cosmic radiation, so that at the time of fall the production of cosmogenic radionuclides ceases. Those that are present in the meteorite, however, continue to decay with their characteristic half-lives. This condition provides a basis for calculating the terrestrial age of the meteorite. To make a precise determination, an investigator must know what the production rate of a particular radionuclide was at the time of fall, and this datum is, of course, unknown for meteorites that fell in the distant past. To overcome this difficulty, investigators can measure the decay rate of a radioactive nuclide that fell on a known date, and by comparison estimate the probable time of fall of another meteorite of the same type. In this procedure, however, there is the assumption that the production rate in space of a particular radionuclide was identical for both meteorites and that both were equally shielded in space. Investigators improve this unsatisfactory method by measuring the decay activity of one or more radionuclides having very long half-lives: aluminum-26, chlorine-36, beryllium-10, or manganese-53. If the cosmic ray age of the meteoroid is much greater than the half life such radionuclides will have become "saturated"—that is, the decay rate will equal the production rate—and therefore investigators can estimate production rates by comparison of several decay rates with a reasonable degree of accuracy. The steady-state rates depend upon the amount of shielding.

Efforts to estimate terrestrial ages did not receive a high priority in the 1950s and 1960s, because of the greater interest in ascertaining formation and exposure ages. Furthermore, the half-lives of some of the more useful radioactive nuclides, such as manganese-53, were not well known.

Antarctic meteorite finds, however, have recently stimulated experimental activity in determining terrestrial ages. The substantial numbers of meteorites that have been gathered in widely separated areas of the continent can, if their terrestrial ages are known, provide valuable glaciological data on the minimum age and the motion of parts of the Antarctic ice sheet. By 1982 scientists had estimated the ages of over a hundred Antarctic meteorites; the two oldest were 540,000 and 690,000 years.[45] One team has extended their experiments to redetermine the terrestrial ages of several previously dated meteoritic irons. They found, for example, that the terrestrial age of the Tamarugal iron, estimated to have fallen about 1,200 years ago, was actually about 3.6 million years.[46] With the present activity, one may expect more advances in the calculation of more precise terrestrial ages.

Geologists, when examining rocks, are now aware that they may contain fossil meteorites. Swedish scientists recently reported the find of a chondrite fragment, possibly a H type, in an Ordovician limestone, and estimated its age to be about 463 million years. As noted above, H-type chondrites that have fallen recently show cosmic-ray exposure ages that peak at about 4 million years, so this find gives evidence of a H-type meteorite parent body that broke up much earlier.[47] In any event, estimated terrestrial ages indicate that meteorites have been falling to earth for millenia, so we now turn to current ideas about their origin.

THE ORIGINS OF METEORITES

John A. Wood, in his 1963 status report on the physics and chemistry of meteorites, observed that their study had "generated an astonishing diversity of hypotheses," and that there were "perhaps as many opinions as to the origin of meteorites as there are students of meteorites."[48] Fortunately, during the past two decades, the number of hypotheses has not increased in the same proportion as the number of meteoriticists. There does appear to be general agreement about the occurrence of major processes and events, although there is a considerable variety of interpretations as to just what occurred. The situation is neither unfortunate nor surprising. It reflects the explosive growth of the field and the application of new research techniques, which have increased rather than diminished the numbers of complex problems. The following sketch, which omits technical details, is intended to indicate in what areas and for what reasons different opinions exist.

There is a general consensus that about 4.6 billion years ago, the Solar System began to form from the gravitational collapse of a nebula of unknown extent, which contained a mass of gas and dust slightly larger than the present mass of the Solar System. In the process, the energy release elevated the temperature within the nebula, and high velocities within the cloud caused sufficient mixing to ensure chemical homogeneity. Primordial matter then began to condense from the gas-dust mixture. Because the abundances of the nonvolatile elements in the Sun and in type 1 carbonaceous chondrites are about the same, the primordial matter was probably of solar composition (fig. 52). The accretion of the matter resulted in the formation of small bodies, many of which were swept up by the planets or ejected from the Solar System. At a very early stage, heating occurred within some of the smaller bodies, which caused chemical differentiation and probably metamorphism. A number of these bodies, which either formed in the asteroid belt or came there as a result of perturbations, collided, and the fragments were eventually propelled

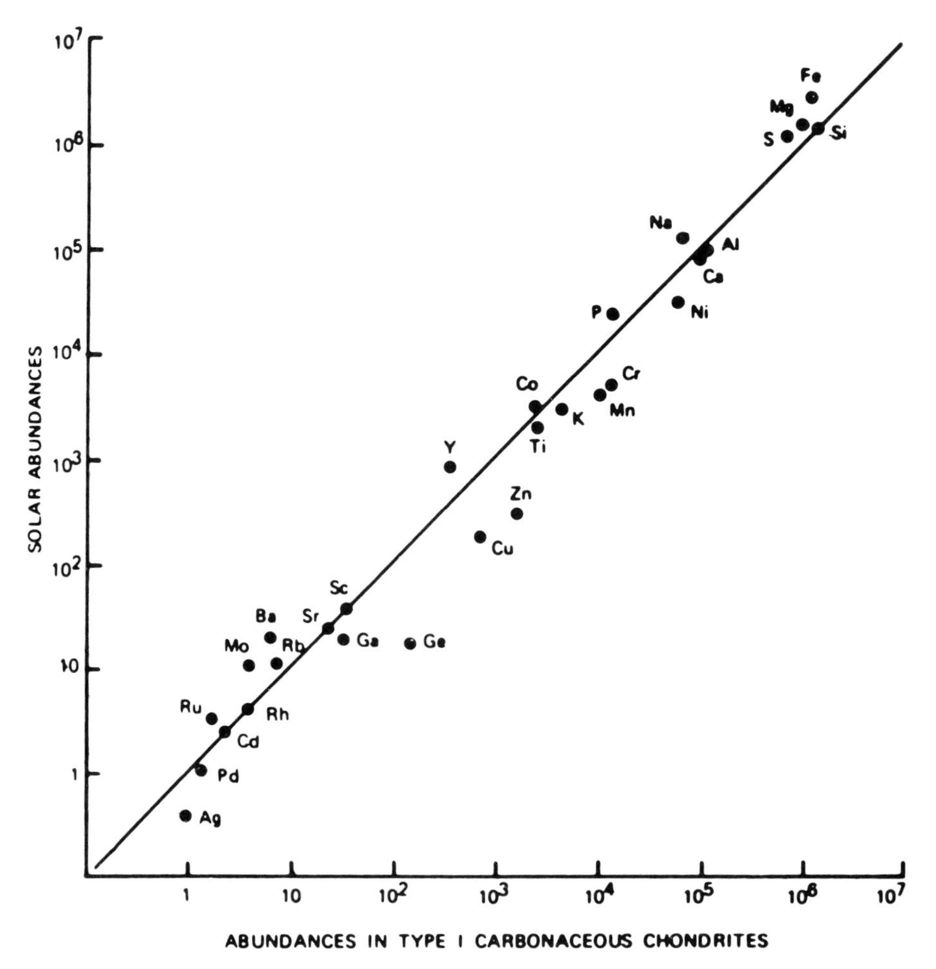

Fig. 52. Comparison of elemental abundances (normalized to Si = 10^6 atoms) in type 1 carbonaceous chondrites with those in the Sun. From Mason (1971). Reprinted with permission of Gordon & Breach Science Publishers.

into Earth-crossing orbits. It is in attempting to fill in the details of this highly superficial picture that many puzzling problems arise.

With respect to the condensation of the nebula, the goal of meteoriticists is to elucidate the details of the process by which each of the eight classes of chondrites formed from the solar dust-gas, in order to account for the differences in chemical composition and petrologic types and to explain the presence of chondrules. The most direct method assumes that chemical and thermal equilibrium prevailed in the nebula, which permits the framing of models to predict the sequence of the formation of condensates from the cooling cloud of dust-gas at pressures of about 10^{-3} atmospheres.[49] Such models provide the bases for interpreting the observed differences among the chondrites, but they also give rise to spirited disagreements. Three examples amply demonstrate how interpretations differ: the observed depletion of volatile elements in type 3 to type 6 ordinary chondrites; the existence of calcium-aluminum-rich inclusions or aggregates (CAIs) in some carbonaceous chondrites; and the origin of chondrules.

Ordinary chondrites have abundances of refractory elements, which condense between 1800 and 1400° K, that are similar to the abundances of these elements in type 1 carbonaceous chondrites. However, the abundances of volatile elements in these chondrites are depleted relative to the type 1 carbonaceous chondrites, and the amount of depletion increases fairly consistently from petrologic type 3 to type 6. To account for this, a number of scientists favor the idea that these volatiles were expelled during a process of thermal metamorphism; the higher the temperature, the greater the release of volatiles and the more homogenization and recrystallization. However, some inconsistencies in the occurrence and distribution of the volatile elements in these chondrites lead others to argue against this explanation. For example, Fredriksson and his coworkers, after a study of several H-type chondrites, concluded that there was "no compelling evidence for the assumption" that any chondrite ever experienced "severe long term thermal metamorphism."[50] A second interpretation attributes the volatile depletion to incomplete condensation in the nebula. The grains of the type 6 to type 3 chondrites condensed at progressively lower temperatures, which were still higher than the condensation temperature of the volatile elements, and then these aggregates, while cooling, failed to maintain equilibrium with the gas. A third hypothesis is that the chondritic material condensed and accreted into large chunks at high temperatures, causing the embedded grains to lose contact with the gas, leaving them poor in volatiles, and that the recrystallization of the accreted material occurred while at elevated temperatures.[51] Each interpretation has its adherents, and none is patently erroneous.

CAIs attracted widespread attention when quite a number were found in the Allende carbonaceous meteorite. Clarke and his coworkers at the Smithsonian Institution reported that "they are quite unlike any composition that has yet been recorded from stony meteorites."[52] Subsequent Rb/Sr dating indicated that they formed at least 10 million years before other chondrites.[53] Then, in 1973, Robert N. Clayton and his group at Chicago's Fermi Institute detected oxygen isotope anomalies in the Allende CAIs, which could not be explained by fractionation processes in the nebula, and suggested that they were presolar interstellar dust grains.[54] Later, they found the isotope anomalies in other classes of carbonaceous chondrites, and Clayton proposed that a nearby exploding supernova had injected them into the solar nebula a few million years before it condensed, and too late for thorough mixing with the nebula dust.[55] The existence of the anomalies and Clayton's hypothesis has stimulated substantial research. Milton Blander at Argonne National Laboratory, for example, found it difficult to conceive of a mechanism for transferring millimeter-sized material from a supernova into the solar nebula.[56] Blander, who has studied kinetic effects such as nucleation constraints and slow reactions that might influence the formation of nebular condensates, pointed out that diffusion might account for the oxygen isotope anomalies. He thought it possible that the precursors of the CAIs formed at high temperatures in an outer nebular region that was rich in oxygen-16, and then were transferred by convection or gravitational forces to an inner region, where they were intimately mixed with material having "normal" oxygen isotopes. Diffusion of oxygen between the two kinds of material then produced the observed anomalies. He emphasized that his suggestion required a "layered" or inhomogeneous solar nebula. This question of the source of the CAIs is by no means settled.

A century or more ago, Reichenbach, Sorby, Tschermak, and Brezina all had different ideas as to the origin of chondrules (see chap. 5). Particularly since the 1960s, knowledge of the mineralogy, petrology, and chemistry of chondrules has increased substantially (pl. 4). Because there is an overlap in the chemical composition and in texture among the chondrules, meteoriticists cling to the idea that a basic process of formation occurred, with minor variations that can account for the differences. Hypotheses proposed in recent years include formation (1) by direct condensation from the nebula, (2) by volcanic activity on parent bodies, (3) by impact of solids within the nebula, (4) by impact on the surfaces of parent bodies, or (5) by the melting in the nebula of preexisting solids.[57] All face criticism. Direct condensation models suffer from the fact that liquid condensates having the chemical composition of most chondrules would be unstable at the presumed temperatures and pressures, compelling advocates to resort to imaginative theories to account for the existence of the chondrules. If produced by volcanic action, chondrules would be compositionally nonchondritic; furthermore, there should also be lavalike chondrites, which are absent. Impact models have difficulty in explaining the great abundance of chondrules, the oxygen isotope heterogeneities, and the very old ages. Proponents of the idea that chondrules were preexisting solids in the nebula which were subsequently melted must account for the process that led to substantial chemical fractionation of the dust in the nebula, as well as the source of heat that caused the melting. Chondrule formation, then, remains a vexing problem for cosmochemists.

Meteoriticists are probably expending as much thought and effort on models of the accretion, size, and early history of parent bodies as they are on the condensation process. One leading theory is that the particles of primordial matter drifted to the central plane of the solar nebula at a time of diminished turbulence. Cohesion occurred, and because the material agglomerated in loose clusters, gentle collisions resulted in continued accretion.[58] Later impacts between bodies up to 100 km in diameter probably aided compaction and consolidation and some mixing of material. The age data indicate that only a few million years after condensation commenced, these parent bodies experienced heating that was sufficient to produce melting in some and varied degrees of metamorphism of the chondritic material in others. The source of this heat is still a question. Heat from a superluminous sun would not be adequate to produce melting at necessary depths of bodies in the asteroid belt. The abundances of long-lived radionuclides would be insufficient to cause melting in bodies 500 km in diameter or less—the largest possible size of a meteorite parent body. The release of gravitational energy during formation or of kinetic energy during accretion also appear inadequate. Another possibility is the action of a solar (T-Tauri) wind or plasma, which would induce electric currents and cause heating. Calculations show that this mechanism would be sufficient to cause melting in bodies up to 250 km in diameter at the Earth's distance from the Sun, but that there would be insufficient heat to accomplish this in the asteroid belt. The preferred source at present is heating produced by the radioactive decay of aluminum-26, which has a half-life of 740,000 years. The decay product of aluminum-26 is magnesium-26, and although enough of it to produce melting has been found in a few refractory inclusions of the Allende meteorite, there remains serious doubt as to whether the mean nebular abundance of aluminum-26 overall was adequate.[59] However, it is certain that heat was provided from some source.

There is evidence that the heating of chondrite-like material of some parent bodies caused at least partial melting and produced most of the achondrites, stony irons, and irons. Detailed studies reveal links between certain classes, which indicate that they may have formed in the same parent body.[60] The aubrites, for example, have similar oxygen isotope ratios to the enstatite chondrites, and may have been produced by the partial melting and metamorphism of the enstatite chondrite parent body. The basaltic achondrites (eucrites and howardites) may have been produced by the partial melting and fractional crystallization of chondritic material that was poor in volatiles. The mesosiderites appear to be a mechanical mixture of metal and howarditic or eucritic material. Some of the irons, for example Group IIIA,IIIB, definitely formed in the core of parent bodies. Others, such as Group IA,IB, show signs of incomplete melting, indicating that they were never a part of a stable nickel-iron, iron-sulfide magma. The majority of the pallasites constitute material that was at the interface between the core and the silicate layer of a parent body, and their texture gives evidence that metal intruded into the olivine layer.

Scientists recently have given a great deal of attention to the peculiar and very rare SNC achondrites: the shergottites, the nakhlites, and the sole chassignite.[61] Mineralogically and chemically, they resemble magmatic cumulates produced on the Earth, which suggests that they formed on very large parent bodies. All display young ages that apparently date recrystallization: the shergottites, about 650 million years; the nakhlites, about 1.3 billion years; and the chassignite, about 1.4 billion years. The rubidium-strontium age of the shergottites—180 million years—indicates a major shock event at that time. The content of noble gases and the nitrogen isotope ratios are very similar to Martian rocks.[62] Although the evidence points to a Martian origin of the shergottites, dynamicists have yet to agree on the impact mechanism that could have ejected meteorites from the Martian surface, given the gravitational field of Mars.[63] However, the fact that U.S. and Japanese scientists have definitely identified two Antarctic meteorites as having a lunar origin lends further support to the surmise that the SNC meteorites formed on a large planet and were blasted from its surface into an Earth-crossing orbit.[64]

Estimates of the size of parent bodies changed considerably after 1960. Brown and Patterson, Urey and Craig, and Ringwood followed Daly in considering that the planet or planets from which meteorites came were of lunar size or larger. They thought that considerable pressure at the core was necessary to form the Widmanstätten structure and to produce diamonds and other high-pressure minerals. Alternatively, O. J. Schmidt, a Soviet scientist, had in 1944 suggested that the accretion of the planetisimals in the asteroid belt had ceased because of Jupiter's perturbations, and thereafter the opposite process of collision and disintegration had commenced.[65] Other Soviet scientists in the 1950s vigorously supported Schmidt's idea. B. Y. Levin investigated further the problem of collisional breakups, and A. A. Yavnel suggested that each of five chemical groups into which he separated meteorites came from a different small planet.[66] Then in 1957 Urey made calculations showing that the core of a lunar-sized body would not cool to the temperature indicated by the Widmanstätten structure in the time permitted by the age of the Solar System.[67] In 1960, Robert A. Fish, Gordon G. Goles, and Edward Anders at the Fermi Institute reported that, given an extinct source of radioactive heat, all meteorites could have been produced in asteroids with maximum diameters of about 500 km. Their calculations indicated reasonable time scales for the operation of the processes of segregation and differentiation, and they suggested that diamonds might be produced by impact instead of by high pressure within a large planet. They postulated that the radioactive disintegration of aluminum-26 had provided the necessary heat, but they listed a half-dozen other possible sources of radioactive heat.[68]

Further studies appeared to strengthen the connection between asteroids and meteorites. Investigations determined that the Widmanstätten structure would not form in the cores of parent bodies above 12,000 atmospheres, which placed an upper limit of a 1,600-km diameter on their size. Whereas early cooling rate estimates indicated a maximum parent body diameter of 600 km, a recent revision suggests that meteoritic iron parent bodies need only be a few tens of kilometers in diameter.[69]

Meanwhile, in the 1950s, a few astronomers began to study the asteroids more intensively.[70] At that time, observers had identified over 1,600 asteroids, which were numbered and whose orbital elements were known with reasonable certainty. They had also accumulated other data. In 1867 Daniel Kirkwood had reported gaps between the orbits of the 80 main belt asteroids then known, attributing them to the correspondence of these orbital periods with certain fractions of the period of Jupiter (see chap. 5). These gaps became more prominent as more asteroids were discovered. Also, about 1920, Kiyotsuyu Hirayama, a Japanese astronomer, had identified five asteroid groups, which he termed "families." The asteroids of each family had similar eccentricities, inclinations, and orbital semi-major axes, and Hirayama postulated that the members of each family had once composed a single body that had broken apart (fig. 53). Some families had many members; Flora, for example, had 81, and Eos had 38 members. As for size, Edward E. Barnard at the Lick and Yerkes observatories had, beginning in the 1890s, made micrometric measurements on photographic plates of the larger asteroids. He estimated that Ceres, the largest, had a diameter of 800 km, Pallas 500 km, and Vesta 400 km. Early twentieth-century observers estimated the sizes of the smaller asteroids on the basis of brightness; the diameters of about a dozen were over 150 km, and most were less than 75 km. Astronomers also knew, because of the fluctuations in brightness, that many asteroids rotated, and by the mid-1930s they were aware that the orbits of several asteroids—Eros, Amor, Apollo, Adonis, and Hermes—either approached closely or intersected the orbit of Earth. In 1929 N. T. Bobrovnikoff at the Lick Observatory observed color differences among a few asteroids and obtained photographic spectra that showed distinct color variations. He and other astronomers in the early 1930s attempted to correlate asteroid spectra with the reflective properties of meteorites. Unfortunately, they neglected to prepare adequately the surfaces of the meteorites, so that their experiments were not successful.

Asteroid studies advanced on several fronts about 1950. The International

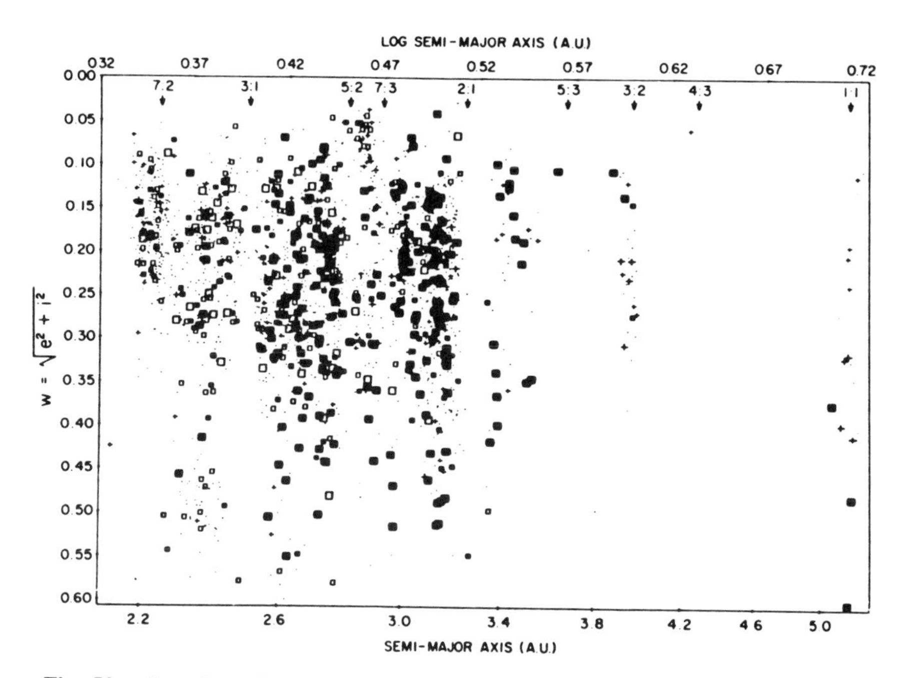

Fig. 53. Overview of the asteroid belt. Nearly all numbered asteroids are plotted as a function of the (log) semi-major axis and a quantity related to departure from an in-plane circular orbit. Kirkwood gaps in the main belt population are correlated with Jupiter commensurabilities indicated at the top. Prominent clusters of identical symbols are the more populous, homogenous Hirayama families: Themis (a = 3.14, w = 0.15); Eos (a = 3.01, w = 0.19); Koronis (a = 2.88, w = 0.06); and Flora (a = 2.21, w = 0.16). Reprinted with permission from article by C. R. Chapman in *Asteroids,* ed. T. Gehrels. Tucson: University of Arizona Press, 1979.

Astronomical Union established the Minor Planet Center at the Cincinnati Observatory (moved to the Smithsonian Astronomical Observatory at Cambridge in 1978), and in 1952 coordinated its activities with the Institute of Theoretical Astronomy in Leningrad, which published an annual catalog of asteroid orbital data. Ernst J. Öpik, an Estonian astronomer who had for three decades studied the physics of meteor phenomena, began to analyze the long-range effects of planetary perturbations on asteroid orbits and also the consequences of collisions between them. Gerard P. Kuiper at the Yerkes Observatory and Tom Gehrels at the University of Arizona initiated the Yerkes-McDonald Survey of asteroids, which from 1950 to 1952 employed photoelectric photometry to search for asteroids, taking over 2,400 photographs over 40° of ecliptic latitude.

The space program increased interest in the asteroids because of projected fly-bys, but the emergence of the powerful computers required for the Apollo flights probably contributed more to knowledge of asteroids than did studies of possible missions. The computational capabilities soon permitted calculation of planetary perturbation effects on asteroidal orbits over time spans that had previously eluded dynamicists. Icarus, the asteroid 14.4-km in diameter which passed within about 7 million km of the Earth in 1968, highlighted the probability of past Earth-asteroid collisions and stimulated intensive study of those classes of asteroids whose orbits intersect or nearly intersect that of the Earth. At the end of the 1970s, Eugene M. Shoemaker and his coworkers at the California Institute of Technology reported that the total population of these classes, whose diameters range from 1 to 10 km, was about 1,300, and that there was a probability that 3 to 4 would collide with the Earth every million years.[71]

The first results of multiple camera arrays, which were established to take photographs automatically of bright meteors and fireballs, appeared in the 1960s. Computation of the orbit of the H-5 chondrite, whose fall on 7 April 1959 at Pribram, Czechoslovakia, was recorded by the Czech All-Sky Network, revealed that its apelion was just outside of the main asteroid belt (fig. 54).[72] The orbit of the Lost City, Oklahoma H-5 chondrite, which fell 3 January 1970, was computed from the photographs of the now-defunct Prairie Network, which demonstrated that it passed into the asteroid belt.[73] Similarly, the Canadian Meteorite Observatory and Recovery Project photographed the fall of the L-6 chondrite on 5 February 1977 at Innisfree, Alberta, and computations determined that it also passed into the main belt.[74]

The revival in 1969 of Bobrovnikoff's idea of comparing the spectra of asteroids and meteorites, but with more sophisticated techniques, yielded startling results.[75] Thomas B. McCord and his coworkers at MIT compared the spectrum of Vesta with the spectra of meteorites, and concluded that its surface was composed of material similar to meteoritic basaltic achondrites (eucrites). Over the next few years McCord and Clark R. Chapman collaborated in measuring the spectra of about 100 asteroids. Other observers took up the study, so that by the late 1970s about 400 asteroids had been classified by spectral types, each corresponding to a meteoritic analog or designated as unclassifiable. C-type asteroids, whose spectra resemble those of carbonaceous chondrites, comprise about 74 percent of those in the main belt; S-type, whose

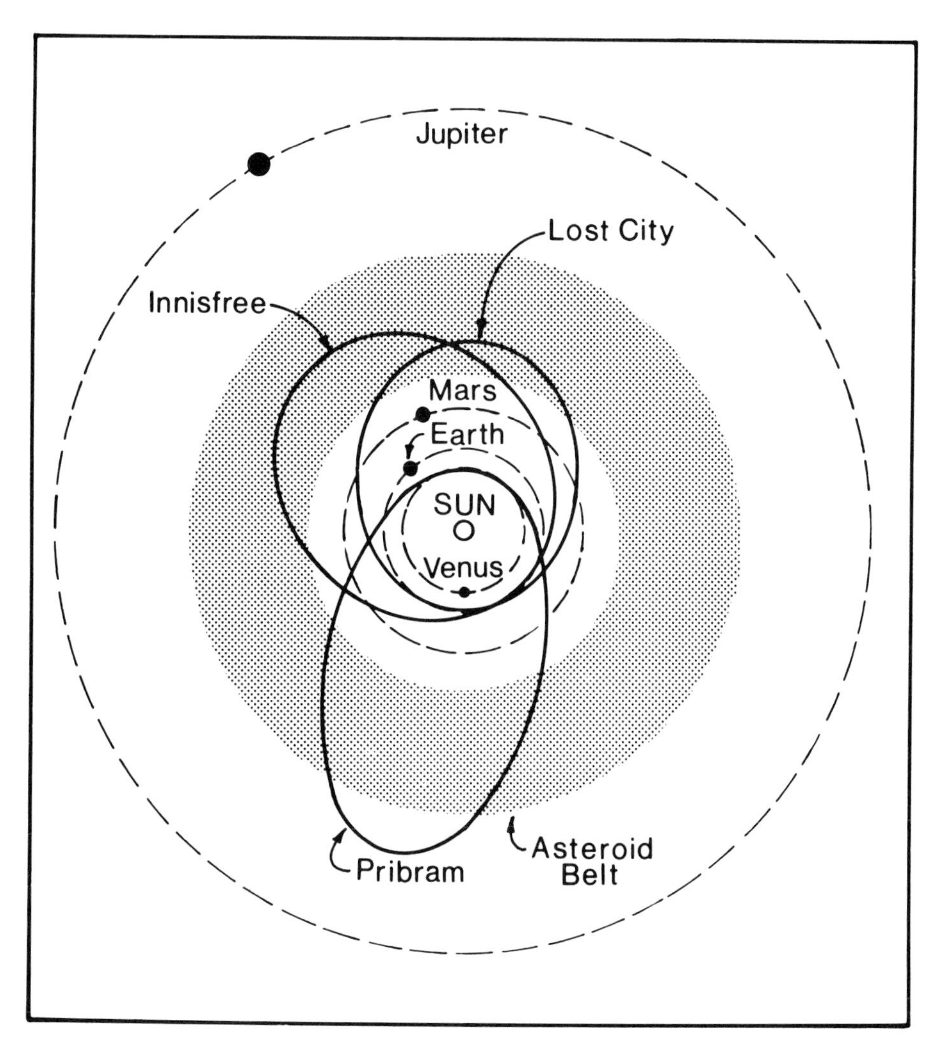

Fig. 54. Orbits of the Pribram, Lost City, and Innisfree meteorites, calculated from photographic data. Redrawn with alterations from a figure furnished by John T. Wasson, with permission.

spectra are similar to those of ordinary chondrites or stony irons, about 16 percent; and the other types make up the remaining 10 percent (table 8).

B. H. Zellner of the University of Arizona has emphasized that the classification of spectral types, the mineralogical interpretation, and the meteoritic identification are three distinct steps.[76] Nevertheless, these findings were puzzling. The vast majority of the meteorite falls in collections: more than 80 percent are ordinary chondrites; less than 5 percent are carbonaceous chondrites; and just over 1 percent are stony irons. If the interpretation of reflectivity studies of the main-belt asteroids is correct, then the Earth is receiving a very biased sample from that area. The carbonaceous chondrites are more likely to be destroyed during entry into the Earth's atmosphere, but this fact could not explain the discrepancy entirely.

TABLE 8
DESIGNATION OF ASTEROID TYPES

Type	Albedo	Spectrum	Mineralogy	Meteoritic analogues
C	low	relatively flat, weak features	silicates plus opaques (carbon)	carbonaceous chondrites
S	moderate	reddish Fe^{2+} absorptions	silicates plus metal	stony irons (H chondrites?)
M	moderate	slightly reddish, featureless	metal, or metal plus neutral silicates	nickel-irons enstatite chondrites
E	high	flat featureless	neutral silicates	enstatite achondrites
R	moderate to high	red, strong features	Fe^{2+} silicates	various or unknown (ordinary chondrites?)
U	various	unusual	various	various or unknown (certain achondrites)

Note: Reprinted with permission from article by B. Zellner in *Asteroids,* ed. T. Gehrels. Tucson: University of Arizona Press, 1979.

There was another problem that scientists had studied from the late 1960s—namely, how fragments of main-belt asteroids achieved an Earth-crossing orbit. George W. Wetherill of the Carnegie Institution of Washington proposed in 1977 that most meteorites were associated with the Apollo and Amor classes of Earth-crossing asteroids. The orbits of these asteroids pass through the main belt, where they experience collisions, with many of the fragments being scattered along their orbits. Wetherill calculated that about 400 tons of these pieces enter the Earth's atmosphere each year, about three-fourths of this amount undergoing disintegration during fall. The flux of bright meteors entering the Earth's atmosphere, as measured by camera networks; the falls on the Moon recorded by seismometers; and the recovery rates of meteorites tend to confirm Wetherill's estimate. Also, a few reflectivity studies of a number of Apollo-Amor objects indicated that some are similar to ordinary chondrites and some to carbonaceous chondrites.[77]

Still, a paradox existed. Cosmic-ray exposure data indicated that ordinary chondrites in collections had ages of just a few million years or even less, which gave evidence that they had been separated from their parent bodies relatively recently. Yet Wetherill's studies showed that collision fragments of main-belt asteroids at the 3:1 Kirkwood gap, which were propelled within the orbit of Mars, would be perturbed into an Earth-crossing orbit only after about 100 million years, during which time most would be destroyed. There was no correspondence between these time scales.

In 1983, however, Jack Wisdom of MIT determined that Jupiter's perturbations produced what he termed "a chaotic zone" at the 3:1 Kirkwood gap, where the initial conditions of a fragment's entry yielded a 50 percent chance that its orbit could extend to the Earth's orbit directly on a time scale of about 1 million years. Following this result, Wetherill returned to the problem and in 1985 published his findings. To begin with, about 50 percent of the asteroids that are in orbits adjacent to and on each side of the 3:1 gap are S-type, corresponding to ordinary chondrites. Wetherill calculated that when the cluster of fragments from the chaotic zone were near perihelion at one astronomical unit, the Earth's perturbations altered their orbits so that, over time, some would collide with Earth or Venus, some would be ejected from the Solar System because of Jupiter's additional perturbations, and some would be returned to the asteroid belt. Further, Wetherill found that the orbital elements and velocities of those of the cluster that became Earth-crossing would cause more meteorites to fall on Earth during the afternoon and evening hours—the same percentage, in fact, as has been observed.

Wetherill also suggested that other chaotic zones may exist at Kirkwood gaps, and mentioned in particular the 4:1 gap near the region of the Flora family. These asteroids had previously been considered a source of differentiated meteorites. Wetherill pointed out that because of resonances, larger bodies approximately 20 km in diameter could, as a result of collisions, be thrown into a Mars-crossing orbit, and some could eventually become Earth-crossing.[78] This proposal requires much more study. However, Wisdom's and Wetherill's work with respect to the chaotic zone and the fate of fragments entering it does represent a major advance in accounting for the falls of ordinary chondritic meteorites.

Earlier, Wetherill exposed another problem. There is a continual loss of large Apollo objects over time by virtue of collisions with Earth, Venus, the Moon, Mercury, and Mars, or by ejection from the Solar System as a result of planetary perturbations. He calculated that about fifteen new Apollos more than 1 km in diameter must be supplied every million years to account for the present population. The probability that the collision of two main-belt asteroids could place an object of that size directly into an Earth-crossing orbit is very low, since the velocity required for such a transfer is 6 km per second, and this energy requirement would cause melting and disintegration during a collision. Less intense collisions near a few of the Kirkwood gaps could result in the transfer of fragments into a Mars-crossing and eventually into an Earth-crossing orbit, but Wetherill's computations led him to believe that this mechanism would supply only one of the fifteen Apollos needed every million years. His conclusion was that extinct comets are the source of the other Apollos.

Certain evidence supports Wetherill's hypothesis. In 1983 the Infrared Astronomy Satellite detected an object, tentatively designated as minor planet TB1983, which was at first considered to be an asteroid because of its sharp outline on a photographic plate. Later, however, astronomers found that it shared the same orbit as the dust that creates the Gemenid meteor showers each December. Also, studies in 1984 of the asteroid Oljato revealed magnetic disturbances in its wake, which indicate the presence of a cometary tail. However, current theories of the origin of comets place their source at distances so far from the Sun that the meteorites in our collections could not have formed there; moreover, scientists believe that the nuclei of comets are so friable that they could not survive atmospheric passage. But astronomers admit that there is little certainty as to where comets originated and that the composition of their nuclei is speculative.[79]

One way to improve understanding of the origin of meteorites would be to explore the asteroids and to sample the surfaces of several different types. A number of scientists and engineers have set up models of asteroidal tours extending up to about five years, and have calculated the type of spacecraft and propulsion systems required for such endeavors.[80] Perhaps such missions will be realized during the next few decades.

MATTERS OF LIFE AND DEATH

The controversies of the 1870s and 1880s as to whether meteorites contained germs of life or fossil organisms were short-lived (see chap. 5). The question arose again in the early 1930s, when Charles B. Lipman, bacteriologist and graduate dean at the University of California, Berkeley, announced that sixteen sterilized specimens of eight meteorites placed in 24- to 48-hour cultures had produced rods and coccoid cells.[81] There were no carbonaceous chondrites among his samples; six were ordinary chondrites, one was a howardite, and one was an iron octahedrite. Lipman had sterilized the surfaces of the meteorites by immersing them in a 30 percent solution of hydrogen peroxide, and he insisted that his method prevented any contamination. In answer to any possible objection that the bacteria resembled terrestrial types too closely, Lipman stated that there was no valid reason why bacteria similar to ours might not have evolved on other planets or other systems in space.

Articles in the *New York Times,* reporting that Lipman had found "life in meteors," aroused members of the scientific community, although a few days after they appeared Lipman notified the press that his results did not prove that life had come to Earth from the cosmos.[82] Michael A. Farrell, a Yale bacteriologist, wrote that Lipman's technique and interpretation was open to serious question. Previously, he noted, Lipman had found bacteria in coal and in Precambrian rock, and he pointed out that the meteorites could have been contaminated while they were in contact with the ground, during handling, on the shelves of a museum, or during crushing prior to the experiments. "Lipman's excursions into the field of life beyond this globe," he concluded, "must be considered as a flight of imagination through space."[83] Harvey H. Nininger, the most enthusiastic of twentieth-century meteorite recoverers, was similarly unimpressed with Lipman's claim. It was doubtful, he stated, that air contamination could be prevented, since a freshly fallen meteorite would naturally absorb bacteria.[84] Sharat K. Roy, curator of geology at Chicago's Field Museum of Natural History, duplicated Lipman's experiments, and found, as Lipman had, rods and coccoid cells. However, Roy identified these as *Staphylococcus albus* and *Bacillus subtilis,* both frequent laboratory contaminants.[85]

The *New York Times,* knowing the public's interest in theories of the origins of life, kept abreast of the flow of scientific articles and in 1936 published a summary headed: "Are meteorites alive? A controversy whether they come bearing bacteria."[86] Lipman finally yielded some ground. He discounted Roy's experimental results because Roy had used much smaller meteorite samples than he had; thus the findings were not comparable. However, Lipman now stated that he did not intend to present any theory about the extraterrestrial origin of life. Rather, he was mainly interested in the survival of bacteria for long periods of time, and he had just presented the factual evidence of having found bacteria in meteorites.[87] Roy subsequently answered Lipman's criticisms, and there the matter ended.[88]

Twenty-five years later, the question of the presence of life in meteorites reappeared. In March 1961 Bartholomew S. Nagy and Douglas J. Hennessy of Fordham University and Warren G. Meinschein of Esso Research presented a paper at the New York Academy of Sciences reporting on their mass spectrometric analysis of the Orgueil carbonaceous chondrite. The hydrocarbons in the meteorite, they wrote, "resemble in many important aspects the hydrocarbons in the products of living things and sediments on earth. Based on these preliminary studies, the composition of the hydrocarbons in the Orgueil meteorite provide evidence for biogenic activity." This statement was sufficiently cautious, in that it appeared to call for further studies, but the authors also suggested that "biogenic processes occur and that living forms exist in regions of the universe beyond the earth."[89] The *New York Times* also quoted the scientists as saying: "We believe that wherever this meteorite originated something lived."[90]

Eight months later, George Claus, a microbiologist at New York University Medical Center, and Nagy reported that they had found "organized elements" in the Orgueil and Ivuna carbonaceous chondrites.[91] There were five distinct types, and from their morphological, optical, and staining results, the authors concluded that they were "possible remnants of organisms" (fig. 55).

Many scientists were skeptical about Nagy's claims. To begin with, the Fischer-Tropsch and several other processes had for two decades been producing a variety of synthetic hydrocarbons: engine fuels, waxes, and raw materials for fats and soaps. In 1953, Stanley L. Miller, one of Urey's students at the University of Chicago, had synthesized in the laboratory several complex hydrocarbons, including amino acids, by means of an electrical discharge in a mixture of hydrogen, water vapor, methane, and ammonia. Urey, in fact, had proposed the occurrence of such an abiogenic process in the primordial nebula to account for the presence of organic compounds in meteorites.[92] Also, George G. Mueller at University College London, during an analysis of the carbonaceous material of the Cold Bokkeveld meteorite, did not detect any optically active organic compounds that would have indicated the presence of biogenic material.[93] Further, current theory held that biogenic material evolved in the presence of liquid water, and the condition of the inorganic constituents in carbonaceous meteorites gave evidence that there was no running water where they originated.

The interpretation that the observed microstructures were fossil organisms was also suspect. Philip Morrison, a physicist then at Cornell University, offered a thoughtful alternative interpretation, assuming the validity of the observations. Under suitable environmental conditions, he wrote, inorganic matter can assume intricate patterns, such as the beautiful dendritic and platelike forms exhibited by snowflakes. The morphology of organic mixtures could also display complexity without it being assumed that living processes were or had been in operation. The "organized elements," in Morrison's view,

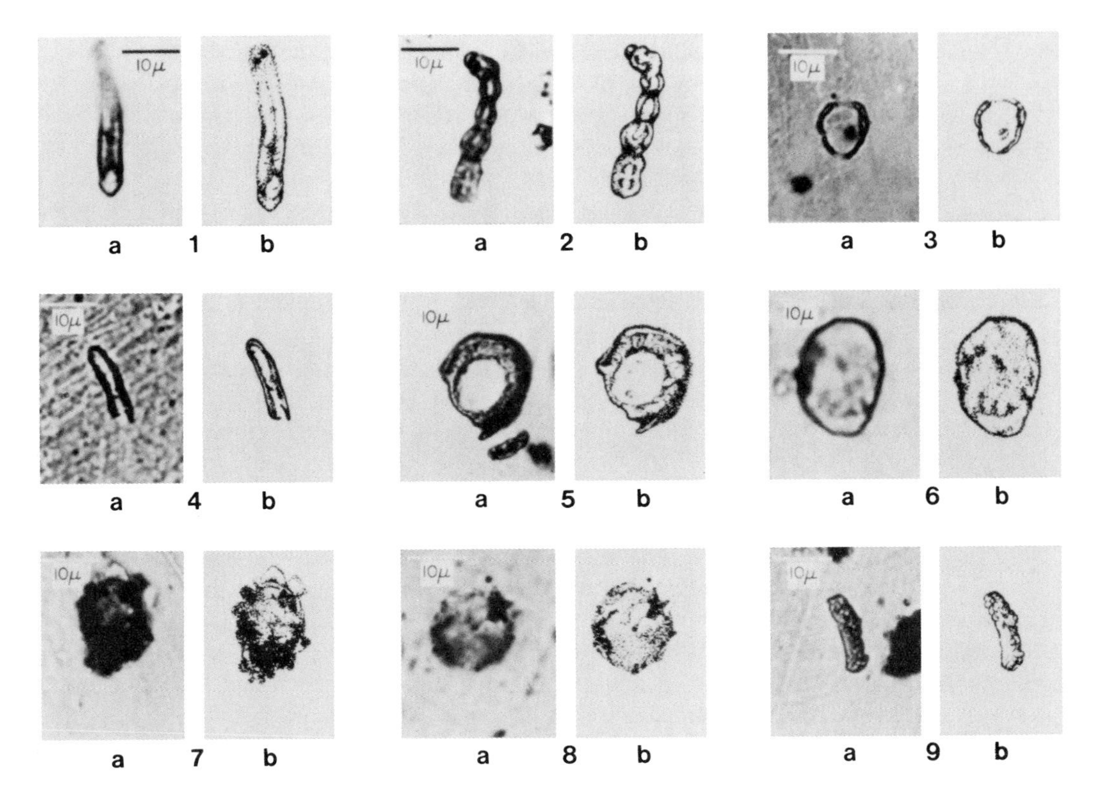

Fig. 55. Photomicrographs and drawings of nine untreated Orgueil meteorite microstructures analyzed with the electron microprobe. 6a, b; 7a, b; and 8a, b consisted of silicates; the remainder appear to be limonite. Reprinted by permission from *Nature* 198:123. Copyright © 1963 Macmillan Journals Limited.

would not be microfossils but "carbonaceous snowflakes." Morrison emphasized that his interpretation did not diminish the importance of the presence of the organized elements. Such order on a micron scale might possibly be a precursor of living systems.[94]

Several scientists at the University of Chicago, led by Edward Anders, presented telling criticisms of Nagy's evidence. Frank W. Fitch, a pathologist, Henry P. Schwarcz of the Fermi Institute, and Anders first reported that three classes of Nagy's organized elements were troilite and magnetite globules, sulfur and hydrocarbon droplets, and apparent mineral grains.[95] What was involved, then, were subjective differences in judgment concerning the morphology of the particles. After exchanging samples with Nagy and Claus, Anders and Fitch published a followup article.[96] They reported that they found ragweed pollen grains, perhaps a furnace ash sphere, and perhaps starch grains. The high density of most particles having morphological simplicity indicated that they were predominantly inorganic. Morphologically complex particles, they stated, were very rare. Some fluoresced in ultraviolet light; some took on a biological stain; and many showed a resemblance to airborne contaminants. Somewhat later, Ryoichi Hayatsu at the Fermi Institute failed to detect in an

Orgueil specimen the optical activity found by Nagy, and suggested that the organic compounds in it were synthesized by abiogenic processes.[97]

Nagy and his coworkers, however, did receive scientific support. Frederick D. Sisler of the U.S. Geological Survey sterilized samples of the Murray carbonaceous chondrite in a germ-free laboratory, and reported that some particle-like forms appeared in the culture media after several months.[98] Frank L. Staplin, a micropaleontologist with Imperial Oil Company of Canada, found several cellular forms in the Orgueil meteorite, which he interpreted as microfossils.[99] At Victoria University, New Zealand, Michael H. Briggs and G. Barrie Kitto detected complex organic microstructures in the Mokoia carbonaceous chondrite, which they judged were compatible either with a biogenic or abiogenic origin. J. Desmond Bernal, the eminent English crystallographer, was also inclined to believe that the organized elements were organic microfossils.[100]

It was Harold Urey, however, who, though expressing skepticism, assisted Nagy the most. He arranged a research position for Nagy at the University of California, San Diego, to enable him to continue his research. In 1966 Urey published a comprehensive review of all the evidence presented by Nagy and others that indicated that carbonaceous meteorites carried biogenic material. "If the materials . . . were of terrestrial origin," Urey wrote, "it would be firmly suggested that the materials were of biological origin and indigenous to the samples."[101] However, Urey also viewed Nagy's findings, if valid, as supportive of his own hypothesis concerning meteorites and the Moon. In 1959 Urey had suggested that some meteorites had been ejected from the lunar surface as a result of the impacts of large meteorites, and in 1962 he proposed that all chondrites came from the Moon.[102] In his 1966 review, he postulated that carbonaceous chondrites containing primitive life forms might have originally been terrestrial material that had been transferred to the Moon, when, according to one hypothesis of the Moon's origin, the Moon escaped from the Earth, or, according to another, the Moon was captured by the Earth and caused terrestrial matter to be flung far into space. The Apollo lunar landings, Urey thought, should recover rock samples that would reveal a great deal about the history of the Solar System and the origin of life, so that it was premature to reject out of hand evidence of fossil life in meteorites. If life had, in fact, existed in meteorites, and if meteorites did not come from the Moon, Urey concluded, it would be necessary to develop more complex theories for the origin of meteorites.[103]

At the time, the major consequence of the controversy about the possible presence of fossil organisms in meteorites was that NASA officials took elaborate precautions to make certain that the recovered lunar rocks were isolated and handled as if they contained "high hazard pathogens." The Apollo astronauts were quarantined at the Lunar Recovery Laboratory, specially constructed at Houston, while scientists performed tests to ensure that they had not brought back with them any organisms that posed a threat to the Earth's biosphere.[104] Meticulous examinations did not detect any fossil life forms in the lunar samples, and the Apollo crews did not recover any carbonaceous chondrites.[105]

When the manned lunar missions came to an end, the controversy about organized elements in meteorites also faded into the background. Nagy, in 1975, published a detailed review and critique of all of the experiments that provided evidence for the presence of biogenic material in meteorites, as well as the refutations. In it, he displayed some ambivalence, but he concluded that he thought it was unlikely that the organized elements were extraterrestrial microfossils.[106] In 1982, however, Nagy and a coworker reported the discovery in optical activity tests that some of the amino acids in the Murchison carbonaceous chondrite were partially racemic, although they emphasized that the sample, despite elaborate precautions, might have been contaminated.[107] The question of the presence of biogenic material in meteorites, then, may well enjoy a revival.

On the morning of 30 June 1908, a tremendous explosion occurred in the atmosphere above the Stony Tunguska River basin in central Siberia.[108] The remoteness of the unpopulated area, the onset of World War I, and the Russian Revolution prevented scientific exploration of the phenomenon until the early 1920s. Then, Leonid A. Kulik, curator of the Meteorite Department of the Mineralogical Museum of the Soviet Academy of Sciences at Leningrad, made inquiries which revealed that people over a wide area had seen a fireball, heard loud explosions, and felt blasts of hot air, and that seismometer instruments as far away as Tiblisi had recorded an earthquake. In 1927 Kulik located the site of the explosion, which was evident because of the widespread devastation of the forest, and two years later he led the first of numerous Soviet scientific expeditions to determine the cause of the Tunguska event.

In Great Britain in 1908, at five widely separated locations, new instruments for recording variations in atmospheric pressure, called microbarographs, went into operation, and these simultaneously recorded oscillations that were not associated with normal atmospheric changes. The owner of one instrument, when reading a popular account in 1929 of Kulik's investigations at Tunguska, noted the coincidence of the date of the explosion and the date of the instrumental oscillations, and he brought this fact to the attention of F. J. W. Whipple, director of the Kew Observatory. Whipple found that the aerial and seismometer records coincided, and then there was also recollection that for days after the explosion the night sky over northern Europe and Asia had been luminescent because clouds of particles at altitudes of several hundred kilometers reflected rays of the sun beneath the horizon. Whipple estimated that five thousand times as much energy was expended in creating air waves as went to produce seismic waves, and he suggested that a comet had entered the atmosphere and disintegrated explosively above the Earth.[109]

Kulik's initial reports mentioned the existence of many shallow craters in the immediate area of the explosion and also the recovery by peasants of pieces of iron, which led to the supposition that a large iron meteorite had broken apart above the site. Later studies, particularly by E. L. Krinov, discounted this.[110] The blast leveled about 2,000 km^2 of forest, and the directions of the fallen trees indicated that the explosion occurred at an altitude of several kilometers (fig. 56). A detailed study of the neighboring swampland showed that there was no connection between the observed swamp holes and

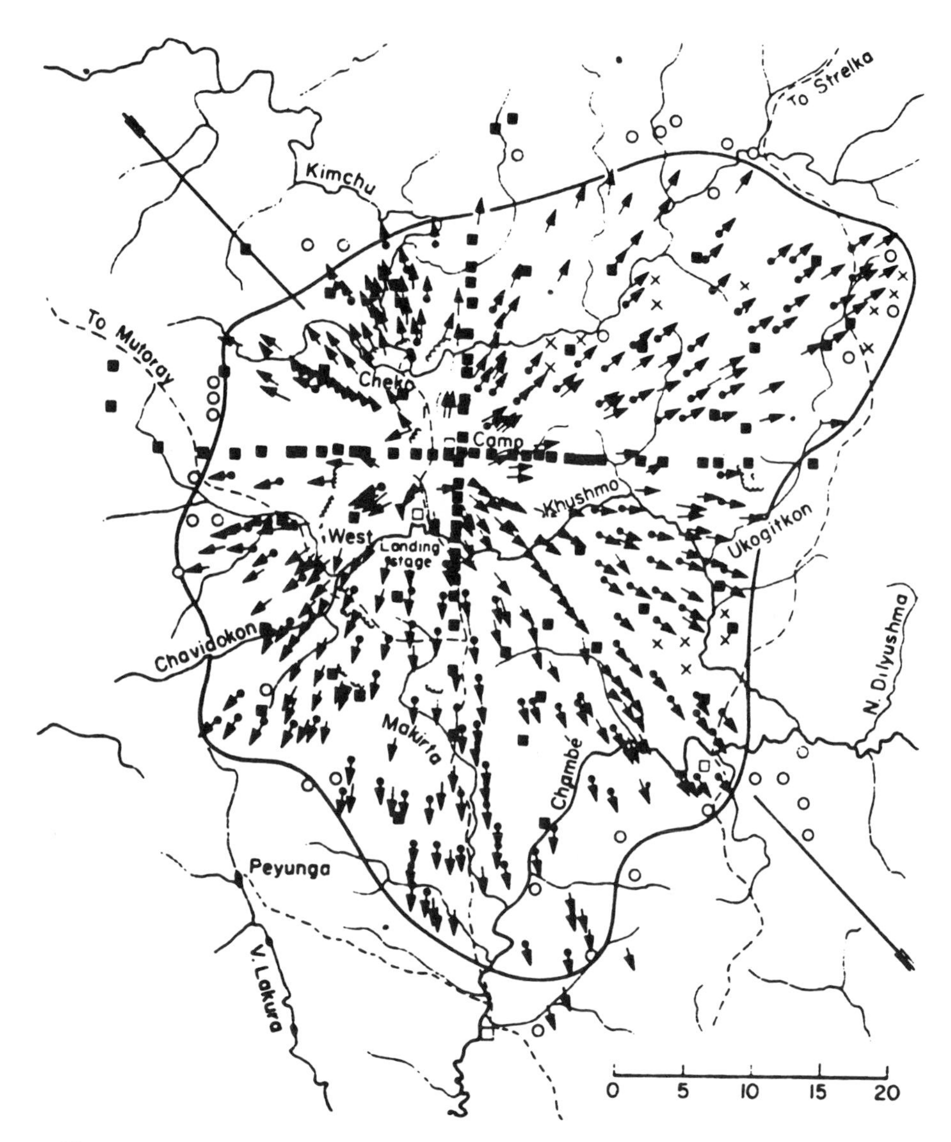

Fig. 56. Map of the forest destruction at Tunguska. Arrows indicate the mean direction of flattened trees. From Krinov (1966). Reprinted with permission of Pergamon Press.

the explosion. Painstaking soil research involving density and magnetic separations recovered numerous black metallic spherules, less than 0.10 mm in diameter, which were irregularly distributed in the area due to wind conditions. Krinov and other Soviet scientists concluded that "the nucleus of a small easily decomposed comet that entered the Earth's atmosphere" caused the Tunguska explosion. Other recent and more imaginative explanations were that the Tunguska meteor was a "black hole" of antimatter, which was explosively annihilated upon contact with "normal" matter, or that it was a nuclear-powered alien spacecraft that went out of control and disintegrated over the Tunguska basin.[111]

Beginning in the early 1950s, geologists identified an increasing number of terrestrial craters as probably having been formed by the impact of giant meteorites; the total is now about seventy. Canadian scientists of the Department of Energy, Mines and Resources, led by C. S. Beals, dominion astronomer, have been particularly active, employing aerial photographs to detect circular features with raised rims (fig. 57; table 9). That the depression at Sudbury, Ontario, was originally a crater 50 km in diameter and 3 km deep is intriguing because large deposits of nickel, iron, and copper are located at the site. Subsequent deformation and metamorphism enlarged the crater to its present size of 100 km in diameter. One theory is that the meteorite impact caused magmatic activity, which brought the metals close to the Earth's surface.[112] In addition to aerial surveys, structural, magnetic, and seismic

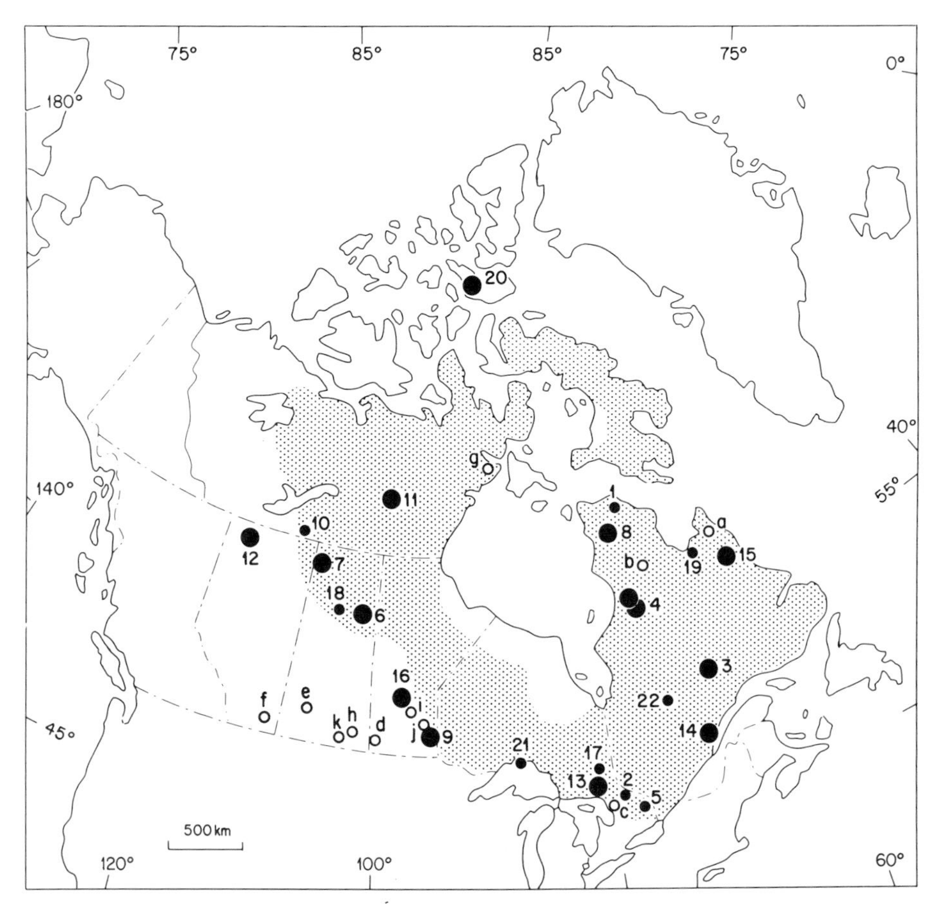

Fig. 57. Map of Canada showing the locations of probable impact craters (solid circles), where evidence of shock metamorphism has been found. Possible impact structures (open circles) are sites that possess a number of characteristics compatible with a meteorite impact origin, but where shock metamorphism has not been recognized, most likely because of deep erosion or because the material is buried beneath a thick sequence of later sedimentary rocks or many meters of water. Courtesy Earth Physics Branch, Energy, Mines and Resources, Ottawa, Canada.

TABLE 9

Canadian Impact Structures, January 1984

		Map ref.	Name	Diam. (km)	Age (m.y.)
A	Probables	1	New Quebec	3.2	~ 5
	(solid circles)	2	Brent	3.8	450 ± 30
		3	Manicouagan	100	210 ± 4
		4	Clearwater West	32	290 ± 20
			Clearwater East	22	290 ± 20
		5	Holleford	2	550 ± 100
		6	Deep Bay	12	100 ± 50
		7	Carswell	37	485 ± 50
		8	Lac Couture	8	385 ± 45
		9	West Hawk Lake	2.7	100 ± 50
		10	Pilot Lake	6	~ 440
		11	Nicholson Lake	12.5	< 400
		12	Steen River	25	95 ± 7
		13	Sudbury	140	1850 ± 3
		14	Charlevoix	46	360 ± 25
		15	Mistastin	28	38 ± 4
		16	Lake St. Martin	23	225 ± 40
		17	Wanapitei	8.5	37 ± 2
		18	Gow Lake	5	< 250
		19	Lac La Moinerie	8	~ 400
		20	Haughton	20	~ 20
		21	Slate Islands	30	~ 350
		22	Ile Rouleau	4	< 300
B	Possibles	a	Merewether	0.2	~ 0.01
	(open circles)	b	Lac Kakiattukallak	6	600
		c	Skeleton Lake	3.5	450 ± 50
		d	Hartney	6	150 ± 30
		e	Elbow	8	~ 80
		f	Eagle Butte	10	~ 40
		g	Meen Lake	4	< 600
		h	Dumas	4	~ 70
		i	Highrock Lake	2.9	< 435
		j	Poplar Bay	3	100 ± 50
		k	Viewfield	2.5	~ 200

Note: Courtesy of the Earth Physics Branch, Energy, Mines and Resources, Ottawa, Canada; m.y. = millions of years.

indications and evidence of shock metamorphism in nearby rocks aid geologists in confirming that a crater resulted from meteorite impact. For example, coesite is a dense polymorph of silica that is produced at pressures between about 25 and 100 kilobars, and stishovite is an even more dense polymorph of silica that has been subjected to pressures of over 100 kilobars. Both phases are reliable indicators of metamorphism caused by intense impact. E. C. T. Chao and his coworkers in the early 1960s first recognized both in the Coconino sandstone at Meteor Crater.[113]

Lunar explorations gave scientists much information on impact craters—in particular, their ages, which had previously only been estimated by crater density distributions. Radiometric dating of the rocks brought back by the Apollo astronauts revealed that the number of large meteorites striking the Moon and producing craters in the first half billion years of Solar System history was about one hundred times the present flux. By extension, geologists assume that the Earth in that period was subject to the same heavy bombardment. Evidences of these very early terrestrial craters no longer exist. The Sudbury crater is one of the oldest; an estimate of its age is 1.8 billion years.[114]

The investigations of impact craters as well as studies of lunar rocks also yielded new insights on the origin of tektites. There was a consensus in the early 1960s that the absence of any cosmogenic nuclides in tektites ruled out any extraterrestrial source other than the Moon. Thereafter, scientists detected coesite and an impact metamorphic product of zirconium silicate, called baddeleyite, in certain tektites, and also iron-nickel spherules in philippinites, providing evidence that they might be impact glass ejected from meteorite craters. Then, they identified the Ries crater in southeast Germany as the parent crater of the moldavite-strewn field 300 km away, and Lake Bosumtwi as the parent crater of the Ivory Coast tektite field, also about 400 km distant. Radiometric dating of the craters and of the nearby tektites indicated that in both cases the craters and the tektites were formed at the same time. Also, the Apollo missions did not return any silica-rich rocks from the Moon, which cast considerable doubt on a lunar origin of tektites. Thus the majority of geochemists came to view tektites as terrestrial impact glass.[115]

John A. O'Keefe at NASA/Goddard Space Flight Center, however, strongly disagreed with this assessment. One of his major arguments against the impact theory of origin was that the "instantaneous formation of dense, water-free glass from ordinary sandstones . . . seems indefensible." Further, he pointed out that the problem of how tektites were launched from their supposedly parent craters had not been solved. His own view was that tektites were ejected from lunar volcanoes.[116]

Despite the lack of unanimity among scientists concerning the origin of tektites, in the 1970s there was sufficient evidence that the impact of giant meteorites had in minor ways altered the Earth's surface. The major changes, geologists agreed, were due to processes that operated gradually over time. Thus, they offered alternative theories for the ending, 65 million years ago, of the Cretaceous geological period (the age of dinosaurs) and the beginning of the Tertiary period (the age of mammals), during which transition an estimated 70 percent of the living species in the oceans and on land became extinct. The last great retreat of the oceans from the continents, accompanied by climatic changes; the reversal of the Earth's magnetic field; or radiation from a supernova were possible explanations of the transition.

The hypothesis for the Cretaceous-Tertiary (K/T) transition, published in early 1980 by a group at the University of California, Berkeley, led by Luis W. Alvarez, startled geologists and paleontologists alike.[117] An asteroidal body about 10 km in diameter, they postulated, had crashed into the Earth. The impact injected into the atmosphere pulverized rock having a total mass about sixty times as large as the object. Part of the dust remained in the stratosphere and blanketed the entire planet for several years, causing prolonged darkness and suppressing photosynthesis, and resulting in the mass extinction of species. Their evidence for the catastrophe was the much higher than normal concentrations of iridium found at three K/T boundary sites in Italy, Denmark, and New Zealand. Further, the chemical composition of the boundary clay, enriched by stratospheric dust, differed substantially from that of the clay that was mixed with the Cretaceous and Tertiary limestones. Alvarez and his associates admitted that they were unable to determine the site of the asteroidal impact, but pointed out that the chances were two out of three that it had crashed into an ocean. They also suggested that four other major extinctions in the past 570 million years might have been caused by the impact of asteroids.

The Alvarez article stimulated worldwide scientific activity and resulted in a flood of publications that has not yet ceased. The hypothesis raised three separate but related issues. First, did the iridium enrichment of the K/T boundary material at sites that had at the time been subject to the vagaries of ocean currents constitute sufficient evidence of the collision and disintegration of a large extraterrestrial object? Second, could such an impact cause the extinction of a huge number of both land and water species, including dinosaurs, by the mechanism that the Alvarez group suggested? Third, was there any evidence to support the suggestion that asteroid impacts had resulted in other extinctions of life?

By 1984 scientists in several different laboratories had examined K/T boundary cores from fifty different sites worldwide, and had found iridium enrichment in all but two.[118] Studies also revealed enrichment of the clay in noble metals, which were present in cosmic abundances.[119] Moreover, two of the sites giving evidence of the iridium anomaly were in continental sedimentary rocks of the Raton Basin in northeastern New Mexico, which were deposited under freshwater conditions.[120] This finding weakened any contention that ocean currents might be responsible for the anomaly. Also, Bruce F. Bohor and his colleagues at the U.S. Geological Survey, Denver, reported a most significant determination.[121] They made a meticulous mineralogical examination of the K/T boundary claystone at Brownie Butte in east central Montana. In the quartz grains that they separated out, they discovered shock metamorphic features that were directly analogous to those found in rocks associated with meteorite impact craters, and they also saw indications of stishovite, the high-pressure silica polymorph. Subsequently, they found shock features in quartz grains from several other sites.[122]

Much of the discussion after 1980 centered on the possible processes that the impact of an asteroidal body would trigger, thus causing extinction. Here, the differences in points of view were considerable. Paleontologists pointed out that the extinction of species was gradual, some not dying out for a few million years, which ruled out the idea of a sudden catastrophic ending.[123] Even scientists who reported iridium anomalies remarked that the hypothesis of the Alvarez group was flawed, because the darkness they postulated could have lasted for no more than a few weeks; thus photosynthesis would have been suppressed for a much shorter time than three years.[124] Kenneth J. Hsü at the

Swiss Federal Institute of Technology, in collaboration with about twenty other scientists from a dozen research institutions in Europe and the U.S., suggested in 1982 the sequence of events that the impact possibly triggered.[125] Immediate extinction of species at the ocean surfaces probably resulted either from cessation of photosynthesis for a limited time, due to atmospheric dust; from poisoning of marine plankton by cyanide from a comet (which Hsü favored), or from other poisonous heavy metals; or from the flash heating of the atmosphere during the entry of the large body. The absence of plankton caused a break in the ocean food chain and resulted in a substantial increase in atmospheric carbon dioxide during the next 50,000 years. Excess atmospheric carbon dioxide caused a temperature rise, and thermal stress both in the oceans and on land led to the extinction of many species over the ensuing tens of thousands of years. In support of this schema, the Hsü group provided evidence from one site that there had been a dramatic decrease of calcium carbonate in the sediments just above the K/T boundary and that it returned to normal values about 350,000 years after the beginning of the Tertiary period. Also, anomalies in oxygen isotope values indicated that there had been a temperature increase during the transition.

Such reports did not convince some geologists and paleontologists. Charles B. Officer and Charles L. Drake of Dartmouth College, for example, insisted in 1983 that studies of other boundary cores indicated that the calcium carbonate decrease was apparently not a worldwide event. The fossil sequence at various boundary sites showed, also, that the transition from the Cretaceous to the Tertiary occupied a longer span of time in some places than in others. In another article, they argued that the geological evidence favored the deposition of iridium over a period of up to 100,000 years, which indicated that the enrichment probably resulted from extended and intense volcanic activity. The shock metamorphic features found in the boundary material by Bohor's group, they thought, could have been due to overpressure that was associated with the magmatic activity.[126]

Meanwhile, several scientists were attempting to determine whether iridium enrichments occurred simultaneously with other extinctions. In 1982 R. Ganapathy, a research scientist at J. T. Baker Chemical Co., reported finding a high iridium content in boundary material deposited 34 million years ago, separating the Eocene and Oligocene epochs. At that time, Ganapathy wrote, five major species of Radiolaria became extinct; also, the time coincided with the age of North American micro-tektites, several billion tons of which have been found in strewn fields in the south and east of the United States.[127] However, other scientists did not detect any iridium anomalies in samples taken across two late Cambrian boundaries, both of which were associated with species' extinctions.[128]

Despite sharp criticism, the continued confirmation of the existence of iridium enrichment at the K/T boundary has led a number of geologists and geochemists to believe that a large extraterrestrial mass—in all likelihood a comet—collided at that time with the Earth. There is not a consensus, however, as to what processes the event set in motion that led to the extinction of numerous species. Nor is the present evidence sufficient to conclude that the impacts of comets or asteroids resulted in species extinctions at other times, although statisticians are now at work attempting to correlate extinctions with recurrent catastrophic events.[129]

If the impact of giant meteorites was instrumental in bringing about mass extinctions, it follows that they influenced the course of the evolution of life on our planet. By the same token, the presence of complex amino acids in meteorites has led some scientists to conjecture that meteorites could have brought life prototypes to Earth. One fact is certain, however: scientists will be investigating both hypotheses for many decades to come.

Notes

1. DISBELIEF

1. Berwerth (1916), 311–312 and (1918), 1–2 attributed the disbelief to the Christian interpretation of the fall of stones as omens of God's wrath or of approaching evil, which eighteenth-century scientists countered by denying falls altogether. Paneth (1940), 1–2 placed the onus of blame on the French Academy of Sciences, which, while trying to eradicate superstition, refused to credit well-authenticated falls. McCall (1973), 17 stated that there was "considerable justification for this scepticism" because of the human proclivity to embroider tales.
2. See Menzel (1976), 17, who, in support of his belief that *Physics Today* should not publish material concerning UFOs, wrote: "I predict that 'Party-Liners' [UFO adherents] will send vigorous protests, perhaps citing the notorious boo-boo of the French Academy in 1790, when they refused to recognize the cosmic origin of meteorites—a decision they reversed only 13 years later." Menzel's statement about the French Academy is erroneous in several respects. For UFOs and the present scientific community, see Sagan and Page, eds. (1971) and Cowen (1980).
3. Westrum (1978), 461–493. The abstract reads: "In making decisions about the reality of alleged anomalous events, scientists are likely to weigh both the a priori plausibility of what is alleged and the credibility of the reports which reach them. The present paper is an attempt to examine the anomaly reporting processes which led to the scientific recognition of the reality of meteorites in the eighteenth century. It is shown that scientists fail to make realistic assumptions about anomaly reporting, and that this failure affects the accuracy of decisions made about anomalies. The treatment of reports about alleged anomalous events is further shown to be related to the scientific community's concerns about protecting its internal processes from external interference. The recognition of meteorites took place only when the savants of the eighteenth century 1) found a way of evaluating the reports, 2) devised a theory to explain them, and 3) received unimpeachable eyewitness testimony of their occurrence."
4. Sambursky (1962), 112, 160; Pallulla, *DSB* I:267–281.
5. Lee (1952), 29–33, 233–235.
6. Corcoran (1971), 75–77.
7. Rackham (1938), I:239.
8. Hicks (1925), II:641.
9. Woodward (1695), 208–209.
10. I. Newton (1952), 380.
11. Debus (1965).

12. Lemery (1700).
13. Dortous de Mairan (1733). Briggs (1967) gave a complete account of the theories of Halley, Dortous de Mairan, and Euler.
14. Canton (1753), 357–358.
15. Labaree (1959–1971), 4:488–489.
16. Hooke (1665), 243.
17. Hall and Hall, eds. (1962), 374–377.
18. I. Newton (1952), 368.
19. Brewster (1837), I:203–208.
20. Chladni (1794), 56–57.
21. Rackham (1938), I:285–286.
22. Dryden, trans., and Clough, reviser (1932), 531–532. All the accounts give evidence that a meteorite fell, although no fragment survives. Burnet (1961), 252 dated the fall in 468–467 B.C., corresponding to Chinese records of a cometary appearance, thought to be Halley's comet, in 467 B.C.
23. Lee (1952), 55.
24. Ibid., 287. Adams (1954), 81 stated that Aristotle believed that stones were generated in the clouds, and were "cast down upon the earth with great violence and hence are called thunderbolts." But there must be some error, because Lee's translation does not include any such passage. Aristotle wrote that the forcible ejection of the dry exhalation from a condensing cloud of moist air produced thunder (*brontes*). A thunderbolt (*keraunos*) was the result, if a "large quantity of wind of fine texture is squeezed out." He then distinguished between two kinds of thunderbolts: one which moved so rapidly through an object that it neither burned nor blackened it; the other moving more slowly so that it scorched the object but still did not burn it. In the *Meteorologica,* Aristotle used the word thunderbolts (*keraunos*) in seven places, in most instances in the same context as firewinds, hurricanes, and whirlwinds. Not once in these passages does the word stone (*lithos*) occur, which is the word used in the account of the Aegospotami fall.
25. Fraustadt and Prescher, trans. and eds. (1956), III:142.
26. Holmyard and Mandeville, eds. and trans. (1927), 23–26.
27. Fraustadt and Prescher, trans. and eds. (1956), III:162.
28. Aldrovandi (1648), 23. Aldrovandi's lectures date from about 1600, and his mineralogical writings were collected and published in 1648.
29. Boodt (1744), 36. His *Gemmarum et Lapidum Historia* was first published in 1609, and was reprinted in many editions and translations until the mid-eighteenth century.
30. Adam and Tannery, eds. (1897–1913), VI:310.
31. Rohault (1672), II:367–369. Gronberg (1772) gave an account of the various hypotheses concerning thunderstones, and Cogels (1907) published a comprehensive bibliography.
32. Lemery (1700), 106–107.
33. Alembert and Diderot, eds. (1757).
34. Wallerius (1778), II:609. The first edition was published in 1763 in Berlin by F. Nicolai.
35. Labaree (1959–1971), 14:57.
36. Fougeroux de Bonderoy et al. (1777).

37. Gesner (1565), 58–70.

38. Balfour (1929), 41–42; Heizer (1962), 260–261.

39. Jussieu (1723).

40. Mahudel (1745).

41. Romé de l'Isle (1767), II:331; Haüy (1801), IV:89–93.

42. Rackham (1938), I:283–285.

43. Brewster (1819).

44. Ehrenberg (1839).

45. Frérét (1746). Descartes attempted to explain the fall of milk, blood, small animals, etc., by stating that matter dropped from the clouds merely looked like such material (Adam and Tannery, eds., 1897–1913, VI:321).

46. Celis (1788). See also Alembert and Diderot, eds. (1757).

47. Buchwald (1975), I:38–40.

48. Blagden (1784). Keay (1980) summarized twentieth-century theories.

49. Halley (1714–1716*a*), 162.

50. Pringle (1759).

51. Le Roy (1771).

52. Joule (1848).

53. Bertholon (1787), II:14–30.

54. Clap (1781).

55. Blagden (1784), 224–230.

56. Olmsted (1834), 155.

57. Rittenhouse (1786).

58. Barham (1718); Chladni (1794), 28–29.

59. Hutchison et al. (1977), 71.

60. Stütz (1790).

61. Chladni (1819*b*), 256.

62. Howard (1802), 200.

63. Romé de l'Isle (1783), 167; Haüy (1801), III:65, IV:3–6.

64. Hermstadt (1797); Howard (1802), 188–189.

65. Proust (1799).

66. Fouchy's 1769 memoir (1772) mentioned Aire-sur-la-Lys.

67. Berwerth (1916), 312; Paneth (1940), 3–4; McCall (1973), 22; Westrum (1978), 464; Bigot de Morogues (1812), 166–168.

68. See Guerlac, *DSB* VIII:66–68; Rappaport (1967); and Meldrum (1933), 419 for details of Lavoisier's early career.

69. Fougeroux de Bonderoy et al. (1777). The chemists did measure the quantities involved in their wet analysis, and the calculations yielded 8.5 percent sulfur, 36 percent iron, and 55.5 percent vitrifiable earth.

70. Kobell (1864), 134–136.

71. Howard (1802), 187.

72. Withering (1790).

73. Gauthier de la Peyronne, trans. (1794), VI:346–356. See also Valmont de Bomare (1762), II:132–133; Charpentier (1778), 343–344; Bergman (1783), 95; Cronstedt (1788), II:720–722; Gallitzen (1792), 87; and Haüy (1801), IV:1–4.

74. Güssman (1785), (1803).
75. Vandermonde et al. (1786). For a complete account, see C. S. Smith (1964).
76. Bigot de Morogues (1812).
77. Blumhof (1816), 311–312; Westrum (1978), 488.
78. Buchwald (1975), I:38–40.
79. Halliday et al. (1984), on the basis of photographic observations, estimate that 8 events per year occur in an area of 10^6 km^2.
80. Drapiez, ed. (1837), I:81.
81. For example, Paneth (1940), 3: "Leading in this scepticism was the French Academy of Sciences, then the highest known authority in matters scientific"; M. Romig (1966), 13: "It is well known that the Academy maintained in the face of overwhelming evidence that the fall of stones from the sky was impossible . . . before Biot's report"; and Westrum (1978), 469: "For the verdict of the *Académie* was widely viewed as being the official viewpoint of science itself, not just that of a single academy."
82. Hahn (1971), 145.
83. Fouchy (1772).
84. Stütz (1790).
85. Tata (1794).
86. Saint-Amans (1802); Bertholon (1791).
87. *JBRS* (1754–1757), XXII:290–300.
88. Gibbon (1932), I:126.
89. Chladni (1794), 36–37.
90. Stepling (1754).
91. *Bulletin des sciences de la Société Philomatique* (1791), No. 1.
92. Lalande (1802).
93. Bingley (1796).
94. Westrum (1978), 479–485.

2. ACCEPTANCE

1. Thomson, G. [William] (1804).
2. Chladni (1799), 332; Brewster (1832*b*); Hoff (1832), 222–223; Baldanza (1965), 214.
3. Hamilton (1795).
4. Levi-Donati (1975).
5. Tata (1794), (1804).
6. Olbers to Gauss, 24 June 1824, in Schilling, ed. (1894–1909), II:321.
7. Chladni (1803), 323–324; (1819*b*), 7–8.
8. Chladni (1794), 1–2.
9. Chladni (1819*b*), 7–12; Lamétherie (1798*a*), 185–186; Fulda (1799).
10. Edward King (1796); Chladni (1798*a*).
11. Lamétherie (1798*b*).
12. Coquebert, trans. (1802–03).

13. Chladni (1801); Pictet (1801).

14. G. A. Deluc (1803).

15. BMNH, Sowerby Archives; BMNH, Meteorite file—Wold Cottage.

16. Edward King (1796).

17. Bingley (1796).

18. Anon. (1796), *Gentlemen's Magazine*.

19. Southey (1797).

20. Baudin (1798).

21. BMNH, Meteorite file—Wold Cottage.

22. Howard (1802), 175–179.

23. See D. W. Sears (1976) for Howard's biography.

24. Quoted in D. W. Sears (1976), from *Copy Minute Book of the Royal Society* 37 (1800–1802), 1 December 1800.

25. Pictet (1801).

26. On p. 204 of Howard's report (1802), Bournon stated that Born obtained the "Bohemian iron" as a gift from the Freiberg Academy. Charpentier (1778), 343–344 described a piece of "native iron" in the collection of the Freiberg Academy, which was apparently a stony iron. The locations of the Steinbach finds—Johanngeorgenstadt and Rittersgrün—are adjacent to the border between Saxony and Bohemia, and in the wars and treaties of the eighteenth century involving Austria, Prussia, and Saxony, borders constantly shifted. Bournon's description of the "Bohemian iron" makes it clear that it was a stony iron. Since no other meteorite finds in Bohemia fit the description, one can legitimately infer that the "Bohemian iron" was a piece of Steinbach. The only remaining problem is what happened to the piece examined by Bournon and Howard; it apparently did not go to the British Museum with the Greville (Born) collection.

27. D. W. Sears (1976) pointed out Howard's detection of thermoluminescence.

28. Howard (1802). D. W. Sears (1976) argued that the chemical analyses of meteorites—particularly Howard's—were crucial in the scientific acceptance of meteorites. They were important, but I believe that the increasing number of reported falls was more significant.

A long paragraph in Bournon's original manuscript, which followed his description of Krasnojarsk, was deleted from the published report. Bournon thought that the iron had a "powerful attraction" for water because small drops appeared on its surface, even though it lay on a table in a warm room close to a fire. The drops had a salty taste, "slightly styptic," and not the "Inky taste which is peculiar to Sulphate of Iron." A corresponding paragraph in Howard's original manuscript was also deleted. After he placed the iron in distilled water for a few hours, Howard analyzed the solution and found that iron chloride was present (microfilm of Royal Society original mss., BMNH, Mineralogy Department).

The question of whether or not iron chloride—the mineral called lawrencite—exists in meteorites has provoked controversy since C. T. Jackson (1838) observed that grass-green drops appeared on the surface of the Lime Creek iron. C. U. Shepard (1842), E. Harris (1859), and C. L. Buchner (1859) held that the chlorine was added by immersion in groundwater. However, J. Lawrence Smith (1855*b*) and (1874) claimed that he found solid iron chloride in the Tazewell and Deep Springs irons. Reichenbach (1857*b*) accepted Smith's claim, and J. W. Mallet (1871) rejected it. G. Daubrée (1877) found iron chloride in the Ovifak iron, unaware that it was a pseudometeorite, and named the mineral "lawrencite" in honor of Smith. More recently, Buchwald (1975) has termed lawrencite "a cosmic ghost" that probably does not exist, at least in iron meteorites. But K. Keil (1968), 6968 provided microprobe evidence for the presence of lawrencite in the Indarch enstatite chondrite. (I am grateful to Dr. Roy S. Clarke, Jr., for giving me, in a private communication, the gist of the arguments of contemporary meteoriticists for and against the existence of lawrencite.)

29. There is also a question about Howard's Siratik sample. Buchwald (1975) examined many fragments of Siratik in museum collections and found that they could be divided into two groups: (1) reheated and hammered pieces of original meteoritic hexahedrites containing about 5.6 percent nickel; and (2) cast-iron pieces having a nickel content of less than 0.10 percent, indicating a terrestrial origin. Buchwald included in the latter group the two specimens of Siratik, now in the British Museum (Natural History), which were available to Howard, and which Buchwald described as "completely mutilated." Was there a third piece, now lost, which contained the 5 percent nickel reported by Howard? Or was Howard's analysis inaccurate? According to later analyses of Krasnojarsk, Howard's determination of 17 percent nickel was much too high, which casts some doubt on his Siratik result.

30. Drée (1803*a*).

31. Lamétherie (1803*c*).

32. Biot (1807), 226.

33. J. A. Deluc (1801).

34. G. A. Deluc (1801–02).

35. Patrin (1801).

36. Pictet (1802).

37. Patrin (1802).

38. Bournon (1803).

39. Patrin (1803).

40. Biot (1802*a*).

41. Drée (1803*d*). Charles Barthold (1800), a professor at the École centrale of the Haut-Rhin, had analyzed the Ensisheim stone. He reported finding 42 percent silica, 20 percent iron, 17 percent alumina, 14 percent magnesia, 2 percent sulfur, and 2 percent lime. He concluded that it was an iron-bearing argillaceous stone, which was the result of the decomposition of the primitive rocks of a nearby mountain from which it had become detached. Howard censured Barthold for not testing for nickel, and he also questioned the presence of such a large amount of alumina. When Fourcroy read this in Howard's report (1802), he immediately analyzed the stone and found 2.4 percent nickel and no alumina. Vauquelin confirmed Fourcroy's results.

42. Sage (1803*a*); Vauquelin (1802–03); Tonnelier-Breteuil (1802); Drée (1803*a*), (1803*b*), (1803*c*), (1803*d*).

43. G. A. Deluc (1803). Substantially the same memoir appeared earlier in *JM* 13 (1802).

44. Lambotin (1803); Brewster (1832*b*).

45. Fourcroy (1803).

46. Crosland (1967), 214.

47. Biot (1807), (1803*c*).

48. For more details about the analyses of Woodhouse and Silliman, see Greene and Burke (1978), 24–26, 97–98.

49. Mathews (1908), 210–211.

50. Ibid., 212.

51. Schofield (1966), 320.

52. These letters are in the Ellicott papers, Library of Congress. I am grateful to Professor John C. Greene for furnishing me the citations while his book *American Science in the Age of Jefferson,* Ames: Iowa State University Press, 1984, was still in draft.

53. S. L. Mitchill, *A Discourse on the Character and Services of Thomas Jefferson,* 35–36, New York: G. & C. Carvill, 1826.

54. Lipscomb (1905), 11:440–442.

55. Bigot de Morogues (1812), 319; Haüy (1822), III:557; Brewster (1819), (1832*b*).

56. Lavoisier (1790), 30–32.

57. Ibid., 30.

58. Lamétherie (1798*a*).

59. Ibid., (1803*a*).

60. Gilbert (1803).

61. Brewster (1832*b*).

62. Soldani (1808); Forster (1823); Wrede (1803); Serres de Mesplès (1813); Séguin (1814); Beauford (1802); Salverte (1803); Murray (1819); Marichal (1812).

63. Marvin (1978), (1979). Two members of the committee, Robert Hare and Joseph Cloud, were members of the committee that recommended publication of the Silliman-Kingsley memoir two years previously, so that they accepted meteorite falls. Hare had invented the oxy-hydrogen blowpipe, and it is probable that he thought the ignition of hydrogen gas would not produce the result postulated by Spalding.

64. Atkinson (1819).

65. Ritter (1803), (1804), (1805).

66. Egen (1822); Butler (1829).

67. Hoff (1835).

68. Shepard (1848*c*).

69. Anon. (1854), *Southern Literary Messenger;* Mrs. G. S. Silliman (1859).

70. J. L. Smith (1855*b*).

71. Egen (1822), 376. A number of commentators have reported that the atmospheric hypothesis had a very short life span (e.g., Berwerth 1916, 315). On the contrary, it enjoyed scientific support for over a half century.

72. Mrs. G. S. Silliman (1859). I am grateful to Dr. Michele Aldrich for informing me of the existence of this privately printed book, which was reprinted in the *Edinburgh New Philosophical Journal* (1862), 16:227–248, and also for pointing out that Mrs. Gold Selleck Silliman was Hepsa Ely Silliman. Mrs. Silliman was obviously well acquainted with contemporary chemical theory, and cited Benjamin Silliman's chemistry text in her treatise.

73. Forster (1823); Ideler (1832). Ideler edited *Aristotelis Meteorologia,* 2 vols., Leipzig, 1824–1826.

74. Home (1972) gave the details of this activity. P. M. Terzago in 1664 offered the explanation that a stone falling from the air, which allegedly killed a Franciscan monk in 1650 in Milan, had been ejected from the Moon. Late eighteenth- and early nineteenth-century scientists did not cite Terzago's report. See the journal *Cosmos* (1850), 1:121.

75. Lichtenberg to W. Herschel, June 1787, *GCL* IV:708–709.

76. Home (1972); Maskelyne (1794).

77. Lichtenberg to W. Olbers, 8 February 1783, *GCL* IV:843–844. Humboldt (1872), I:118 stated in a footnote that Chladni's book appeared two months before the Siena fall.

78. Olbers to Gauss, 25 December 1802, in Schilling, ed. (1894–1909), II:121.

79. Olbers (1803*a*).

80. Chladni (1819*b*), 7–8; Benzenberg (1834), 1.

81. Lamétherie (1803*c*), 26–27.

82. Laplace (1802).

83. Biot (1802*a*), (1803*a*), (1803*b*); Poisson (1803); Hutton (1812).

84. Gauss to Olbers, 4 January 1803, in Schilling (1894–1909), II:123.

85. Olbers (1803*a*), (1803*b*).

86. Chladni (1803).

87. Drée (1803*c*); Ende (1804); Klaproth (1803*a*); Kratter (1825), 78.

88. Chladni (1805*b*), 260–271.

89. Chladni (1808).

90. Blumhof (1816).

91. Chladni (1818*b*).

92. Laplace (1878–1912), 6:254, Book IV, Ch. V.

93. Ibid., 6:483, Book V. Ch. VI.

94. Brewster (1820).

95. J. L. Smith (1855*b*), 165–166.

96. Brandes (1805).

97. Olbers (1837*a*), 55.

98. Poisson (1837).

99. The title of Benzenberg's book (1834) gives a hint of its peculiar content. Both Chladni and Brandes, who would have been able to refute Benzenberg, were dead when the book was published.

100. Berzelius (1834).

101. J. L. Smith (1855*a*), 165–174.

102. Gould (1859).

103. Brewster (1832*b*).

104. Silliman and Kingsley (1809).

105. Day (1810). Neither Day nor Silliman and Kingsley mentioned the theory of cosmic origin, which testifies to the low esteem in which it was held at the time.

106. Farey and Bevan (1813).

107. Chladni (1817*a*), 101; (1823), 257.

108. Silliman Family Papers, Yale University Library, "Lecture Notes on Meteorites," by Benjamin Silliman, 57.

109. Day (1810), 167.

110. Bowditch (1815). He thought that the evidence "strongly" favored the Earth satellite theory, but that the similarity of meteorites weighed against it.

111. Chladni (1818*a*), (1818*b*). The statement appeared in *Taschenbuch für die gesammte Mineralogie* (1817), 11:256.
112. Olbers (1837*a*).
113. J. L. Smith (1855*b*), 340–342.
114. Drée (1803*c*), 426.
115. Brewster (1832*b*).
116. Chladni (1803), 327.
117. Benzenberg (1834), 5–11; Olbers (1837*a*). The observations and calculations are reported in German *Meilen,* which I have converted to kilometers.
118. Lichtenberg to Benzenberg, 3 November 1798, *GCL* IV:998–999.
119. Chladni (1803), 325–326.
120. Chladni (1817*b*).
121. Ibid., (1817*a*).
122. Ibid., (1818*a*).
123. Olbers (1837*a*); Bessel (1839).
124. Gilbert (1805).
125. Chladni (1805*b*), 272–273.
126. Bigot de Morogues (1812), 319; Haüy (1822), III:557.
127. Brewster (1832*a*).
128. Olbers (1837*a*).

3. MATHEMATICAL ASTRONOMY AND STATISTICS

1. Olmsted (1834), 363–364.
2. *The Old Countryman,* New York, 20 November 1833; Rev. 6:12–17.
3. Olmsted's estimate was 207,840; Arago estimated 240,000.
4. Denison Olmsted Papers, Yale University Library, "Lecture Notes," Box 10, Folder 296, p. 215.
5. Olmsted (1836*b*).
6. Olmsted (1834). Accounts of Olmsted's "explanation" differ. C. S. Lyman (1868) does not mention Olmsted's misconceptions, while Simon Newcomb (1868), 42–43 viewed the fate of Olmsted's hypothesis "as a warning to philosophers . . . to assign no cause which is not a real phenomenon." Curiously, Olivier (1925), 27, while noting that a number of Olmsted's conclusions were erroneous, wrote that Olmsted's view that the nebulous body had its aphelion near the orbit of the Earth "proved absolutely correct."
7. Denison Olmsted Papers, Yale University Library, "Lecture Notes," Box 10, Folder 296, p. 214.
8. Twining (1834), (1835).
9. Bache (1835*a*).
10. Bache (1835*b*).
11. Bache (1836); Olmsted (1836*a*).
12. Humboldt (1819); Ellicott (1804); Olbers (1837*a*).
13. Walker (1843), 120.

14. Olbers (1837*a*).

15. Olbers (1837*a*); Walker (1843), 139–140. Locke reported his observations in two letters to the *Cincinnati Daily Gazette,* so Arago and Quetelet can hardly be blamed for not knowing about his discovery.

16. Olbers (1837*a*).

17. Biot (1837).

18. Olmsted (1837).

19. Denison Olmsted Papers, Yale University Library, "Lecture Notes," Box 10, Folder 286, pp. 218 ff. Walker (1843), writing in 1841, stated that he believed that Olmsted still held the theory. Olmsted would undoubtedly have been appalled to know that he inspired Edgar Allan Poe to write the following poem:

MONODY ON DOCTOR OLMSTED—IF DEAD—DAMNED

If this prime ass is with his brother worms—
Forgive the license of the clASSic terms!—
I hope they'll make a "black-board" of his face
And make it minus all but its grimace,
That devilish simper which he frequent wore
When, with the glibness of a saint, he swore,
That, "His last work was plainer far than dirt"—
As tainting too, thou muddy-headed squirt!
Plain—dead level—flat—they are the same—
At him, ye worms! till appetite is lame.

The "Monody" was found in 1938, and one of Poe's editors thinks that Poe composed it in early 1846, after Olmsted had received an honorary L.L.D. from New York University, at which occasion Poe was asked to read a poem. Because the "Monody" was written in a minute hand that Poe used frequently in the 1830s, however, the editor concedes it may have been composed at a time when Poe learned that Olmsted had fixed the height of the nebulous cloud at precisely 2,238 miles above the Earth on the basis of inexact data. Poe, the editor states, disliked pretentious scientists, and he surmises that Poe was inebriated at the time of the poem's composition. See Thomas O. Mabbott, ed., *Collected Works of Edgar Allan Poe,* Belknap Press, 1962, I:500–501.

20. Olbers (1838).

21. Lovering (1839).

22. Quetelet apparently began to have doubts that showers were cosmic phenomena; see H. A. Newton (1863*a*).

23. Olbers (1837*a*).

24. Herrick (1839*a*).

25. Herrick (1839*b*).

26. Herrick (1838*b*).

27. Walker (1843), 109.

28. Ibid., 88–93.

29. Ibid., 95–96.

30. Olmsted (1834).

31. Hitchcock (1834).

32. Olmsted (1834).

33. Olbers (1837*a*).

34. Agassiz (1840).

35. SIA, Charles U. Shepard Papers, Box 2, Folder 15.

36. Fort (1919), particularly pp. 14, 47, and chaps. IV and XXVII.

37. Quetelet (1839); Herrick (1841); Chasles (1841); E. C. Biot (1848); Needham (1959), 184, 425, 430, 433. Edouard Biot died in 1850 at the age of 47; Needham considered that his early death was a distinct loss to the history of science in China.

38. H. A. Newton (1863*a*), 149.

39. Ibid., (1864), 378–381.

40. Ibid., (1864); Kirkwood (1867), 22, thought H. A. Newton's idea of a short period orbit was wrong, because none "of nearly 100 known bodies which revolve about the sun in orbits of small eccentricity has a retrograde motion."

41. Olivier (1925), 33. For Walker's value, see Walker (1843), Table I.

42. H. A. Newton (1862).

43. *BAAS* (1868), 396; Schiaparelli (1867).

44. Olivier (1925), 33. Schiaparelli's complaint seems justified. For example, Newcomb (1868), 48 wrote: "Soon after the discovery of the connection of the November meteoroids with Tempel's comet [1866 I] was made known, an Italian astronomer, M. Schiaparelli, was fortunate enough to identify the orbit of the August meteoroids with that of a telescopic comet discovered in 1862." Thus, just a year later, Newcomb reversed the sequence of the discoveries.

45. *BAAS* (1868), 394–396.

46. *BAAS* (1868), 397; Le Verrier (1867); Schiaparelli (1867); Oppolzer (1866–67); Peters (1866–67); Oppolzer (1867*a*).

47. Adams (1867).

48. Weiss (1866–67); Arrest (1867), Galle (1867).

49. Lockyer (1890), 147.

50. N. Nordenskiöld (1823).

51. Shepard (1850*a*), 154–157.

52. Greg (1854), 332–333.

53. Kesselmeyer (1860); Prior (1923); Hey (1966); Buchwald (1975), II:764.

54. Chladni (1819*b*), 93; Olbers (1837*a*). Schreibers based his estimate on the fact that in the 26 years from 1790 through 1815 there were 10 reported falls in France. He calculated that the area within which the falls occurred was 1/1800th of the Earth's surface. Hence, the total number of falls in the period was 18,000, equivalent to 700 per year. See Benzenberg (1834), 39.

55. Greg (1854), 331.

56. Reichenbach (1858*d*).

57. SIA, C. U. Shepard Papers, undated manuscript, Box 2, Folder 15.

58. Greg (1854), 334–341.

59. *BAAS* (1860), 26.

60. Mason (1962*b*), 24–25.

61. Haidinger (1867*a*), (1867*b*).

62. Halliday et al. (1984).

63. In this discussion, I have depended on Buchwald (1975), I:5–32.

64. Prevost (1803*b*). The passage is from Lucretius, *De rerum natura,* VI:177.

65. Biot (1803*d*). The other relevant passages are Lucretius, *De rerum natura,* V:305; Ovid, *Metamorphoses,* Book 2, xi, and Book 14, xvii; and Virgil, *Aeneid,* IX:486.

66. Pictet (1803).

67. Prevost (1803*c*).

68. Day (1810), 173.

69. Higgins (1818), (1819).

70. Buchner (1859), 169.

71. Fox (1969) gave a detailed account of Mollet's work and of the development of the fire piston.

72. Benzenberg (1811), quoted in Haidinger (1861), 353.

73. Grotthus (1821), 343–345.

74. Joule (1848). Joule states in the article that he presented his conclusions at a lecture in Manchester on 28 April 1847.

Joule did not give the bases of his calculations. However, it seems certain that he employed the formula $F = mv^2 = \rho A/g\ v^2$, where ρ is the density of the air, A is the resisting area of the air, g is the acceleration of gravity, and v is the velocity of the air. If we assume an altitude of 50 miles, the density of air at sea level to be .075 pounds per cubic foot, and g (corrected for altitude) to be 31.42 feet per second^2, then:

$$F = \frac{.01 \cdot .075 \cdot .25}{31.42} \cdot 18 \cdot 18 \cdot 5280 \cdot 5280 = 53292.$$

A slight change, either in ρ or in g, will yield Joule's value. To arrive at the total BTU, Joule used the formula:

$$\frac{F \cdot s}{782} = \frac{51600 \cdot 20 \cdot 5280}{782} = 6967980.$$

For a good discussion of *vis viva,* see Thomas S. Kuhn (1959). I have been unable to determine just when meteorite scientists began to use $mv^2/2$ instead of mv^2. Lyman (1868), 139 is the first scientist I have found to use $mv^2/2$ in the context of meteorites.

75. Joule and Thomson (1857).

76. Tyndall (1857).

77. Daubrée (1864), 8, quoted in Haidinger (1869), 283.

78. Haidinger (1861).

79. J. L. Smith (1861).

80. Story-Maskelyne and Lang (1863).

81. Haidinger (1869).

82. Haidinger did refer later to the publications of Joule and Thomson (1857) and of Tyndall (1857). See Haidinger (1863).

83. Schiaparelli (1868); *BAAS* (1868), 410. See also M. Romig (1966).

84. Olivier (1925).

85. Paneth (1940), 12–13; Öpik (1933); Wylie (1940); Buchwald (1975), I:15.

86. A. S. Herschel to L. Fletcher, 5 April 1881, BMNH, Meteorite file—Middlesbrough.

87. Herschel (1881*b*).

4. NINETEENTH-CENTURY FOUNDATIONS OF METEORITE ANALYSIS

1. A sample of about 150 chemists and mineralogists who studied meteorites disclosed that about 37 percent were German, with the balance about equally divided between Austria-Hungary, France, Great Britain, the United States, and the Scandinavian countries. Two-thirds were university or college professors, and almost all the rest were museum curators, geologists holding governmental positions, or mineral collectors or dealers. About half held a Ph.D. degree or the equivalent.
2. Klaproth (1812). Levi-Donati and Clarke (1983) reported that a 14-gram specimen labeled Brianza in the Mineralogical Museum of the University of Parma, Italy, is an artificial iron, not a brecciated chondritic stone, as listed in Hey (1966).
3. Gillet de Laumont (1815).
4. For example, a copper nugget reported to have fallen in 1931 at Eaton, Weld County, Colorado, was discredited in 1973. In contrast, the Oktibbeha County iron meteorite, found in the mid-nineteenth century in an Indian burial mound in Mississippi, was considered doubtful because of its very high nickel content (60 percent), but has been proved to be genuine in the last decade; see Reed (1972) and Kracher and Willis (1981).
5. See J. L. Smith (1855*a*), 157–158.
6. Haüy (1801), III:405.
7. Haüy (1801), II:210–221.
8. Fletcher (1901), 8, 18.
9. Stromeyer's report is printed as an appendix to Rose (1825).
10. J. L. Smith (1855*a*), 152. However, there were errors. Shepard (1829) made three specific-gravity determinations of samples of "crysolite" present in the Richmond meteorite, and calculated a mean specific-gravity value that was lower than that of any of the samples. Thomas Davies, a laboratory assistant at the British Museum (Natural History), made long-division errors in calculating the specific gravities of the Cachiyuyal and Kadonah meteorites.
11. Reichenbach (1859*c*), 359–361.
12. Story-Maskelyne and Lang (1863), 48.
13. J. L. Smith (1870).
14. BMNH, Sir Lazarus Fletcher, Box 1—meteorite analysis notebooks.
15. Berzelius (1834), 9–10.
16. J. L. Smith (1853).
17. J. L. Smith (1871*a*).
18. See Jeffery (1970), 73; and W. T. Hall (1935), 225 ff.
19. Berzelius (1834), 127–129; Hutchison et al. (1977), 127.
20. Baumhauer (1845).
21. In 1866 Rammelsberg, among others, thought that the sulfide dissolution was negligible. Cited in A. A. Moss et al. (1968), 104.
22. Fletcher (1901), 10.
23. Berzelius (1834), 140–141.
24. Rose (1825).

25. In the German original Rose used the word "augite," whereas the French translation, Rose (1826) changed it to "pyroxene." Similarly, Rose's "chrysolite" and "olivine" became "peridot" in the French translation. While these changes reflect the insistence of French mineralogists at the time that the nomenclature introduced by Haüy be adhered to, they caused problems later when the various pyroxenes were more clearly distinguished.

26. Shepard (1846).

27. Reichenbach (1857*b*).

28. For a description of Sorby's early work, see D. W. Humphries, "Sorby: The Father of Microscopical Petrography," in Cyril S. Smith, ed. (1965), 17–41.

29. BMNH, Victor von Lang box—diary and crystallographic notebook.

30. Story-Maskelyne (1870).

31. G. Kurat and L. Pabst (1974).

32. G. Tschermak (1885*a*); see also Wood and Wood, trans. (1964).

33. Sorby (1886), 140.

34. Sorby (1877), 497.

35. Cohen (1894–1905), I:27; see also Perry (1944), 1–2.

36. McGucken (1969) described the development of spectroscopy in the nineteenth century in great detail.

37. Wöhler (1861).

38. Engelbach (1862). Lithium is one of the least abundant elements in meteorites.

39. Berman (1937).

40. Doelter (1914); Laurent (1846); Frémy (1856).

41. For example, the text by William Foster, *Inorganic Chemistry for Colleges* (New York: Van Nostrand, 1929), described the theory on pp. 469–472.

42. Burke (1966), 114–125.

43. Ibid., 131.

44. Rose (1852), 9.

45. Doelter (1914), 66.

46. Berzelius (1834), 30–32. Berzelius stated that Nordenskiöld had given him a fragment containing only one recognizable mineral and that the groundmass was not included. He described the mineral as a partially clear, white, foliated mineral that appeared crystalline on the surface and crumbled easily. It contained black points that were magnetic and dissolved without any emission of gas or odor. He compared the crumbled pieces to the fragile powder of vitreous feldspar. The sample that he analyzed weighed only 1.215 grams, of which 1.136 grams dissolved in aqua regia followed by sodium carbonate. He did not report the length of time that these reagents acted on the sample. His results were:

	Entire mass	Percent dissolved	Oxygen content	
Silica	0.425	37.411		19.44
Magnesia	0.344*	32.922	12.74	19.54
Iron oxide	0.325	28.610	6.51	
Manganese oxide	0.009	0.793	0.17	
Alumina	0.003	0.264	0.12	

	Entire mass	Percent dissolved	Oxygen content
Copper oxide, tin oxide, potassium and sodium	trace	trace	
Undissolved	0.079		
	1.215	100.00	

*This is obviously a typographical error and should be 0.374 to arrive at the correct total weight and percent of dissolved magnesia.

If one takes the formula for silica as SiO_3 and uses Berzelius's values of atomic weights—0 = 8.013; Si = 22.221; Mg = 12.689; Fe = 27.181; Mn = 27.716; Al = 12.716—then the oxygen content is exactly as reported. One can only guess where the error occurred.

47. Rose (1837–42).

48. Tschermak (1872), (1883).

49. Laugier (1821).

50. Rose (1825), 176–177.

51. Shepard (1846), 381.

52. Rammelsberg (1848).

53. Ibid., (1843).

54. Bonsdorff (1822).

55. Harris (1859), 32, 49. Rose (1864*b*), 106 thought that it was capricious to postulate the presence of unidentified minerals from the results of an analysis of a mixture.

56. Shepard (1846), 381.

57. Ibid., (1848*c*), 416.

58. Rose (1864*b*), 117–121; Cohen (1894), I:270–271.

59. J. L. Smith (1855*a*).

60. Kenngott (1856); Cohen (1894–1905), I:271.

61. J. L. Smith (1864). Rose (1864*b*), 117 f. called the mineral shepardite, but this name was never used.

62. Story-Maskelyne (1870); Rammelsberg (1870), 76.

63. Schreibers (1820), 70–72. Widmanstätten in 1807 sold his family's printing business in Graz and moved to Vienna, where he became director of the collection of technological artifacts being assembled by Emperor Francis I. Schreibers wrote that Widmanstätten repeated the process on sections of the Toluca, Elbogen, Lenarto, and Krasnojarsk meteorites, and that, although "tiresome duties" prevented Widmanstätten from publishing his finds, he did make them known orally. Schreibers printed several illustrations of Widmanstätten patterns, including a striking one in color of the Krasnojarsk pallasite, in his 1820 volume. The meteorites themselves were used as printing blocks. Karl A. Neumann, professor of chemistry at Prague, mentioned Widmanstätten's discovery in a report on the Elbogen iron, and postulated that a difference in nickel content caused the difference in the color of the alternating lamellae (Neumann, 1812).

64. Thomson's article appeared in Italian in the *Atti dell' Academia delle Scienze di Siena* in 1808. R. T. Gunther (1939) located this publication in writing a biographical sketch of Thomson, and M. H. Hey (1939) and F. A. Paneth (1960) both pointed out that Thomson deserved credit for at least the simultaneous discovery of the structure. C. S. Smith (1962) noted that the article bore the date 1804, and about 1965 Marjorie Hooker found that it had been published earlier in French, in the October and November 1804 issues of the *Bibliothèque Britannique;* see M. Hooker (1974); C. D. Waterston (1965), 133, n. 53. Roy S. Clarke, Jr. (1977) and Clarke and Goldstein (1978) brought Thomson's distinct priority to the attention of a wider audience. For Thomson's biography, see Waterston (1965).

65. Thomson (1804).

66. Biot (1803*c*).

67. Gillet de Laumont (1815). C. S. Smith (1962) provided an extensive critique.

68. Schreibers (1820), 84.

69. Berzelius (1833). Clarke and Goldstein (1978) gave full details.

70. Berzelius (1834).

71. Cohen (1894–1905), I:121.

72. Clarke and Goldstein (1978).

73. Cohen (1894–1905).

74. Shepard (1847).

75. Haidinger (1847), 581.

76. J. G. Neumann (1849).

77. Sadebeck (1875).

78. Buchwald (1975), I:90.

79. Glocker (1848), 336.

80. J. L. Smith (1855*a*).

81. Reichenbach (1861).

82. Rammelsberg (1870).

83. Meunier (1869), 31–34.

84. Davidson (1891).

85. Cohen (1894–1905), I:96–105.

86. Fletcher (1908).

87. Sorby (1877).

88. Ibid., 287.

89. Howard (1802), 182, 190–191, 211.

90. Proust (1805).

91. Rose (1825).

92. Shepard (1829).

93. Berzelius (1834).

94. Rammelsberg (1849), 445; Fischer (1848), 593.

95. W. S. Clark (1852).

96. Silliman and Hunt (1846), 343.

97. Bergemann (1849), 407–408.

98. J. L. Smith (1855*b*), 156.

99. Reichenbach (1861), 115.

100. Tschermak (1871), 189, 192–194; Brezina (1880). Later, in 1885, Brezina regretted his decision. Schreibers, Brezina wrote, had in fact made the first observation of these lamellae, but Schreibers wrote in such an unreadable style that his discovery was completely overlooked. Now, he [Brezina], had named the feature in honor of Schreibers's bitterest enemy. See Brezina (1885), iv–v. Reichenbach's antagonism not only to Schreibers but also to other Vienna mineralogists (Partsch, Hörnes, and Haidinger) is described in chap. 6. Schreibers's observation was made in his description of the Hraschina meteorite (1820), 1–9, but it is certainly not clear from the text that Schreibers was describing iron sulfide, a point noted by Cohen (1894–1905), I:189n.

101. Reichenbach (1862), 620–636.

102. Haidinger (1863).

103. Rose (1864*b*), 39, 40, 139, 140.

104. Rammelsberg (1870).

105. Meunier (1869).

106. Cohen (1894–1905), I:188.

107. Ibid., I:189.

108. Ibid., II:247–257.

109. Clarke and Goldstein (1978), 73–74.

110. Berzelius (1833). This article appeared originally in 1832 in the Swedish, and is cited in Clarke and Goldstein (1978), 3.

111. Berzelius (1834).

112. Shepard (1846), 380.

113. Ibid., 383.

114. Cohen (1894–1905), I:210.

115. Patera and Haidinger (1849). Haidinger proposed that the mineral Shepard termed "schreibersite" be called "shepardite," but his suggestion was not accepted.

116. Fischer (1848).

117. Shepard (1848*b*).

118. W. S. Clark (1852).

119. Shepard (1853).

120. Cohen (1894–1905), I:122.

121. J. L. Smith (1855*a*), 153. Smith based the formula on two analyses:

	No. 1	No. 2
Iron	57.22	56.53
Nickel	25.82	28.00
Phosphorus	13.92	14.86

The formula Ni_2Fe_4P, he wrote, would yield the following:

Phosphorus	1 atom	15.47 percent
Nickel	2 atoms	29.17 percent
Iron	4 atoms	55.36 percent

It is not clear from these figures what atomic weights Smith used. Modern atomic weights give 8.3 percent P; 31.6 percent Ni; and 60.8 percent Fe. Berzelius's (1836–37) atomic weights (P = 15.717; Ni = 29.622; Fe = 27.181) would give 8.5 percent P; 32.2 percent Ni; and 59.2 percent Fe. It is, of course, possible that Smith erred in the calculation.

122. Bergemann (1856).

123. Reichenbach (1857*b*), 619.

124. Ibid., (1861), 485 ff.

125. Ibid., (1862), 148–151.

126. Rose (1864*b*), 48–55.

127. Tschermak (1874).

128. Flight (1882).

129. Cohen (1894–1905), I:121–124.

130. N. Nordenskiöld (1823), 333–334.

131. Laugier (1821), 272.

132. Partsch (1843*b*). Moritz Hörnes brought Partsch's system up to date in 1862, and retained the anomalous and normal categories.

133. Laugier (1827) analyzed the Renazzo carbonaceous chondrite and concluded it was unusual only because it contained less nickel and sulfur.

134. Shepard (1846), (1847), (1848*c*).

135. Boisse (1850).

136. Ibid., 162 ff.

137. Reichenbach (1859*b*).

138. Cohen (1894–1905), III:7.

139. Rose (1863*b*), (1864*b*).

140. Tschermak (1872), (1883).

141. Brezina (1885), 156.

142. Tschermak (1884).

143. Story-Maskelyne (1863).

144. Ibid., (1875), 504.

145. Shepard (1867).

146. Daubrée (1867).

147. Ibid., (1879), 486–502.

148. For example, Lacroix (1927*a*), 419–420.

149. Meunier (1885).

150. Brezina (1885), 160; Cohen (1894–1905), III:19.

151. Brezina (1885), (1904).

5. LATE NINETEENTH-CENTURY METEORITE THEORIES

1. H. A. Newton (1886), 734.

2. Drake (1957), 49–50.

3. W. Herschel (1789), (1791); Halley (1714–1716*b*). In 1791 Herschel described what he called a "planetary nebula," in which a self-luminous "shining fluid" surrounded a central star. He postulated that the fluid might be responsible for the regeneration of the central star.

4. Huggins (1864).

5. Cappocci de Belmonte (1840).

6. S. G. Brush, "John James Waterston," *DSB* (1976), 14:185.

7. Fletcher, BMNH, *An Introduction to the Study of Meteorites with a List of Meteorites Represented in the Collection*. London: British Museum (Natural History), 1890, 51. Altogether, Fletcher published twelve of these catalogs, beginning in 1881 and ending in 1914.

8. Reichenbach (1857*a*), (1857*b*).

9. Ibid., (1858*c*).

10. Ibid., (1859*c*), 372–373.

11. Ibid., (1858*d*).

12. Ibid., (1859*e*).

13. Ibid., (1859*d*), 307–311.

14. Ibid., (1859*e*), 464–465; (1860*a*).

15. Haidinger (1861).

16. BMNH, Herschel file, Herschel to Story-Maskelyne, 16 February 1876.

17. Boussingault (1861); Graham (1866–67).

18. Ansdell and Dewar (1886).

19. Kirkwood (1867), 106–111.

20. Boisse (1850), 162.

21. There had been several successful attempts to produce minerals artificially: James Hall's synthesis of calcite in 1805; Pierre Berthier's production of pyroxene in 1825; and Henri de Sénarmont's work on quartz in 1851. Daubrée seems to have been the first to attempt to synthesize meteorites.

22. BMNH, Herschel file, Herschel to Story-Maskelyne, 26 August 1867.

23. Daubrée (1866).

24. Ibid., (1879), II:474, 579–614.

25. Fouqué and Michel-Lévy (1881).

26. Tschermak (1876).

27. Ball (1879).

28. Tschermak (1907).

29. Zöllner (1873).

30. Wright (1875*a*), (1875*b*).

31. Ibid., (1875*c*).

32. Ibid., (1876).

33. Sorby (1864*a*).

34. Ibid., (1865).

35. Ibid., (1864*b*).

36. Ibid., (1866).

37. Ibid., (1877).

38. Story-Maskelyne (1875).

39. Ibid., BMNH manuscript.

40. Prior (1916).

41. Meadows (1972) combined an excellent biography of Lockyer and a balanced critique of his scientific work.

42. Meadows (1972), 176.
43. Tait (1869).
44. Meadows (1972), 177.
45. Lockyer (1887), 153–154.
46. Ibid., (1887), 155; Lockyer (1890), 63–68, 125–147.
47. Meadows (1972), 180.
48. Lockyer (1887), 128–129, 140–141.
49. Ibid., (1890), 237–241.
50. Ira S. Bowen in 1927 determined that the nebulium spectrum was due to ionized oxygen and nitrogen atoms.
51. Lockyer (1887), 150; (1890), 325.
52. Tait's theory of 1879 is described in Lockyer (1890), 227–229.
53. The mass of the Earth is 5.98×10^{27} grams. Tait assumed the mass of a comet was 6.17×10^{24} grams. King (1976), 293 states that in 1969 Z. Sekanina estimated that the mass of the nuclei of 23 short-period comets averaged 10^{17} to 10^{18} grams.
54. G. H. Darwin (1888). Meadows (1972), 190 implied that Lockyer asked Darwin to make the analysis.
55. Lockyer (1890), 325; G. H. Darwin (1888), 15.
56. Bredikhin (1890).
57. Tennyson, "God and the Universe" (1892), quoted in Meadows (1972), 207.
58. Meunier (1887).
59. Meadows (1972), 185.
60. Högbom (1901).
61. Flammarion (1864), 186 ff.
62. Kahl (1971), 294. Helmholtz delivered the lecture at Heidelberg and at Cologne in spring 1871.
63. W. Thomson (1871). Thomson and Helmholtz were correspondents although they apparently arrived at the same hypothesis independently. Arrhenius (1908) and Becquerel (1924) reported that in 1821 Count Eleanore de Montivault suggested that life came to Earth on meteorites from lunar volcanoes, and that in 1865 H. E. Richter, a German physician, postulated that the carbon in meteorites demonstrated the presence of organic life on other celestial bodies.
64. Nagy (1975), chap. 2 published selected excerpts from the reports of several of these chemists.
65. Berzelius (1834), 113–123, 143. Nagy did not excerpt Berzelius's final conclusion. Monge et al. (1806) and Thenard (1806) made reports on the Alais meteorite. Both reported carbon and water, but said nothing about organic matter.
66. Wöhler (1858); Wöhler and Hörnes (1859); Wöhler (1859); Wöhler (1860).
67. Wöhler and Hörnes (1859). Kobell (1864) 686, 688 listed both substances under the general class of bitumen or asphalt, containing carbon and hydrogen. He gave the formula for ozocerite as CH—carbon 85.75 percent, hydrogen 14.29 percent; and for scheerite C^2H^4, carbon 75 percent, hydrogen 25 percent.
68. E. Harris (1859), 20, 42.
69. Haüy (1801), III:321.
70. Clöez (1864).
71. M. Crosland, "Pierre Eugène Marcellin Berthelot," *DSB* II:67.

72. Berthelot (1868).

73. Ibid., (1866).

74. S. P. Thompson (1910), II:607.

75. L. Huxley, ed., (1918), II:126–127.

76. Zöllner (1873), xxv–xxvi.

77. Flight (1877), 714; Helmholtz (1874), 212.

78. Flight, ibid. Allen Thomson (1877), lxxv, in his presidential address to the British Association in 1877, dismissed William Thomson's hypothesis on the same grounds as Zöllner.

79. S. P. Thompson (1910), II:607, 1102.

80. Arrhenius (1908); Becquerel (1924); Figuier (1872).

81. Flight (1877), 718.

82. Gümbel (1878).

83. O. Hahn (1879).

84. O. Hahn (1880). Hahn wrote that he had 360 thin sections of Knyahinya, a L5 chondrite; 6 of l'Aigle, a L6 chondrite; 6 of Pultusk, a H5 chondrite; and 1 to 3 thin sections of 17 other meteorites. None of those identified in his photographs were carbonaceous meteorites. Hahn and Weinland focused on the structure or morphology of the forms viewed in the meteorites and made no mention of chemical composition. Nagy (1975), 75, 613 refers to Hahn's hypothesis but does not point out that Hahn's thin sections were from ordinary chondrites. Nagy suggests that Hahn may have been looking at fossils from terrestrial sedimentary rocks, but it is evident from the title and content of Hahn's 1880 work that he was looking at chondrules. At the time, Reichenbach's collection of meteorites had been bequeathed to Tübingen and was under Quenstadt's supervision. It is quite probable that Quenstadt, who was a paleontologist, would have known whether terrestrial rocks had become intermixed with the meteorites.

85. Weinland (1881*a*), (1881*b*), (1882).

86. Vogt (1882).

87. Rolle (1884).

88. Anon. (1882), *Popular Science Monthly*.

89. Birgham (1881).

6. CURATORS AND COLLECTORS

1. Chladni (1819*b*), 5. The falls listed by Chladni were Thüringen (1581), Zwickau (1647), Miskolz (1559), Führen (1654), Bern (1693), and Verona (1668). On the basis of the evidence, Hey, ed. (1966) considered the Fünen and Verona (Vago) falls to have been genuine.

2. Westrum (1978), 472.

3. Romé de l'Isle (1767), II:7.

4. VNHM, Acquisitions Register, 1888, Lot LVIII, #3756. The pieces weighed 26 grams and 7 grams, and at the time Brezina was unsure whether they were fragments of the Vago or the Albareto fall.

5. BMNH, Meteorite File, "Limerick." Letter from John Collins, Holly Park, Kilcoman, Limerick, 11 June 1946.

6. Hey, ed. (1966), 435–436; Hutchison et al. (1977), 214.

7. Schreibers (1820), v, footnote; Chladni (1819*b*), 90.

8. Fitzinger (1868). Brezina (1885) stated that the cabinet had acquired eight meteorites prior to Schreibers's appointment, but he did not list them.

9. Fitzinger (1856).

10. VNHM, Misc. Correspondence, B. Silliman, Jr., to Partsch, 11 August 1842.

11. Ibid., W. E. Logan to Partsch, 9 July 1855.

12. Partsch (1843*a*), 70–72.

13. VNHM, Misc. Correspondence, C. U. Shepard to Partsch, 11 May 1847.

14. Fitzinger (1856).

15. Berwerth (1918), 23–24.

16. Brezina (1885), 152.

17. Haidinger (1859*a*), 16.

18. For biographical information on Reichenbach, see Berwerth (1913–14), Engelhardt (1977), Partington (1972), IV:401–404.

19. Hofmann, ed., (1888), I:249, 254. Wöhler wrote to Liebig on 30 December 1844 that "our names have been brought into disrepute" because of the publication, but the matter did not appear to have damaged their reputations. The article was entitled: "Untersuchung über den Magnetismus und damit verwandte Gegenstände," *Annalen der Chemie,* 53 (1845), 1–270.

20. Reichenbach (1859*c*).

21. Haidinger (1859*b*).

22. Reichenbach (1860*a*), footnote, 373–374.

23. Engelhardt (1977) appended the documents pertaining to the gift. Shepard visited Vienna in 1862 and wrote to Maskelyne: "The collection of Reichenbach is astonishingly fine. I had no conception of it before the view. In some respects, it fairly rivals the Imperial Cabinet" (BMNH, Shepard to Maskelyne, 17 September 1862).

24. BMNH, Hörnes file, Hörnes to Maskelyne, 7 July 1861.

25. Ibid., Reichenbach file, Reichenbach to Maskelyne, 29 March 1867.

26. W. Campbell Smith (1969); Hey, ed., (1966), lix. The letter presenting the Krasnojarsk specimen from J. de Stehlin, Counsellor of State to the Tsar, is in the BMNH "Krasnojarsk" file.

27. John Ross (1819), 88, Appendix III, lxxxix; Sabine (1819).

28. BMNH, "Otumpa" file, unsigned note dated 14 August 1872. A decade earlier a piece had been cut from the mass, and according to records a pair of pistols were made from it and sent as a gift to U.S. President James Monroe. The pistols are displayed in the James Monroe Law Office, Museum and Memorial Library at Fredericksburg, Virginia. Buchwald tested a bleb from one of the pistols with a microprobe and found it contained less than 0.3 percent nickel, and Daniel J. Melton of the U.S. Geological Survey independently found a negligible amount of nickel in spectroscopic tests. Buchwald believes that the gunsmith found it too difficult to work the meteoritic iron and substituted wrought iron; see Buchwald (1975), II:373–374.

29. A. Russell (1952).

30. BMNH, Heuland file, Heuland to Konig, 23 August 1843.

31. Ibid., Heuland to Konig 17 October 1843.

32. Ibid., Heuland to Konig, 19 October 1843.

33. Unfortunately, the file does not disclose the price paid. A note appended to the letter of 19 October 1843 indicates that Heuland's cost for the three meteorites was £468, of which that attributable to the Walker County iron was £285. The exchange rate at the time was $4.44, so that his cost was approximately $1,265. An undated letter in the handwriting of Henry Heuland, given to George T. Prior by A. G. F. Gregory in May 1926, reads:

> Meteoric iron, from a mass which originally weighed 165 Pounds, was discovered in the autumn of 1832 by Mr. Wiley Speaks, whilst on a deer hunting excursion in the North East Corner of Walker County, Alabama, United States. From its weight he supposed it to be silver, and had it directly conveyed to his house, where it remained till 1842 when it came into the possession of Professor Gerard Troost of Nashville, Tennessee, who allowed him 1320 Dollars for it—about £300, and who again ceded it to Mr. Heuland with other meteoric irons from divers localities of Tennessee in the United States. (BMNH, "Walker Co." file)

Troost, in his 1845 report on the meteorite, stated that only a few fragments had been cut from the main mass before it came into his possession. If Troost paid $1,320 for the meteorite, he just about recovered his cost. However, Troost retained several pieces, which are now in various collections.

34. Buchwald (1975), III:1277 states that Reichenbach labeled his piece of the Walker County iron "Claiborne," and exchanged several fragments of it under that name. Also, see Buchwald (1975), II:500–501 concerning the different names attached to the Cosby's Creek iron.

35. BMNH, Troost file, Troost to Heuland, 2 February 1846.

36. Ibid.

37. Ibid., Heuland file, Heuland to Konig, 6 April 1846.

38. Ibid., Heuland to Konig, 6 July 1846.

39. Ibid., Heuland to Konig, 12 October 1846.

40. Ibid., Heuland to Konig, 2 February 1847.

41. Ibid., Heuland to Konig, 3 June 1848.

42. Ibid., Heuland to Konig, 15 November 1848.

43. Ibid., Shepard file, C. U. Shepard, Amherst, 26 October 1824.

44. Ibid., Shepard to Konig, 30 July 1845.

45. Ibid., Troost file, Troost to Heuland, 2 February 1846.

46. Rose (1863*b*).

47. For the history of the Durango meteorite, see Buchwald (1975), II:550.

48. MNH, early catalogs.

49. Buchner (1863), notes that Paris had 53 localities.

50. Lacroix (1927*a*), 411.

51. Ibid., 447, 448, 451.

52. MNH Catalog. Charcas is no. 382, Lucé is no. 277.

53. For Greg's biography, see G. F. Herbert Smith (1907).

54. In 1855 Greg had 37 irons and 40 stones in his meteorite collection, including a few duplicates. See SIA, C. U. Shepard collection, Box 1, Folder 21, Greg to Shepard 18 October 1855.

BMNH, Mineralogy Department, Fletcher file. Krantz sent the entire collection to the British Museum; his price was £686.9.0, plus freight charges.

Maskelyne and Greg declined to purchase several meteorites, and the final outcome was as follows:

Returned to Krantz—19 specimens	£ 163.9.3
Bought by the British Museum–37 specimens	£ 321.3.0
Bought by Greg for the British Museum—27 specimens	£ 58.3.3
Bought by Greg for himself—47 specimens	£ 102.4.9
Bought by the British Museum for payment in 1863—1 specimen, a sizable piece of Ocatitlan (Toluca)	£ 64.16.0
	£709.16.3

55. BMNH, Mineralogy Department, Fletcher file. Krantz to Maskelyne, 27 March 1861.

56. Shepard (1861).

57. BMNH, Smith file, Smith to Maskelyne, 13 September 1862.

58. Ibid., Shepard file, Shepard to Maskelyne, 17 September 1862.

59. BMNH, Catalogue file.

60. BMNH, Exchange of Duplicates file. Several of the exchanges involved a substantial number of specimens. For example, in 1862 and 1863, the museum received from Yale: Denton County, Walker County, Pittsburg, Salt River, Weston, and Richmond. It gave: Arva [Magura], Segowlie, Butsura, Durala, Umbala, Shalka, Akbarpur, Futtehpur, Yatoor, and Newstead [a pseudometeorite]. From Berlin, it received 14 specimens in exchange for 11; and from Vienna, it received 12 specimens and gave 13.

61. BMNH, "Cranbourne" file; also Buchwald (1975), II:508.

62. The meteorites were: Durala, Bustee, Shalka, Dhurmsala, Jamkheir, Shytal, Supuhee, Udipi, and Lodran. One sheaf of correspondence illustrating this approach is worth quoting as a timeless example of bureaucracy in action. The covering letter, dated 11 November 1867, was signed by Herman Merivale, India Office, and sent to the museum:

> Sir:
>
> I am directed by the Secretary of State for India to inform you that copies of your letter dated the 17th of July last . . . regarding the presentation to the British Museum of Aerolites of importance obtainable in British India, were transmitted to the three Presidencies in that country with a view to meeting as far as practicable, the wishes of the Trustees of the Museum, and that a reply has just been received from the Madras Government, a copy of which and the instructions issued to the several Collectors in that Presidency, is herewith forwarded for the information of the Trustees.
>
> I am also directed to state that no reply has yet reached Sir Stafford Northcote from the Governments of India or Bombay, but that when such is received it will be duly communicated to you.

The attachments read:

> Public Letter from Fort Saint George
> Dated 4th October, No. 20 of 1867
>
> In acknowledging the receipt of your Despatch of the 9th August last, No. 29, we have the honour to transmit for your information, a copy of the

instructions which we have issued to the several Collectors in this Presidency, with reference thereto.

No. 192. Order thereon, 23 September 1867, No. 1184

The Governor in Council directs that copies of the foregoing Despatch, and of its enclosures, be sent to the Collectors of the seven Districts in this Presidency, who will report, without delay, whether there are any Meteorites procurable in their respective Districts, and will furnish, with reference to any that may be, or may have been, brought to their notice, as much information, of the nature of that indicated in the memorandum appended to Sir C. Wood's Despatch recorded in the Proceedings of Government under date the 9th April 1863, as they may be able to procure.

2. It is understood that Meteorites have recently fallen in the Districts of Kurnool and Cuddapah. The Collectors of these Districts will report on this point, and will explain why the instructions conveyed in Sir C. Wood's Despatch . . . have not been observed.

63. BMNH, "Veramin" and "Ogi" files. In return for the Ogi specimen, the museum presented "a set of casts of Assyria antiquities, with the translation of inscriptions."

64. Silliman Family Papers, Yale University Library, Letter, Nevil Story-Maskelyne to B. Silliman, Jr., 25 August 1862.

65. Haidinger (1867*c*).

66. BMNH, Tschermak file.

67. Ibid., Daubrée and Meunier files.

68. George Kish, "(Nils) Adolf Erik Nordenskiöld," *DSB* X:148.

69. BMNH, A. E. Nordenskiöld file, Nordenskiöld to Maskelyne, 23 August 1864.

70. Ibid., Nordenskiöld to Maskelyne, 13 March 1869.

71. Ibid., Nordenskiöld to Maskelyne, 4 November 1870.

72. Ibid., Nordenskiöld to Maskelyne, 25 November 1870 and 24 January 1871.

73. Buchwald (1975), II:413; Hey (1966), 365.

74. Hlawatsch (1909). Brezina assumed responsibility for the meteorites in 1878, and in 1885 became head of the mineralogical-petrographical division.

75. Miers (1921).

76. BMNH, "Estherville" file: Birge to BMNH, 16 September 1879; Brezina to Maskelyne, 28 October 1879; Brezina to Maskelyne, 3 November 1879; Brezina to Fletcher, 6 April 1881; also miscellaneous notes on the cutting process.

77. VNHM, Inv. Nr. M/345/: Brezina to Herr Hofrath, 23 August 1878; Stürtz to Brezina, 16 September 1878. I am unable to determine whether Stürtz's specimen was actually a piece of the Augusta County iron or whether it was a piece of Staunton. Buchwald has been somewhat successful in determining the location of portions of these two meteorites, both of which were sold in the late nineteenth century under the name, Augusta County. See Buchwald (1975), II:277–278, III:1166.

78. VNHM, Miscellaneous documents: authorization by the Imperial Museum to pay the U.S. Consul at Vienna for the account of W. E. Hidden of New York, dated 4 February 1881. BMNH, Brezina file: Brezina to Fletcher, 20 August 1881; note by Fletcher, 10 October 1881; Brezina to Fletcher, 11 October 1881; Fletcher to Brezina (copy), 14 October 1881.

79. BMNH, Brezina file, Brezina to Fletcher, 18 October 1881.

80. Ibid., Brezina to Fletcher, 22 April 1886.

81. Ibid., December 1882, July 1884; note on letter of Brezina to Fletcher, 6 June 1891.

82. Ibid., Brezina to Fletcher, 2 March 1886.

83. VNHM, Inv. Nr. M/345/: "Erwerbung für die Sammlung Min-Petr. Abteilung;" "Uebersicht der Acquisitions . . . im Jahre 1887;" "Uebersicht der Acquisitions . . . im Jahre 1888;" also Register of Acquisitions, 1887 and 1888.

84. The 5.110-kg Misteca iron was purchased from the Bonn dealer, August Krantz, for 355 florins. Brezina sent it to the Krenkle Bros. machine shop at Waldkirche, near Freiberg im Breisgau, where it was sectioned into a total of 29 pieces, ranging in weight from 39 to 431 grams. The total loss in cutting was 1,511 grams or about 30 percent. Brezina assigned values to these sections, ranging from 5 to 40 florins, and sold or exchanged all but 7 pieces.

The description in the register of the Kendall County iron presents several problems. Brezina wrote that it was "originally an irregular egg-shaped block 20 cm. high and 15 cm. thick," that it was purchased from F. E. Grothaus of San Antonio, Texas, for 293.08 florins (equivalent to $111.69), and that the "iron meteorite had fallen ? and was found (according to the report of Grothaus) in Kendall County, 4–5 miles from San Antonio, 29°30 N 98°25 W, Bexar County, Texas, USA a few years before 1887." He then described the 8.95-kg piece, which is still in the Vienna collection, stating that the polished and etched face displayed a characteristic "hexahedral breccia with many 1–5 mm. green, fewer hazelnut to walnut green, grains." Sectioning the larger half yielded 70 pieces, but Brezina did not mention by whom or where the cutting was performed, nor the loss sustained. The actual site of the find is more puzzling. The Kendall County line is approximately 23 English miles from the center of San Antonio, so that a site 4–5 miles north of San Antonio would be located within Bexar County. Friedrich E. Grothaus, however, emigrated from Germany in 1848, and in 1854 settled in Comfort, Texas in northwestern Kendall County. In 1887 he was a bookkeeper in San Antonio. It is possible that when he made the report to Brezina, he used German "Meilen," which when converted to English miles would place the find location about 23.5 miles north of San Antonio, just inside the Kendall County line.

George F. Kunz of New York, a mineralogist and consultant to Tiffany & Co. on gems and precious stones, sold the Eagle Station, Catorze (Charcas-Descubridora), and Nelson County meteorites to the Vienna museum for 11,472 florins (approximately $4,400.) The Eagle Station pallasite came to the museum in 5 pieces, the 35.650-kg main mass and 4 small pieces totaling 127 grams. Brezina set a value of 6,000 florins on all these pieces. Trenkle Bros. of Waldkirche cut the main mass into 60 pieces with a loss of about 3.8 kg. Brezina retained the largest piece, weighing 15.195 kg, for the Vienna collection, and placed its value at 2,500 florins. The 41.530-kg Catorze mass was not sectioned, except for a 78-gram piece, which Brezina inspected microscopically. He mentioned that: "At one place in an artificially produced furrow, there are remains of a copper celt, with which someone evidently long ago attempted to split off a piece of the iron, thus producing the furrow." He set a value of just over 3,400 florins on the two pieces. The Nelson County meteoritic iron was cut into 100 pieces, the largest being a 17.2-kg endpiece, and Brezina set a total value of 1,970 florins on the iron.

Brezina purchased the Hex River Mountains and Kokstad irons from Professor

Daniel P. Hahn of Capetown, paying 1,600 florins for the former and 1,700 florins for the latter. When cut, they yielded a total of 79 pieces. Brezina retained for the Vienna collection a 31.2-kg piece of Hex River Mountains and a 38.6-kg specimen of Kokstad.

There were numerous other meteorites that Brezina acquired and sectioned for sale from about 1890, cutting the iron meteorites at the museum with a motor-driven endless steel wire, which reduced the loss considerably.

85. BMNH, Brezina file, Brezina to Fletcher, 27 April 1891; Bement file, Bement to Fletcher, 11 September 1893.

86. VNHM, Inv. Nr. M/345/: "Brezina Anschuldigungen, Disziplinaruntersuchung, Pensionierung," 30 August 1896.

87. BMNH, Mineralogy Department, Exchange of Duplicates, 1859–1919; Hlawatsch (1909). Berwerth (1918), 51 wrote that Brezina was such an enthusiastic collector that he overspent the museum's budget repeatedly, and that since this policy was not feasible in the long run, he resigned.

88. For example, in 1896 Warren Foote, a Philadelphia dealer, offered Fletcher the 522-lb Sacramento Mountains iron (the entire mass except for a small specimen he had promised to Bement) for a price of $1,200 delivered in London. Fletcher declined the offer, stating that he would be glad to receive a slice later on, that he was not anxious to be the sole possessor of a meteorite or to prevent other people from having slices, and that in any case there was not much more room in the museum for big meteorites (BMNH, Foote file, note on letter from A. Foote to Fletcher, 13 October 1896). Foote did finally cut slices from the mass, and Fletcher's cost for a 8.75-kg endpiece was $175 (ibid., Foote to Fletcher, 27 April 1897 and July 1897).

89. Barcena (1876).

90. Fletcher (1890); see Buchwald (1975), passim.

91. BMNH, "Crumlin" file.

92. Narendra (1978).

93. BMNH, Smith file, Smith to Maskelyne, 1 March 1870.

94. Ibid., Smith to Maskelyne, 9 September 1863. Smith enclosed a letter from a J. Berrien Lindsley, possibly an official at the University of Nashville, which provided the information on the Troost collection.

95. Ibid., Smith to Maskelyne, 3 May 1875.

96. Ibid., Smith to Maskelyne, 8 September 1875.

97. Ibid., Smith to Maskelyne, 28 September 1878.

98. J. L. Smith (1884); Huntington (1888).

99. BMNH, Shepard file, Shepard to Fletcher, 16 December 1880.

100. SIA, Shepard Papers, Box 1, Folder 8; Box 2, Folder 1.

101. Ibid., Box 2, Folder 5; BMNH, Shepard file.

102. Ibid., Box 1, Folder 8; Mason (1975).

103. Dupree (1957), 79–90.

104. Mason (1975).

105. Henderson and Dole (1964) made a comprehensive report on the Port Orford meteorite. There is a "Copy of Dr. Evans' Journal" in the Division of Meteorites, Smithsonian Institution. Whether the Port Orford meteorite exists is open to question. Why Evans, an experienced field geologist who made careful notes about other matters, would fail to make even a passing reference in his journal to finding the meteorite is a puzzle. However, he had no reason to prevaricate in his correspondence with Jackson. It is possible that failing health may have impaired his memory. Buchwald (1975), III:1393–1400 mentioned that specimens of the Imilac pallasite from Chile were widely distributed in the second quarter of the nineteenth century, and one wonders whether a sample that had reached the west coast of North America found its way into Dr. Evans's mineral box. But recent analyses apparently demolish this speculation. Scott, *GCA* (1977), 41:359–360 reported trace element data on the metal in Imilac, Port Orford, and Krasnojarsk, and believed that the differences between the germanium levels in Imilac and Port Orford clearly differentiate the two. A. M. Davis (Ph.D. thesis, Yale University, 1977) supported Scott's results. However, a specimen of the Imilac pallasite, found in 1974 and now at the Smithsonian Institution, has a fusion crust that is strikingly similar to that of the Port Orford sample at the Smithsonian. The evidence for the existence of the meteorite, in terms of statements about it in the early 1860s and of modern investigations, is conflicting. I am grateful to Roy S. Clarke, Jr., for providing me with the information he has gathered on the Port Orford meteorite and with the references to it just listed.

106. BMNH, meteorite file "?Trenton," Henry to Maskelyne, 11 March 1861. A note on this letter reads "?Trenton," but the content clearly refers to the Port Orford meteorite.

107. Dupree (1957), 85–86.

108. BMNH, meteorite file "Butcher Irons," Baird to Maskelyne, 3 May 1870.

109. VNHM, Inv. Nr. M/345/: Maskelyne to Tschermak, 27 May 1870.

110. Buchwald (1975), III:1236–1238.

111. Mason (1975). A letter from Joseph Henry to J. Lawrence Smith, dated 20 June 1865, mentions the Tucson meteorite. Henry reported that he could not send Smith a section because of the difficulty experienced in sawing the meteorite (letter from "Secretary Outgoing Correspondence, No. 1, page 13" furnished by Roy S. Clarke, Jr.).

112. Pierson (1873); Monnig (1939).

113. BMNH, meteorite file "Casas Grandes," Ward to Fletcher, 22 April 1901, 29 May 1901; Merrill to Fletcher, 10 June 1901.

114. Mason (1975).

115. BMNH, Bement file—in particular, Bement to Fletcher, 30 October 1894; 18 October 1897; 22 October 1897; and Fletcher's notes on these letters.

116. Ibid., Bement to Fletcher, 15 March 1898, 10 August 1899.

117. Hovey (1912). J. P. Morgan, Jr., gave the collection to the museum. Twenty-two of the nearly 300 meteorites were new specimens.

118. Buchwald (1975), II:410 ff. gave a comprehensive summary of the explorations for the Cape York meteorite, including his own experiences in locating and retrieving the 7.8-kg Savik II and the 20-ton Agpalilik pieces in the early 1960s.

119. Peary (1914), 145–149, 533–618.

120. Lange (1958).

121. H. A. Ward (1904*d*).

122. Kohlstedt (1980); R. Ward (1948).

123. BMNH, meteorite file "Otumpa," C. Hicks to Fletcher, 24 June 1889.

124. R. Ward (1948), 255.

125. UR, File Drawer 8, note of Avery Coonley.

126. Ibid., File Drawers 8 and 9.

127. Ibid.

128. Ibid.

129. Private communication from Edward J. Olsen, Curator of Mineralogy, Field Museum of Natural History.

130. SIA, Merrill Papers, George L. English to Merrill, 20 October 1926, Box 23, Folder 5.

131. BMNH, meteorite file "Yorktown," Bailey to Gregory, 16 March 1885.

132. Wülfing (1894).

133. Ibid., (1895).

134. Ibid., (1897).

135. Cohen (1899).

136. Wülfing (1899).

137. H. A. Ward (1904*b*), (1904*c*).

138. Foote (1913).

139. SIA, Merrill Papers, Box 17, Folder 9, Merrill to Farrington, 22 October 1913; 26 October 1913; 5 November 1913; Farrington to Merrill, 25 October 1913; 31 October 1913.

7. FOLKLORE, MYTH, AND UTILITY

1. Virgil, *Aeneid,* II:687–706; *Acts of the Apostles,* 19:35; Lenain de Tillemont (1693–1739), III, Part I:265–283.

2. Andrée (1901), 404; Alois (1905), 75; Anon. (1953), 295; Bartsch (1879–80), II:202; Carmen (1955), 101; Drechsler (1903–1906), II:135; Fogel (1915), 81; Grimm (1875–1878), II:602; Hughes (1949), 298; Koch (1961), 78; Radford and Radford (1948), 219; Sartori (1906), 207; Hand, ed. (1964), 192.

3. Holmberg (1922–23), 21; Urbes (1898), 152.

4. Lütolf (1862), 322.

5. Reiser (1897–1903), II:436.

6. Vicuna Cifuentes (1915), 251; Radoza (1970), I:168.

7. Bjorkman (1973), 98, 102; Hughes (1949), 288.

8. Shakespeare, *King Henry IV, Part I,* I, i:10–11, and V, i:20–23; *Julius Caesar,* II, ii, 19, 31–32.

9. Bjorkman (1973).

10. Alois (1905), 75; Bartsch (1879–80), II:202; Höhn (1913), 313; Holmberg (1922–23), 109; Meier (1852), 506; Meyer (1900), 516; Schlosser (1891), 389; Schullerus (1912), 218.

11. Birlanger (1874), I:395.

12. Grimm (1875–1878), II:602.

13. Hurt (1900), 27; Schullerus (1912), 158; Urbes (1898), 151.

14. Amersbach (1891–1893), II:33; Basset (1899), 579–580; Grimm (1875–1878), II:602; Gundel (1912), 95; Saucier (1956), 138; Simon (1960), 123; Spech (1935), 19; Wehrham (1910), 208.

15. J. D. Clark (1966), 23; Koch (1961), 84.
16. Hand, ed. (1964), 193.
17. Carmer (1940), 362.
18. Anon. (1884), I:216; Espinosa (1910), 414; Dähnhardt (1907), 134.
19. Basset (1899), 98; Radford and Radford (1948), 225.
20. Gundel (1912), 251.
21. Anon. (1886), 843; Drechsler (1903–1906), II:135; Hendriks (1966), 30; Meyer (1900), 516; Montell (1966), 83.
22. Alois (1905), 75; Basset (1899), 578–579; Hockner (1854), 418; Morel and Walter (1972), 106.
23. Vicuna Cifuentes (1915), 250.
24. Millington and Maxfield (1906), 210.
25. Bjorkman (1973), 92; Bartsch (1879–80), II:202; V. D. Chamberlain (1982).
26. Anon. (1867), 780.
27. Alois (1905), 75; Birlanger (1861–62), I:190.
28. Cockayne (1864–1866), 233–287.
29. Basset (1899), 656.
30. Dähnhardt (1907), 135.
31. Bjorkman (1973), 107–109; Jastrow (1912), II:692.
32. Gundel (1912), 84–95; Schlosser (1891), 289.
33. Bartsch (1879–80), II:202.
34. Mannhardt (1858), 424.
35. Andrée (1901), 389.
36. Bartsch (1879–80), I:256–260.
37. Birlanger (1861–62), I:189–190.
38. Drechsler (1903–1906), II:135.
39. Bjorkman (1973), 94–95.
40. Peary (1914), 559.
41. UR, Henry Augustus Ward Papers, File Drawer 8.
42. Botley (1970), 304–305.
43. Bächtold (1915), 207.
44. Mendoza (1952), 42.
45. BMNH, meteorite file "Jhung."
46. Buchwald (1975), III:814.
47. BMNH, meteorite file "Tenham."
48. Bird (1820), 156–158.
49. Falconet (1729), 519 wrote that baetyls were stones that had fallen from heaven in a fireball and that the god Coelus gave the name of his son, Betul, to the stones. Prevost (1803*a*), 115–118 stated that baetyls were stones representing Cybele, mother of the gods, and that priests wore the stones next to their breasts. Wainwright (1934*a*), 32–46 wrote that baetyls were sacred meteorites and that the word had a Semitic origin, from Beth-el or "home of God."
50. Wainwright (1928), (1930), (1931), (1932*b*), (1933), (1934*b*). The quotation is from (1931), 189.
51. Forbes (1950), 428.

52. H. A. Newton (1897), 14; Farrington (1900), 202.

53. Wainwright (1931), 189n.

54. Farrington (1900), 200; Wainwright (1935), 43.

55. Brezina (1904), 214. Wainwright (1933), 44 thought that the mention of the bull's head linked Astarte to both Ammon and Zeus, who were frequently depicted as having the head of a bull.

56. Dietz and McHone (1974), note, 176–177.

57. Partsch (1857). The following descriptions relies on Partsch, Dietz, and McHone (1974), Thomsen (1980), and the authorities the two latter cite.

58. Dietz and McHone (1974), 175.

59. Thomsen (1980), 89.

60. Buchwald (1975), III:838.

61. Kinnicut (1884), 381–384; Buchwald (1975), II:656–658.

62. Grogan (1948); Buchwald (1975), II:635.

63. Buchwald (1975), II:776.

64. Ibid., II:399.

65. Ibid., II:431.

66. Ibid., III:878.

67. Dorsey (1906), 61–62; V. D. Chamberlain (1982).

68. Mallet (1884); Buchwald (1975), III:1305.

69. W. F. Butler (1875), 304–305; Coleman (1886), 97.

70. Peary (1914), 611.

71. BMNH, meteorite file "Ogi."

72. Grube (1910), 95–96.

73. Brezina (1887), 69–70.

74. Saxton (1870); BMNH, meteorite file "Nedagolla."

75. Lacroix (1927*b*).

76. Brezina (1887), 69–70.

77. BMNH, meteorite file "Kota-Kota."

78. Ibid., "Uwet."

79. Ibid., "Queen's Mercy."

80. Merian (1864); H. A. Newton (1897), 3. Joshua's rout of the army of the Amorite kings may also be an example of this sort. After Joshua's attack at Gilgal, his opponents fled, and when the enemy reached the descent at Bath-horon, the "Lord cast great stones from the sky upon them all the way to Azekah, so that they died" (Josh. 10:11). Bosler (1945) thought that this event could have been a meteorite shower of the magnitude of that at Pultusk in 1868.

81. Needham (1959), III:419.

82. John Wasson noted and informed me of this coincidence.

83. Balfour (1929), 37–49.

84. Wainwright (1931), 192–194; (1932*b*), 170–172.

85. Needham (1959), III:434.

86. The following account relies on Andrée (1889), 30–41; Tylor (1889), II:166, 263–264; Balfour (1929), 37–49; and Gardner (1942), 90.

87. Narendra (1978), 11.

88. Bjorkman (1973), 126–127; Waldbaum (1980), 69–79.
89. Andrée (1884), 2. I have been unable to locate the cited source.
90. For example, Bradley Stoughton (1934), 2. Although Stoughton does not state explicitly that the fabrication of meteoritic iron led to the smelting of iron, his text and the accompanying illustration imply that this was the case.
91. Wainwright (1932*a*).
92. Ibid., 15.
93. Forbes (1950), 428.
94. Forbes (1954), 594.
95. Bjorkman (1973), 114.
96. Gettens et al. (1971).
97. Chung (1979).
98. Wasson and Sedwick (1969), as cited by Buchwald (1975), II:657.
99. Grogan (1948).
100. Buchwald (1975), II:636–637.
101. BMNH, Maskelyne manuscript, Ch. III; Farrington (1915), 35.
102. Bjorkman (1973), 110–113.
103. Tuckerman (1944). This meteorite might be the same one that Avicenna reported could not be forged (see chap. 1).
104. Sowerby (1820).
105. Berwerth (1907).
106. Cohen (1894–1905), I:62–63 summarized all of the above trials.
107. Ibid., 64.
108. Merrill (1929*a*), 103–104.
109. Buchwald (1975), I:162–164.
110. Ibid., 41–43. Foshag (1941), 141 referred to this condition, but he did not point out clearly that terrestrial reheating was the cause. He said: "the granulation of hexahedrites by secondary heating accounts for the so-called 'low nickel ataxites.'"
111. Howard-White (1963), 93.
112. Stodart and Faraday (1820); L. P. Williams (1965); Yeo and Miller (1965).
113. Howard-White (1963), 94–95.
114. Buchwald (1975), II:584.

8. NEW DIRECTIONS 1900–1950

1. BMNH, Prior file.
2. A. E. Nordenskiöld (1878).
3. Wahl (1910).
4. Prior (1913).
5. Prior (1914).
6. Merrill (1918).
7. SIA, Merrill Papers; Prior to Merrill, 19 November 1917, Box 20, Folder 14.
8. Ibid., Merrill to Prior, 18 December 1917, and 6 August 1918.

9. Urey and Craig (1953). They also included one analysis by Hillebrand and two by Penfield. Three analyses performed for Merrill by W. Tassin were also accepted.
10. SIA, Merrill Papers, Merrill to S. Meunier, 15 June 1915, Box 19, Folder 14.
11. Ibid., Merrill to Prior, 6 August 1918, Box 20, Folder 14.
12. Ibid., Prior to Merrill, 9 October 1918.
13. Prior (1920).
14. Ibid., 56n.
15. Lacroix (1927*a*), (1927*b*).
16. Prior (1916), 41.
17. Prior (1920), 52.
18. SIA, Merrill Papers, Merrill to Prior, 18 December 1917, Box 20, Folder 14.
19. Merrill (1920).
20. Ibid., (1921).
21. Ibid., (1929*b*).
22. Ibid., (1928).
23. Wahl (1910), 94–95.
24. Doelter (1914); Berman (1937).
25. Berman (1937), 344.
26. Bragg (1930).
27. Perry (1944), 37; A. D. Romig and Goldstein (1980), 1155. I have relied on Perry in the following discussion of the early phase diagrams.
28. Buchwald (1975), III:1055.
29. Perry (1944), 39. The controversy about taenite continued for more than twenty years. In 1938 John Buddhue postulated two forms of taenite, nickel-rich and nickel-poor, with subsequent changes in the phase diagram.
30. Ibid., 38.
31. J. Young (1926), especially 638, 640.
32. Belaiew (1924).
33. Mehl and Derge (1937); Mehl (1965).
34. Mehl (1965), 255, 263; Buchwald (1975), I:115.
35. Clarke and Goldstein (1978), 5.
36. Perry (1944).
37. T. C. Chamberlin (1901).
38. SIA, Merrill Papers, Prior to Merrill, 19 November 1917, Box 20, Folder 14.
39. Olivier (1925), 261.
40. Trowbridge (1918); (1924), 38–39.
41. Niessl (1917), 18.
42. Hoffmeister (1925).
43. Manian et al. (1934).
44. W. M. Foote (1912).
45. Lindemann and Dobson (1922).
46. Paneth (1940), 13.
47. Manian et al. (1934).
48. Watson (1939).

49. Paneth (1940), 12.
50. Farrington (1901), 532.
51. Cohen (1894–1905), II:290.
52. Harkins (1917).
53. Ibid., 876–879.
54. Noddack and Noddack (1930).
55. H. N. Russell (1929).
56. C. Oftedahl, *DSB* (1976), 14:58–59.
57. E. D. Goldberg, *DSB* (1972), 5:456–459.
58. Goldschmidt (1922).
59. Adams and Washington (1924).
60. Washington (1926).
61. Goldschmidt (1929), (1930).
62. Goldschmidt and Peters (1932*a*), (1932*b*).
63. Brown (1949); Brown and Goldberg (1949).
64. Daly (1943).
65. Ibid., 413.
66. Ibid., 416n.
67. W. Thomson (1862).
68. Ibid., (1899).
69. Badash (1968) gave a comprehensive review of the early history of radioactive dating techniques. I have relied on his account.
70. Ibid.
71. Ramsay and Soddy (1903); Boltwood (1904); Rutherford (1905).
72. Boltwood (1905).
73. Rutherford (1906), 189.
74. Badash (1968).
75. Russell (1921).
76. Strutt (1906); Quirke and Finkelstein (1917).
77. E. G. Spittler, *DSB* (1974), 10:288–289.
78. Urry (1933).
79. Ibid.
80. Urry (1937).
81. Paneth (1939), (1940), 20.
82. Foshag (1941), 43; R. D. Evans (1943).
83. Arral et al. (1942).
84. Bauer (1947).
85. Huntley (1948).
86. Bauer (1948).
87. Nier (1935); Brewer (1935); Smythe and Hemmendinger (1937).
88. A. E. Foote (1891).
89. BMNH, "Canyon Diablo" file. Foote to Fletcher, 4 December 1891. Foote suggested that Fletcher pay only transportation charges on a large mass and display it in the expectation that donors could come forward and provide funds for its purchase. Fletcher declined, and purchased a small specimen for £10.

90. Nininger and Nininger (1950).

91. SIA, Merrill Papers, Box 17, Folders 15–16; G. K. Gilbert (1896).

92. Brandon Barringer (1964) gave an account of the attempts to locate the mass and of the controversies that arose.

93. D. M. Barringer (1905).

94. Tilghman (1905).

95. Merrill (1908).

96. Darton (1910).

97. D. M. Barringer (1914).

98. D. M. Barringer (1924).

99. B. Barringer (1964).

100. E. M. Shoemaker (1963). Comparing the crater size with those produced in nuclear bomb tests, and assuming a velocity of 15 km/sec, Shoemaker estimated that the size of the meteorite was 63,000 tons. Earlier estimates ranged between 5,000 and 3,000,000 tons. See also Elbert King (1976), 90–91.

101. D. M. Barringer, Jr. (1928). Subsequently, drill tests proved that the original crater floor was about 30 meters below the rim and had been filled with drifting sand. Over 100 drill holes were sunk to bedrock without encountering an obstruction, indicating that the meteorite must have suffered complete disintegration and/or volatilization. Other smaller craters adjacent to the main crater did yield meteoritic fragments. See Buchwald (1975), III:937–938.

102. Spencer (1933*a*).

103. Later in 1933, Spencer and Hey suggested that the two Nejed irons, brought to London in 1887 and 1893, were identical to the fragment recovered by Philby. Buchwald (1975), III:1269–1271 supported this identification. Spencer also reported on what was then known about the Tunguska event, which occurred in 1908 in central Siberia. We defer discussion about this until chap. 9, because it was not until the early 1960s that Soviet scientists completed a comprehensive survey.

104. Melton (1938). The origin of the Carolina Bays still constitutes a problem for geologists.

105. G. K. Gilbert (1892).

106. D. M. Barringer (1924). Also, A. C. Gifford (1924) postulated that lunar craters were explosion craters produced by meteorite impacts.

107. Spencer (1933*a*), 242.

108. Dietz (1946).

109. Daly (1946).

110. LaPaz (1949).

111. Barnes (1963). I have relied on Barnes for the history of the discovery of tektite-strewn fields. The nineteenth-century reports he cited are: Joseph Mayer, *Böhmischen Gesell. Wiss. Abh. Jahrg.*, 1787, 265 (southern Bohemia moldavites); Charles Darwin, *Geological Observations on the Volcanic Islands,* London, 1844 (australites); P. Van Digk, *Jaarb. Mijn. Ned. Indie,* (1879), 2:225 (billitonites); and R. D. M. Verbeek, *Proc. K. Akad. Wet.,* (1897), 5:421, (javanites). The word "tektite" derives from the Greek root meaning "molten."

112. Mason (1962*b*), 201.

113. Wahl's belief was described by Michel (1922), 315–317.

114. Ibid., 266, 318–319.

115. Ibid., 321–323. Michel (1922), 320–321 reported that C. E. Tilley (1922) showed that when the refractive indices of tektites and various types of natural glasses were plotted against their respective densities, the tektites could be easily differentiated.

116. Michel (1922), 318.

117. Spencer (1933*b*).

118. Paneth (1940), 24–25.

9. CONTEMPORARY METEORITE RESEARCH AND THEORIES

1. Keil and Fredriksson (1981).

2. Clarke et al. (1970), 1–2.

3. Marvin and Mason, eds. (1980), (1982); Yanai (1979).

4. Marvin (1983).

5. Brown and Patterson (1947*a*), (1947*b*), (1948).

6. Urey and Craig (1953).

7. Of the 192 analyses discarded, 14 were superseded by others, 23 were duplicates, and 48 were incomplete. The 94 superior analyses were by no means all recent. The numbers performed by decade were:

1861–1870	6	1911–1920	11
1871–1880	8	1921–1930	15
1881–1890	5	1931–1940	20
1891–1900	6	1941–1950	11
1901–1910	10	1951	2

8. In fact, Urey and Craig wrote that the diameter of the body would be about 1700 km or larger, but they gave this figure as the diameter of the Moon instead of its radius. They stated, however, that the calculated density of the L-type parent body was similar to that of the Moon. Six years later, Urey did state specifically that the "primary" objects were of lunar size, and identified "secondary" objects resulting from breakup with the asteroids.

9. Wiik (1956).

10. Ringwood (1961).

11. Mason (1960).

12. Mason (1962*a*).

13. Keil and Fredriksson (1964).

14. Mason (1963*a*).

15. Mason (1967*a*).

16. Van Schmus and Wood (1967).

17. Dodd et al. (1975); Wasson (1985).

18. Goldberg et al. (1951).

19. Lovering et al. (1957).

20. Wasson (personal communication).

21. Wasson (1967).

22. Wasson and Kimberlin (1967).

23. Ibid., (1985).
24. Dodd (1981).
25. Anders (1963) wrote a comprehensive review of age determination studies in the 1950s and early 1960s, and Wasserburg and Burnet (1969) gave additional information.
26. Ibid.
27. Herzog and Pinson (1956).
28. Patterson (1956).
29. Dodd (1981).
30. Anders (1963); D. W. Sears (1978).
31. Herzog and Pinson (1956).
32. Anders (1963), 428.
33. Dodd (1981), 165.
34. Papanastassiou and Wasserburg (1969).
35. Wasserburg et al. (1969); Sanz and Wasserburg (1969). Wetherill et al. (1973) found an inclusion in the Allende meteorite that had a much lower initial ratio, indicating that the formation or isotopic equilibration of the previously measured chondrites occurred many millions of years later.
36. D. W. Sears (1978), 129–130; Dodd (1981), 166–167.
37. Papanastassiou and Wasserburg (1969), 375.
38. Wasserburg and Burnet (1969), 468.
39. Paneth et al. (1952).
40. Begemann et al. (1957).
41. I have relied on Anders (1963) for this discussion of methods.
42. Wasson (1974).
43. Wasson (1985).
44. Wetherill (1977).
45. Evans et al. (1982).
46. Nishiizumi et al. (1984), abstract. Anders (1963) cited the previously reported age of the Tamarugal iron.
47. Thorslund and Wickman (1981).
48. Wood (1963).
49. Some scientists question the assumption of thermal and chemical equilibrium. Alfvén et al. (1977) stated that the role of plasma and hydromagnetic processes in the primordial nebula has been minimized in the prevailing theory. They believed that these processes played a fundamental role in the chemical differentiation of the nebula and in the formation of the planets.
50. Fredriksson et al. (1975).
51. Dodd (1981).
52. Clarke et al. (1970), 48.
53. Wetherill et al. (1973); Gray et al. (1973).
54. Clayton et al. (1973).
55. Clayton et al. (1977); Clayton (1977); Schramm and Clayton (1978).
56. Blander (1979).
57. Wood and McSween (1977); Dodd (1981); Taylor et al. (1983).
58. Cameron (1975), (1979).

59. Wasson (1985).

60. Dodd (1981).

61. Ibid.

62. Bogard and Johnson (1983).

63. Kerr (1983).

64. Kerr (1984*b*).

65. Schmidt (1945).

66. Mason (1962*b*).

67. Ibid., 197.

68. Fish et al. (1960). Although most scientists thereafter favored the theory that meteorites originated in asteroidal size bodies, this view was not unanimous. In 1972 M. W. Ovenden, astronomer at the University of British Columbia, on the basis of dynamical calculations revived the hypothesis that a planet about 90 times the mass of the Earth once existed between Mars and Jupiter. He added, however, that possibly it was always in the shape of a ring. Disintegration and dispersal of all pieces, except what remained in the asteroid belt, occurred about 16 million years ago. Later, T. C. Van Flandern of the U.S. Naval Observatory stoutly defended this idea before a rather hostile audience, and suggested that the missing material became long-period comets. See Ovenden (1972) and Van Flandern (1977). In the late 1950s, unanimity of opinion that meteorites came exclusively from within the Solar System still did not exist. L. LaPaz (1958) calculated that about as many meteorites came from interstellar space as came from within the Solar System.

69. Dodd (1981); Narayan and Goldstein (1985).

70. I have relied on Baker (1938), Chapman (1975), and Gehrels (1979) as sources relative to the studies of asteroids prior to and during the 1950s.

71. Shoemaker et al. (1979).

72. Ceplecha (1961).

73. McCrosky (1970); McCrosky et al. (1971); Clarke et al. (1971).

74. Halliday et al. (1978).

75. Chapman (1975).

76. Zellner (1979).

77. Wetherill (1977).

78. Ibid., (1985).

79. Wetherill (1979); Waldrop and Kerr (1983); Kerr (1985). Jan Hendrick Oort, the Dutch astronomer who first proposed the existence of a cloud of comets, thought it originated from minor planets that escaped from the asteroid belt at an early stage in the history of the Solar System. He also speculated that the cloud might conceivably be part of the remnants of a disrupted planet (Oort 1950). One of the most recent theories places the source at between 300 and 3,000 astronomical units distant from the Sun (Delsemme 1977).

80. Morrison and Niehoff (1979).

81. Lipman (1932).

82. *New York Times,* 31 January 1933, p. 19, col. 8; ibid., 5 February 1933, part 8, p. 6, col. 1–4; ibid., 10 February 1933, p. 19, col. 6.

83. Farrell (1933).

84. Nininger (1933).

85. Roy (1935).
86. *New York Times,* 28 June 1936, part 10, p. 9, col. 3–5.
87. Lipman (1936).
88. Roy (1937).
89. Nagy et al. (1961); Mason (1963*b*).
90. *New York Times,* 17 March 1961.
91. Claus and Nagy (1961).
92. Urey (1952).
93. Mueller (1953).
94. Morrison (1962). C. S. Smith (1981) has made several important studies of the structures displayed by inorganic matter.
95. Fitch et al. (1962).
96. Anders and Fitch (1962).
97. Hayatsu (1964).
98. Mason (1963*b*).
99. Urey (1966).
100. Urey et al. (1962).
101. Urey (1966).
102. Urey (1959); Urey et al. (1962).
103. Urey (1966).
104. McLane et al. (1967).
105. Elbert King (1976).
106. Nagy (1975), 684.
107. Engel and Nagy (1982).
108. I have relied on Spencer (1933*a*) and Krinov (1966) for details of the Tunguska event.
109. Whipple (1930); Spencer (1933*a*).
110. Krinov (1966).
111. Baxter and Atkins (1976); Stoneley (1977).
112. Elbert King (1976); Wasson (1985).
113. Elbert King (1976).
114. Ibid.; Wasson (1985).
115. Elbert King (1976); Wasson (1985).
116. O'Keefe (1976), 182, 196–197.
117. L. Alvarez et al. (1980).
118. W. Alvarez et al. (1984).
119. Ganapathy (1980).
120. Orth et al. (1981).
121. Bohor et al. (1984).
122. Kerr (1984*c*).
123. Archibald and Clemens (1982); Officer and Drake (1983).
124. Kyte et al. (1980).
125. Hsü et al. (1982).
126. Officer and Drake (1983), (1985).

127. Ganapathy (1982). Ganapathy calculated that the impacting object had a diameter of 3 km and a mass of 50 billion tons. Ganapathy (1983) also reported detecting an iridium enrichment in an Antarctic ice core, whose age coincided with the time of the Tunguska event.

128. Orth et al. (1984). These boundaries, separated by about 10 million years, are about 525 million years old. They concluded that the lack of an iridium anomaly at one site did not rule out the possibility of the extinction as a result of the impact of an extraterrestrial body, believing that such an event could only be excluded if no iridium anomalies were detected at several other sites. The conclusion is significant, because it demonstrates that the theory has become well-entrenched in a very short period of time.

129. Kerr (1984*a*).

References

ABBREVIATIONS

AAAS	*Proceedings of the American Association for the Advancement of Science*
ACP	*Annales de Chimie et de Physique*
AJS	*American Journal of Science*
AN	*Astronomische Nachrichten*
ANSP	*Proceedings of the Academy of Natural Sciences, Philadelphia*
AP	*Annalen der Physik*
AS	*Mémoires de l'Académie Royale des Sciences, Paris*
AWB	*Abhandlungen der königlichen Akademie der Wissenschaften zu Berlin*
BAAS	*Report of the British Association for the Advancement of Science*
BB	*Bibliothèque Britannique, Genève*
BMNH	British Museum (Natural History) Archives
CR	*Académie des Sciences, Paris, Comptes Rendus*
DSB	*Dictionary of Scientific Biography*
EPSL	*Earth and Planetary Science Letters*
GCA	*Geochimica et Cosmochimica Acta*
GCL	*Georg Christoph Lichtenberg, Schriften und Briefe,* W. Promies, ed., 4 vols. München, 1967–1972
HMAR	*Histoire et Mémoires de l'Académie Royale des Inscriptions et Belles Lettres, Paris*
JBRS	*Journal Book of the Royal Society of London*
JM	*Journal des Mines*
JP	*Journal de Physique, de Chimie, et d'Histoire Naturelle*
MM	*The Mineralogical Magazine*
MNH	Muséum d'Histoire Naturelle, Paris, Archives
NAS	*Proceedings of the National Academy of Sciences, U.S.*
PM	*Philosophical Magazine and Journal of Science*
PR	*The Physical Review*
PRS	*Proceedings of the Royal Society of London*
PT	*Philosophical Transactions of the Royal Society of London*
SAW	*Sitzungsberichte der kaiserlichen königlichen Akademie der Wissenschaften zu Wien, Mathematisch-naturwissenschaftliche Klasse*

SIA	Smithsonian Institution Archives
UR	Henry Augustus Ward Papers, Department of Rare Books and Special Collections, Rush Rhees Library, The University of Rochester
VNHM	Naturhistorische Museum Wien, Mineralogisch-Petrographische Abteilung, Archives

ARCHIVAL NOTES

British Museum (Natural History)

The Mineralogy Library retains five types of archival material on meteorites:

1. *Meteorite files,* which contain correspondence and other information related to individual meteorites. These are listed under the meteorite name, and sometimes under the former name (e.g., Otumpa instead of Campo del Cielo).
2. *Files of incoming correspondence,* listed alphabetically under the name of the writer. Some letters contain information about meteorites.
3. *The diaries, laboratory and other notebooks, and reports* of each scientist employed in the Mineralogy Department from about 1860 forward. These are kept in individual boxes.
4. Almost all of the *meteorite catalogues* published by the museum have been retained.
5. Boxes containing several hundred *reprints of journal articles* on meteorites, which date from about 1860.

Apart from the Acquisitions Register, the Mineralogy Department has two pertinent files. One is a list of the exchange of duplicate meteorite specimens from the 1860s forward. The other is a file of notes made by Sir Lazarus Fletcher, evidently compiled to aid him in preparing his editions of the museum catalogue.

Smithsonian Institution

The Smithsonian Institution Archives hold the George P. Merrill Collection and the Charles U. Shepard Collection, both of which are particularly rich in material pertaining to meteorites.

The Division of Meteorites, U.S. National Museum of Natural History, has the F. A. Paneth Collection of rare books and about 1,200 reprints of journal articles on meteorites, and also the Maskelyne Collection (bound) of about 175 reprints of journal articles.

Department of Rare Books and Special Collections, Rush Rhees Library, The University of Rochester

The Henry Augustus Ward Papers contain a large amount of material on meteorites. In addition to Ward's correspondence, account books, and financial papers, there is much information on the Ward-Coonley meteorite collection, descriptions and photographs of meteorites, newspaper clippings, biographical material, and several hundred reprints of journal articles.

Naturhistorische Museum, Wien, Mineralogisch-Petrographische Abteilung

The department has files on its scientific personnel (e.g., von Schreibers, Mohs, and Partsch), some of which date from the early 1800s. There are also files of miscellaneous correspondence having to do with meteorites. The Acquisitions Registers, and particularly the entries made by Brezina in the late 1800s, carry much detail with respect to purchases, gifts, and exchanges of meteorites. The original meteorite thin sections, prepared by Tschermak, are retained in the department offices.

BOOKS AND ARTICLES

Adam, Charles, and Paul Tannery, eds.
1897–1913 *Oeuvres de Descartes*. 12 vols. Paris.

Adams, Frank Dawson.
1954 *The Birth and Development of the Geological Sciences*. New York: Dover.

Adams, John C.
1867 "On the orbit of the November meteors." *Royal Astronomical Society (Monthly Notices)* 27:247–252.

Adams, Leason H., and Henry S. Washington
1924 "The distribution of iron in meteorites and in the earth." *Journal of the Washington, D.C. Academy of Sciences* 14:333–340.

Agassiz, Louis
1840 "On Animals found in Red Snow." *BAAS,* Part II:143.

Aldrovandi, Ulisse
1648 *Museum Metallicum*. Bologna.

Alembert, Jean L. R. d', and Denis Diderot, eds.
1757 "Pierres de Foudre." *Dictionnaire Raisonée des Sciences, des Arts, et des Métiers* VII:214.

Alfvén, H., G. Arrhenius, and D. A. Mendis
1977 "The Role of Plasma in the Primeval Nebula." In *Comets, Asteroids, Meteorites*. Ed. A. H. Delsemme. 561–568. Toledo, Ohio: University of Toledo.

Alois, John
1905 *Sitte, Brauch, und Volksglaube in deutschen Westböhmen*. Prag.

Alvarez, Luis W., Walter Alvarez, Frank Asaro, and Helen V. Michel
1980 "Extraterrestrial Cause for Cretaceous-Tertiary Extinction." *Science* 208:1095–1108.
1981 "Asteroid Extinction Hypothesis." *Science* 211:648–656.

Alvarez, Walter, Erle G. Kauffman, Finn Surlyk, Luis W. Alvarez, Frank Asaro, and Helen V. Michel
1984 "Impact Theory of Mass Extinctions and the Invertebrate Fossil Record." *Science* 223:1135–1141.

Amersbach, Karl
1891–1893 *Sage und Märchen bei Grimmelshausen*. 2 vols. Baden Baden.

Anders, Edward
1963 "Meteorite Ages." In *The Moon, Meteorites and Comets*. Ed. Bar-

bara M. Middlehurst and Gerard P. Kuiper. Chicago: University of Chicago Press.

Anders, Edward, and Frank Fitch

1962 "Search for Organized Elements in Carbonaceous Chondrites." *Science* 138:1392–1399.

Andrée, Richard

1884 *Die Metalle bei den Naturvölken*. Leipzig: Veit.

1889 *Ethnographische Parallen und Vergleich*. Leipzig: Veit.

1901 *Braunschweiger Volkskunde*. Braunschweig.

Anonymous

1796 "Two Stones that Fell from the Sky." *Gentleman's Magazine* 66:1007–1008.

1824 "Notice on the Malleable Iron of Louisiana." *AJS* 8:218–225.

1834 "The Meteoric Shower." *New England Magazine* 6:47–54.

1854 "Meteorites." *Southern Literary Messenger* 20:286–294.

1860 "Report on Observations of Luminous Meteors." *BAAS*, 26.

1864 "Meteoric Stones." *All the Year Round* 10:488–491.

1867 "Comets and Shooting Stars." *Every Saturday* 41:780–783.

1868 "Papers Bearing on Meteoric Astronomy." *BAAS*, 393–417.

1879 "Earth-born Meteorites." *Cornhill Magazine* 40:446–464.

1882 "J. L. Smith, Organic Remains in Meteorites." *Popular Science Monthly* 20:568–569.

1884 *Biblioteca de las Tradiciones Populares Españoles*. Madrid.

1886 "New Orleans Superstitions." *Harper's Weekly* 30:843.

1953 "Collecteanea." *Folklore* 64:286–301.

Ansdell, Gerrard, and James Dewar

1886 "On the Gaseous Constituents of Meteorites." *PRS* 40:552.

Arago, Dominique F.

1836 "Sur les étoiles filantes." *ACP* 61:176–181.

Archibald, J. David, and William A. Clemens

1982 "Late Cretaceous Extinctions." *American Scientist* 20:377–385.

Arral, W. J., R. B. Jacobi, and F. A. Paneth

1942 "Meteorites and the Age of the Solar System." *Nature* 149:235–238.

Arrest, Heinrich L. d'

1867 "Ueber einige merkwürdige Meteorfälle beim Durchgange der Erde durch die Bahn des Biela'schen Cometen." *AN* 69 (Nr. 1633), col. 7–10.

Arrhenius, Svante

1908 *Worlds in the Making*. New York and London: Harper and Brothers.

Atkinson, Henry

1819 "On Hypotheses Proposed for Explaining the Origin of Meteoric Stones." *PM* 54:336–342.

Bache, Alexander D.

1835*a* "Meteoric Observations made on or about the 13th of November, 1834." *AJS* 27:335–338.

1835*b* "Replies to a Circular in relation of an Unusual Meteoric Display on the 13th Nov. 1834, addressed by the Secretary of War to the Military Posts of the United States, with other Facts relating to the same Question." *AJS* 28:305–309.

1836 "Observations upon the Facts recently presented by Prof. Olmsted, in relation to Meteors, seen on the 13th of Nov. 1834." *AJS* 29:383–398.

Bächtold, Hans
1915 "Mitteilungen aus dem schweizerischen Soldatenleben." *Schweizerisches Archiv für Volkskunde* 19:201–264.

Badash, Lawrence
1968 "Rutherford, Boltwood, and the Age of the Earth." *AAAS* 112:157–169.

Baker, Robert H.
1938 *Astronomy*. New York: D. Van Nostrand Co., Inc.

Baldanza, B.
1965 "Italian Meteorites." *MM* 35:214.

Balfour, Henry
1929 "Concerning Thunderbolts." *Folklore* 40:37–49.

Ball, Robert S.
1879 "Speculations on the Source of Meteorites." *Nature* 19:493–495.

Barcena, Mariano
1876 "On certain Mexican Meteorites." *ANSP* 28:122–126.

Barham, Henry
1718 "Relation of a Fiery Meteor seen by him in Jamaica, to strike into the Earth." *PT* 30:837–838.

Barnes, Virgil E.
1963 "Tektite Strewn Fields." In *Tektites*. Ed. John A. O'Keefe. 25–50. Chicago: University of Chicago Press.

Barringer, Brandon
1964 "Daniel Moreau Barringer (1860–1929) and His Crater (the Beginning of the Crater Branch of Meteoritics." *Meteoritics* 2:183–199.

Barringer, Daniel M.
1905 "Coon Mountain and its Crater." *ANSP* 57:861–886.
1909 *Meteor Crater (formerly called Coon Mountain or Coon Butte) in Northern Central Arizona*. Delivered before the National Academy of Sciences. Privately published.
1914 "Further Notes on Meteor Crater, Arizona." *ANSP* 66:556–566.
1924 "Further Notes on Meteor Crater in Northern Central Arizona." *ANSP* 76:275–278.

Barringer, Daniel M., Jr.
1928 "A New Meteor Crater." *ANSP* 80:307–311.

Barthold, Charles
1800 "Analyse de la Pierre de Tonnerre." *JP* 50:169–176.

Bartsch, Karl
1879–80 *Sagen, Märchen und Gebräuche in Mecklenburg*. 2 vols. Wien.

Basset, René
1899 "Folk-Lore Astronomique." *Revue des traditions populaires* 14:579–657.

Baudin, N.
1798 "Account of Fiery Meteor seen in Gascony, July 24, 1790 with Observations on Fireballs and Shooting Stars by Chladni." *PM* 2:225–231.

Bauer, Carl A.
1947 "Production of Helium in Meteorites by Cosmic Radiation." *PR* 72:354–355.
1948 "The Absorption of Cosmic Radiation in Meteorites." *PR* 74:225–226.

Baumhauer, Edouard H. von
1845 "Ueber den mutmasslichen Ursprung der Meteorsteine, nebst einer Analyse des Meteorsteins welcher am 2 Juli 1843 in der Provinz Utrecht gefallen ist." *AP* 64:465–502.

Baumhauer, Edouard H. von, and F. Seelheim
1862 "Über eine für einen Meteorsteine gehaltene Gesteinmasse." *AP* 116:189–190.

Baxter, John, and Thomas Atkins
1976 *The Fire Came By*. Garden City, N.Y.: Doubleday & Co., Inc.

Bayen, N.
1791 "Observations communiqué à M. Bayen: sur un globe de feu qui a paru dans ces contrées dans la nuit." *Bulletin des sciences de la Société Philomatique,* No. 1.

Beauford, William
1802 "Some Conjectures respecting the Origin of Stones which have been observed to fall from the Clouds." *PM* 14:148–151.

Beck, Ludwig
1879 "Das Meteoreisen in technische und culturgeschichtliche Beziehung." *Archiv für Anthropologie* 12:293–314.

Becquerel, Paul
1924 "La vie terrestre provient-elle d'un autre monde?" *Bulletin de la société astronomique de France* 38:393–417.

Begemann, Friedrich, J. Geiss, and D. C. Hess
1957 "Radiation Age of a Meteorite from Cosmic-Ray-Produced He^3 and H^3." *PR* 107:540–542.

Belaiew, N. T.
1923 *Crystallisation of Metals*. London: University of London Press.
1924 "On the Genesis of Widmanstätten structure in Meteorites and in Iron Nickel and Iron Carbon Alloys." *MM* 20:173–185.

Beloch, Julius
1900 "Die Bevölkerung Europas zur Zeit der Renaissance." *Zeitschrift für Socialwissenschaft* 3:765–786.

Benzenberg, Johann F.
1811 *Briefe geschrieben auf einer Reise durch die Schweiz im Jahre 1810*. Düsseldorf.
1834 *Die Sternschnuppen sind Steine aus den Mondvulkanen, die einer Durchmesser von 1 bis 5 Fuss haben, welche bei 8000 Fuss Geschwindigheit in 1 Seconde NICHT wieder auf den Mond zurückkommen und die dann mit Millionen um die Erde herumlaufen*. Bonn: Eduard Weber.

Bergemann, Carl W.
1849 "Über das Meteoreisen von Zacatecas." *AP* 78:406–413.
1856 "Untersuchungen von Meteoreisen." *AP* 100:245–260.

Bergman, Torbern
1783 *Outlines of Mineralogy*. Translated by William Withering, M.D.

Birmingham: Piercy and Jones.

Berman, Harry

1937 "Constitution and Classification of the Silicates." *American Mineralogist* 22:342–408.

Berthelot, Pierre-Eugène-Marcellin

1866 "Sur l'origine des carbures et des combustible minéraux." *ACP* 9:481–483.

1868 "Sur la matière charbonneuse des météorites." *CR* 67:849.

Bertholon, Pierre

1777 "Mémoire dans lequel on donne un nouveau moyen de se préserver du tonnerre, après avoir prouvé que la foudre s'élève souvent de la terre." *JP* 10:179–196.

1787 *De l'Électricité des Météors*. 2 vols. Lyon.

1791 "Observations d'un globe de feu." *Journal des Sciences Utiles* 4:224–228.

Bertrand, Louis

1801 "Fer natif de Sibérie. Discussion sur son origine, dans une lettre de M. Bertrand aux redacteurs." *BB* 17:433–441.

Berwerth, Friedrich M.

1906 *Andreas Xaver Stütz*. Wien.

1907 "Javanische Waffen mit Meteoreisen-pamor." *Mineralogische und petrologische Mitteilungen* 26:506–507.

1912 "Meteoriten." In *Handwörterbuch der Naturwissenschaften*. Ed. F. Korschott. VI:845–862. Jena.

1913–14 "Karl Ludwig Freiherr v. Reichenbach." *Mineralogische und petrologische Mitteilungen* 32:153–169.

1916 "On the Origin of Meteorites." *Annual Report of the Smithsonian Institution,* 311–320.

1917 "Können die Tektite als Kunstprodukte gedeutet werden?" *Zentralblatt für Mineralogie,* 240–254.

1918 "Die Meteoritensammlung der naturhistorischen Hofmuseums als Born der Meteoritenkunde." *SAW* 127:1–81.

Berzelius, Jöns Jacob

1833 "Untersuchung einer bei Bohumiltz in Böhmen gefunden Masse." *AP* 27:118–132.

1834 "Über Meteorsteine." *AP* 33:1–32, 113–148.

1836–37 "On Meteoric Stones." *Edinburgh Philosophical Journal* 22:1–7.

Bessel, Friedrich W.

1839 "Vorläufige Nachricht über eine die Berechnung der Sternschnuppen betreffende Arbeit." *AP* 47:525–527.

Bigot de Morogues, Pierre M.-S.

1812 *Mémoire historique et physique sur les chutes des pierres tombées sur la surface de la terre à diverses époques*. Orléans: Jacob Ainé.

Bingley, W.

1796 "Stones fallen from the Air a Natural Phenomenon." *Gentlemen's Magazine* 66, Part 2:726–728.

Biot, Edouard C.

1848 "Catalogue géneral des Étoiles Filantes et des autres Météores observées en Chine pendant 24 siècles depuis le 7ème siècle av. J.C." *Mémoires présentée par divers savants à l'Académie Royale des sciences (Paris)* 10:129–352, 415–422. (Presented 21 May and 26 July, 1841.)

Biot, Jean B.

1802*a* "Note sur des substances pierreuses d'une nature particulière, que l'un assure être tombée sur la terre." *Bulletin des sciences de la Société Philomatique* 66:139–140.

1802*b* "Sur les substances minérales prétendus tombées du ciel, et nouvellement analyses par MM. Howard et Bournon." *Bulletin des sciences de la Société Philomatique* 68:153–156.

1803*a* "Hypothese La Place's über den Ursprung der meteorischen Steine, vorgetragen und erörtert." *AP* 13:358–370.

1803*b* "Lettre de M. Biot à l'occasion de celle de M. Deluc sur les pierres tombées." *BB* 21:315–320.

1803*c* "Lettre à M. Pictet." *BB* 23:394–405.

1803*d* "Lettre." *BB* 24:82–85.

1807 "Relation d'un Voyage fait dans le Département de l'Orne, pour constater la réalité d'un météore observé á l'Aigle le 6 floréal an 11." *Mémoires de la classe des sciences, mathématique et physique, L'Institut National de France,* 224–266. (Read 17 July 1803.)

1837 "The November Meteors." *AJS* 32:181–183.

Bird, G.

1820 "Letter." *PM* 56:156–158.

Birgham, Francis

1881 "The Discovery of Organic Remains in Meteor Stones." *Popular Science Monthly* 20:83–87.

Birlanger, Anton

1861–62 *Volkstümliches aus Schwaben.* 2 vols. Freiburg.

1874 *Aus Schwaben: Sagen, Legenden, Aberglauben, usw.* 2 vols. Wiesbaden.

Bjorkman, Judith Kingston

1973 "Meteors and Meteorites in the Ancient Near East." *Meteoritics* 8:91–132.

Blagden, Charles

1784 "An account of some late Fiery Meteors, with Observations." *PT* Part I:201–232.

Blander, Milton

1979 "Non-Equilibrium Effects on the Chemistry of Nebular Condensates: Implications for the Planets and Asteroids." In *Asteroids.* Ed. Tom Gehrels. 809–821. Tucson: University of Arizona Press.

Blumhof, J. G. L.

1816 "Nachträge zu Chladni's neuen chronologischen Verzeichnisse herabgefallener Stein-und Eisenmassen, mit einen Zusatze von Dr. Chladni." *AP* 53:307–312.

Bogard, D. D., L. Husain, and R. J. Wright

1976 "^{40}Ar-^{39}Ar dating of collisional events in chondrite parent bodies." *Journal of Geophysical Research* 81:5664–5678.

Bogard, D. D., and P. Johnson

1983 "Martian Gases in an Antarctic Meteorite?" *Science* 221:651–654.

Boguslawski, Georg von

1867*a* "Notiz über die kosmisches Theorie der Feuermeteore." *AP* 130:165–168.

1867*b* "Das November Phaenomenon der Sternschnuppen in seinen einzelnen Erscheinungen, von den ältesten Zeiten bis 1866." *AP* 130:471–493.

Bohor, Bruce F., E. E. Foord, P. J. Modreski, and D. M. Triplehorn
1984 "Mineralogic Evidence for an Impact Event at the Cretaceous-Tertiary Boundary." *Science* 224:867–869.

Boisse, A.
1850 *Recherches sur l'histoire, la nature et l'origine des Aérolithes.* Aveyron: Rodez.

Boltwood, Bertram B.
1904 "On the Ratio of Radium to Uranium in some Minerals." *AJS* 118:97–103.
1905 "The Origin of Radium." *PM* 9:599–613.

Bonsdorff, P. A. von
1822 "Forsök att bestämma sammansättningen af de mineralier, hvilka kristallisera i Amphiboleus form." *ACP* 20:5–34.

Boodt, Anselm Boece de
1744 *Le parfait ioaillier, ou Histoire des pierreries.* Lyon.

Bosler, Jean
1945 "Sur une averse de météorites mentionée dans la Bible." *Astronomie* 59:47.

Botley, Cecelia
1970 "A comet and meteors in the early 16th century." *Journal of the British Astronomical Association* 80:304–305.

Bouche-Leclerq, Auguste
1899 *L'Astrologie grecque.* Paris.

Bournon, Jacques-Louis Comte de
1803 "Lettre de M. le comte de Bournon à M. Delamétherie, en réponse à la critique de M. Patrin à l'égard des pierres tombées de l'atmosphere." *JP* 56:294–304.

Boussingault, Jean B.
1861 "Sur la présence de l'azote dans un fer météorique." *ACP* 63:336–343.
1868 "Analyse d'une fonte chromifère. Dosage du carbone dans la fonte, le fer, et l'acier." *CR* 66:873–877.

Bowditch, Nathaniel
1815 "An estimate of the height, direction, velocity, and magnitude of the meteor that exploded over Weston, in Connecticut, December 14, 1807. With methods of Calculating observations made on such Bodies." *Memoirs of the American Academy of Arts and Sciences* 3, Part 2:213–236.

Brady, L. F.
1931 "A new iron meteorite from Pojoaque, New Mexico." *AJS* 21:178.

Bragg, William Lawrence
1930 "The Structure of Silicates." *Zeitschrift für Kristallographie* 74:237–305.

Brandes, Heinrich W.
1805 "Einige kritische Bemerkungen über Höfe, Ringe, Nebensonnen, Fata Morgana, usw." *AP* 19:370–371.

Bredekhin, Fedor A.
1890 "Abstract." *Nature* 42:20.

Brewer, A. Keith
1935 "Further evidence for the existence of K^{40}." *PR* 48:640.

Brewster, David
1819 "Account of Meteoric Stones, Masses of Iron, and Showers of Dust, Red Snow, and other Substances, which have fallen from the Heavens from the earliest period down to 1819." *Edinburgh Philosophical Journal* 1:223–235.
1820 "Meteorology." *Edinburgh Philosophical Journal* 3:197.
1832*a* "Astronomy." *Edinburgh Encyclopedia* II:602–606.
1832*b* "Meteorites." *Edinburgh Encyclopedia* XIII:126–158.

Brewster, David, ed.
1837 *Letters of Euler on Different Subjects in Natural Philosophy addressed to A German Princess*. 2 vols. New York: Harper and Bros.

Brezina, Aristides
1880 "Über die Reichenbach'schen Lamellen in Meteoreisen." *Denkschriften der Akademie der Wissenschaften, Wien* 43:13–16.
1885 *Die Meteoritensammlung des k.k. mineralogischen Hofkabinetes in Wien am 1. Mai 1885*. Wien: Alfred Hölder.
1887 "Urgeschichtliche Meteorite." *Mitteilungen der anthropologische Gesellschaft Wien* 17:69–70.
1896 *Die Meteoritensammlung der k.k. Naturhistorischen Hofmuseums am 1. Mai 1895*. Wien.
1904 "The arrangement of collections of meteorites." *Proceedings of the American Philosophical Society* 43:211–247.

Briggs, J. Morton, Jr.
1967 "Aurora and Enlightenment; Eighteenth Century Explanations of the Aurora Borealis." *Isis* 58:491–503.

Brown, Harrison S.
1949 "A Table of Relative Abundances of Nuclear Species." *Reviews of Modern Physics* 21:625–634.
1953 *A Bibliography on Meteorites*. Chicago: University of Chicago Press.
1957 "The Age of the Solar System." *Scientific American* 196, 4:80–94.

Brown, Harrison S., and Edward Goldberg
1949 "The Neutron Pile as a Tool in Quantitative Analysis: The Gallium and Palladium Content of Iron Meteorites." *Science* 109:347–353.

Brown, Harrison S., and Claire Patterson
1947*a* "The Composition of Meteoritic Matter. I. The Composition of the Silicate Phase of Stony Meteorites." *Journal of Geology* 55:405–411.
1947*b* "The Composition of Meteoritic Matter. II. The Composition of Iron Meteorites and of the Metal Phase of Stony Meteorites." *Journal of Geology* 55:508–510.
1948 "The Composition of Meteoritic Matter. III. Phase Equilibria, Genetic Relations, and Planet Structure." *Journal of Geology* 56:85–111.

Buchner, Chr. Ludwig Otto
1859 *Die Feuermeteore, inbesondere die Meteoriten, historisch und naturwissenschaftlich betrachtet*. Giessen: J. Riker.
1863 *Die Meteoriten in Sammlungen*. Leipzig: W. Engelmann.

Buchwald, Vagn F.
1975 *Handbook of Iron Meteorites*. 3 vols. Berkeley, Los Angeles, London: University of California Press.

Burke, John G.
1966 *Origins of the Science of Crystals*. Berkeley and Los Angeles: University of California Press.

Burnet, John
1961 *Early Greek Philosophy*. New York: Meridian.

Burney, William
1821 "On the Appearances of Meteors as Prognosticated by Wind and Rain." *PM* 58:127–130, 198–200.

Butler, ?
1829 "Origin of Aerolites." *Monthly Magazine* 8 NS:111.

Butler, William Francis
1875 *The Great Lone Land: A Narrative of Travel and Adventure in the North-West of America*. London.

Cajori, Florian, ed.
1962 *Sir Isaac Newton's Mathematical Principles*. Berkeley and Los Angeles: University of California Press.

Cameron, A. G. W.
1975 "The Origin and Evolution of the Solar System." *Scientific American* 233, 3:32–41.
1979 "On the Origin of Asteroids." In *Asteroids*. Ed. Tom Gehrels. 992–1007. Tucson: University of Arizona Press.

Canfield, Frederick A.
1923 *The Final Disposition of Some American Collections of Minerals*. Dover, N.J.

Canton, John
1753 "Electrical Experiments, With an Attempt to Account for their Several Phenomena." *PT* 48:357–358.

Cappocci di Belmonte, Ernesto
1840 "Sur la périodicité des aérolithes." *CR* 11:357–360.

Carmen, Roy
1955 *La Litterature orale en Gaspene*. Ottawa.

Carmer, Carl
1940 *Listen for the Lonesome Drum*. New York.

Carpenter, R. R.
1945 "A Legal Treatise on Meteorites: or, Did you ever see a 'Tomanowas'?" *Popular Astronomy* 53:186–192, 238–245.

Cassidy, William A., Luisa M. Villar, Theodore E. Bunch, Truman P. Kohman, and Daniel J. Milton
1965 "Meteorites and Craters of Campo del Cielo Argentina." *Science* 149:1055–1064.

Cassidy, William A., and Louis A. Rancitelli
1982 "Antarctic Meteorites." *American Scientist* 70:156–164.

Celis, Don Rubin de
1788 "An Account of a Mass of Native Iron found in South America." *PT* 78:37–42, 183–189.

Ceplecha, Z.
1961 "Multiple Fall of Pribram Meteorites Photographed." *Bulletin of the Astronomical Institute of Czechoslovakia* 12:21–47.

Chamberlain, V. D.
1982 "The Skidi Pawnees." Abstract of paper read to the Meteoritical Society, 1982.

Chamberlain, Thomas C.
1901 "On a Possible Function of Disruptive Approach in the Formation of Meteorites, Comets, and Nebulae." *Journal of Geology* 9:369–392.
1924 "Review of 'Study of Fundamental Problems of Geology.'" *Yearbook of the Carnegie Institution* 23:270–282.
1929 *The Two Solar Families: The Sun's Children*. Chicago: University of Chicago Press.

Chant, C. A.
1943 "Meteoritic Iron in Ancient Egypt." *Journal of the Royal Astronomical Society of Canada* 37:127.

Chapman, Clark R.
1975 "The Nature of Asteroids." *Scientific American* 232, 1:24–33.
1979 "The Asteroids: Nature, Interrelations, Origin, and Evolution." In *Asteroids*. Ed. Tom Gehrels. 25–60. Tucson: University of Arizona Press.

Charpentier, Johann F. W.
1778 *Mineralogische Geographie der Chursachsischen Lande*. Leipzig: S. L. Crusius.

Chasles, Michel
1841 "Catalogue d'apparitions d'étoiles filantes pendant six siècles de 538 à 1123." *CR* 12:499–509.

Chladni, Ernst F. F.
1794 *Ueber den Ursprung der von Pallas gefundenen und anderer ihr ähnlicher Eisenmassen*. Riga: J. F. Hartknoch.
1798*a* "Observations on a Mass of Iron found in Siberia by Professor Pallas and on other Masses of the like Kind, with Some Conjectures respecting their Connection with certain Natural Phenomena." *PM* 2:1–8.
1798*b* "Account of a remarkable fiery Meteor seen in Gascony on the 24th of July 1790 by M. Baudin, Professor of Philosophy at Pau. With some Observations on Fireballs and Shooting Stars." *PM* 2:225–231.
1799 "Observations on Fire-Balls and Hard Bodies which have fallen from the Atmosphere." *PM* 2:331–335.
1801 "Observations sur une masse de fer, trouvée en Sibérie par le Prof. Pallas, et sur d'autres masses du même genre, avec quelques conjectures sur les rapports avec certaines phénomènes naturels." *BB* 16:73–88.
1803 "Chronologisches Verzeichniss der mit einen Feuermeteor niedergefallen Stein-und Eisenmassen, nebst einigen Bemerkungen." *AP* 15:307–328.
1805*a* "Berichtigung einen angeblichen Meteorstein betreffend." *AP* 19:243.
1805*b* "Einige Kosmologische Ideen, die Vermehrung oder Verminderung der Masse eines Weltkörpers betreffend." *AP* 19:257–281.
1808 "Beiträge zu den Nachrichten von Meteorsteinen." *AP* 29:375–383.
1817*a* "Ueber die sprungweise gehende Bewegung mancher Feuerkugeln,

nebst einigen Folgerungen." *AP* 55:91–101.
1817*b* "Ueber einige von Himmel gefallene Materien, die von den gewöhnlichen Meteorsteinen verschieden sind." *AP* 55:249–276.
1818*a* "Ueber Sternschnuppen von dem Dr. Benzenberg." *AP* 58:289–302.
1818*b* "Verschiedene physikalische Bermerkungen." *AP* 59:1–11.
1818*c* "Ueber Dinge, die sich in dem Weltraume befinden, und von den bekannten Weltkörpern verschieden sind." *AP* 59:87–94.
1819*a* "Vierte Fortsetzung des Verzeichnisses der von Himmel gefallen Massen, nebst Bemerkungen über einige schon bekannte Massen dieser Art, und Beiträgen zur Geschichte hierher gehörender Meteore." *AP* 60:238–254.
1819*b* *Über Feuer-Meteore und über die mit denselben herabgefallenen Massen*. Wien.
1823 "Neue Beiträge zur Kenntnis der Feuermeteore und der herabgefallenen Massen." *AP* 75:229–257.

Chung, Li
1979 "Studies on the Iron Blade of a Shang Dynasty Bronze Yüeh-Axe Unearthed at Kao-Ch'eng, Hopei, China." *Ars Orientalis* XI:259–289.

Clap, Thomas
1781 *Conjectures upon the Nature and Motion of Meteors*. Norwich, Conn.: John Trumbull.

Clark, J. D.
1966 "North Carolina Superstitions." *North Carolina Folklore* 14:10–23.

Clark, William S.
1852 *On Metallic Meteorites—Inaugural Dissertation*. Göttingen: Kaestner.

Clarke, Roy S., Jr.
1977 "William Thomson (1761–1806) A Neglected Meteoriticist." *Meteoritics* 12:194–195.

Clarke, Roy S., Jr., and Eugene Jarosewich
1969 "Classification and Bulk Chemical Composition of the Campo del Cielo, Argentina, Meteorite." *Meteoritics* 4:162.

Clarke, Roy S., Jr., Eugene Jarosewich, Brian Mason, Joseph Nelen, Manuel Gómez, and Jack R. Hyde
1970 *The Allende, Mexico, Meteorite Shower*. Washington, D.C.: Smithsonian Institution Press.

Clarke, Roy S., Jr., Eugene Jarosewich, and Joseph Nelen
1971 "The Lost City, Oklahoma, Meteorite: An Introduction to Its Laboratory Investigation and Comparisons with Pribram and Ucera." *Journal of Geophysical Research* 76:4135–4143.

Clarke, Roy S., Jr., and Joseph I. Goldstein
1978 *Schreibersite Growth and Its Influence on the Metallography of Coarse-Structured Iron Meteorites*. Smithsonian Contributions to the Earth Sciences, No. 21. Washington, D.C.: Smithsonian Institution Press.

Clarke, W. A.
1836 "On the Origin of Shooting Stars." *AJS* 30:369–370.

Claus, George, and Bartholomew Nagy
1961 "A microbiological examination of some carbonaceous chondrites." *Nature* 192:594–596.

Clayton, Robert N.
1977 "Interstellar Material in Meteorites: Implications for the Origin and Evolution of the Solar Nebula." In *Comets, Asteroids, Meteorites*. Ed. A. H. Delsemme. 335–341. Toledo, Ohio: University of Toledo.

Clayton, Robert N., Lawrence Grossman, and Toshiko K. Mayeda
1973 "A Component of Primitive Nuclear Composition in Carbonaceous Meteorites." *Science* 182:485–488.

Clayton, Robert N., Naoki Onuma, Lawrence Grossman, and Toshiko K. Mayeda
1977 "Distribution of the Pre-solar Component in Allende and Other Carbonaceous Chondrites." *EPSL* 34:209–224.

Cloëz, Stanislas
1864 "Analyse chimique de la pierre météorique d'Orgueil." *CR* 59:37–40.

Cockayne, T. O., ed.
1864–66 *Leechdoms, Wortcunning and Starcraft*. London.

Cogels, Paul
1907 *Céranies et Pierres de Foudre: Histoire et Bibliographie*. Anvers: J. van Hille—de Baeker.

Cohen, Emil W.
1894–1905 *Meteoriten-Kunde*. 3 vols. in one. Stuttgart.
1899 "Über den Wülfing'schen Tauschwerth der Meteoriten im Vergleich mit dem Handelspreisen." *Mittheilungen aus dem naturwissenschaft Vereinigung für Neu-Vorpommern und Rügen* 31:50–62.

Coleman, A. F.
1886 "A Meteorite from the Northwest (Iron Creek)." *Transactions of the Royal Society of Canada* 4, Sec. 3:97.

Coquebert, Eugène, trans.
1802–03 "Réflexions sur l'origine des diverses Masses, der fer natif, et notamment de celle trouvée par Pallas en Sibérie." Traduites de l'Allemand de M. Chladni. *JM* 15:286–320, 446–485.

Corcoran, Thomas H., ed. and trans.
1971 *Seneca Naturales Quaestiones*. 2 vols. Cambridge, Mass.: Loeb Library.

Coulvier-Gravier, Remi Armand
1859 *Recherches sur les Météores*. Paris: Mallet-Bachelier.

Coulvier-Gravier, Remi Armand, and J. F. Saigey
1847 *Recherches sur les Étoiles Filantes*. Paris: Hachette.

Cowen, Robert C.
1980 "UFOs: Fact and Frivolity." *Technology Review* 82:6–7.

Cronstedt, Axel F.
1788 *An Essay Towards a System of Mineralogy*. Translated by Gustav von Engestrom. 2 vols. London: John Hyacinth de Magellan.

Crosland, Maurice
1967 *The Society of Arcueil*. London: Heineman.

Dähnhardt, Oskar
1907 *Natursagen*. Leipzig.

Daly, Reginald A.
1943 "Meteorites and an Earth Model." *Bulletin of the Geological Society of America* 54:401–456.
1946 "Origin of the Moon and Its Topography." *Proceedings of the Amer-*

ican Philosophical Society 90:104–119.

Darnton, Robert
1968 *Mesmerism and the End of the Enlightenment in France*. Cambridge, Mass.: Harvard University Press.

Darton, Nelson Horatio
1910 "A Reconnaissance of Parts of Northwestern New Mexico and Northern Arizona." *Bulletin of the U.S. Geological Survey*, No. 435.

Darwin, George H.
1888 "On the mechanical conditions of a swarm of meteorites and on theories of cosmogony." *PRS* 45:3–16.

Daubrée, Gabriel-Auguste
1864 "Complément d'Observations sur la chute de météorites qui a eu lieu 14 Mai 1864 aux environs d'Orgueil." *Nouvelles Archives du Muséum d'Histoire Naturelle* 3:1–19.
1866 "Expériences synthétiques relatives aux météorites. Rapprochements auxquels elles conduisent, tant pour la formation de ces corps planétaires que pour celle du globe terrestre." *CR* 62:200–206, 369–375, 660–674.
1867 "Classification adoptée pour la collection de météorites du Muséum." *CR* 65:60–63.
1877 "Observations sur la structure intérieure d'une des masses de fer natif d'Ovifak." *CR* 84:69.
1879 *Études Synthétiques de Géologie Expérimentale*. 2 vols. Paris: Dunod.
1893 "Observations sur les conditions qui paraissent avoir présidé à la formation des météorites." *CR* 116:345–347.

Davidson, John M.
1891 "Analyses of Kamacite, Taenite and Plessite from the Welland Meteoric Iron." *AJS* 192:64–66.
1899 "Platinum and Iridium in Meteoric Iron." *AJS* 207:4.

Day, Jeremiah
1810 "A View of the Theories which have been proposed to explain the Origin of Meteoric Stones." *Memoirs of the Connecticut Academy of Arts and Sciences* 1:163–174.

Debus, Allen
1965 *The English Paracelsians*. London: Oldbourne.

Delsemme, A. H.
1977 "The Origin of Comets." In *Comets, Asteroids, Meteorites*. Ed. A. H. Delsemme. 453–467. Toledo: University of Toledo.

Deluc, Guillaume-Antoine
1801–02 "Observations sur la Masse de Fer de Sibérie." *JM* 11:213–220.
1802 "Sur la Masse de Fer de Sibérie, et sur les Pierres supposées tombées de l'Atmosphère." *JM* 13:92–107.
1803 "Nouvelles Considérations sur la Masse de Fer de Sibérie et sur les Pierres supposées tombées du Ciel sur la Terre." *BB* 23:78–112.

Deluc, Jean-André
1801 "Pierres tombées." *BB* 18:272–298.

Denning, W. F.
1880 "The August Meteors." *Popular Science Monthly* 18:178–190.
1888 "A History of the August Meteors." *Nature* 38:392–394.

Dietz, Robert S.
1946 "The Meteoritic Impact Origin of the Moon's Surface Features." *Journal of Geology* 54:359–375.

Dietz, Robert S., and John McHone
1974 "Kaaba Stone: Not a Meteorite, Probably an Agate." *Meteoritics* 9:173–179.

Dodd, Robert T.
1981 *Meteorites: A Petrological-chemical Synthesis*. Cambridge: Cambridge University Press.

Dodd, Robert T., J. E. Grover, and G. E. Brown
1975 "Pyroxenes in the Shaw (L-7) Chondrite." *GCA* 39:1585–1594.

Doelter, Cornelio A.
1914 "Konstitution der Silicate." *Handbuch der Mineralchemie* 2:61–109.

Doolittle, M. H.
1878 "Aerolithic Disturbance of Planetary Motions." *Bulletin of the Philosophical Society of Washington, D.C.* 2:190–192.

Dorsey, George A.
1906 *The Pawnee: Mythology*. Washington, D.C.: Carnegie Institution.

Dortous de Mairan, Jean J.
1733 *Traité Physique et Historique de l'Aurore Boréale*. Paris.

Drake, Stillman, trans.
1957 *Discoveries and Opinions of Galileo*. Garden City, N.Y.: Doubleday Anchor.

Drapiez, Pierre A. J., ed.
1837 *Dictionnaire Classique des Sciences Naturelles*. Bruxelles: Meline, Cans.

Drechsler, Paul
1903–06 *Sitte, Brauch und Volksglaube*. 2 vols. Leipzig.

Drée, Etienne Marquis de
1803*a* "Account and Description of a Stone which fell from the Clouds in the Commune of Sales, near Ville-Franche, in the Department of the Rhone." *PM* 16:217–224.
1803*b* "Recherches sur les masses minérales dites tombées de l'atmosphère sur notre globe." *JP* 56:380–389.
1803*c* "Suite de recherches sur les masses minérales dites tombées de l'atmosphère sur notre globe." *JP* 56:405–427.
1803*d* "On the Stones said to have fallen at Ensisheim, in the Neighborhood of Agen, and at other Places." *PM* 16:289–298.

Dryden, John, trans., and Arthur H. Clough, reviser
1932 *Plutarch: The Lives of the Noble Grecians and Romans*. New York: Modern Library.

Duflos, Adolph F.
1849 "Chemische Zerlegung das Meteoreisenmasse von Seeläsgen." *AP* 74:61–66.

Duflos, Adolph F., and Nicolas W. Fischer
1847 "Analyse des Braunauer Meteoreisens und Fortsetzung der begonnen Analyse." *AP* 72:575–580.

Dufresne, E. R., and Edward Anders
1963 "Chemical Evolution of the Carbonaceous Chondrites." In *The Moon,*

Meteorites, and Comets. Ed. Barbara M. Middlehurst and Gerard P. Kuiper. 496–526. Chicago: University of Chicago Press.

Dupree, A. Hunter
1957 *Science in the Federal Government*. Cambridge, Mass.: Harvard University Press.

Egen, P. N. C.
1822 "Versuch eines Beweisses, dass wahrscheinlich die Feuermeteore atmosphärischen Ursprungs sind." *AP* 72:375–422.

Ehrenberg, C. G.
1839 "Über das in Jahre 1686 in Curland vom Himmel gefallene Meteorpapier." *AP* 46:187–188.

Ellicott, Andrew
1804 "Account of an extraordinary Flight of Meteors (commonly called shooting stars)." *Transactions of the American Philosophical Society* 6:28–29.

Ende, Ferdinand A. von
1804 *Über Massen und Steine, die aus dem Monde auf die Erde gefallen sind*. Braunschweig.

Engel, Michael, and Bartholomew Nagy
1982 "Distribution and enantiomeric composition of amino acids in the Murchison meteorite." *Nature* 296:837.

Engelbach, Theophilus
1862 "Lithium und Strontium im Meteorstein vom Capland." *AP* 116:512.

Engelhardt, Wolf Freiherr von
1977 *Mineralogie, Geologie und Paläontologie an der Universität Tübingen von den Anfängen bis zur Gegenwart*. Tübingen: Mohr.

Erman, Georg Adolph
1839 "Ueber einige Thatsachen, welche wahrscheinlich machen, dass die Asteroïden der Augustperiods sich im Februar, und die der Novemberperiode im Mai eines jeden Jahres zwischen der Sonne und der Erde, auf dem Radius Vector der letzeren, befinden." *AP* 49:582–601.

Espinosa, Aurelio M.
1910 "New Mexico Spanish Folklore." *Journal of American Folklore* 23:395–418.

Evans, J. C., J. H. Reeves, and L. A. Rancitelli
1982 "Aluminum 26: Survey of Victoria Land Meteorites." In *Catalog of Meteorites from Victoria Land, Antarctica, 1978–1980*. Ed. Ursula Marvin and Brian Mason. 70–74. Smithsonian Contributions to the Earth Sciences, No. 24. Washington, D.C.: Smithsonian Institution Press.

Evans, Robley D.
1943 "Measurements of the Age of the Solar System." *Geological Series of the Field Museum of Natural History* 7:79–98.

Falconet, ?
1729 "Sur les Baetyles." *HMAR* VI:513–532. (Read Sept. 1722.)
1756 "Dissertation sur la Pierre de la Mère des Dieux." *HMAR* XXIII:213–241. (Read April 1750.)

Farey, John, and Benjamin Bevan
1813 "On the Connexion between Shooting Stars and large Meteors."

Nicholson's Journal of Natural Philosophy 34:298–300.

Farrell, Michael A.
1933 "Living Bacteria in Ancient Rocks and Meteorites." *American Museum Novitiates,* No. 645:1–3.

Farrington, Oliver C.
1900 "The Worship and Folk-Lore of Meteorites." *Journal of American Folklore* 13:199–212.
1901 "The Constituents of Meteorites." *Journal of Geology* 9:522–532.
1906 "Professor Henry A. Ward." *Science* 24:61, 153–154.
1915 *Meteorites: Their Structure, Composition, and Terrestrial Relations.* Chicago.

Farrington, Oliver C., comp.
1907 *Analyses of Iron Meteorites.* Chicago: Field Museum of Natural History.

Fechner, Gustav T.
1876 *Erinnerungen an die letzen Tage der Odlehre und ihres Urhebers.* Leipzig: Breitkopf und Härtel.

Figuier, Louis
1872 *The Tomorrow of Death.* Translated by S. R. Crocker. Boston.

Fireman, Edward L.
1970 "The Lost City Meteorite—A Deep Space Probe for Cosmic Rays." *Sky and Telescope* 39:6.

Firsoff, V. A.
1963 *Life Beyond the Earth: A Study in Exobiology.* London: Hutchinson & Co.

Fischer, Nicolas W.
1848 "Schluss der Untersuchung des Braunauer Meteoreisens." *AP* 73:590–594.

Fish, Robert A., Gordon G. Goles, and Edward Anders
1960 "The Record in Meteorites. III. On the Development of Meteorites in Asteroidal Bodies." *Astrophysical Journal* 132:243–258.

Fitch, Frank, H. P. Schwarcz, and Edward Anders
1962 "Organized Elements in Carbonaceous Chondrites." *Nature* 193:1123–1125.

Fitzinger, L. J.
1856 *Nekrolog, Paul Partsch.* Wien.
1868 *Geschichte der kais, kön. Hof-Naturalien Cabinetes zu Wien.* Wien.

Flammarion, Camille
1864 *La Pluralité des Mondes Habités.* Paris: Didier et Cie.

Fletcher, Lazarus
1890 "On the Mexican Meteorites with especial regard to the supposed occurrence of wide-spread meteoritic showers." *MM* 9:91–178.
1901 "On the Meteoric Stones which fell near Zomba, British Central Africa, on January 25th, 1899; with notes on the chemical analyses of such bodies." *MM* 13:1–37.
1904 *Introduction to the Study of Meteorites.* London: HMSO.
1908 "On the possible existence of a Nickel-Iron Constituent (Fe_5Ni_2) in both the Meteoric Iron of Youndegin and the Meteoric Stone of Zomba." *MM* 15:147–152.

Flight, Walter

1876 "Accounts of Aerolites and Aerolitic Meteors, and Accounts of recent Researches on Them." *BAAS* 46:168–170.

1877 "Meteorites and the Origin of Life." *Eclectic Magazine* 89:711–718.

1882 "Report of an Examination of the Meteorites of Cranbourne in Australia, of Rowton in Shropshire, and of Middlesbrough in Yorkshire." *PT* 173:885–897.

1887 *A Chapter in the History of Meteorites*. London: Delau & Co.

Fogel, E. M.

1915 *Beliefs and Superstitions of the Pennsylvania Germans*. Philadelphia.

Foote, Albert E.

1891 "A New Locality for Meteoric Iron with a Preliminary Notice of the Discovery of Diamonds in the Iron." *AAAS* 40:279–283.

Foote, Warren M.

1912 "Preliminary note on the shower of meteoric stones near Holbrook, Arizona, July 19, 1912, including a reference to the Perseid swarm of meteors visible from July 11 to August 22." *AJS* 134:437–456.

1913 "Factors in the exchange value of meteorites." *Proceedings of the American Philosophical Society* 52:516–543.

Forbes, R. J.

1950 *Metallurgy in Antiquity*. London: E. J. Brill.

1954 "Extracting, Smelting, and Alloying." In *A History of Technology*. Ed. Charles Singer et al. I:572–599. London: Oxford University Press.

Forster, Thomas

1823 *Researches about Atmospheric Phaenomena*. London.

Fort, Charles

1919 *The Book of the Damned*. New York: Horace Liveright.

Foshag, W. F.

1941 "Problems in the Study of Meteorites." *American Mineralogist* 26:137–144.

Fouchy, Jean Paul Grandjean de

1772 "Trois faits singuliers du même genre." *AS* (1769):20–21.

Fougeroux de Bonderoy, Auguste-Denis, Louis-Claude Cadet de Gassicourt, and Antoine Laurent de Lavoisier

1777 "Rapport fait à l'Académie Royale des Sciences, par MM. Fougeroux, Cadet, & Lavoisier d'une observation communiquée par M. l'Abbé Bachelay sur une pierre qu'on prétend être tombée du Ciel pendant un orage." *JP* II:251–255.

Fouqué, Ferdinand, and Auguste Michel-Lévy

1881 "Expériences synthétiques à la reproduction artificielle des météorites," *CR* 93:674–675.

Fourcroy, Antoine F. de

1803 "Memoir on the stones which have fallen from the Atmosphere, and particularly near Laigle in the Department of l'Orne, on the 26th of April last." *PM* 16:299–305.

Fox, Robert

1969 "The Fire Piston and Its Origins in Europe." *Technology and Culture* 10:355–370.

Fraustadt, Georg, and Hans Prescher, trans. and eds.
1956 *Georgius Agricola: Ausgewählte Werke*. Berlin: Deutscher Verlag der Wissenschaft.

Fredriksson, Kurt, Ananda Dube, Eugene Jarosewich, Joseph A. Nelen, and Albert F. Noonan
1975 "The Pulsora Anomaly: A Case Against Metamorphic Equilibration in Chondrites." In *Mineral Sciences Investigations 1972–1973*. Ed. George S. Switzer. 41–53. Smithsonian Contributions to the Earth Sciences, No. 14. Washington, D.C.: Smithsonian Institution Press.

Frémy, Edmond
1856 "Recherches sur les silicates." *CR* 43:1146–1149.

Frérét, Nicolas
1746 "Reflections sur les prodiges rapportez dans les Anciens." *HMAR* 4:411–436. (Read Feb. 1717.)

Fulda, F. C.
1799 "Observations on Fire Balls." *PM* 3:66–76. (Read at the Physical Society, Göttingen, 7 Dec. 1796.)

Galle, Johann G.
1867 "Ueber den muthmasslichen Zusammenhang der periodischen Sternschnuppen des 20 April mit den ersten Cometen des Jahres 1861." *AN* 69 (Nr. 1635), col. 33–36.

Gallitzen, D. de
1792 *Traité ou Description abrègée méthodique des Mineraux*. Maestricht: J. P. Roux.

Ganapathy, R.
1980 "A Major Meteorite Impact on the Earth 65 Million Years Ago: Evidence from the Cretaceous-Tertiary Boundary Clay." *Science* 209:921–923.
1982 "Evidence for a Major Meteorite Impact on the Earth 34 Million Years Ago: Implications for Eocene Extinctions." *Science* 216:885–888.
1983 "The Tunguska Explosion of 1908: Discovery of Meteoritic Debris near the Explosion Site and at the South Pole." *Science* 220:1158–1161.

Gardner, G. R.
1942 "British Charms, Amulets, and Talismans." *Folk-Lore* 53:96.

Gauthier de la Peyronne, C., trans.
1794 *Voyages du Professeur Pallas dans Plusiers Provinces de l'Empire de Russe*. 8 vols. Paris: Maradan.

Gehrels, Tom
1979 "The Asteroids: History, Surveys, Techniques, and Future Work," in *Asteroids*. Ed. Tom Gehrels. 3–24. Tucson: University of Arizona Press.

Genth, Frederick A.
1854 "On a new Meteorite from New Mexico." *AJS* 67:239–240.

Gesner, Conrad
1565 *De Rerum Fossilium, Lapidum et Gemmarum Figuri*. Zurich.

Gettens, Rutherford, Roy S. Clarke, Jr., and W. T. Chase
1971 *Two Early Chinese Bronze Weapons with Meteoritic Iron Blades*.

Washington, D.C.: Freer Gallery of Art, Occasional Papers Vol. 4, No. 1.

Gibbon, Edward
1932 *Decline and Fall of the Roman Empire*. 3 vols. New York: Modern Library.

Gifford, A. C.
1924 "The Mountains of the Moon." *New Zealand Journal of Science and Technology* 7:129–142.

Gilbert, Grove Karl
1892 "The Moon's Face: A Study of the Origins of Its Features." *Bulletin of the Philosophical Society of Washington, D.C.* 12:241–292.
1896 "The Origin of Hypotheses, Illustrated by the Discussion of a Topographic Problem." *Science* 3:1–13.

Gilbert, Ludwig W.
1803 "Hypothese über der Ursprung der Meteorsteine von Joseph Izarn." *AP* 15:437–462.
1805 "Der neue Harding'sche Planet Juno." *AP* 19:129–131.

Gillet de Laumont, Francois P. R.
1815 "Sur un Aërolith tombée en Moravie, et sur un masse de fer natif tombée en Bohème." *JM* 38:232–237.

Gilvarry, J. J., and J. E. Hill
1956 "The Impact of Large Meteorites." *Astrophysical Journal* 124:610–622.

Girtanner, Christoph
1792 *Anfangsgründe der antiphlogischen Chemie*. Berlin.

Glocker, F.
1848 "Über die krystallinische Structur des Eisens." *AP* 73:332–336, 593.

Goldberg, Edward, A. Uchiyama, and Harrison Brown
1951 "The Determination of Nickel, Cobalt, Gallium, Palladium and Gold in Iron Meteorites." *GCA* 2:1–25.

Goldschmidt, Victor M.
1922 "Über die massenverteilung im Erdinnern, verglichen mit der Structur gewissen Meteoriten." *Naturwissenschaften* 10, Heft 42:918–920.
1929 "The Distribution of the Chemical Elements." *Proceedings of the Royal Institution of Great Britain* 26:73–86.
1930 "Geochemische Verteilungsgesetze und kosmische Häufigkeit der Elements." *Naturwissenschaften* 18:599–1013.

Goldschmidt, V. M., and C. Peters
1932*a* "Zur Geochemie des Berylliums." *Nachrichten von der Gesellschaft der Wissenschaften zu Göttingen,* Part 2: 360–376.
1932*b* "Zur Geochemie der Edelmetalle." *Nachrichten von der Gesellschaft der Wissenschaften zu Göttingen,* Part 2: 377–401.

Gopalan, K., and George W. Wetherill
1968 "Rubidium-Strontium Age of Hypersthene (L) Chondrites." *Journal of Geophysical Research* 73:7133–7136.
1971 "Rubidium-Strontium Studies on Black Hypersthene Chondrites: Effects of Shock and Reheating." *Journal of Geophysical Research* 76:8484–8492.

Gould, Benjamin A.
1859 "On the Supposed Lunar Origin of Meteorites." *AAAS* 13:181–187.

Graham, Andrew, A. W. R. Bevan, and Robert Hutchison
1985 *Catalogue of Meteorites.* Tucson: University of Arizona Press.

Graham, Thomas
1866–67 "On the Occlusion of Hydrogen Gas by Meteoric Iron." *PRS* 15:502–503.

Gray, C. M., D. A. Papanastassiou, and G. J. Wasserburg
1973 "The Identification of Early Condensates from the Solar Nebula." *Icarus* 20:213–239.

Greene, John C., and J. G. Burke
1978 *The Science of Minerals in the Age of Jefferson.* Philadelphia: The American Philosophical Society.

Greg, Robert P.
1854 "Observations on Meteorites or Aerolites, considered geographically, statistically, and cosmically, accompanied by a Complete Catalog." *PM* 8:329–342, 449–463.
1855 "1. Description of a new Meteoric Iron from Chile, containing native Lead. 2. Fall of a large Mass of Meteoric Iron at Corrientes in South America." *PM* 10:12–15.
1860 "A Catalogue of Meteorites and Fireballs from A.D. 2 to A.D. 1860." *BAAS* (1860):48–118.
1861 "Footnote." *PM* 22:456–458.

Grew, Nehemiah
1673 "On the Nature of Snow." *PT* VIII:5193.

Grimm, J. W.
1875–78 *Deutsche Mythologie.* 3 vols. Berlin: Meyer.

Grogan, Robert M.
1948 "Beads of Meteoric Iron from an Indian Mound near Havana, Illinois." *American Antiquity* 13:302–305.

Gronberg, N. A.
1772 "Dissertation sur la pierre de tonnerre." *JP* 2:555–560.

Grotthus, Theodor V.
1821 "Untersuchung eines in Kurland im Dönaberg'schen Kreise, am 30 Juni (12 Juli) 1820 herabgefallenen Meteorsteins." *AP* 67:337–370.

Grube, Wilhelm
1910 *Religion und Kultus der Chinesen.* Leipzig.

Gümbel, Carl W.
1878 "Ueber die in Bayern gefundenen Steinmeteoriten." *Sitzungberichten der math, phys. Classe d. K. b. Akademie d. Wissenschaften, München* 101:14–72.

Gundel, Wilhelm
1893 "Sternschnuppen." *Paulys Real-Encyclopädie der classischen Altertumswissenschaft* 3:2439–2442. Stuttgart.
1912 *Sterne und Sternbilder im Glauben des Altertums und der Neuzeit.* Bonn.

Gunther, R. T.
1939 "Dr. William Thomson, F.R.S., a Forgotten English Mineralogist, 1761–c. 1806." *Nature* 143:667–668.

Güssman, Francisco
1785 *Lithophylacium Mitisianum: Dissertatione Praeuia et Observa-*

tionibus Perpetuis Physico Mineralogicis Explicatum. Viennae: Typis Iosephi Nobili de Kurzbeck.

1803 *Über die Steinregen*. Wien.

Hahn, Otto

1879 *Die Urzelle: nebst dem Beweis dass Granit, Gneiss, Serpentin, Talk, Gewisse Sandsteine, auch Basalt, endlich Meteorstein und Meteoreisen aus Pflanzen bestehen*. Tübingen: Laupp'schen.

1880 *Die Meteorite (Chondrite) und ihre Organismen*. Tübingen: Laupp'schen.

Hahn, Roger

1971 *The Anatomy of a Scientific Institution*. Berkeley, Los Angeles, London: University of California Press.

Haidinger, Wilhelm K.

1847 "Über das Meteoreisen von Braunau." *AP* 72:580–582.

1859*a* "Der Meteorit von Kakowa bei Orowitza." *SAW* 34:11–21.

1859*b* "Freiherr K. v. Reichenbach gegen Dr. M. Hörnes. Sendschriften an Hrn Prof. Dr. J. C. Poggendorf." *AP* 108:174–178.

1859*c* "Der Meteoreisenfall von Hraschina bei Agram am 26 Mai 1751." *SAW* 35:361–390.

1861*a* "Considerations on the Phenomena attending the Fall of Meteorites on the Earth." *PM* 22:349–361, 442–458.

1861*b* "Das Meteor von Quenggouk in Pegu, und die Ergebnisse des Falles daselbst am 27 Dec. 1857." *SAW* 44:637–642.

1863 "Die Meteorit von Albareto im k.k. Hof-Mineralien Cabinete, von Jahre 1766, und der Troilit." *SAW* 47:283–298.

1866 "Der Meteorsteinfall am 9 Juni 1866 bei Knyahinya, nächst Nagy-Berenzna in Ungher Comitate." *SAW* 54:200–205.

1867*a* "Die Tageszeiten der Meteoritenfälle verglichen." *SAW* 55:131–144, 189–194.

1867*b* "Die Localstunden von 178 Meteorfällen." *SAW* 55:651–658.

1867*c* "Die Meteoriten des k.k. Hof-Mineraliencabinetes am 1 Juli 1867, und der Fortschritte seit 7 Jänuar 1859." *SAW* 56:175–184.

1869 "On the Phenomena of Light, Heat, and Sound accompanying the Fall of Meteorites." *AJS* 98:280–285.

Hall, A. Rupert, and Marie B. Hall

1962 *Unpublished Scientific Papers of Isaac Newton*. Cambridge: Cambridge University Press.

Hall, William T.

1935 *Textbook of Quantitative Analysis*. New York: John Wiley & Sons, Inc.

Halley, Edmund

1714–16*a* "An Account of several extraordinary Meteors or Lights in the Sky." *PT* 29:159–164.

1714–16*b* "An Account of Several Nebulae or lucid Spots like Clouds, lately discovered among the Fixt Stars by help of the Telescope." *PT* 29:390–392.

1719 "An Account of the extraordinary Meteor seen all over England, on the 19th of March 1718/9 with a Description of the uncommon Height thereof." *PT* 30:978–990.

Halliday, Ian, Alan T. Blackwell, and Arthur A. Griffin

1978 "The Innisfree Meteorite and the Canadian Camera Network." *Jour-*

nal of the Royal Astronomical Society of Canada 72:15–39.
1984 "The Frequency of Meteorite Falls on the Earth." *Science* 223:1405–1407.

Halliday, Ian, and Arthur A. Griffin
1982 "A Study of the Relative Rates of Meteorite Falls on the Earth's Surface." *Meteoritics* 17:31–46.

Hamilton, William
1795 "An Account of the late Eruption of Mount Vesuvius." *PT* 85:73–116.

Hand, Wayland D., ed.
1964 *The Frank C. Brown Collection of North Carolina Folklore,* Vol. 7. Durham, N.C.

Harkins, William D.
1917 "The Evolution of the Elements and the Stability of Complex Atoms." *Journal of the American Chemical Society* 39:856–879.

Harkins, William D., and S. B. Stone
1925 "The isotopic Composition of the Element Chlorine in the Meteorites." *NAS* 11:643–645.

Harris, Elijah P.
1859 *The Chemical Constitution and Chronological Arrangement of Meteorites: Inaugural Dissertation.* Göttingen: Kaestner.

Harris, John
1704 *Lexicon Technicum.* London.

Haüy, René Just
1801 *Traité de Minéralogie.* 5 vols. Paris.
1822 *Traité de Minéralogie.* 5 vols. Paris.

Hayatsu, Ryoichi
1964 "Orgueil Meteorite: Organic Nitrogen Contents." *Science* 146:1291–1293.

Heilbron, John L.
1977 *Electricity in the 17th and 18th Centuries.* Berkeley, Los Angeles, London: University of California Press.

Heizer, Robert F.
1962 "The Background of Thomsen's Three-Age System." *Technology and Culture* 3:259–266.

Helmholtz, Hermann von
1874 "On the Use and Abuse of the Deductive Method in Physical Science." *Nature* 11:149–151, 211–212.

Henderson, Edward P., and Hollis M. Dole
1964 "The Port Orford Meteorite." *The ORE BIN* 26:113–130.

Hendriks, G. D.
1966 *Mirrors, Mice, and Mustaches.* Austin.

Hermstadt [Hermbstaedt], Sigismund F.
1797 "Observations sur les Attractions prochaines." *ACP* 22:107–108.

Herrick, Edward C.
1838*a* "On the Shooting Stars of August 9th and 10th 1837, and on the Probability of the Annual Occurrence of a Meteoric Shower in August." *AJS* 33:176–180.
1838*b* "Further Proof of an Annual Meteor Shower in August, with Remarks

on Shooting Stars in general." *AJS* 33:354–364.

1839*a* "On the Meteoric Shower of April 20, 1803, with an Account of Observations made on or about 20th April, 1839." *AJS* 36:358–363.

1839*b* "Report on the Shooting Stars of August 9th and 10th, 1839, with other Facts relating to the frequent Occurrence of a Meteoric Display in August." *AJS* 37:325–338.

1841 "Contributions towards a History of the Star-Showers of Former Times." *AJS* 40:349–365.

1861 "Shooting Stars in November, 1860." *AJS* 51:137.

Herschel, Alexander S.

1881*a* "The Progress of Meteor Spectroscopy." *Nature* 24:507–508.

1881*b* "On the Fall of an Aerolite near Middlesbrough, Yorkshire on March 14, 1881." *BAAS* (1881):296–302.

Herschel, William

1789 "Catalogue of a second Thousand of new Nebulae and Clusters of Stars; with a few introductory Remarks on the Construction of the Heavens." *PT* 79:212–255.

1791 "On Nebulous Stars, properly so called." *PT* 81:71–88.

Herzog, Leonard F., and William H. Pinson, Jr.

1956 "Rb/Sr Age, Elemental and Isotopic Abundance Studies of Stony Meteorites." *AJS* 254:555.

Hey, Max H.

1939 "History of the Widmanstätten Structure." *Nature* 143:764.

Hey, Max H., ed.

1966 *Catalogue of Meteorites*. London: British Museum (Natural History).

Hicks, R. D., trans.

1925 *Diogenes Laertius*. 2 vols. London.

Higgins, William

1811 "Description and Analysis of a Meteoric Stone, which fell in the County of Tipperary in Ireland in the month of August 1810." *PM* 38:262–268.

1818 "Account of a Shower of Meteoric Stones, which fell in the County of Limerick." *PM* 51:355.

1819 "Bericht von dem Steinregen, welcher sich am 10 Sept. 1813 in der Grafschaft Limerick in Irland ereignet hat." *AP* 60:233–237.

Hinrichs, Gustavus

1875 "The Great Iowa Meteor." *Popular Science Monthly* 7:588–596.

Hitchcock, Edward

1834 "On the Meteors of Nov. 13, 1833." *AJS* 25:354–363.

Hlawatsch, C.

1909 *Nekrolog und Schriftverzeichnis von Dr. Aristides Brezina*. Wien.

Hockner, N.

1854 "Tempelnsagen." *Zeitschrift für deutsche Mythologie und Sittenkunde* 2:413–420.

Hoff, Karl E. A. von

1832 "Neue Beiträge zu Chladni's Verzeichnissen von Feuermeteoren und herabgefallen Massen." *AP* 24:221–242.

1835 "Neue Beiträge zu Chladni's Verzeichnissen von Feiermeteoren und herabgefallen Massen." *AP* 34:339–370.

Hoffmeister, C., ed.
1925 "G. von Niessl: Katalog der Bestimmungsgrössen für 611 Bahnen grosser Meteore." *Denkschriften der Akademie der Wissenschaften, Wien* 100:1–70.

Hofmann, A. W., ed.
1888 *Aus Justus Liebig's und Friedrich Wöhler's Briefwechsel in den Jahre 1829–1873*. 2 vols. Braunschweig.

Högbom, A. G.
1901 "Eine meteorstatistische Studie." *AJS* 162:399–400.

Höhn, H.
1913 "Sitte und Brauch bei Tod und Begräbnis." *Mitteilungen über volkstümliche Überlieferung in Württemberg* 7:311.

Holmberg, Uno
1922–23 "Der Baum des Lebens." *Annales Academiae Scientarum Fennicae*, Série B, XVI:40.

Holmyard, E. J., and D. C. Mandeville, eds. and trans.
1927 *Avicennae: De Congelatione et Conglutinatione Lapidum, being sections of the Kitâb Al-Shifâ*. Paris: Paul Geuthner.

Holzhayer, J. B.
1872 "Osliann." *Verhandlungen der Gelehrten Estonischen Gesellschaft* 7:2–10.

Home, Roderick W.
1972 "The Origin of Lunar Craters: An 18th Century View." *Journal of the History of Astronomy* 3:1–10.

Hooke, Robert
1665 *Micrographia*. London.

Hooker, Marjorie
1974 "William Thomson." *Collected Abstracts: International Association: Ninth General Meeting,* 72. West Berlin and Regensburg.

Hoppe, Günter
1979 *Über der kosmischen Ursprung der Meteorite und Feuerkugeln (1794)*. Leipzig: Akademische Verlagsgesellschaft.

Hovey, Edmund O.
1912 "New Accessions of Meteorites." *The American Museum Journal* 12:257.

Howard, Edward
1802 "Experiments and observations on certain Stony and Metalline Substances, which are said to have fallen on the Earth; also on various kinds of native Iron." *PT* 92, Part I:168–212.

Howard-White, F. B.
1963 *Nickel, An Historical Review*. London.

Hsü, Kenneth J., Q. He, Judith A. McKenzie, Helmut Weissert, Katharina Perch-Nielsen, Hedy Oberhänsli, Kerry Kelts, John LaBrecque, Lisa Tauxe, Urs Krähenbühl, Stephen F. Percival, Jr., Ramil Wright, Anna Marie Karpoff, Nikolai Petersen, Peter Tucker, Richard Z. Poore, Andrew M. Gombos, Kenneth Pisciotto, Max F. Carman, Jr., and Edward Schreiber
1982 "Mass Mortality and Its Environmental Consequences." *Science* 216:249–256.

Huggins, William
1864 "On the Spectra of some of the Nebulae." *PRS* 13:492–493.

Hughes, Gwladys
1949 "Folk Beliefs and Customs in an Hawaiian Community." *Journal of American Folklore* 62:288–298.

Humboldt, Alexander von
1819 "Fireballs." *PM* 53:312–314.
1872 *Cosmos*. Translated by E. C. Otté. 4 vols. New York: Harper & Bros.

Huntington, Oliver W.
1886 "On the Crystalline Structure of Iron Meteorites." *Proceedings of the American Academy of Arts and Sciences* 13:478–498.
1888 "Catalogue of all Recorded Meteorites, with a Description of the Specimens in the Harvard College Collection." *Proceedings of the American Academy of Arts and Sciences* 23:37–110.

Huntley, H. E.
1948 "Production of Helium by Cosmic Rays." *Nature* 161:356.

Hurt, J.
1900 *Über estonische Himmelskunde*. Leipzig.

Hutchison, Robert, A. W. R. Bevan, and J. M. Hall
1977 *Appendix to the Catalogue of Meteorites*. London: British Museum (Natural History).

Hutton, Charles
1812 *Tracts on Mathematical and Philosophical Subjects*. 3 vols. London.

Huxley, Leonard, ed.
1918 *Life and Letters of Sir Joseph Dalton Hooker*. 2 vols. London.

Ideler, Julius L.
1832 *Über der Ursprung der Feuerkugeln und des Nordlichts*. Berlin: Enslin.

Izarn, Joseph
1803 *De Pierres Tombées du Ciel, ou Lithologie Atmosphèrique*. Paris.

Jackson, Charles T.
1838 "Chemical Analysis of Meteoric Iron from Claiborne, Clarke Co., Alabama." *AJS* 34:332–337.

Jackson, C. T., and A. A. Hayes
1845 "Remarks on the Alabama Meteoric Iron, with a Chemical Analysis of the Drops of green Liquid which exude from it," and "On the same Subject, with Remarks on the Origin of the Chlorine found in the Alabama Iron, and a Description of New Methods employed in the Analysis of Meteoric Irons." *AJS* 48:145–156.

Janos, Leo
1980 "Timekeepers of the Solar System." *Science 80* I, 4:44–55.

Jastrow, Morris
1912 *Die Religion Babyloniens und Assyriens*. 3 vols. Giessen.

Jeffery, P. G.
1970 *Chemical Methods of Rock Analysis*. Oxford: Pergamon Press.

Johnson, Samuel
1755 *A Dictionary of the English Language*. 2 vols. London.

Jones, A. T.
1922 "The Temperature of Meteorites." *Science* 56:169–172.

Joule, James P.
1848 "On Shooting Stars." *PM* 32:349–351.

Joule, James P., and W. Thomson
1857 "On the Thermal Effects of Fluids in Motion." *PM* 13:286–292.

Jussieu, Antoine de
1723 "De l'Origine et des Usages de la Pierre de Foudre." *AS*, 6–9.

Kahl, Russell
1971 *Selected Writings of Hermann von Helmholtz*. Middletown, Conn.

Kaushal, Sushil K., and George W. Wetherill
1969 "Rb^{87}-Sr^{87} Age of Bronzite (H group) Chondrites." *Journal of Geophysical Research* 74:2717–2726.

Keay, Colin S. L.
1980 "Anomalous Sounds from the Entry of Meteor Fireballs." *Science* 210:11–19.

Keil, Klaus
1968 "Mineralogical and Chemical Relationships among Enstatite Chondrites." *Journal of Geophysical Research* 73:6945–6976.

Keil, Klaus, and Kurt Fredriksson
1964 "The Iron, Magnesium and Calcium Distribution in Coexisting Olivines and Rhombic Pyroxenes of Chondrites." *Journal of Geophysical Research* 69:3487–3515.
1981 "Electron Probe Microanalysis." In *The Encyclopedia of Mineralogy*. Ed. Keith Frye. 143–148. Stroudsburg, Pa.: Hutchinson Ross.

Kenngott, Gustav A.
1856 "Enstatite." *AJS* 71:200.

Kerr, Richard A.
1983 "A Lunar Meteorite and Maybe Some from Mars." *Science* 220:285–289.
1984*a* "Periodic Impacts and Extinctions Reported." *Science* 223:1277–1279.
1984*b* "Second Lunar Meteorite Identified." *Science* 224:274.
1984*c* "An Impact but No Volcano." *Science* 224:858.
1985 "Could an Asteroid be a Comet in Disguise?" *Science* 227:930–931.

Kesselmeyer, P. A.
1860 *Über den Ursprung der Meteorsteine*. Frankfurt am Main: Brönner.

Keysner, Blanche W.
1958 "Divining the Future." *Keystone Folklore Quarterly* 3:5–8.

King, Edward
1796 *Remarks Concerning Stones Said to Have Fallen From the Clouds, Both in These Days and in Ancient Times*. London.

King, Elbert A.
1976 *Space Geology: An Introduction*. New York: John Wiley & Sons, Inc.

Kinnicut, Leonard P.
1884 "Report on the Meteoric Iron From the Altar Mounds in the Little Miami Valley, Ohio." *Seventeenth Report, Peabody Museum of Archaeology and Ethnology,* 381–384. Cambridge, Mass.

Kirkwood, Daniel
1859 "On the Possible Intersection of the Orbits of Mars and Certain of the Asteroids." *AJS* 77:335.
1860 "On the Nebular Hypothesis." *AJS* 80:161–181.
1864 "On Certain Harmonies of the Solar System." *AJS* 88:1–18.

1867 *Meteoric Astronomy: A Treatise on Shooting Stars, Fireballs, and Aerolites*. Philadelphia: J. B. Lippincott & Co.

Klaproth, Martin H.

1803*a* "Ueber meteorische Stein und Metallmasse." *AWB*, 21–43.

1803*b* "Verhandlungen, die Analyse und den Ursprung meteorischer Stein-und-Metallmassen betreffend." *Allgemeines Journal der Chemie* I:1–36, 466–468.

1812 "Chemische Untersuchung zweier Gediegen-Eisen-Massen." *Journal für Chemie und Physik* 5:1–5.

Kobell, Franz von

1864 *Geschichte der Mineralogie*. München: Cottaschen.

Koch, William E.

1961 *Kansas Folklore*. Lincoln: University of Nebraska Press.

Kohlstedt, Sally Gregory

1980 "Henry A. Ward: The Merchant Naturalist, and American Museum Development." *Journal of the Society for the Bibliography of Natural History* 9:647–661.

Kracher, Alfred, and John Willis

1981 "Composition and Origin of the Unusual Oktibbeha County Iron Meteorite." *Meteoritics* 16:239–246.

Kratter, H.

1825 *Versuch einer Entwicklung der Grundbegriffe die Meteorsteine und Darstellung der vorzüglichen Hypotheses, ihren Ursprung betreffend*. Wien: Walleshauser.

Krinov, E. L.

1966 *Giant Meteorites*. Translated by J. S. Romankiewicz. New York: Oxford University Press.

1974 "The Committee on Meteoritics of the USSR Academy of Sciences." *Meteoritics* 9:141–144.

Kuhn, Thomas S.

1959 "Energy Conservation as an Example of Simultaneous Discovery." In *Critical Problems in the History of Science*. Ed. M. Clagett. 321–356. Madison: University of Wisconsin Press.

Kurat, Gero, and Lisa Pabst

1974 *Katalog der Meteoriten-Präparatesammlung der Mineralogisch-Petrographischen Abteilung des Naturhistorischen Museums Wien*. Sonderpublikation Nr. 1/1974, Min-pet. Abt.: Naturhistorischen Museum Wien.

Kyte, Frank T., Zhiming Zhou, and John T. Wasson

1980 "Siderophile-enriched Sediments from the Cretaceous-Tertiary Boundary." *Nature* 288:651–656.

Labaree, Leonard

1959–71 *The Papers of Benjamin Franklin*. 15 vols. New Haven: Yale University Press.

Lacroix, Alfred

1927*a* "Les Météorites tombées en France et dans ses Colonies et conservées au Muséum National d'Histoire Naturelle, avec Remarques sur la Classification des Météorites." *Bulletin de Muséum National d'Histoire Naturelle* 33:411–455.

1927*b* "Le Fer météorique de l'Oasis de Tamentit dans le Touat." *CR* 184:1217–1220.

1927*c* "La Composition et la Structure du Fer météorique de Tamentit." *CR* 185:315–317.

Lalande, Jerome de

1802 "Lettre sur les Pierres de Foudre." *JP* 55:451–453.

Lambotin, ?

1803 "Sur des Pierres tombées de l'Atmosphère." *JP* 56:458–461.

Lamétherie, Jean C. de

1798*a* "Note sur la Nature des Globes de Feu qui tombent de l'Atmosphère, & sur le Fer Natif." *JP* 47:185–186.

1798*b* "Discours Préliminaire." *JP* 48:17–18.

1802 "Discourse Préliminaire: Des Globes de Feu, ou des Pierres tombées du Haut de l'Atmosphère sous Formes de Globes de Feu." *JP* 56:23–25.

1803*a* "Lithologie atmosphèrique." *JP* 56:441–458.

1803*b* "Sur les Pierres tombées à Aigle, Département de l'Orne." *JP* 56:428.

1803*c* "Discours Préliminaire: Des Substances minérales tombées de l'Atmosphère." *JP* 58:25.

Lange, E. F.

1958 "Oregon Meteorites." *Oregon Historical Quarterly* 59:105–115.

LaPaz, Lincoln

1946 "On the Ownership of Recovered Meteorites." *Popular Astronomy* 54:93–95.

1949 "The Craters on the Moon." *Scientific American* 181, 120–24.

1958 "The Effects of Meteorites upon the Earth (including its Inhabitants, Atmosphere, and Satellites)." *Advances in Geophysics* 4:217–350.

Laplace, Pierre-Simon

1802 "Laplace an von Zach." *Monatliche Korrespondenz,* September. 277.

1878–1912 *Oeuvres complètes de Laplace*. 14 vols. Paris.

1951 *A Philosophical Essay on Probabilities*. Translated by F. W. Truscott and F. L. Emory. New York: Dover.

Larkin, Henry

1880 "Planets, Moons, and Meteorites." *British Quarterly Review* 71:143–156.

Laugier, André

1821 "Analyse de la Pierre météorique de Juvénas." *ACP* 19:264–273.

1827 "Note sur la Composition chimique de la Pierre météorique de Ferrare." *ACP* 54:139–142.

Laurent, Auguste

1846 "Sur les Silicates." *CR* 23:1055–1058.

Lavoisier, Antoine L. de

1790 *Elements of Chemistry*. Translated by Robert Kerr. London: Creech.

Lee, H. D. P., trans.

1952 *Aristotle Meteorologica*. Cambridge, Mass.: Loeb Library.

Lemery, Nicolas

1700 "Explication physique et chimique des Feux souterrains, des Tremblemens de Terre, des Ouragans, des Éclairs, & du Tonnerre." *AS,* 101–110.

Lenain de Tillemont, Louis S.
1693–1739 *Histoire des Empereurs*. 6 vols. Brussels: E. H. Friex.

Le Roy, Jean-Baptiste
1771 "Mémoire sur le Météore ou Globe de Feu, observé au mois de Juillet dernier, dans une grande partie de la France." *AS*, 668–692.

Le Verrier, Urban-J.-J.
1867 "Sur les Étoiles filantes de 13 Novembre et du 10 Août." *CR* 64:94–99.

Levi-Donati, G. R.
1975 "La Polemica sulla 'piogetta di sassi' del 1794." *Physis* 17:94–111.

Levi-Donati, G. R., and Roy S. Clarke, Jr.
1983 "Brianza: An Example of Wrongly Believed Chondrite." *Meteoritics* 18:165.

Lindemann, F. A., and G. M. B. Dobson
1922 "A Theory of Meteors, and the Density and Temperature of the Outer Atmosphere to which it Leads." *PRS* A102:411–437.

Lipman, Charles B.
1932 "Are there Living Bacteria in Stony Meteorites?" *American Museum Novitiates,* No. 588:1–19.
1936 "Bacteria in Meteorites." *Popular Astronomy* 44:442–446.

Lipscomb, Andrew A., ed.
1905 *The Writings of Thomas Jefferson*. 20 vols. Washington, D.C.

Littman, Enno
1905 "Sternsagen und Astrologisches am Nordabessinier." *Archiv für Religionswissenschaft* 11:298–319.

Lockyer, J. Norman
1887 "Researches on the Spectra of Meteorites." *PRS* 43:117–156.
1890 *The Meteoritic Hypothesis*. London: Macmillan.

Loomis, Elias
1835 "On Shooting Stars." *AJS* 28:95–104.

Looritz, Oskar
1951 *Grundzüge des estonischen Volksglaubens*. 3 vols. Lund.

Lovering, J.
1839 "Meteoric Observations." *AJS* 35:323–328.

Lovering, John F., Walter Nichiporuk, Arthur Chodos, and Harrison Brown
1957 "The Distribution of Gallium, Germanium, Cobalt, Chromium, and Copper in Iron and Stony Iron Meteorites in Relation to Nickel Content and Structure." *GCA* 11:263–278.

Lütolf, Alois
1862 *Sagen, Bräuche, Legenden aus fünf Orten, Lucern, Uri, Schwiz, Unterwalden, und Zug*. Lucerne.

Lyman, C. S.
1868 "Meteoric Astronomy, and the New Haven Contributions to Its Progress." *New Englander* 27:129–163.

McCall, G. J. H.
1973 *Meteorites and their Origins*. Newton Abbot: David and Charles.

McCrosky, Richard E.
1970 "The Lost City Meteorite Fall." *Sky and Telescope* 39:2–6.

McCrosky, Richard E., A. Posen, G. Schwartz, and C.-Y. Shao

1971 "Lost City Meteorite—Its Recovery and a Comparison with Other Fireballs." *Journal of Geophysical Research* 76:4090–4108.

McGucken, William

1969 *Nineteenth-Century Spectroscopy*. Baltimore: Johns Hopkins Press.

McLane, James C., Elbert A. King, Jr., Donald A. Flory, Keith A. Richardson, James P. Dawson, Walter W. Kemmerer, and Bennie C. Wooley

1967 "Lunar Receiving Laboratory." *Science* 155:525–529.

McSween, Harry Y., and Steven M. Richardson

1977 "The Composition of Carbonaceous Chondrite Matrix." *GCA* 41:1145–1161.

Mahudel, Nicolas

1745 "Sur les prétendues Pierres de Foudre." *HMAR* 12:163–169. (Read in 1734.)

Mallet, John W.

1871 "On Three Masses of Meteoric Iron, from Augusta Co., Virginia." *AJS* 102:11–14.

1884 "On a Mass of Meteoric Iron from Wichita County, Texas." *AJS* 128:285–288.

Manian, S. H., H. C. Urey, and W. Bleakney

1934 "An Investigation of the Relative Abundance of Oxygen Isotopes 0^{16}: 0^{18} in Stone Meteorites." *Journal of the American Chemical Society* 56:2601–2609.

Mannhardt, Wilhelm

1858 *Germanische Mythen*. Berlin.

Marichal, G. A.

1812 *Mon Opinion sur la Formation des Aérolithes*. Paris: Didot.

Marschall, A. Franz

1852 *Nekrolog des k.k. Hofrathes Carl Ritter v. Schreibers*. Wien.

Marsden, Brian G.

1979 "The Work of the Minor Planet Center." In *Asteroids*. Ed. Tom Gehrels. 77–83. Tucson: University of Arizona Press.

Marvin, Ursula B.

1978 "Theoretical Meteoritics and Early American Referees (1810)." *Meteoritics* 13:555.

1979 "An Attempt to Explain the Origin of Meteoric Stones by Lyman Spaulding." *Meteoritics* 14:193–205.

1983 "Extraterrestrials have Landed on Antarctica." *New Scientist,* 17 March:710–715.

Marvin, Ursula B., and Brian Mason, eds.

1980 *Catalog of Antarctic Meteorites, 1977–1978*. Smithsonian Contributions to the Earth Sciences, No. 23. Washington, D.C.: Smithsonian Institution Press.

1982 *Catalog of Meteorites from Victoria Land, Antarctica, 1978–1980*. Smithsonian Contributions to the Earth Sciences, no. 24. Washington, D.C.: Smithsonian Institution Press.

Maskelyne, Nevil

1794 "An Account of an Appearance of Light, like a Star, seen lately in the dark Part of the Moon." *PT* 84:429–440.

Mason, Brian H.
1960 "The Origin of Meteorites." *Journal of Geophysical Research* 65:2965–2970.
1962*a* "The Classification of Chondritic Meteorites." *American Museum Novitiates,* No. 2085:1–20.
1962*b* *Meteorites.* New York: John Wiley & Sons, Inc.
1963*a* "Olivine Composition in Chondrites." *GCA* 27:1011–1023.
1963*b* "Organic Matter from Space." *Scientific American* 208, 3:43–49.
1967*a* "Olivine Composition in Chondrites—A Supplement." *GCA* 31:1100–1103.
1967*b* "Meteorites." *American Scientist* 55:429–455.
1975 "Mineral Sciences in the Smithsonian Institution." In *Mineral Science Investigations, 1972–1973.* Ed. George S. Switzer. 1–10. Smithsonian Contributions to the Earth Sciences, No. 14. Washington, D.C.: Smithsonian Institution Press.

Mason, Brian, ed.
1971 *Handbook of Elemental Abundances in Meteorites.* New York: Gordon & Breach Science Publishers.

Mathews, Catherine Van Cortlandt
1908 *Andrew Ellicott: His Life and Letters.* New York: Grafton.

Maxwell, James Clerk
1859 "On the Stability of the Motion of Saturn's Rings." *Monthly Notices of the Astronomical Society* 19:297–304.
1857–62 "On Theories of the Constitution of Saturn's Rings." *Proceedings of the Royal Society of Edinburg* 4:99–101.

Meadows, A. J.
1972 *Science and Controversy: A Biography of Sir Norman Lockyer.* London: Macmillan.

Mehl, Robert F.
1965 "On the Widmanstätten Structure." In *The Sorby Centennial Symposium on the History of Metallurgy.* Ed. C. S. Smith. 243–269. New York: Gordon and Breach.

Mehl, Robert F., and G. Derge
1937 "Studies Upon the Widmanstätten Structure, VIII. The Gamma-Alpha Transformation in Iron-Nickel Alloys." *Transactions of the American Institute of Mining and Metallurgical Engineers* 125:482–500.

Meier, Ernst H.
1852 *Deutsche Sagen, Sitten und Gebräuche aus Schwaben.* Stuttgart.

Meldrum, A. N.
1933 "Lavoisier's Early Work in Science." *Isis* 19:330–363; 20:396–425.

Melton, F. A.
1938 "Possible late Cretaceous Origin of the Carolina 'Bays.'" *Bulletin of the Geological Society of America* 49:1954.

Mendoza, V. T. and V. R. R. de
1952 *Folklore de San Pedro Piedre Gorde.* Mexico.

Menzel, D. H.
1976 "Case Against UFOs." *Physics Today* 29:13–15.

Menzies, A. W. E.
1925 "Isotopic Composition and Atomic Weight of Chlorine in Meteorites." *Nature* 116:643–644.

Merian P.
1864 "Ueber den Meteorsteinfall zu Ensisheim." *AP* 122:182–186.

Merrill, George P.
1908 "The Meteor Crater of Canyon Diablo, Arizona: Its History, Origin and Associated Meteoric Irons." *Smithsonian Institution Miscellaneous Collections* 50:461–498.
1916 *Handbook and Descriptive Catalogue of the Meteorite Collections in the United States National Museum*. United States National Museum Bulletin 94. Washington, D.C.: Smithsonian Institution.
1918 "Report on Researches on the Chemical and Mineralogical Composition of Meteorites with especial reference to Their Minor Constituents." *Memoirs of the National Academy of Sciences* 14:1–29.
1919 "The Percentage Number of Meteorite Falls and Finds Considered with reference to their Varying Basicity." *Proceedings of the National Academy of Sciences. U.S.* 5:37–39.
1920 "The Cumberland Falls, Whitley County, Kentucky Meteorite." *Proceedings of the U.S. National Museum* 57:97–105.
1921 "On Metamorphism in Meteorites." *Bulletin of the Geological Society of America* 32:395–416.
1928 "Concerning the Origin of Metal in Meteorites." *Proceedings of the U.S. National Museum* 73:1–7.
1929*a* "The Story of Meteorites." In *Minerals from Earth and Sky*. Part I. Ed. G. P. Merrill and W. F. Foshag. Smithsonian Scientific Series, vol. 3. New York.
1929*b* *Composition and Structure of Meteorites*. U.S. National Museum Bulletin, No. 149. Washington, D.C.: Smithsonian Institution.

Meunier, Stanislaus
1867 *Géologie Comparée. Étude Descriptive, Théorique et Expérimentale sur les Météorites*. Paris: Bureau du Cosmos.
1869 "Recherches sur la composition et la structure des météorites." *ACP* 17:21–35, 60–70.
1885 "Géologie Comparée—sur la Classification et l'Origine des Météorites." *CR* 101:728.
1887 "Les météorites et l'analyse spectrale." *CR* 105:1095–1097.
1898 *Guide dans la Collection de Météorites avec le Catalogue des Chutes representées au Muséum*. Paris: Muséum d'Histoire Naturelle.

Meyer, Elard H.
1900 *Badisches Volkleben im 19 Jahrhundert*. Strassburg.

Michel, Hermann
1922 "Fortschritte in der Meteoritenkunde seit 1900." *Fortschritte in der Mineralogie, Kristallographie und Petrographie* 7:245–326.

Middlehurst, Barbara M., and Gerard P. Kuiper, eds.
1963 *The Moon, Meteorites and Comets*. Chicago: University of Chicago Press.

Miers, Henry A.
1921 "Biographical Notice of Sir Lazarus Fletcher (1854–1921)." *MM* 19:181–192.

Miller, A. M.
1923 "Meteorites." *Scientific Monthly* 17:435–440.

Millington, W. H., and B. L. Maxfield
1906 "Philippine (Visayan) Superstitions." *Journal of American Folklore* 19:205–211.

Monck, W. H. S.
1890 "Our Astronomical Column." *Nature* 42:90.

Monge, G., J. Fourcroy, C. Berthollet, J. B. Biot, and L. Vauquelin
1806 "Extrait d'un Rapport fait à l'Institut, sur l'Aërolithe tombé près de Valence, Arrondissement d'Alais." *BB* 32:277–281.

Monnig, O. E.
1939 "How the Casas Grandes, Chihuahua, Mexico, Meteorite got to Washington, D.C." *Popular Astronomy* 47:152.

Montell, Lynwood
1966 "Death Beliefs from the Kentucky Foothills." *Kentucky Folklore* 12:81–86.

Morel, Robert, and Suzanne Walter
1972 *Dictionnaire des Superstitions*. Paris.

Moore, Carleton B.
1971 *Meteorites*. Boston: Houghton Mifflin.

Morrison, David, and John Nieshoff
1979 "Future Exploration of the Asteroids." In *Asteroids*. Ed. Tom Gehrels. 227–250. Tucson: University of Arizona Press.

Morrison, Philip
1962 "Carbonaceous 'snowflakes' and the origins of life." *Science* 135:663–664.
1971 "The Nature of Scientific Evidence: A Summary." In *UFOs—A Scientific Debate*. Ed. C. Sagan and T. Page. 276–290. Ithaca: Cornell University Press.

Moss, A. A., M. H. Hey, C. J. Elliott, and A. J. Easton
1968 "Methods for the Chemical Analysis of Meteorites: II. The Major and Some Minor Constituents of Chondrites." *MM* 36:101–119.

Mueller, George G.
1953 "The Properties and Theory of Genesis of the Carbonaceous Complex within the Cold Bokevelt Meteorite." *GCA* 1:1–10.

Muir, A., ed.
1958 *V. M. Goldschmidt: Geochemistry*. Oxford: Clarendon Press.

Murphey, B. F., and A. O. Nier
1941 "Variations in the Relative Abundance of the Carbon Isotopes." *PR* 59:771–772.

Murray, John
1819 "On Aerolites." *PM* 54:37–43.

Nagata, Takesi
1979 *Proceedings of the Fourth Symposium on Antarctic Meteorites*. Tokyo: Memoirs of National Institute for Polar Research, Special Issue No. 15.

Nagy, Bartholomew
1967 "The Possibility of Extraterrestrial Life: Ultra-microchemical Analyses and Electron Microscopic Studies of Microstructures in

Carbonaceous Meteorites." *Review of Paleobotany and Palynology* 3:237–242.

1975 *Carbonaceous Meteorites*. Amsterdam: Elsevier Scientific.

Nagy, Bartholomew, D. J. Hennessy, and W. G. Meinschein

1961 "Mass Spectroscopic Analysis of the Orgueil Meteorite: Evidence for Biogenic Hydrocarbons." *Annals of the New York Academy of Sciences* 93:25–35.

Nagy, Bartholomew, K. Fredriksson, H. C. Urey, G. Claus, C. A. Andersen, and J. Percy

1963 "Electron Probe Microanalysis of Organized Elements in the Orgueil Meteorite." *Nature* 198:121–125.

Narayan, C., and J. I. Goldstein

1985 "A Major Revision of Iron Meteorite Cooling Rates—An Experimental Study of the Growth of the Widmanstätten Pattern." *GCA* 49:397–410.

Narendra, Barbara

1978 "The Peabody Museum Meteorite Collection: A Historic Account." *Discovery* 13:11–23.

Needham, Joseph

1959 *Science and Civilization in China. Vol. 3. Mathematics and the Sciences of the Heavens and the Earth*. Cambridge: Cambridge University Press.

Neumann, J. G.

1849 "Über die krystallinische Struktur des Meteoreisens von Braunau." *Naturwissenschaftliche Abhandlungen Wien* 3:45–56.

Neumann, K. A.

1812 "Der verwünschte Burggraf in Ellbogen in Böhmen, ein Meteorolit." *AP* 42:197–209.

Newcomb, Simon

1868 "Meteoric Showers." *North American Review* 107:38–50.

Newton, Hubert A.

1862 "An Account of Two Meteoric Fireballs, observed in the United States, Aug. 2, and Aug. 6, 1860, with Computation of their Paths." *AJS* 83:335–348.

1863*a* "Evidence of the Cosmical Origin of Shooting Stars derived from the Dates of Early Star Showers." *AJS* 86:145–149.

1863*b* "Procession and Periodicity of the November Star-shower." *AJS* 86:300–301.

1864 "The Original Accounts of the Displays in Former Times of the November Star-shower, together with a Determination of the Length of Its Cycle, Its Annual Period, and the Probable Orbit of the Group of Bodies around the Sun." *AJS* 87:377–389; 88:53–61.

1879 "Relation of Meteorites to Comets." *Nature* 18:315–317, 340–342.

1886 "Meteorites, Meteors, and Shooting Stars." *Popular Science Monthly* 29:733–747.

1891 "Fireball in the Madonna di Foligno." *AJS* 141:235–238.

1897 "The Worship of Meteorites." *AJS* 203:1–14.

Newton, Isaac

1952 *Opticks*. Based on the 4th ed., London, 1730. New York: Dover.

Nier, Alfred O.
1935 "Evidence for the Existence of an Isotope of Potassium of Mass 40." *PR* 48:283–284.

Niessl, Gustav von
1917 "The Determination of Meteor-Orbits in the Solar System." Translated by C. Abbe. *Smithsonian Miscellaneous Collections* 66, 16:1–35.

Nininger, Harvey H.
1933 "Concerning Bacteria in Meteorites." *Popular Astronomy* 41:214–215.

Nininger, Harvey H., and A. D. Nininger
1950 *The Nininger Collection of Meteorites: A Catalog and a History.* Winslow, Ariz.: American Meteorite Museum.

Nishiizumi, K., D. Elmore, and J. R. Arnold
1984 "Cosmogenic Nuclides in Peculiar Meteorites." *Meteoritics* 19:283 (abstract).

Noddack, Ida, and Walter Noddack
1930 "Die Haufigkeit der chemischen Elements." *Naturwissenschaften* 18:757–764.

Nordenskiöld, Adolf E.
1878 "On the Composition and Common Origin of Certain Meteorites." *Nature* 18:510–511.

Nordenskiöld, Nils
1823 "Mineralogical Description of some Aërolites which fell near Wiborg." *Edinburgh Philosophical Journal* 9:333–334.

Officer, Charles B., and Charles L. Drake
1983 "The Cretaceous-Tertiary Transition." *Science* 219:1383–1390.
1985 "Terminal Cretaceous Environmental Events." *Science* 227:1161–1167.

O'Keefe, John A.
1976 *Tektites and Their Origins.* Amsterdam: Elsevier.

Olbers, Heinrich Wilhelm M.
1803*a* "Ueber die von Himmel gefallenen Steine." *Monatliche Korrespondenz* 7:146–160.
1803*b* "Letter from Dr. Olbers of Bremen to Baron von Zach on the Stones which have fallen from the Heavens." *PM* 15:289–293.
1805 "Entdeckung eines beweglischen Sterne, den man gleichfalls für einem zwischen Mars und Jupiter sich aufhaltenden planetarischen Körper halten kann." *Astronomische Jahrbach,* 108.
1829 "Auszug aus einem Schreiben, die am 26 September 1829 beobachtete Feuerkugel betreffend." *AN* 8:15–16.
1830 "Noch etwas über die am 26. September 1829 gesehene Feuerkugel." *AN* 8:159–162.
1837*a* "Die Sternschnuppen." *Schumacher's Jahrbuch für 1837,* 36–64.
1837*b* "Noch etwas über Sternschnuppen, als Nachtrag." *Schumacher's Jahrbuch für 1837,* 278–282.
1838 "Die Sternschnuppen im August 1837." *Schumacher's Jahrbuch für 1838,* 317–330.
1839*a* "November-Beobachtungen von Sternschnuppen 1838 in Bremen." *AN* 16:177–180.

1839*b* "Sternschnuppen-Beobachtung im August 1839." *AN* 16:385–386.

Olivier, Charles P.

1925 *Meteors*. Baltimore: Williams & Wilkins.

Olmsted, Denison

1834 "Observations on the Meteors of November 13, 1833." *AJS* 25:363–411; 26:137–174.

1836*a* "On the Causes of the Meteors of November 13, 1833." *AJS* 29:376–383.

1836*b* "Remarks on Shooting Stars, in reply to Rev. W. A. Clarke." *AJS* 30:370–376.

1837 "On the Meteoric Shower of Nov. 1836." *AJS* 31:386–395.

Oort, Jan H.

1950 "The Structure of the Cloud of Comets Surrounding the Solar System and A Hypothesis Concerning Its Origin." *Bulletin of the Astronomical Institute of the Netherlands* 11:91–110.

Öpik, Ernst

1933 "Meteorites and the Age of the Universe." *Popular Astronomy* 41:71.

Oppolzer, Theodor von

1866 "Ueber die Bahn des Cometen I 1866 (Tempel)." *SAW* 53:247–257.

1866–67 "Bahnbestimmung des Cometen I, 1866." *AN* 68 (nr. 1624), col. 241–250.

1867*a* "Schreiben des Herrn Th. Opolzer an den Herausgeber." *AN* 68 (Nr. 1629), col. 333–334.

1867*b* "Ueber die Bahn des Cometen III, 1862." *AN* 69 (Nr. 1638), col. 81–88.

Orth, Charles J., James S. Gilmore, Jere D. Knight, Charles L. Pillmore, Robert H. Tschudy, and James E. Fassett

1981 "An Iridium Anomaly at the Palynological Cretaceous-Tertiary Boundary in Northern New Mexico." *Science* 214:1341–1343.

Orth, Charles J., Jere D. Knight, Leonard R. Quintana, James S. Gilmore, and Alison R. Palmer

1984 "A Search for Iridium Abundance Anomalies at Two Late Cambrian Biomere Boundaries in Western Utah." *Science* 223:163–165.

Ovenden, M. W.

1972 "Bode's Law and the Missing Planet." *Nature* 239:508–509.

Owen, E. A., and A. H. Sully

1939 "The Equilibrium Diagram of Iron-Nickel Alloys." *PM* 27:614–636.

Pai, D. K., C. Tuniz, R. K. Moniot, T. H. Kruse, and G. F. Herzog

1980 "Beryllium 10 in Australasian Tektites: Evidence for a Sedimentary Precursor." *Science* 218:787–789.

Paneth, Friedrich A.

1939 "The Age of Meteorites." *Royal Astronomical Society, Occasional Notes,* No. 5:57–64.

1940 *The Origin of Meteorites*. London: Oxford University Press.

1954 "Die Heliummethode zur geologischen Altensbestimmung und das Alter der Eisenmeteorite." *Zeitschrift für Elektrochemie* 58:567–573.

1960 "The Discovery and Earliest Reproduction of the Widmanstätten Structure." *GCA* 18:176.

Paneth, Friedrich A., P. Reasbeck, and K. I. Mayne

1952 "Helium 3 Content and the Age of Meteorites." *GCA* 2:300–303.

Papanastassiou, Dimitri A., and Gerald J. Wasserburg
1969 "Initial Strontium Isotopic Abundances and the Resolution of Small Times Differences in the Formation of Planetary Objects." *EPSL* 5:361–376.

Partington, J. R.
1972 *A History of Chemistry*. Vol. IV. London: Macmillan.

Partsch, Paul
1843*a* *Kurze Uebersicht der im k.k. Hof-Mineralien-Kabinette zu Wien zur Schau gestellten acht Sammlungen*. Wien: Prandel.
1843*b* *Die Meteoriten oder von Himmel gefallenen Steine und Eisenmassen im k.k. Hof-Mineralien Kabinette zu Wien*. Wien.
1857 "Über den schwarzen Stein in der Kaaba zu Mekka." *Denkschriften der Akademie der Wissenschaften, Wien* 13:1–5.

Patera, Adolphe, and W. K. Haidinger
1849 "Meteorite of Arva." *AJS* 58:439–440.

Patrin, Eugène M. L.
1801 "Météores Lumineux." *BB* 18:204–224.
1802 "Considérations sur les masses de pierres et de matières metalliques qu'on suppose tombée de l'Atmosphère." *JP* 55:376–393.
1803 "Lettre de M. L. Patrin à J. C. Delamétherie à l'Occasion des Pierres tombées de l'Atmosphère." *JP* 56:392.

Patterson, Claire
1956 "Age of Meteorites and the Earth." *GCA* 10:230–237.

Peary, Robert E.
1914 *Northward Over The "Great Ice."* New York: F. A. Stokes.

Perry, Stuart H.
1944 *The Metallography of Meteoric Iron*. U.S. National Museum Bulletin, No. 184. Washington, D.C.: Smithsonian Institution.

Peters, Christian F. W.
1866–67 "Bemerkung über den Sternschnuppenfall von 13 November und 10 August 1866." *AN* 68 (Nr. 1626), col. 287–288.

Phipson, T. L.
1867 *Meteors, Aerolites, and Falling Stars*. London: Lovell Reeve.

Pictet, Marc A.
1801 "Pierres Météoriques: Details sur les Pierres de ce Genre vues chez E. Howard." *BB* 17:414.
1802 "Notice des Rédacteurs." *BB* 20:89–93.
1803 "Sur l'échauffement des projectiles par leur frottement contre l'air." *BB* 23:331–336.

Pierson, W. M.
1873 "Correspondence Relative to the Discovery of a Large Meteorite in Mexico." *Annual Report of the Smithsonian Institution,* 419–422.

Plummer, J. J.
1881 "On the Nature and Supposed Origin of Meteorites." *Good Words* 22:850–855.

Podosek, Frank A.
1970 "Dating of Meteorites by the High-Temperature Release of Iodine-Correlated Xe^{129}." *GCA* 34:341–365.

Poisson, Simon D.
1803 "Sur les substances minérales qu l'en suppose tombées du ciel sur la

terre." *Bulletin de sciences de la Société Philomatiques* 3:180–184.
1837 *Recherches sur la Probabilité des Jugemens*. Paris: Bachelier.

Pollack, James B.
1983 "Environmental Effects of an Impact-Generated Dust Cloud: Implications for the Cretaceous-Tertiary Extinction." *Science* 219:287–289.

Prätorius, Johann
1666 "Anthropodemus, Plutonicus des ist Neue Weltbeschreibung von allerly wunderbaren Menschen." *Handwörterbuch des deutschen Aberglaubes* VIII:117.

Prevost, Pierre
1803*a* "Pierre tombée du Ciel: Note du même sur les Traditions Anciennes de ce Phénomène." *BB* 23:115–118.
1803*b* "Sur les Pierres tombantes." *BB* 23:320–331.
1803*c* "Pierres tombées." *BB* 24:188–203, 295–296, 393–397.
1804 "Aérolithes." *BB* 25:85–88.
1805 "Beobachtungen der Feuerkugel am 8ten März 1798 und Bemerkungen über dieses Phänomen." *AP* 19:220–234.

Pringle, John
1759 "Some Remarks upon the Several Accounts of the Fiery Meteor (which appeared on Sunday the 26th of November, 1758) and upon Other Such Bodies." *PT* 51, Part 1:259–274.

Prior, George T.
1913 "On the Remarkable Similarity in Chemical and Mineral Composition of Chondritic Meteoric Stones." *MM* 17:33–38.
1914 "The Meteorites of Uwet, Kota Kota, and Angela: Redeterminations of Nickel and Iron in the Baroti and Wittekrantz Meteoric Stones." *MM* 17:127–134.
1916 "On the Genetic Relationship and Classification of Meteorites." *MM* 18:26–44.
1920 "The Classification of Meteorites." *MM* 19:51–63.
1923 *Catalogue of Meteorites in the British Museum*. London.

Pröhle, Heinrich
1859 *Sagen des Ober-Harzes*. Leipzig.

Promies, Wolfgang, ed.
1967–72 *Georg Christoph Lichtenberg, Schriften und Briefe*. 4 vols. München: Carl Hanser.

Proust, Joseph L.
1799 "Sur le Fer Natif du Pérou." *JP* 49:148–149.
1805 "Sur le Fer Natif, et sur du Nouveau Sulfures de Manganèse." *JP* 61:272.

Pruett, J. H.
1939 "The Willamette, Oregon, Meteorite in History." *Popular Astronomy* 47:148–150.
1942 "Death of the Discoverer of the Willamette, Oregon, Siderite." *Contributions, Society for Research on Meteorites* 3:79.

Puckett, N. N.
1926 *Folk Beliefs of the Southern Negro*. Chapel Hill: University of North Carolina Press.

Quetelet, Lambert A. J.
1839 *Catalogue des Principles Apparations d'Étoiles Filants*. Brussels.

Quirke, T. T., and L. Finkelstein
1917 "Measurement of the Radioactivity of Meteorites." *AJS* 194:237–242.

Rackham, H., trans.
1938 *Pliny Natural History*. Cambridge, Mass.: Loeb Library.

Radoza, F. D. y
1970 *Dictionary of Philippine Folk Beliefs and Customs*. 4 vols. Cagayan de Oro City.

Radford, Edwin, and M. A. Radford
1948 *Encyclopaedia of Superstitions*. London.

Rammelsberg, Karl F.
1843 "Ueber die Bestandtheile der Meteorsteine." *AP* 60:130–139.
1848 "Über die Zusammensetzung des Meteorsteine von Juvenas, und seinen Gehalt an Phosphorsäure und Titansäure." *AP* 73:585–590.
1849 "Über die chemische Zusammensetzung des Meteoreisens von Seeläsgen." *AP* 74:444–446.
1870 "Die chemische Natur der Meteoriten." *AWB*, 75–160.

Ramsey, William, and Frederick Soddy
1903 "Experiments in Radioactivity, and the Production of Helium from Radium." *Nature* 68:354–355.

Rappaport, Rhoda
1967 "Lavoisier: Geologic Activities 1763–1792." *Isis* 58:375–385.

Reed, S. J. B.
1972 "The Oktibbeha County Iron Meteorite." *MM* 38:623–626.

Reichelt, H.
1913 "Die Steinerne Himmel." *Indogermanische Forschungen* 32:23–57.

Reichenbach, Karl L. von
1857*a* "Über den Meteoriten von Hainholz." *AP* 101:311–313.
1857*b* "Zum Meteoriten von Hainholz." *AP* 102:618–621.
1858*a* "Über die Rinde der meteorische Eisenmassen." *AP* 103:637–646.
1858*b* "Über die Rinden der Meteorsteine." *AP* 104:473–482.
1858*c* "Die Meteoriten und die Kometen nach ihren gegenseitigen Beziehungen." *AP* 105:438–460.
1858*d* "Über die Anzahl der Meteoriten und Betrachtungen über ihre Rolle im Weltgebäude." *AP* 105:551–563.
1859*a* "Die meteorischen Kügelchen der Capitän Callum." *AP* 106:476–490.
1858*b* "Anordnung und Einteilung der Meteoriten." *AP* 107:155–182.
1859*c* "Über die chemische Beschaffenheit der Meteoriten." *AP* 107:353–374.
1858*d* "Über das Gefüge der Steinmeteoriten." *AP* 108:291–311.
1859*e* "Über die Zeitfolge und die Bildungen der näheren Bestandtheile der Meteoriten." *AP* 108:452–465.
1860*a* "Meteoriten in Meteoriten." *AP* 111:353–386.
1860*b* "Meteoriten und Sternschnuppen." *AP* 111:387–401.
1861 "Über das innere Gefüge der näheren Bestandtheile des Meteoreisens." *AP* 114:99–132, 250–274, 477–485, 485–491.
1862 "Über die näheren Bestandtheile des Meteoreisens." *AP* 115:148–156, 620–636; 116:576–591.
1864 "Die Sternschnuppen in ihren Beziehungen zur Erdoberfläche." *AP* 123:368–374.

1865*a* "Geschichte der Meteoriten von Blansko." *AP* 124:213–234.
1865*b* "Die schwarzen Linien und Ablösungen in den Meteoriten." *AP* 125:308–325.

Reinsch, P.
1872 "Über die mikroskopische Structur des Krähenberger Meteoriten und über die Anfertigung mikroskopischer Präparate von Meteorsteinen." *Tageblatt der Versammlungen deutscher Naturforscher und Ärtze, Leipzig,* 133.

Reiser, Karl A.
1897–1903 *Sagen, Gebräuche und Sprichwörten des Allgäus.* 2 vols. Kempten.

Richardson, Robert S.
1965 "The Discovery of Icarus." *Scientific American* 212, 4:106–115.

Ringwood, Albert E.
1961 "Chemical and Genetic Relationships among Meteorites." *GCA* 24:159–197.

Rittenhouse, David
1786 "Observations on the Account of a Meteor." *Transactions of the American Philosophical Society* 2:173–176.

Ritter, Johann W.
1803 "Einiges über Nordlichter und deren Periode, und über den Zusammenhang des Nordlights mit dem Magnetismus und des Magnetismus mit den Feuerkugeln, dem Blitze und der Elektrizität." *AP* 15:206–226.
1804 "Ideen über Feuerkugeln und Meteorsteine, die Perioden ihrer Erscheinung, und die Übereinstimmung mit denen der Norlichter und Gewitter, und ụber den daraus hervorgehenden tellurischen Ursprung der gefallener Stein-und Eisenmassen." *AP* 16:221–241.
1805 "Noch einiges über Nordlichter und Feuerkugeln, und auffallende meteorologische Wahrnehmung am 20sten Nov. 1804." *AP* 19:235–242.

Roe, Arthur
1975 "The C. U. Shepard Mineral Collection and the Two Drs. Shepard." *The Mineralogical Record* (Sept.–Oct.):253–257.

Rohault, Jacques
1672 *Traité de Physique.* 2 vols. Amsterdam.

Rolle, Friedrich
1884 *Die hypothetischen Organismen-Reste in Meteoriten.* Wiesbaden: J. F. Bergmann.

Romé de l'Isle, Jean B. L.
1767 *Catalogue Systèmatique et Raisonée des Curiosités de la Nature et de l'Art, qui Composent le Cabinet de M. Davila.* 3 vols. Paris: Briasson.
1783 *Cristallographie.* Paris.

Romig, A. D., Jr., and J. I. Goldstein
1980 "Determination of the Fe-Ni and Fe-Ni-P Phase Diagrams at Low Temperatures (700 to 300° C.)." *Metallurgical Transactions A* IIA:1151–1159.

Romig, Mary F.
1966 "The Scientific Study of Meteors in the 19th Century." *Meteoritics* 3:11–25.

Romig, Mary F., and Donald L. Lamar
1964 "Strange Sounds from the Sky." *Sky and Telescope* 28:214.

Rose, Gustav
1825 "Ueber die in Meteorsteinen vorkommenden krystallisarten Mineralien." *AP* 80:173–192.
1826 "Sur les Minéraux cristallisés qui se trouvent dans les Aérolithes." *ACP* 31:81–100.
1837–42 *Mineralogisch-geologische Reise nach den Ural.* 2 vols. Berlin.
1852 *Das Krystallo-Chemische Mineralsystem.* Leipzig.
1861 "Ueber das Vorkommen von krystallisirten Quarz in dem Meteoreisen von Xiquipilco in Mexico." *AP* 188:184–188.
1863*a* "Über zwei neue Meteoritenfälle nach den Mittheilungen, die Hr. Prof. Grewingk gemacht hat." *AP* 120:619–623.
1863*b* "Systematisches Verzeichnis der Meteoriten in den mineralogischen Museum der Universität zu Berlin." *AP* 118:419–423.
1864*a* "Über das angebliche Meteoreisen von Pompeji in der Chladnischen Meteoritensammlung." *AP* 123:374–377.
1864*b* "Beschreibung und Einteilung der Meteoriten auf Grund der Sammlung im mineralogischen Museum zu Berlin." *AWB,* 23–161.

Ross, John
1819 *A Voyage of Discovery made under the Orders of the Admiralty in His Majesty's Ships Isabella and Alexander for the Purpose of Exploring Baffin's Bay and . . . a Northwest Passage.* London: John Murray.

Roy, Sharat K.
1935 "The Question of Living Bacteria in Meteorites." *Field Museum of Natural History, Geological Series* 6:179–198.
1937 "Additional Notes on the Question of Living Bacteria in Stony Meteorites." *Popular Astronomy* 45:499–504.

Russell, Arthur
1952 "John Henry Heuland." *MM* 29:395–405.

Russell, Henry N.
1921 "A Superior Limit to the Age of the Earth's Crust." *PRS* 99A:84–86.
1929 "The Composition of the Sun's Atmosphere." *Astrophysical Journal* 70:11–82.

Rutherford, Ernest
1905 "Present Problems in Radioactivity." *Popular Science Monthly* 67:5–34.
1906 *Radioactive Transformations.* New York: Charles Scribners Sons.

Sabine, Edward
1819 "Notes on Meteoric Iron used by the Esquimaux of the Arctic Highlands." *Quarterly Journal of Literature, Science and the Arts* 6:369; 7:72–94.

Sadebeck, Alexander
1875 "Studien aux dem mineralogischen Museum der Universität Kiel." *AP* 156:554–563.

Sagan, Carl, and T. Page, eds.
1971 *UFOs—A Scientific Debate.* Ithaca: Cornell University Press.

Sage, Balthazar G.
1803*a* "Sur une Pierre qu'on dit tombée du Ciel dans les Environs de

Villefranche, Département du Rhône." *JP* 56:314–315.
1803*b* "Examen comparé de la Pierre Météorique d'Aigle et de celle de Villefranche." *JP* 57:70–73.

Saint-Amans, Jean F. B.
1802 "Lettre." *BB* 20:85–89.

Salverte, Eusebius
1803 "Conjectures on the Stones which have Fallen from the Atmosphere." *PM* 15:354–359.

Sambursky, S.
1962 *The Physical World of Late Antiquity*. New York: Basic Books.

Sanz, H. G., and Gerald J. Wasserburg
1969 "Determination of an Internal ^{87}Rb-^{87}Sr Isochron for the Olivenza Chondrite." *EPSL* 6:333–345.

Sartori, P.
1906 "Zur Volkskunde des Regierungsbezirke Minden." *Zeitschrift des Vereins für rheinische und westphälische Volkskunde* 7:200–210.

Saucier, Corinne
1956 *Traditions de la Paroisse des Avoyelles en Louisiana*. Philadelphia.

Saxton, G. H.
1870 "Letter from Colonel Saxton Describing the Fall of Nedogolla." *Proceedings of the Asiatic Society of Bengal, Calcutta,* 64–65.

Scherman, L.
1896 "Die Sterne im indogermanischen Seelenglauben." *Am Urquelle-Monatschrift für Volkskunde* 6:5–9.

Schiaparelli, Giovanni V.
1867 "Sur les Étoiles Filantes, et spécialement sur l'Identification des Orbites des Essaims d'Août et de Novembre avec celles des Comètes de 1862 et de 1866." *CR* 64:598–599.
1868 "Sulla Velocità della Meteore Cosmiche nel Loro Movimento a Traverso dell'Atmosfera Terrestra." *Rendiconti del Reale Instituto Lombardo* 1:1–9.

Schilling, C., ed.
1894–1909 *Wilhelm Olbers, sein Leben und seine Werke*. 2 vols. Berlin: Springer.

Schlosser, A.
1891 "Volksmeinung und Volks-beglaube aus der Deutschen Steiermark." *Germania* 36:380–408.

Schmidt, O. J.
1945 "A Meteoric Theory of the Origin of the Earth and Planets." *Nature* 156:185. (Abstract.)

Schofield, Robert E., ed.
1966 *A Scientific Autobiography of Joseph Priestley (1733–1804)*. Cambridge, Mass.: MIT Press.

Schönwerth, F. X. von
1857–59 *Aus der Oberpfalz Sitten und Sagen*. Augsburg.

Schramm, D. N., and R. N. Clayton
1978 "Did a Supernova Trigger the Formation of the Solar System?" *Scientific American* 239, 4:124–130.

Schreibers, Carl von
1820 *Beyträge zur Geschichte und Kenntnis meteorischer Stein und Metalmassen, und der Erscheinungen, welche deren Niederfall zu begleiten pflegn (als Nachtrag zu Herrn Chladni's neuestem Werke über Feuer-Meteore).* Wien: J. G. Heubner.
1836 "Über die neulichst bei Magdeburg zufällig aufgefundene problematische Metallmasse." *Neues Jahrbuch für Mineralogie, Geologie, und Paläontologie,* 497–498.

Schullerus, P.
1912 "Glaube und Brauch bei Tod und Begräbnis der Romänen in Harbechtale." *Zeitschrift des Vereins für Volkskunde* 22:156–164.

Sears, Derek W.
1976 "Edward Charles Howard and an Early British Contribution to Meteoritics." *Journal of the British Astronomical Association* 86:133–139.
1978 *The Nature and Origin of Meteorites.* New York: Oxford University Press.

Sears, Derek W., and H. Sears
1977 "Sketches in the History of Meteoritics 2: The Early Chemical and Mineralogical Work." *Meteoritics* 12:27–46.

Sears, Paul M.
1965 "Notes on the Beginnings of Modern Meteoritics." *Meteoritics* 2:293–299.

Séguin, Armand
1814 "Summary Ideas on the Probabilities of the Origin of Aerolites." *PM* 44:212–215.

Serres de Mesplès, Marcel P. T. de
1813 "Observations sur la Chute des Pierres, ou sur les Aérolithes." *ACP* 85:262–308.

Shepard, Charles U.
1829 "A Mineralogical and Chemical Description of the Virginia Aerolite." *AJS* 16:191–203.
1842 "Analysis of Meteoric Iron from Cocke County, Tennessee, with Some Remarks about Chlorine in Meteoric Iron Masses." *AJS* 43:354–363.
1846 "Report on Meteorites." *AJS* 52:377–392.
1847 "Report on Meteorites." *AJS* 54:74–87.
1848*a* "An Account of the Meteorite of Castine, Maine." *AJS* 56:251–253.
1848*b* "Observations on Rammelsberg's Analysis of the Juvenas Meteoric Stone, and on the Conclusion of Fischer's Examination of the Braunau Meteoric Iron." *AJS* 56:346–349.
1848*c* "Report on Meteorites." *AJS* 56:402–447.
1850*a* "Account of Three New American Meteorites and Geographical Distribution of Such Bodies Generally." *AAAS* 3:147–157.
1850*b* "On Meteorites." *AJS* 60:127–129.
1851 "On Meteorites." *AJS* 61:36–40.
1853 "Notice on the Meteoric Iron found near Senaca River, Cayuga County, N.Y." *AJS* 65:363–366.
1861 "Catalogue of the Meteoric Collection of Charles Upham Shepard,

deposited in the Cabinet of Amherst College, Mass." *AJS* 81:456–459.

1867 "New Classification of Meteorites with an Enumeration of Meteoric Species." *AJS* 93:22–28.

Shoemaker, Eugene M.

1963 "Impact Mechanics at Meteor Crater, Arizona." In *The Moon, Meteorites, and Comets*. Ed. B. M. Middlehurst and G. P. Kuiper. Chicago: University of Chicago Press.

Shoemaker, Eugene M., J. G. Williams, E. F. Helin, and R. F. Wolfe

1979 "Earth Crossing Asteroids: Orbital Classes, Collision Rates with Earth, and Origin." In *Asteroids*. Ed. Tom Gehrels. 253–282. Tucson: University of Arizona Press.

Silliman, Benjamin, and J. L. Kingsley

1809 "Memoir on the Origin and Composition of the Meteoric Stones which fell from the Atmosphere, in the County of Fairfield, and State of Connecticut, on the 14th of December 1807." *Transactions of the American Philosophical Society* 6:323–345.

Silliman, Benjamin, Jr., and T. S. Hunt

1846 "On the Meteoric Iron of Texas and Lockport." *AJS* 52:370–376.

Silliman, Mrs. Gold S.

1859 *On the Origin of Aerolites*. New York: W. C. Bryant. Reprinted in *Edinburgh New Philosophical Journal*, 1862, 16:227–248.

Simon, G. H.

1960 "Beliefs common in Ceylon." *Western Folklore* 19:123.

Smith, Cyril S.

1962 "Note on the History of the Widmanstätten Structure." *GCA* 26:971–972.

1964 "The Discovery of Carbon in Steel." *Technology and Culture* 5:149–175.

1981 *A Search for Structure*. Cambridge, Mass.: MIT Press.

Smith, Cyril S., ed.

1965 *The Sorby Centennial Symposium on the History of Metallurgy*. New York: Gordon & Breach.

Smith, G. F. H.

1907 "Robert Philips Greg." *MM* 14:268–271.

Smith, John Lawrence

1853 "New and Ready Methods of Determining the Alkalis in Minerals." *AJS* 65:234–243; 66:53–61.

1854 "Reexamination of American Minerals, Part IV." *AJS* 68:372–381.

1855*a* "Lecture on Meteoric Stones." *Annual Report of the Smithsonian Institution*, 151–174.

1855*b* "Memoir on Meteorites. A Description of Five New Meteoric Irons, with Some Theoretical Considerations on the Origin of Meteorites based on their Physical and Chemical Characters." *AJS* 69:322–343.

1861 "The Guernsey County (Ohio) Meteorites—A Complete Account of the Phenomenon Attending their Fall with a Chemical Analysis of them." *AJS* 81:87–98.

1864 "Bishopville Meteoric Stone." *AJS* 88:225–226.

1870 "Description and Analysis of the Franklin County Meteoric Iron; with Remarks on the Presence of Copper and Nickel in Meteoric Irons;

the Method of Analyzing the Same; and the Probability of Lead in the Tarapaca Iron having Originally been Foreign to that Mass." *AJS* 99:331–335.

1871*a* "On the Determination of the Alkalis in Silicates by Ignition with Carbonate of Lime and Sal-ammoniac." *AJS* 101:269–275.

1871*b* "Mineralogical and Chemical Composition of the Meteoric Stone that Fell near Searsmont, Maine, May 21, 1871." *AJS* 102:200–201.

1873 "A Description of the Victoria Iron seen to Fall in South Africa in 1862, with Some Notes on Chladnite or Enstatite." *AJS* 105:107–110.

1874 "On a Mass of Meteoric Iron of Howard Co., Ind., with Some Remarks on the Molecular Structure of Meteoric Iron, and a Notice Concerning the Presence of Solid Proto-Chloride of Iron in Meteorites." *AJS* 107:391–395.

1876*a* "Research on the Solid Carbon Compounds in Meteorites." *AJS* 111:388–395, 433–442.

1876*b* "Aragonite on the Surface of Meteoric Iron, and a New Mineral (Daubréelite) in the Concretions of the Interior of the Same." *AJS* 112:107–110.

1878 "On the Composition of the New Meteoric Mineral Daubréelite, and Its Frequent if not Universal Occurrence in Meteoric Irons." *AJS* 116:270–272.

1879 "Mémoire sur la Fer Natif du Groenland et sur la Dolérite qui le renferme." *ACP* 16:452–505.

1884 *Original Researches in Mineralogy and Chemistry*. Ed. J. B. Marvin. Louisville, Ky.: J. P. Morton.

Smith, S. W. J., A. A. Dee, and J. Young

1928 "The Mode of Formation of Neumann Bands. Part I, The Mechanism of Twinning in the Body-centered Cubic Lattice. Part II, The Evidence that the Bands are Twins. Part III, The Movement from which the Twinning Results." *PRS* A, 121:477–514.

Smith, W. C.

1969 "A History of the First Hundred Years of the Mineral Collection in the British Museum with particular Reference to the Work of Charles Konig." *Bulletin of the British Museum (Natural History), Historical Series* 3, 8:235–259.

Smithson, James L. M.

1811 "On the Composition of Zeolite." *PT* 11:171–177.

Smythe, W. R., and A. Hemmendinger

1937 "The Radioactive Isotope of Potassium." *PR* 51:178–182.

Soldani, Ambrogio

1808 *Storia di quelle Bolidi, che hanno da sé scaglioto Pietre alla Terra*. Siena: Atti della Acc. dei Fisiocritici.

Sorby, Henry Clifton

1864*a* "On the Conclusion to be Drawn from the Physical Structure of Some Meteorites." *BAAS*, Part 2:70.

1864*b* "On the Microscopical Structure of Meteorites." *PRS* 13:333–334.

1865 "On the Microscopical Structure of Meteorites, with a Note by Professor Brayley." *BAAS*, Part I:139–142.

1866 "On the Physical History of Meteorites in Connection with the Nebular Theory." *Sheffield Literary and Philosophical Society*, 10.

1868 "On Spherules in Meteorites." *BAAS*, 418.

1877 "On the Structure and Origin of Meteorites." *Nature* 15:495–498.

1886 "On the Application of Very High Power to the Study of the Microscopical Structure of Steel." *Journal of the Iron and Steel Institute* 1:140–144.

1887 "On the Microscopical Structure of Iron and Steel." *Journal of the Iron and Steel Institute* 2:255–288.

Southey, Robert

1797 *Letters Written During a Short Residence in Spain and Portugal.* Bristol.

Sowerby, James

1820 "Particulars of the Sword of Meteoric Iron Presented by Mr. Sowerby to Emperor Alexander of Russia." *PM* 55:49–52.

Spech, F. G.

1935 "Penobscot Tales and Religious Beliefs." *Journal of American Folklore* 48:1–107.

Spencer, Leonard J.

1933*a* "Meteorite Craters as Topographic Features of the Earth's Surface. *Geographical Journal* 81:227–248.

1933*b* "Origin of Tektites." *Nature* 131:117–118.

1936 "George Thurland Prior, 1862–1936." *Obituary Notices of the Royal Society of London,* 2, 5:151–159.

1938 "The Kaalijärv Meteorite from the Estonian Craters." *MM* 25:75–80.

1949 "A List of Catalogues of Meteorite Collections." *MM* 28:471–478.

Stepling, Joseph

1754 *De pluvis lapidea anni 1753 ad Strkow et ejus causis.* Prague.

Stodart, James, and Michael Faraday

1820 "Experiments on the Alloys of Steel made with a View to Its Improvement." *Quarterly Journal of Science* 10:354.

Stoneley, Jack

1977 *Cauldron of Hell: Tunguska.* New York: Simon & Schuster.

Story-Maskelyne, M. H. Nevil

1868–69 "Preliminary Notice of the Mineral Constituents of the Breitenbach Meteorite." *PRS* 17:370–372.

1870 "On the Mineral Constituents of Meteorites." *PT* 160:189–214.

1871 "On the Mineral Constituents of Meteorites." *PT* 161:359–367.

1875 "Some Lecture Notes upon Meteorites." *Nature* 12:485–487, 504–507, 520–523.

Story-Maskelyne, M. H. Nevil, and Victor Lang

1863 "Aerolitics." *PM* 25:46–58.

Stoughton, Bradley

1934 *The Metallurgy of Iron and Steel.* 4th ed. New York and London: McGraw Hill Book Company, Inc.

Stromeyer, Friedrich

1833 "Chemische Untersuchung der unlängst bei Magdeburg entdecken und für Meteoreisen gehaltenen Eisenmasse." *AP* 28:551–566.

Strutt, Robert J.

1906 "On the Distribution of Radium in the Earth's Crust, and on the Earth's Internal Heat." *PRS* 77:479.

Stütz, Andreas X.

1790 "Ueber einige vorgeblich vom Himmel gefallene Steine." *Bergbaukunde* 2:398–409.

Suess, Franz E.
1916 "Können die Tektite als Kunstprodukte gedeutet werden?" *Zentralblatt für Mineralogie*, 569–578.

Suess, Hans
1939 "Die Radioactivität des Kaliums als Mittel zur Bestimmung des relativen Alters der Elemente in Meteoriten." *Naturwissenschaften* 27:702–704.

Tait, Peter G.
1869 "On Comets." *Proceedings of the Royal Society of Edinburgh* 6:553–555.

Tata, Domenico
1794 *Memoria sulla pioggia di pietre avvenuta nelle campagna sanese a di 16. di Giugno di questo corrente anno*. Naples.
1804 "Memoria sulla pioggia di pietre, etc. Mémoire sur la pluie de pierres tombées dans la campagne de Siene, le 16 juin 1794, par l'abbé Dominico Tata." *BB* 25:240–267.

Taylor, G. Jeffrey, Edward R. D. Scott, and Klaus Keil
1983 "Cosmic Setting for Chondrule Formation." In *Chondrules and their Origins*. Ed. Elbert A. King. 262–278. Houston, Tex.: Lunar and Planetary Institute.

Taylor, W. J.
1857 "Examination of a Nickel Meteorite from Oktibbeha County, Mississippi." *AJS* 24:293–295.

Thenard, Louis J.
1806 "Analyse d'une Pierre tombée de l'Atmosphère à Valence, Arrondissement Alais, le 15 Mars 1806." *BB* 32:203–209.

Thompson, Sylvanus P.
1910 *The Life of William Thomson, Baron Kelvin of Largs*. 2 vols. London.

Thomsen, Elsebeth
1980 "New Light on the Origin of the Holy Black Stone of the Ka'ba." *Meteoritics* 15:87–91.

Thomson, Allen
1877 "Address." *BAAS*, lxviii–xcviii.

Thomson, G. [William]
1804 "Essai sur le Fer Malléable trouvé en Sibérie par le Prof. Pallas." *BB* 27:135–154, 209–229.

Thomson, William
1862 "On the Secular Cooling of the Earth." *Transactions of the Royal Society of Edinburgh* 23:157–169.
1871 *Address of Sir William Thomson, Knt. L.L.D., F.R.S. President*. London: Taylor and Francis.
1899 "The Age of the Earth as an Abode Fitted for Life." *Science* 9:665–674, 704–711.

Thorslund, Per, and Frans E. Wickman
1981 "Middle Ordovician Chondrite in Fossiliferous Limestone from Brunflo, Central Sweden." *Nature* 289:285–286.

Tilghman, Benjamin C.
1905 "Coon Butte, Arizona." *ANSP*, 57:887–914.

Tilley, C. E.
1922 "Density, Refractivity and Composition Relations of Some Natural Glasses." *MM* 19:275–294.

Tonnelier-Breteuil, Louis A. le
1802 "Notice sur plusiers Substances Pierreuses et Métalliques, que l'on dit être tombées du Ciel, et sur Différences Espèces de Fer Natif (Extrait d'une Mémoire lu à la Société Royale du Londres, par Howard et Bournon)." *JM* 13:11–31, 81–91.
1803 "Sur quelques Faits Nouveaux, relatifs aux Substances que l'on dit tombées de l'Atmosphère, communiqués par M. de Bournon." *JM* 13:446–453.

Topham, E.
1797 "Communication." *Gentleman's Magazine* 67:549.

Troost, Gerard
1846 "Description of Three Varieties of Meteoric Iron." *AJS* 52:356–358.

Trowbridge, Charles C.
1918 "Meteor Train Spectra and Probable Erroneous Conclusions of Observers." *PR* 11:484.

Trowbridge, Charles C. (deceased)
1924 "Spectra of Meteor Trains." *Proceedings of the National Academy of Sciences* 10:24–41.

Tschermak, Gustav
1865 "Chemisch-mineralogische Studien. 1. Die Felspathgruppe." *SAW* 50:566–613.
1870 "Ueber einen Feldspath aus dem Narödal und ueber das Mischungsgesetz der plagioklastischen Feldspaths." *SAW* 60:145–149.
1871 "Das Meteoreisen aus der Wüste Atacama." *Denkschriften der Akademie der Wissenschaften, Wien* 31:187–197.
1872 "Die Meteoriten des k.k. mineralogischen Museum am 1 October 1872." *Mineralogische Mittheilungen,* 165–172.
1874 "Das Krystallgefüge des Eisens, inbesondere des Meteoreisens." *SAW* 70:456–458.
1876 "The Formation of Meteorites and Volcanic Agency." *PM* 1:497–507.
1883 "Beitrag zur Classification der Meteoriten." *AJS* 126:411–412. (Abstract.)
1884 "Beitrag zur Classification der Meteoriten." *SAW* 88:347–371.
1885*a* *Die mikroskopische Beschaffenheit der Meteoriten.* Stuttgart.
1885*b* *Lehrbuch der Mineralogie.* Wein: A. Hölder.
1907 "Über das Eintreffen gleichartigen Meteoriten." *SAW* 116:1–35.

Tuckerman, L. B.
1944 "Early Use of Meteoritic Iron in Weapons." *Journal of the Washington Academy of Sciences* 34:163–164.

Twining, Alexander C.
1834 "Investigations respecting the Meteors of Nov. 13th 1833." *AJS* 26:320–352.
1835 "Meteors on the Morning of November 13th 1834." *AJS* 27:339–340.

Tylor, Edward B.
1889 *Primitive Culture.* 2 vols. New York.

Tyndall, John
1857 "Remarks on Foam and Hail." *PM* 13:352–353.

Urbes, W.
1898 "Aberglaube der Slowenen." *Zeitschrift für östereichische Volkskunde* 4:162.

Urey, Harold C.
1952 "On the Early Chemical History of the Earth and the Origin of Life." *Proceedings of the National Academy of Sciences* 38:351–363.
1959 "Primary and Secondary Objects." *Journal of Geophysical Research* 64:1721–1737.
1965 "Meteorites and the Moon." *Science* 147:1262–1265.
1966 "Biological Material in Meteorites—A Review." *Science* 151:157–166.

Urey, Harold C., and Harmon Craig
1953 "The Composition of the Stone Meteorites and the Origin of Meteorites." *GCA* 4:36–82.

Urey, Harold C., F. Fitch, H. P. Schwarcz, E. Anders, M. H. Briggs, G. B. Kitto, J. D. Bernal, B. Nagy, G. Claus, and D. J. Hennessy
1962 "Life-Forms in Meteorites." *Nature* 193:1119–1133.

Urry, W. D.
1933 "Helium and the Problem of Geological Time." *Chemical Reviews* 13:305–343.
1937 "Age Determination of the Iron Meteorites." *Proceedings of the Geological Society of America* 50:317.

Valmont de Bomare, Jacques C.
1762 *Minéralogie, ou Nouvelle Exposition du Règne Minéral*. 2 vols. Paris.

Vandermonde, C. A., C. L. Berthollet, and G. Monge
1786 "Mémoire sur le Fer Considerée dans Ses Différens États Metallique." *AS*, 132–200.

Van Flandern, T. C.
1977 "A Former Major Planet of the Solar System." In *Comets, Asteroids, Meteorites*. Ed. A. H. Delseeme. 475–481. Toledo, Ohio: University of Toledo.

Van Schmus, William R., and John A. Wood
1967 "A Chemical-Petrologic Classification for the Chondritic Meteorites." *GCA* 31:747–765.

Vauquelin, Nicolas-Louis
1802–03 "Mémoire sur les Pierres dites tombées du Ciel." *JM* 13:308–322.

Vicuna Cifuentes, Julio
1915 *Mitos y Supersticiones* . . . Santiago: Imprenta Universitaria.

Vogt, Carl
1882 *Les Prétendus Organismes des Météorites*. Genève: H. Georg.

Wadsworth, M. E.
1883 "Meteoric and Terrestrial Rocks." *Science* 1:127–130.

Wahl, Walter A.
1910 "Beiträge zur Chemie der Meteoriten." *Zeitschrift für Anorganische Chemie* 69:52–96.

Wainwright, G. A.
1912 "Pre-dynastic Iron Beads in Egypt." *Revue Archéologique* 1:255–259.
1928 "The Aniconic Form of Amon in the New Kingdom." *Annales du Service de Antiquités de l'Égypte* 28:175–189.

1930 "The Relationship of Amun to Zeus and his Connexion with Meteorites." *Journal of Egyptian Archaeology* 16:35–38.
1931 "The Emblem of Min." *Journal of Egyptian Archaeology* 17:185–195.
1932*a* "Iron in Egypt." *Journal of Egyptian Archaeology* 18:3–15.
1932*b* "Letopolis." *Journal of Egyptian Archaeology* 18:159–172.
1933 "The Bull Standards of Egypt." *Journal of Egyptian Archaeology* 19:42–52.
1934*a* "Jacob's Bethel." *Palestine Exploration Fund Quarterly Statement* (January), 32–44.
1934*b* "Some Aspects of Amun." *Journal of Egyptian Archaeology* 20: 139–153.
1935 "Amun's Meteorite and Omphaloi." *Zeitschrift für Ägyptische Altertumskunde* 71:41–44.

Waldbaum, Jane C.
1980 "The First Archaeological Appearance of Iron and the Transition to the Iron Age." In *The Coming of the Age of Iron*. Ed. T. A. Wertime and J. D. Muhly. 69–98. New Haven: Yale University Press.

Waldrop, M. Mitchell, and Richard A. Kerr
1983 "IRAS Science Briefing." *Science* 222:916–917.

Walker, Sears C.
1843 "Researches concerning the Periodical Meteors of August and November." *Transactions of the American Philosophical Society* 8:87–140.

Wallerius, Johann G.
1778 *Systema Mineralogicum*. Vienna.

Ward, Henry A.
1900 *The Ward-Coonley Collection of Meteorites*. Chicago.
1902 "Bacubirito or the Great Meteorite of Sinaloa, Mexico." *Proceedings of the Rochester Academy of Sciences* 4:67–74.
1904*a* *Catalogue of the Ward-Coonley Collection of Meteorites*. Chicago.
1904*b* "Great Meteorite Collections: Some Words as to their Composition as Affecting their Relative Values." *Proceedings of the Rochester Academy of Sciences* 4:149–164.
1904*c* "Values of Meteorites: Relative and Individual." *The Mineral Collector* 11:97–115.
1904*d* "The Willamette Meteorite." *Proceedings of the Rochester Academy of Sciences* 4:137–148.

Ward, Roswell
1948 *Henry A. Ward: Museum Builder to America*. Rochester, N.Y.

Washington, Henry S.
1926 "The Chemical Composition of the Earth, of Meteorites, and of the Sun's Atmosphere." *Bulletin of the National Research Council* 11, Part 2:30–32.

Wasserburg, Gerald J., and D. S. Burnet
1969 "The Status of Isotope Age Determinations on Iron and Stone Meteorites." In *Meteorite Research*. Ed. P. M. Millman. 467–479. Dordrecht: D. Reidel.

Wasserburg, Gerald J., Dimitri A. Papanastassiou, and H. G. Sanz
1969 "Initial Strontium for a Chondrite and the Determination of a Metamorphism or Formation Interval." *EPSL* 7:33–43.

Wassidlo, R.
1895 "Der Naturlesen in Munde des Mecklenburger Volker." *Zeitschrift des Vereins für Volkskunde* 5:424–448.

Wasson, John T.
1967 "The Chemical Classification of Iron Meteorites—I. A Study of Iron Meteorites with Low Concentrations of Gallium and Germanium." *GCA* 31:161–180.
1974 *Meteorites: Classification and Properties*. New York and Berlin: Springer-Verlag.
1985 *Meteorites: Their Record of Early Solar-System History*. New York: W. H. Freeman & Co.

Wasson, John T., and Jerome Kimberlin
1967 "The Chemical Classification of Iron Meteorites—II. Irons and Pallasites with Germanium Concentrations Between 8 and 100 ppm." *GCA* 31:2065–2092.

Wasson, John T., and S. P. Sedwick
1969 "Meteoritic Material from Hopewell Indian Burial Mounds: Chemical Data Regarding Possible Sources." *Nature* 222:22–24.

Wasson, John T., and George W. Wetherill
1979 "Dynamical, Chemical and Isotopic Evidence Regarding the Formation Locations of Asteroids and Meteorites." In *Asteroids*. Ed. Tom Gehrels. 926–974. Tucson: University of Arizona Press.

Waterston, Charles D.
1965 "William Thomson (1761–1806), A Forgotten Benefactor." *University of Edinburgh Journal* (Autumn), 122–134.

Watson, Fletcher
1939 "A Study of Fireball Radiant Positions." *Proceedings of the American Philosophical Society* 81:473–491.

Weeks, J. H.
1908 "Notes on Some Customs of the Lower Congo People." *Folk-Lore* 19:408–437.

Wehrham, K.
1910 "Kleinere Mitteilungen." *Zeitschrift des Vereins für rheinische und westphälische Volkskunde* 7:66.

Weinland, David F.
1881*a* "Korallen in Meteorsteinen." *Das Ausland* 54:301–303.
1881*b* "Weiteres über die Tiereste in Meteoriten." *Das Ausland* 54:501–508.
1882 *Ueber die in Meteoriten entdecken Thierreste*. Esslingen: G. Fröhner.

Weiss, Edmund
1866–67 "Bemerkungen über den Zusammenhang zwischen Cometen und Sternschnuppen." *AN* 68 (Nr. 1632), col. 381–384.

Westrum, Ron
1978 "Science and Social Intelligence about Anomalies: The Case of Meteorites." *Social Studies of Science* 8:461–493.

Wetherill, George W.
1977 "Fragmentation of Asteroids and Delivery of Fragments to Earth." In *Comets, Asteroids, Meteorites*. Ed. A. H. Delsemme. 283–295. Toledo, Ohio: University of Toledo.
1979 "Apollo Objects." *Scientific American* 240, 3:54–65.

1985 "Asteroidal Source of Ordinary Chondrites." *Meteoritics* 20:1–22.

Wetherill, George W., R. Mark, and C. Lee-Hu
1973 "Chondrites: Initial Strontium-87/Strontium-86 Ratios and the Early History of the Solar System." *Science* 182:281–283.

Whipple, F. J. W.
1930 "The Great Siberian Meteor." *The Quarterly Journal of the Royal Meteorological Society* 56:287–304.

Whiston, William
1716 *An Account of a Surprizing Meteor Seen in the Air March the 6th, 1716 at Night*. London.

Wiik, H. Birger
1956 "The Chemical Composition of Some Stony Meteorites." *GCA* 9:279–289.

Williams, I. M., and William A. Cassidy
1983 "Catch a Falling Star: Meteorites and Old Ice." *Science* 222:55–57.

Williams, L. Pearce
1965 "Faraday and the Alloys of Steel." In *The Sorby Centennial Symposium on the History of Metallurgy*. Ed. C. S. Smith. 145–162. New York: Gordon and Breach.

Wimperis, H.
1905 "The Temperature of Meteorites." *Nature* 71:81–82.

Winchell, Horace V.
1923 "A Meteoric Career." *Atlantic Monthly* 131:779–786.

Withering, William
1790 "An Account of Some Extraordinary Effects of Lightning." *PT* 80:293–295.

Wöhler, Friedrich
1858 "Ueber die Bestandtheile des Meteorsteines von Kaba in Ungarn," *SAW* 33:205–209.
1859 "Über die Bestandtheile des Meteorsteines von Capland." *SAW* 35:6–8.
1860 "Neuere Untersuchungen über die Bestandtheile des Meteorsteines von Capland." *SAW* 41:565–567.
1861 "Lithien in Meteoriten." *Annalen der Chemie und Pharmacie* 120: 253–254.

Wöhler, Friedrich, and Mauritz Hörnes
1859 "Die organische Substanz im Meteorsteine von Kaba." *SAW* 34:7–8.

Wood, John A.
1963 "Physics and Chemistry of Meteorites." In *The Moon, Meteorites, and Comets*. Ed. B. M. Middlehurst and G. P. Kuiper. 337–401. Chicago: University of Chicago Press.
1964 "The Cooling Rates and Parent Planets of Several Iron Meteorites." *Icarus* 3:429–459.
1968 *Meteorites and the Origin of Planets*. New York: McGraw-Hill.

Wood, John A., and Harry Y. McSween
1977 "Chondrules as Condensation Products." In *Comets, Asteroids, Meteorites*. Ed. A. H. Delsemme. 365–373. Toledo, Ohio: University of Toledo.

Wood, John A., and E. M. Wood, trans.
1964 *Gustav Tschermak: The Miscroscopic Properties of Meteorites.*

Smithsonian Contributions to Astrophysics. Vol. 4, no. 6. Washington, D.C.: Smithsonian Institution.

Woodward, John
1695 *An Essay Toward a Natural History of the Earth*. London.

Wrede, Ernst F.
1803 "Kritische Bermerkungen über die neuere Hypothesen, wodurch man die unter dem Namen der Feuerkugeln bekannten Lufterscheinungen zu erklären sucht." *AP* 14:55–100.

Wright, Arthur W.
1875*a* "Spectroscopic Examination of Gases from Meteoric Iron." *AJS* 109:294–302.
1875*b* "Preliminary Note on an Examination of Gases from the Meteorite of February, 12, 1875." *AJS* 109:459–460.
1875*c* "Examination of Gases from the Meteorite of February 12, 1875." *AJS* 110:44–49, 206–207.
1876 "On the Gases Contained in Meteorites." *AJS* 111:253–262; 112:165–176.

Wülfing, Ernst A.
1894 "Ueber Verbreitung und Werth der in Sammlungen aufbewahrten Meteoriten." *Separatabdruck aus dem Bericht über die XXVII Versammlung des oberrheinischen geologischen Vereins zu Landau am 29 März*.
1895 "Verbreitung und Wert der in Sammlungen aufbewahrten Meteoriten." *Separatabdruck aus den Jahresheften des Vereins für vaterlande Naturkunde in Württemberg*.
1897 *Die Meteoriten in Sammlungen und Ihre Literatur, nebst einem Versuch, den Tauschwert der Meteoriten zu Bestimmen*. Tübingen.
1899 "Über den Tauschwert der Meteoriten." *Neues Jahrbuch für Mineralogie, Geologie und Paläontologie* 2:116–119.

Wylie, Charles C.
1940 "The Orbit of the Pultusk Meteor." *Popular Astronomy* 48:306–312.

Yanai, Keizo
1979 *Catalog of Yamato Meteorites*. Tokyo: National Institute of Polar Research.

Yeo, R. B. G., and O. O. Miller
1965 "A History of Nickel Steels from Meteorites to Maraging." In *The Sorby Centennial Symposium on the History of Metallurgy*. Ed. C. S. Smith. 467–500. New York: Gordon & Breach.

Young, J.
1926 "The Crystal Structure of Meteoric Iron as Determined by X-ray Analysis." *PRS* A, 112:630–641.

Young, John
1872 "Meteors—Seedbearing and Otherwise." *Cornhill Magazine* 25:77–91.

Zellner, B.
1979 "Asteroid Taxonomy and the Distribution of the Compositional Types." In *Asteroids*. Ed. Tom Gehrels. 783–806. Tucson: University of Arizona Press.

Zöllner, Johann K. F.
1873 "Ueber den Zusammenhang von Sternschnuppen und Kometen." *AP* 148:322–329.

Name Index

Persons whose work or activity is described or whose views are quoted or paraphrased are included. In cases of joint authorship, only the name of the first author is listed. Vital statistics are noted for deceased persons, where such information is available in standard biographical sources. Otherwise, the data of relevant publication is listed.

Subject Index

Meteorite Index

A question mark (?) follows a meteorite name when there is not sufficient data to determine that the meteorite actually fell. See Graham et al. (1985) for complete information and relevant literature on each meteorite.

Designer:	Marvin Warshaw
Compositor:	Prestige Typography
Text:	11/12 Times Roman
Display:	ITC Barcelona Book
Printer:	Braun-Brunfield, Inc.
Binder:	Braun-Brunfield, Inc.